T0360527

Solutions of Nonlinear Differential Equations

Existence Results via the Variational Approach

TRENDS IN ABSTRACT AND APPLIED ANALYSIS

ISSN: 2424-8746

Series Editor: John R. Graef
The University of Tennessee at Chattanooga, USA

Published

**Trends in Abstract
and Applied Analysis**
Volume **3**

Solutions of Nonlinear Differential Equations

Existence Results via the Variational Approach

Lin Li
Shu-Zhi Song

Chongqing Technology and Business University, China

World Scientific

NEW JERSEY · LONDON · SINGAPORE · BEIJING · SHANGHAI · HONG KONG · TAIPEI · CHENNAI · TOKYO

Published by

World Scientific Publishing Co. Pte. Ltd.
5 Toh Tuck Link, Singapore 596224
USA office: 27 Warren Street, Suite 401-402, Hackensack, NJ 07601
UK office: 57 Shelton Street, Covent Garden, London WC2H 9HE

Library of Congress Cataloging-in-Publication Data
Names: Li, Lin (Mathematics professor) | Song, Shu-Zhi.
Title: Solutions of nonlinear differential equations : existence results via the variational approach /
    Lin Li (Chongqing Technology and Business University, China) &
    Shu-Zhi Song (Chongqing Technology and Business University, China).
Description: New Jersey : World Scientific, 2016. | Series: Trends in abstract and applied analysis ;
    volume 3 | Includes bibliographical references and index.
Identifiers: LCCN 2016001133 | ISBN 9789813108608 (hc : alk. paper)
Subjects: LCSH: Differential equations, Nonlinear. | Differential equations,
    Nonlinear--Numerical solutions.
Classification: LCC QA372 .L5 2016 | DDC 515/.355--dc23
LC record available at http://lccn.loc.gov/2016001133

British Library Cataloguing-in-Publication Data
A catalogue record for this book is available from the British Library.

Desk Editors: Kwong Lai Fun/Vishnu Mohan

Typeset by Stallion Press
Email: enquiries@stallionpress.com

Printed in Singapore

Lin Li dedicates this volume to his loving wife
Yan-Hui Wang and daughter Rui-Xi Li.

Preface

Since the birth of the Calculus of Variations, it has been realized that when they apply, variational methods can obtain better results than most other methods. Moreover, they apply in a very large number of situations. It was realized many years ago that the solutions of a great number of problems are in effect critical points of functionals.

In this monograph, we look at how variational methods can be used in nonlinear differential equations. In Chapter 1, we provide some prerequisites for this monograph. We collect some knowledge of Sobolev space and variational principle and so forth. Basically, these theories are essentially known and readily available in many books. Well-trained readers may skip this chapter.

Subsequent chapters deal with fourth-order problems, Kirchhoff problems, nonlinear field problems, gradient systems, and variable exponent problems. The chapters are presented in such a way that, except for references to the basic results in Chapter 1, each is essentially a stand-alone entity that can be read with little reference to other chapters. Each problem we develop in this book has its own difficulties. That is why we intend to develop some standard and appropriate methods that are useful and that can be extended to other problems.

Chapter 2 starts by presenting the basic setting for fourth-order problems. In the second section, sufficient conditions for the existence of multiple solutions to nonlinear fourth-order problems are presented. Section 3 is concerned with the existence of infinitely many solutions, while Section 4 presents sufficient conditions for the existence of multiple solutions of quasilinear fourth-order problems with singular term. Last section is semilinear fourth-order equations on $\mathbb{R}^N$.

In Chapter 3, Kirchhoff problems are discussed. After a brief introduction, the next sections parallel those in Chapter 2 by examining sufficient conditions for the existence of radial solution, the existence of multiple solutions of nonhomogeneous problems, and the existence of solutions with superlinear, asymptotically linear and sublinear nonlinearities, respectively. Final section deals with combined nonlinearities.

Nonlinear Field Problems is the focus of Chapter 4. Topics include Schrödinger–Maxwell equations, Klein–Gordon–Maxwell systems and Klein–Gordon equation coupled with Born–Infeld theory. Solutions of superlinear and asymptotically linear are discussed and quasilinear Schrödinger–Maxwell equations also considered. Existence of solutions for Klein–Gordon–Maxwell systems with superlinear and sublinear nonlinearities, sign-changing potential, without odd nonlinearities, partially sublinear nonlinearities are presented. Last section is about Klein–Gordon equation coupled with Born–Infeld theory.

Chapter 5 is devoted to gradient systems. One dimension systems and elliptic systems are discussed separately.

Chapter 6 is mainly concerned variable exponent problems. Solutions of $p(x)$-Laplacian problems, $p(x)$-Laplacian-like problems and $p(x)$-biharmonic problems are presented.

This monograph may be used as a valuable reference for researchers in abstract and applied analysis and related fields. Nevertheless, both the presentation style and the choice of the material make the present book accessible to all newcomers to this modern research field, which lies at the interface between abstract and applied analysis.

We want to thank our colleagues Prof. Martin Schechter of University of California, Irvine, Prof. Vicenţiu Rădulescu of University of Craiova, Prof. Dušan Repovš of University of Ljubljana, Prof. Pasquale Candito, Prof. Robert Livrea of Università degli Studi 'Mediterranea' di Reggio Calabria, Prof. Shapour Heidarkhani of Razi University, Prof. Ghasem Alizadeh Afrouzi of University of Mazandaran, Prof. Ling Ding of Hubei University of Arts and Science, Prof. Shang-Jie Chen of Chongqing Technology and Business University, Ji-Jiang Sun of Nanchang University, Xin Zhong of Chinese Academy of Sciences, Wen-Wu Pan of Sichuan University of Science and Engineering and Ying Li of National University of Defense Technology, whose collaboration on the application of variational methods to nonlinear differential equations have inspired this monograph.

We want to express deep gratitude to our PhD supervisor, Prof. Chun-Lei Tang of Southwest University (Chongqing), for his guidance and professional advice during our studies at Southwest University (Chongqing).

His personality and erudition have positively influenced our evolution and will accompany us for all our future career.

Lin Li and Shu-Zhi Song

Some Notations and Conventions

- $\mathbb{R}$, $\mathbb{Z}$ and $\mathbb{N}$ denote the set of real numbers, integers and nonnegative integers.
- $\mathbb{R}^N$ denotes the usual N-dimensional Euclidean space.
- $\Omega \subset \mathbb{R}^N$ denotes an open set (usually with a smooth boundary) $\partial\Omega$.
- $L^p(\Omega)$, $1 \leq p < \infty$, denotes the space of measurable functions u on Ω with norm $\|u\|_p := \left(\int_\Omega |u|^p dx\right)^{1/p} < \infty$.
- $L^\infty(\Omega)$ denotes the space of measurable functions u with $|u(x)| \leq C$ a.e. in Ω with norm $\|u\|_\infty := \inf\{C \geq 0 : |u(x)| \leq C$ a.e. in $\Omega\}$.
- $\|\cdot\|_X$ (or simply $\|\cdot\|$) denotes the norm in the space X.
- $\langle \cdot, \cdot \rangle_X$ denotes the inner product in X.
- Δu denotes the Laplacian of u, $\sum_{i=1}^{N} u_{x_i x_i}$.
- $C^1(X, \mathbb{R})$ is the space of continuously differentiable functionals on X.
- $H_0^1(\Omega)$ [resp. $H^1(\Omega)$] is the Sobolev space obtained by completion of $C_0^\infty(\Omega)$ [resp. $C^\infty(\bar\Omega)$] in the norm $\|u\| := \left(\int_\Omega [|\nabla u|^2 + |u|^2] dx\right)^{1/2}$.
- $W_0^{1,p}(\Omega)$ [resp. $W^{1,p}(\Omega)$] is the Sobolev space obtained by completion of $C_0^\infty(\Omega)$ [resp. $C^\infty(\bar\Omega)$] in the norm $\|u\| := \left(\int_\Omega [|\nabla u|^p + |u|^p] dx\right)^{1/p}$, $1 \leq p < \infty$.
- $H^1(\mathbb{R}^N)$ [resp. $W^{1,p}(\mathbb{R}^N)$] is the Sobolev space obtained by completion of $C_0^\infty(\mathbb{R}^N)$ in the norm $\|u\| := \left(\int_{\mathbb{R}^3} [|\nabla u|^2 + |u|^2] dx\right)^{1/2}$ [resp. $\|u\| := \left(\int_{\mathbb{R}^N} [|\nabla u|^p + |u|^p] dx\right)^{1/p}$, $1 \leq p < \infty$].
- E denotes the weighted Sobolev space with the norm $\left(\int_{\mathbb{R}^3} [|\nabla u|^2 + V(x)|u|^2] dx\right)^{1/2}$, where $V : C(\mathbb{R}^3, \mathbb{R})$ and bounded from below.

- $\{u_n\}$ denotes a sequence of functions.
- The arrow $\rightharpoonup$ (resp. $\rightarrow$) denotes weak (resp. strong) convergence.
- The symbol $\hookrightarrow$ denotes the embedding is continuous and $\hookrightarrow\hookrightarrow$ denotes the embedding is compact.

Contents

Chapter 1

Preliminaries and Variational Principles

1.1 Sobolev Spaces

Let Ω be an open subset of $\mathbb{R}^N$, $N \in \mathbb{N}$. Define

$$L^p(\Omega) := \{u : \Omega \to \mathbb{R} \text{ is Lebesgue measurable, } \|u\|_{L^p(\Omega)} < \infty\},$$

where

$$\|u\|_{L^p(\Omega)} = \left(\int_\Omega |u|^p dx \right)^{1/p}, \quad 1 \leq p < +\infty.$$

If $p = +\infty$

$$\|u\|_{L^\infty(\Omega)} = \text{ess sup}_\Omega |u| := \inf_{A \subset \Omega, \text{meas}(A)=0} \sup_{\Omega \backslash A} |u|,$$

where meas denotes the Lebesgue measure. Let

$$L^p_{loc}(\Omega) := \{u : \Omega \to \mathbb{R}, u \in L^p(V) \text{ for each } V \subset\subset \Omega\},$$

where $V \subset\subset \Omega \Longleftrightarrow V \subset \bar{V} \subset \Omega$ and $\bar{V}$ is compact. We will in this book denote $\|u\|_{L^p(\Omega)}$ by $\|u\|_p$ or $|u|_p$.

Let $C_c^\infty(\Omega)$ denote the space of infinitely differentiable functions $\phi : \Omega \to \mathbb{R}$ with compact support in Ω. For each $\phi \in C_c^\infty(\Omega)$ and a multiindex $\alpha = (\alpha_1, \ldots, \alpha_N)$ with order $|\alpha| = \alpha_1 + \cdots + \alpha_N$, we denote

$$D^\alpha \phi = \frac{\partial^{\alpha_1}}{\partial x_1^{\alpha_1}} \cdots \frac{\partial^{\alpha_N}}{\partial x_N^{\alpha_N}} \phi.$$

Definition 1.1. Suppose u, $v \in L^1_{loc}(\Omega)$. We say that v is the αth-weak partial derivative of u, written $D^\alpha u = v$ provided

$$\int_\Omega u D^\alpha \phi \, dx = (-1)^{|\alpha|} \int_\Omega v \phi \, dx$$

for all $\phi \in C_c^\infty(\Omega)$.

1

It is easy to check that the αth-weak partial derivative of u, if it exists, is uniquely defined up to a set of measure zero.

Let $C^m(\Omega)$ be the set of functions having derivatives of order $\leq m$ being continuous in Ω ($m = $ integer ≥ 0 or $m = \infty$).

Let $C^m(\bar{\Omega})$ be the set of functions in $C^m(\Omega)$ all of whose derivatives of order $\leq m$ have continuous extension to $\bar{\Omega}$.

Definition 1.2. Fix $p \in [1, +\infty]$ and $K \in \mathbb{N} \cup \{0\}$. The Sobolev space

$$W^{k,p}(\Omega)$$

consists of all $u : \Omega \to \mathbb{R}$ which have α^{th}-weak partial derivatives $D^\alpha u$ for each multiindex α with $|\alpha| \leq k$ and $D^\alpha u \in L^p(\Omega)$.

If $p = 2$, we usually write

$$H^k(\Omega) = W^{k,2}(\Omega), \quad k = 0, 1, 2, \ldots.$$

Note that $H^0(\Omega) = L^2(\Omega)$. We henceforth identify functions in $W^{k,p}(\Omega)$ which agree a.e.

Definition 1.3. If $u \in W^{k,p}(\Omega)$, we define its norm to be

$$\|u\|_{W^{k,p}(\Omega)} := \begin{cases} \left(\sum_{|\alpha| \leq k} \int_\Omega |D^\alpha u|^p dx \right)^{1/p} &, \quad p \in [1, +\infty), \\ \sum_{|\alpha| \leq k} \text{ess sup}_\Omega |D^\alpha u|, & \quad p = +\infty. \end{cases}$$

Definition 1.4. We denote

$$W_0^{k,p}(\Omega)$$

as the closure of $C_c^\infty(\Omega)$ in $W^{k,p}(\Omega)$ with respect to its norm defined in Definition 1.3. It is customary to write

$$H_0^k(\Omega) = W_0^{2,k}(\Omega)$$

and denote by $H^{-1}(\Omega)$ the dual space to $H_0^1(\Omega)$.

The following results can be found in [1].

Proposition 1.1 ([1]). *For each $k = 1, 2, \ldots$ and $1 \leq p \leq +\infty$, the Sobolev space $W^{k,p}(\Omega)$ is a Banach space and so is $W_0^{k,p}(\Omega)$. In particular, $H^k(\Omega)$, $H_0^k(\Omega)$ are Hilbert spaces.*

Proposition 1.2 ([1]). *Let $(X, \|\cdot\|_X)$ and $(Y, \|\cdot\|_Y)$ be two Banach spaces, $X \subset Y$. We say that X is continuously embedded in Y (denoted by $X \hookrightarrow Y$) if the identity $id : X \to Y$ is a linear bounded operator, that is, there is* *a constant* $C > 0$ *such that* $\|u\|_Y \leq C\|u\|_X$ *for all* $u \in X$. *In this case, constant* $C > 0$ *is called the embedding constant. If moreover, each bounded sequence in* X *is precompact in* Y, *we say the embedding is compact, written* $X \hookrightarrow\hookrightarrow Y$.

Definition 1.5. A function $u : \Omega \subset \mathbb{R}^N \to \mathbb{R}$ is Hölder continuous with exponent $\gamma > 0$ if

$$[u]^{(\gamma)} := \sup_{x \neq y \in \Omega} \frac{|u(x) - u(y)|}{|x - y|^\gamma} < \infty.$$

Definition 1.6. The Hölder space $C^{k,\gamma}(\bar{\Omega})$ consists of all functions $u \in C^k(\bar{\Omega})$ for which the norm

$$\|u\|_{C^{k,\gamma}(\bar{\Omega})} := \sum_{|\alpha| \leq k} \|D^\alpha u\|_{C(\bar{\Omega})} + \sum_{|\alpha| = k} [D^\alpha u]^{(\gamma)}$$

is finite. It is a Banach space. We set $C^{k,0}(\bar{\Omega}) = C^k(\bar{\Omega})$.

We have the following embedding results, see [1–3].

Proposition 1.3 ([1–3]). *If* Ω *is a bounded domain in* $\mathbb{R}^N$, *then*

$$W_0^{k,p}(\Omega) \hookrightarrow \begin{cases} L^q(\Omega), & kp < N, 1 \leq q \leq Np/(N - kp), \\ C^{m,\alpha}(\bar{\Omega}), & 0 \leq \alpha \leq k - m - N/p \text{ and} \\ & 0 \leq m < k - N/p < m + 1. \end{cases}$$

Proposition 1.4 ([1–3]). *If* Ω *is a bounded domain in* $\mathbb{R}^N$, *then*

$$W_0^{k,p}(\Omega) \hookrightarrow\hookrightarrow \begin{cases} L^q(\Omega), & kp < N, 1 \leq q \leq Np/(N - kp), \\ C^{m,\alpha}(\bar{\Omega}), & 0 \leq \alpha \leq k - m - N/p \text{ and} \\ & 0 \leq m < k - N/p < m + 1. \end{cases}$$

The following proposition can be found in [4, 5].

Proposition 1.5 ([4, 5]). *The following embeddings are continuous:*

$$H^1(\mathbb{R}^N) \hookrightarrow L^p(\mathbb{R}^N), \quad \text{if } 2 \leq p < \infty, N = 1, 2,$$
$$H^1(\mathbb{R}^N) \hookrightarrow L^p(\mathbb{R}^N), \quad \text{if } 2 \leq p \leq 2^*, N \geq 3,$$

where $2^* := 2N/(N - 2)$ *if* $N \geq 3$ *and* $2^* = +\infty$ *if* $N = 1, 2$, *is called a critical exponent.*

Let $N \geq 3$ and $2^* = 2N/(N-2)$. The space

$$D^{1,2}(\mathbb{R}^N) := \{u \in L^{2^*}(\mathbb{R}^N) : \nabla u \in L^2(\mathbb{R}^N)\}$$

with the inner product

$$\int_{\mathbb{R}^N} \nabla u \cdot \nabla v dx$$

and the corresponding norm

$$\left(\int_{\mathbb{R}^N} |\nabla u|^2 dx\right)^{1/2}$$

is a Hilbert space. We have the following results

Proposition 1.6 ([5]). *The embedding $D^{1,2}(\mathbb{R}^N) \hookrightarrow L^{2^*}(\mathbb{R}^N)$ $(N \geq 3)$ is continuous.*

In particular, the Sobolev inequality holds

$$S := \inf_{u \in D^{1,2}(\mathbb{R}^N), \|u\|_{2^*}=1} \|\nabla u\|_2^2 > 0.$$

Now, we introduce some theories of Lebesgue–Sobolev space with variable exponent. The detailed description can be found in [6–9]. Denote by $S(\Omega)$ the set of all measurable real functions on Ω. Set

$$C_+(\overline{\Omega}) = \left\{p : p \in C(\overline{\Omega}), p(x) > 1, \forall x \in \overline{\Omega}\right\}.$$

For any $p \in C_+(\overline{\Omega})$, denote

$$1 < p^- := \inf_{x \in \overline{\Omega}} p(x) \leq p(x) \leq p^+ := \sup_{x \in \overline{\Omega}} p(x) < \infty.$$

Let $p \in C_+(\overline{\Omega})$. Define the generalized Lebesgue space by

$$L^{p(x)}(\Omega) = \left\{u : u \in S(\Omega), \int_{\Omega} |u(x)|^{p(x)} dx < \infty\right\},$$

then $L^{p(x)}(\Omega)$ endowed with the norm

$$|u|_{p(x)} = \inf\left\{\beta > 0 : \int_{\Omega} \left|\frac{u(x)}{\beta}\right|^{p(x)} dx \leq 1\right\},$$

becomes a Banach space.

Let $a \in S(\Omega)$, and $a(x) > 0$ for a.e. $x \in \Omega$. Define the weighted variable exponent Lebesgue space $L_a^{p(x)}(\Omega)$ by

$$L_a^{p(x)}(\Omega) = \left\{u : u \in S(\Omega), \int_{\Omega} a(x)|u(x)|^{p(x)} dx < \infty\right\},$$

with the norm

$$|u|_{p(x)} = \inf \left\{ \beta > 0 : \int_\Omega a(x) \left| \frac{u(x)}{\beta} \right|^{p(x)} dx \le 1 \right\}.$$

From now on, we suppose that $a \in L^\infty(\Omega)$ and $\operatorname{essinf}_{x \in \Omega} a(x) = a_0 > 0$. Then obviously $L_a^{p(x)}(\Omega)$ is a Banach space (see [10] for details).

The variable exponent Sobolev space $W^{1,p(x)}(\Omega)$ is defined by

$$W^{1,p(x)}(\Omega) = \{ u \in L^{p(x)}(\Omega) : |\nabla u| \in L^{p(x)}(\Omega) \},$$

with the norm

$$\|u\| = |u|_{p(x)} + |\nabla u|_{p(x)}.$$

We set $\rho(u) = \int_\Omega \left(|\nabla u|^{p(x)} + a(x)|u|^{p(x)} \right) dx$.

Proposition 1.7 ([11]). *For all $u \in W_a^{1,p(x)}(\Omega)$, we have*

(i) $\|u\|_a \le 1 \Rightarrow \|u\|_a^{p^+} \le \rho(u) \le \|u\|_a^{p^-}$,
(ii) $\|u\|_a \ge 1 \Rightarrow \|u\|_a^{p^-} \le \rho(u) \le \|u\|_a^{p^+}$.

Remark 1.1. If $N < p^- \le p(x)$ for any $x \in \Omega$, by Theorem 2.2 in [9] and the equivalence of the norms $\| \cdot \|_a$ and $\| \cdot \|$, we deduce that $W_a^{1,p(x)}(\Omega) \hookrightarrow W_a^{1,p^-}(\Omega)$. Since $N < p^-$, it follows that $W_a^{1,p(x)}(\Omega) \hookrightarrow W_a^{1,p^-}(\Omega) \hookrightarrow\hookrightarrow C(\overline{\Omega})$. Defining the norm

$$\|u\|_\infty = \sup_{x \in \overline{\Omega}} |u(x)|,$$

then there exists a constant $k > 0$ such that

$$\|u\|_\infty \le k\|u\|_a, \quad \forall u \in W_a^{1,p(x)}(\Omega).$$

In the remainder of this section, we recall some definitions and basic properties of variable spaces $L^{p(x)}(\mathbb{R}^N)$ and $W^{1,p(x)}(\mathbb{R}^N)$. For a deeper treatment on these spaces, we refer to [8, 9].

Let $p \in L^\infty(\mathbb{R}^N)$, $p^- > 1$. The variable exponent Lebesgue space $L^{p(x)}(\mathbb{R}^N)$ is defined by

$$L^{p(x)}(\mathbb{R}^N) = \left\{ u : \mathbb{R}^N \to \mathbb{R} : u \text{ is measurable and } \int_{\mathbb{R}^N} |u|^{p(x)} dx < +\infty \right\}$$

endowed with the norm $|u|_{p(x)} \left\{ \lambda > 0 : \int_{\mathbb{R}^N} |\frac{u}{\lambda}|^{p(x)} dx \le 1 \right\}$. Then we define the variable exponent Sobolev space

$$W^{1,p(x)}(\mathbb{R}^N) = \{ u \in L^{p(x)}(\mathbb{R}^N) : |\nabla u| \in L^{p(x)}(\mathbb{R}^N) \}$$

with the norm $\|u\| = |u|_{p(x)} + |\nabla u|_{p(x)}$.

Proposition 1.8 ([12]). *Set $\psi(u) = \int_{\mathbb{R}^N}(|\nabla u(x)|^{p(x)} + |u(x)|^{p(x)})dx$. If $u, u_k \in W^{1,p(x)}(\mathbb{R}^N)$, then*

(1) $\|u\| < 1(= 1; > 1) \Leftrightarrow \psi(u) < 1(= 1; > 1)$;

(2) If $\|u\| > 1$, then $\|u\|^{p^-} \leq \psi(u) \leq \|u\|^{p^+}$;

(3) If $\|u\| < 1$, then $\|u\|^{p^+} \leq \psi(u) \leq \|u\|^{p^-}$;

(4) $\lim_{k \to +\infty} \|u_k\| = 0 \Leftrightarrow \lim_{k \to +\infty} \psi(u_k) = 0$;

To study $p(x)$-biharmonic problems, we need some results on the spaces $L^{p(x)}(\Omega)$ and $W^{k,p(x)}(\Omega)$.

Define the generalized Lebesgue space by

$$L^{p(x)}(\Omega) = \left\{ u : \Omega \to \mathbb{R} \text{ measurable and } \int_\Omega |u(x)|^{p(x)}\, dx < \infty \right\},$$

for $p \in C(\overline{\Omega})$ with $p^- > 1$, where

$$p^+ = \max_{x \in \overline{\Omega}} p(x), \quad p^- = \min_{x \in \overline{\Omega}} p(x).$$

Denote

$$p^*(x) = \begin{cases} \dfrac{Np(x)}{N - p(x)} & \text{if } p(x) < N, \\ +\infty & \text{if } p(x) \geq N, \end{cases}$$

$$p_k^*(x) = \begin{cases} \dfrac{Np(x)}{N - kp(x)} & \text{if } kp(x) < N, \\ +\infty & \text{if } kp(x) \geq N, \end{cases}$$

for any $x \in \overline{\Omega}$, $k \geq 1$. We define a norm, the so-called Luxemburg norm, on this space $L^{p(x)}(\Omega)$ by the formula

$$|u|_{p(x)} = \inf \left\{ \beta > 0 : \int_\Omega \left| \frac{u(x)}{\beta} \right|^{p(x)} dx \leq 1 \right\}.$$

We recall that the space $(L^{p(x)}(\Omega), |\cdot|_{p(x)})$ is a Banach space. If $0 < |\Omega| < \infty$ and p_1, p_2 are variable exponents so that $p_1(x) \leq p_2(x)$ a.e. in Ω then the embedding $L^{p_2(x)}(\Omega) \hookrightarrow L^{p_1(x)}(\Omega)$ is continuous.

Proposition 1.9 ([9]). *The space $(L^{p(x)}(\Omega), |\cdot|_{p(x)})$ is separable, uniformly convex, reflexive and its conjugate space is $L^{q(x)}(\Omega)$, where $q(x)$ is the conjugate function of $p(x)$, i.e.,*

$$\frac{1}{p(x)} + \frac{1}{q(x)} = 1$$

for all $x \in \Omega$. For $u \in L^{p(x)}(\Omega)$ and $v \in L^{q(x)}(\Omega)$ we have

$$\left| \int_\Omega u(x)v(x)\, dx \right| \leq \left(\frac{1}{p^-} + \frac{1}{q^-} \right) |u|_{p(x)} |v|_{q(x)}.$$

The Sobolev space with variable exponent $W^{k,p(x)}(\Omega)$ is defined as

$$W^{k,p(x)}(\Omega) = \left\{ u \in L^{p(x)}(\Omega) : D^\alpha u \in L^{p(x)}(\Omega), |\alpha| \leq k \right\},$$

where $D^\alpha u = \frac{\partial^{|\alpha|}}{\partial x_1^{\alpha_1} \partial x_2^{\alpha_2} \dots \partial x_N^{\alpha_N}} u$ with $\alpha = (\alpha_1, \dots, \alpha_N)$ is a multi-index and $|\alpha| = \sum_{i=1}^N \alpha_i$. The space $W^{k,p(x)}(\Omega)$ equipped with the norm

$$\|u\|_{k,p(x)} = \sum_{|\alpha| \leq k} |D^\alpha u|_{p(x)},$$

also becomes a separable and reflexive Banach space. For more details, we refer the reader to [8, 9, 13, 14].

Proposition 1.10 ([9]). *For $p, r \in C(\overline{\Omega})$ such that $r(x) \leq p_k^*(x)$ for all $x \in \overline{\Omega}$, there is a continuous embedding*

$$W^{k,p(x)}(\Omega) \hookrightarrow L^{r(x)}(\Omega).$$

If we replace $\leq$ with $<$, the embedding is compact.

We denote by $W_0^{k,p(x)}(\Omega)$ the closure of $C_0^\infty(\Omega)$ in $W^{k,p(x)}(\Omega)$. The generalized Sobolev space $W^{2,p(x)}(\Omega) \cap W_0^{1,p(x)}(\Omega)$ equipped with the norm

$$\|u\| = \inf \left\{ \beta > 0 : \int_\Omega \left(\left| \frac{\Delta u(x)}{\beta} \right|^{p(x)} - \lambda \left| \frac{u(x)}{\beta} \right|^{p(x)} \right) dx \leq 1 \right\}.$$

Remark 1.2. According to [15], the norm $\| \cdot \|_{2,p(x)}$ is equivalent to the norm $|\Delta \cdot|_{p(x)}$ in the space X. Consequently, the norms $\| \cdot \|_{2,p(x)}, \| \cdot \|$ and $|\Delta \cdot|_{p(x)}$ are equivalent.

Proposition 1.11 ([16]). *Set $\rho(u) = \int_\Omega \left(|\Delta u|^{p(x)} - \lambda |u|^{p(x)} \right) dx$. For u_n, $u \in X$ we have*

(1) $\|u\| < (=; > 1) \Leftrightarrow \rho(u) < (=; > 1)$,
(2) $\|u\| \leq 1 \Rightarrow \|u\|^{p^+} \leq \rho(u) \leq \|u\|^{p^-}$,
(3) $\|u\| \geq 1 \Rightarrow \|u\|^{p^-} \leq \rho(u) \leq \|u\|^{p^+}$,
(4) $\|u_n\| \to 0 \Leftrightarrow \rho(u_n) \to 0$,
(5) $\|u_n\| \to \infty \Leftrightarrow \rho(u_n) \to \infty$.

1.2 Differentiable Functionals

Let E be a Banach space with norm $\| \cdot \|$. Let $U \subset E$ be an open set of E. The dual (or conjugate) space of E is denoted by E^*, i.e., E^* denotes the set of bounded linear functionals on E. Consider a functional $I : U \to \mathbb{R}$.

Definition 1.7. The functional I has a Fréchet derivative $F \in E^*$ at $u \in U$ if

$$\lim_{h \in E, h \to 0} \frac{I(u+h) - I(u) - F(h)}{\|h\|} = 0.$$

We define $I'(u) = F$ or $\nabla I(u) = F$ and sometimes refer to it as the gradient of I at u. Usually, $I'(\cdot)$ is a nonlinear operator. We use $C^1(U, \mathbb{R})$ to denote the set of all functionals which have continuous Fréchet derivative on U. A point $u \in U$ is called a critical point of a functional $I \in C^1(U, \mathbb{R})$, if

$$I'(u) = 0.$$

Definition 1.8. The functional I has a Gâteaux derivative $G \in E^*$ at $u \in U$ if, for every $h \in E$,

$$\lim_{t \to 0} \frac{I(u + th) - I(u)}{t} = G(h).$$

The Gâteaux derivative at $u \in U$ is denoted by $DI(u)$. Obviously, if I has a Fréchet derivative $F \in E^*$ at $u \in U$, then I has a Gâteaux derivative $G \in E^*$ at u and $I'(u) = DI(u)$. But the converse is not true. However, if I has Gâteaux derivatives at every point of some neighborhood of $u \in U$ such that $DI(u)$ is continuous at u then I has a Fréchet derivative and $I'(u) = DI(u)$. This is a straightforward consequence of the Mean Value Theorem.

Let $f(x, t)$ be a function on $\Omega \times \mathbb{R}$, where Ω is either bounded or unbounded. We say that f is a Carathéodory function if $f(x, t)$ is continuous in t for a.e. $x \in \Omega$ and measurable in x for every $t \in \mathbb{R}$.

The following lemmas comes from Zou and Schechter [17].

Lemma 1.1. *Assume $p \geq 1$, $q \geq 1$. Let $f(x, t)$ be a Carathéodory function on $\Omega \times \mathbb{R}$ and satisfy*

$$|f(x, t)| \leq a + b|t|^{p/q}, \qquad \forall (x, t) \in \Omega \times \mathbb{R},$$

where a, $b > 0$ and Ω is either bounded or unbounded. Define a Carathéodory operator by

$$Bu := f(x, u(x)), \quad u \in L^p(\Omega).$$

Let $\{u_k\}_{k=0}^{\infty} \subset L^p(\Omega)$. If $\|u_k - u_0\|_p \to 0$ as $k \to \infty$, then $\|Bu_k - Bu_0\|_q \to 0$ as $k \to \infty$. In particular, if Ω is bounded, then B is a continuous and bounded mapping from $L^p(\Omega)$ to $L^q(\Omega)$ and the same conclusion is true if Ω is unbounded and $a = 0$.

Lemma 1.2. *Assume p_1, p_2, q_1, $q_2 \geq 1$. Let $f(x,t)$ be a Carathéodory function on $\Omega \times \mathbb{R}$ and satisfy*

$$|f(x,t)| \leq a|t|^{p_1/q_1} + b|t|^{p_2/q_2}, \quad \forall (x,t) \in \Omega \times \mathbb{R},$$

where a, $b \geq 0$ and Ω is either bounded or unbounded. Define a Carathéodory operator by

$$Bu := f(x, u(x)), \quad u \in \mathcal{H} := L^{p_1}(\Omega) \cap L^{p_2}(\Omega).$$

Define the space

$$\mathcal{E} := L^{p_1}(\Omega) + L^{p_2}(\Omega)$$

with a norm

$$\|u\|_{\mathcal{E}} = \inf \left\{ \|v\|_{L^{q_1}(\Omega)} + \|w\|_{L^{q_1}(\Omega)} : u = v + w \in \mathcal{E}, v \in L^{q_1}(\Omega), w \in L^{q_2}(\Omega) \right\}.$$

Then $B = B_1 + B2$, where B_i is a bounded and continuous mapping from $L^{p_i}(\Omega)$ to $L^{q_i}(\Omega)$, $i = 1, 2$. In particular, B is a bounded continuous mapping from $\mathcal{H}$ to $\mathcal{E}$.

The following lemma is enough for us to see that the functionals encountered in this book are of C^1.

Lemma 1.3. *Assume $\sigma \geq 0$, $p \geq 0$. Let $f(x,t)$ be a Carathéodory function on $\Omega \times \mathbb{R}$ satisfying*

$$|f(x,t)| \leq a|t|^{\sigma} + b|t|^{p}, \quad \forall (x,t) \in \Omega \times \mathbb{R},$$

where a, $b > 0$ and Ω is either bounded or unbounded. Define a functional

$$I(u) := \int_{\Omega} F(x,u)dx, \text{ where } F(x,u) = \int_0^u f(x,s)ds.$$

Assume $(E, \| \cdot \|)$ is a Sobolev Banach space such that $E \hookrightarrow L^{p+1}(\Omega)$ and $E \hookrightarrow L^{\sigma+1}(\Omega)$. Then $I \in C^1(\Omega)$ and

$$I'(u)h = \int_{\Omega} f(x,u)h dx, \quad \forall h \in E.$$

Moreover, if $E \hookrightarrow\hookrightarrow L^{\sigma}(\Omega)$, $E \hookrightarrow\hookrightarrow L^{p+1}(\Omega)$, then $I' : E \to E^*$ is compact.

1.3   Ekeland's Variational Principle

For any bounded from below, lower semi-continuous functional f, Ekeland's variational principle provides a minimizing sequence whose elements minimize an appropriate sequence of perturbations of f which converges locally uniformly to f. Roughly speaking, Ekeland's variational principle states that there exist points which are almost points of minima and where the "gradient" is small. In particular, it is not always possible to minimize a nonnegative continuous function on a complete metric space. Ekeland's variational principle is a very basic tool, has effective in numerous situations, which has led to many new results and strengthened a series of known results in various fields of analysis, geometry, the Hamilton–Jacobi theory, extremal problems, the Ljusternik–Schnirelmann theory, etc. Its precise statement is as follows.

Theorem 1.1 ([18]). *Let (X, d) be a complete metric space and let f : $X \to (-\infty, \infty]$ be a lower semi-continuous, bounded from below functional with $D(f) = \{u \in X : f(u) < \infty\} \neq \emptyset$. Then for every $\varepsilon > 0$, $\lambda > 0$, and $u \in X$ such that*

$$f(u) \leq \inf_X f + \varepsilon,$$

there exists an element $v \in X$ such that

(a) $f(v) \leq f(u)$;
(b) $d(v, u) \leq 1/\lambda$;
(c) $f(w) > f(v) - \varepsilon\lambda d(w, v)$ for each $w \in X \setminus \{v\}$.

In the particular case $X = \mathbb{R}^N$, we can give the following simple alternative proof to Ekeland's variational principle, due to Hiriart-Urruty [19]. Indeed, consider the perturbed functional

$$g(w) := f(w) + \varepsilon\lambda \|w - u\|, \quad w \in \mathbb{R}^N.$$

Since f is lower semi-continuous and bounded from below, then g is lower semicontinuous and $\lim_{\|w\| \to \infty} g(w) = \infty$. Therefore there exists $v \in \mathbb{R}^N$ minimizing g on $\mathbb{R}^N$ such that, for all $w \in \mathbb{R}^N$,

$$f(v) + \varepsilon\lambda \|v - u\| \leq f(w) + \varepsilon\lambda \|w - u\|. \tag{1.1}$$

By letting $w = u$ we obtain

$$f(v) + \varepsilon\lambda \|v - u\| \leq f(u)$$

and (a) follows. Now, since $f(u) \leq \inf_{\mathbb{R}^N} f + \varepsilon$, we also deduce that $\|vu\| \leq 1/\lambda$.

We infer from relation (1.1) that, for any w,

$$f(v) \leq f(w) + \varepsilon\lambda[\|w - u\| - \|v - u\|] \leq f(w) + \varepsilon\lambda\|w - u\|,$$

which is the desired inequality (c).

Taking $\lambda = \frac{1}{\sqrt{\varepsilon}}$ in Theorem 1.1 we obtain the following property.

Corollary 1.1. *Let (X, d) be a complete metric space and let $f : X \to (\infty, \infty]$ be lower semi-continuous, bounded from below, and let $D(f) = \{u \in X : f(u) < \infty\} \neq \emptyset$. Then for every $\varepsilon > 0$ and every $u \in X$ such that*

$$f(u) \leq \inf_X f + \varepsilon,$$

there exists an element $u_\epsilon \in X$ such that

(a) $f(u_\epsilon) \leq f(u)$;
(b) $d(u_\epsilon, u) \leq \sqrt{\varepsilon}$;
(c) $f(w) > f(u_\epsilon) - \sqrt{\varepsilon}d(w, u_\epsilon)$ for each $w \in X \setminus \{u_\epsilon\}$.

Let $(X, \|\cdot\|)$ be a real Banach space, and let X^* be its topological dual endowed with its natural norm, denoted for simplicity also by $\|\cdot\|$. We denote by $\langle\cdot, \cdot\rangle$ the duality mapping between X and X^*; that is, $\langle x^*, u\rangle = x^*(u)$ for every $x^* \in X^*$, $u \in X$. Theorem 1.1 readily implies the following property, which asserts the existence of almost critical points. In other words, Ekeland's variational principle can be viewed as a generalization of the Fermat theorem which establishes that interior extrema points of a smooth functional are, necessarily, critical points of this functional.

Corollary 1.2. *Let X be a Banach space and let $f : X \to \mathbb{R}$ be a lower semi-continuous functional which is bounded from below. Assume that f is Gâteaux differentiable at every point of X. Then for every $\varepsilon > 0$ there exists an element $u_\varepsilon \in X$ such that*

(i) $f(u_\varepsilon) \leq \inf_X f + \varepsilon$;
(ii) $\|f'(u_\varepsilon)\| \leq \varepsilon$.

Letting $\varepsilon = 1/n$, $n \in \mathbb{N}$, Corollary 1.2 gives rise to a minimizing sequence for the infimum of a given function which is bounded from below. Note, however, that such a sequence need not converge to any point. Indeed, let $f : \mathbb{R} \to \mathbb{R}$ defined by $f(s) = e^{-s}$. Then, $\inf_{\mathbb{R}} f = 0$, and any minimizing sequence fulfilling (a) and (b) from Corollary 1.2 tends to ∞. The following definition is dedicated to handle such situations.

Definition 1.9. (a) A function $f \in C^1(X, \mathbb{R})$ satisfies the Palais–Smale condition at level $c \in \mathbb{R}$ (abbreviated to $(PS)_c$-condition) if every sequence

$\{u_n\} \subset X$, such that $\lim_{n \to \infty} f(u_n) = c$ and $\lim_{n \to \infty} \|f'(u_n)\| = 0$, possesses a convergent subsequence.

(b) A function $f \in C^1(X, \mathbb{R})$ satisfies the Palais–Smale condition (abbreviated to (PS)-condition) if it satisfies the Palais–Smale condition at every level $c \in \mathbb{R}$.

For reader's convenience, we introduce the Cerami condition (abbreviated to $(C)_c$-condition), which was established by Cerami [20].

Definition 1.10. Assume functional Φ is C^1 and $c \in \mathbb{R}$, if any sequence $\{u_n\}$ satisfying $\Phi(u_n) \to c$ and $(1 + \|u_n\|)\|\Phi'(u_n)\| \to 0$ has a convergence subsequence, we say Φ satisfies Cerami condition at the level c.

Combining this compactness condition with Corollary 1.2, we obtain the following result.

Theorem 1.2. *Let X be a Banach space and let f be a function $f \in C^1(X, \mathbb{R})$ which is bounded from below. If f satisfies the $(PS)_c$-condition (or $(C)_c$-condition) at level $c = \inf_X f$, then c is a critical value of f; that is, there exists a point $u_0 \in X$ such that $f(u_0) = c$ and u_0 is a critical point of f, that is, $f'(u_0) = 0$.*

1.4   Minimax Principles

In this section we are interested in some powerful techniques for finding solutions of some classes of stationary nonlinear boundary value problems. These solutions are viewed as critical points of a natural functional, often called the energy associated to the system. The critical points obtained in this section by means of topological techniques are generally nonstable critical points which are neither maxima nor minima of the energy functional.

In many nonlinear problems we are interested in finding solutions as stationary points of some associated energy functionals. Often such a mapping is unbounded from above and below, so that it has no maximum or minimum. This forces us to look for saddle points, which are obtained by minimax arguments.

One of the most important minimax results is the so-called Mountain Pass Theorem, whose geometrical interpretation will be briefly described in the following. Denote by f the function which measures the altitude of a mountain terrain and assume that there are two points in the horizontal plane, e_0 and e_1, representing the coordinates of two locations such that $f(e_0)$ and $f(e_1)$ are the deepest points of two separated valleys. Roughly speaking, our aim is to walk along an optimal path on the mountain from the point $(e_0, f(e_0))$ to $(e_1, f(e_1))$, spending the smallest amount of energy by passing the mountain ridge between the two valleys. Walking on a path $(\gamma, f(\gamma))$ from $(e_0, f(e_0))$ to $(e_1, f(e_1))$ such that the maximal altitude along γ is the smallest among all such continuous paths connecting $(e_0, f(e_0))$ and $(e_1, f(e_1))$, we reach a point L on γ passing the ridge of the mountain which is called a mountain pass point.

Theorem 1.3 ([21]). *Let E be a real Banach space and suppose that $I \in C^1(E, \mathbb{R})$ satisfies the condition*

$$\max\{I(0), I(u_1)\} \leq \alpha < \beta \leq \inf_{\|u\|=\rho} I(u),$$

for some $\alpha < \beta$, $\rho > 0$ and $u_1 \in E$ with $\|u_1\| > \rho$. Let $c \geq \beta$ be characterized by

$$c = \inf_{\gamma \in \Lambda} \max_{0 \leq \tau \leq 1} I(\gamma(\tau)),$$

where $\Lambda = \{\gamma \in C([0,1], E) : \gamma(0) = 0, \gamma(1) = u_1\}$ is the set of continuous paths joining 0 and u_1. Then, there exists a sequence $\{u_k\} \subset E$ such that

$$I(u_k) \to c \geq \beta \text{ and } I'(u_k) \to 0, \text{ as } k \to \infty.$$

Theorem 1.4 ([22]). *Let E be a real Banach space with its dual space E^*, and suppose that $I \in C^1(E, \mathbb{R})$ satisfies*

$$\max\{I(0), I(e)\} \leq \mu < \eta \leq \inf_{\|u\|=\rho} I(u),$$

for some μ, η, $\rho > 0$ and $e \in E$ with $\|e\| > \rho$. Let $c \geq \eta$ be characterized by

$$c = \inf_{\gamma \in \Gamma} \max_{0 \leq \tau \leq 1} I(\gamma(\tau)),$$

where $\Gamma = \{\gamma \in C([0,1], E) : \gamma(0) = 0, \gamma(1) = e\}$ is the set of continuous paths joining 0 and e, then there exists a sequence $\{u_n\} \subset E$ such that

$$I(u_n) \to c \geq \eta \text{ and } (1 + \|u_n\|)\|I'(u_n)\|_{E^*} \to 0, \text{ as } n \to \infty.$$

Recall that $\mathcal{J}$ has a local linking at 0 with respect to the direct sum decomposition $E = E^- \oplus E^+$, if there is $\rho > 0$ such that

$$\begin{cases} \mathcal{J}(u) \leq 0, & \text{for } u \in E^-, \|u\| \leq \rho, \\ \mathcal{J}(u) \geq 0, & \text{for } u \in E^+, \|u\| \leq \rho. \end{cases} \tag{1.2}$$

It is then clear that 0 is a (trivial) critical point of $\mathcal{J}$. Next, we consider two sequences of finite dimensional subspaces

$$E_0^\pm \subset E_1^\pm \subset \cdots \subset E^\pm$$

such that

$$E^\pm = \overline{\bigcup_{n\in\mathbb{N}} E_n^\pm}.$$

For multi-index $\alpha = (\alpha^-, \alpha^+) \in \mathbb{N}^2$ we set $E_\alpha = E_{\alpha^-}^- \oplus E_{\alpha^+}^+$ and denote by $\mathcal{J}_\alpha$ the restriction of $\mathcal{J}$ on E_α. A sequence $\{\alpha_n\} \subset \mathbb{N}^2$ is admissible if, for any $\alpha \in \mathbb{N}^2$, there is $m \in \mathbb{N}$ such that $\alpha \leq \alpha_n$ for $n \geq m$; where for $\alpha, \beta \in \mathbb{N}^2$, $\alpha \leq \beta$ means $\alpha^\pm \leq \beta^\pm$. Obviously, if $\{\alpha_n\}$ is admissible, then any subsequence of $\{\alpha_n\}$ is also admissible.

Definition 1.11 ([23]). We say that $\Phi \in C^1(X)$ satisfies the Palais–Smale condition ((PS) for short), if whenever $\{\alpha_n\} \subset \mathbb{N}^2$ is admissible, any sequence $\{u_n\} \subset E$ such that

$$u_n \in E_{\alpha_n}, \qquad \sup_n \Phi(u_n) < \infty, \qquad \|\Phi'_{\alpha_n}(u_n)\|_{E^*_{\alpha_n}} \to 0 \qquad (1.3)$$

contains a subsequence which converges to a critical point of Φ.

Theorem 1.5 ([24]). *Suppose that $\Phi \in C^1(X)$ has a local linking at 0, Φ satisfies (PS) condition, Φ maps bounded sets into bounded sets and, for every $m \in \mathbb{N}$,*

$$\Phi(u) \to -\infty, \qquad as \ \|u\| \to \infty, \ u \in E^- \oplus E_m^+. \qquad (1.4)$$

Then Φ has a nontrivial critical point.

Definition 1.12. A sequence $\{u_{\alpha_n}\}$ with $\{\alpha_n\}$ admissible and $u_{\alpha_n} \in X_{\alpha_n}$ is said to be a $(Ce)^*_c$ sequence if $\sup I(u_{\alpha_n}) < \infty$ $(I(u_{\alpha_n}) \to c)$ and $(1 + \|u_{\alpha_n}\|)I'_{\alpha_n}(u_{\alpha_n}) \to 0$ as $n \to \infty$. The functional I is said to satisfy the $(Ce)^*_c$ condition if every $(Ce)^*_c$ sequence has a convergent subsequence.

The $(Ce)^*$ condition implies the $(Ce)^*_c$ condition for every $c \in \mathbb{R}$. We get the following developed local linking theorem obtained by the authors in [23].

Theorem 1.6 ([23]). *Suppose X is a Banach space, $X = V \oplus W$ with $\dim V < +\infty$ and $I \in C^1(X, \mathbb{R})$ satisfies the following assumptions:*

(I_1) *I has a local linking at 0;*
(I_2) *I satisfies the $(Ce)^*$ condition;*
(I_3) *I maps bounded sets into bounded sets;*
(I_4) *for every finite dimensional subspace $E \subset W$, $I(u) \to -\infty$, as $\|u\| \to \infty$, $u \in V \oplus E$.*

Then I has at least two critical points.

Here, we recall a linking theorem of Schechter [25]. Let E be a Banach space. The set Φ of mappings $\Gamma(t) \in C(E \times [0,1], E)$ is to have following properties:

(a) For each $t \in [0,1), \Gamma(t)$ is a homeomorphism of E onto itself and $\Gamma(t)^{-1}$ is continuous on $E \times [0,1)$.

(b) $\Gamma(0) = I$.

(c) For each $\Gamma(t) \in \Phi$ there is a $u_0 \in E$ such that $\Gamma(1)u = u_0$ for all $u \in E$ and $\Gamma(t)u \to u_0$ as $t \to 1$ uniformly on bounded subsets of E.

(d) For each $t_0 \in [0,1)$ and each bounded set $A \subset E$ we have

$$\sup_{\substack{0 \le t \le t_0 \\ u \in A}} \{\|\Gamma(t)u\| + \|\Gamma^{-1}(t)u\|\} < \infty.$$

A subset A of E links a subset B of E if $A \cap B = \phi$ and, for each $\Gamma(t) \in \Phi$, there is a $t \in (0,1]$ such that $\Gamma(t)A \cap B \ne \phi$.

Theorem 1.7 ([25]). *Let I be a C^1-functional on H, and let A, B be subsets of H such that A links B and*

$$a_0 := \sup_A I \le b_0 := \inf_B I.$$

Assume that

$$a := \inf_{\Gamma \in \Phi} \sup_{0 \le s \le 1, u \in A} I(\Gamma(s)u)$$

is finite. Then there is a sequence $(u_k) \subset H$ such that

$$I(u_k) \to a, \quad (1 + \|u_k\|)I'(u_k) \to 0.$$

If this sequence (u_k) has a convergent subsequence, it produces a critical point of I.

Now, we introduce the definition of genus.

Definition 1.13. Let X be a real Banach space and A a subset of X. A is said to be symmetric if $u \in A$ implies $-u \in A$. For a closed symmetric set A which does not contain the origin, we define the genus $\gamma(A)$ of A as by the smallest integer k such that there exists an odd continuous mapping from A to $\mathbb{R}^k \setminus \{0\}$. If does not exist such a k, we define $\gamma(A) = \infty$. Moreover, we set $\gamma(\emptyset) = 0$. Let Γ_k denote the family of closed symmetric subsets A of X such that $0 \notin A$ and $\gamma(A) \ge k$.

For the convenience of the readers, we summarize the properties of a genus. We refer the readers to [26] for the proof of the next proposition.

Proposition 1.12 ([26]). *Let A and B be closed symmetric subsets of X which do not contain the origin. Then (i)-(v) below hold.*

(i) If there is an odd continuous mapping from A to B, then $\gamma(A) \leq \gamma(B)$;
(ii) If there is an odd homeomorphism from A onto B, then $\gamma(A) = \gamma(B)$;
(iii) If $\gamma(B) < \infty$, then $\gamma(\overline{A \setminus B}) \geq \gamma(A) - \gamma(B)$;
(iv) If A is compact, then $\gamma(A) < \infty$ and $\gamma(N_\delta(A)) = \gamma(A)$ for $\delta > 0$ small enough;
(v) The n-dimensional sphere S^n has genus equal to $n + 1$.

The following abstract critical point theorem established by Rabinowitz [21].

Theorem 1.8 ([21]). *Let $I \in C^1(E, \mathbb{R})$ be an even functional on a Banach space E. Assume $I(0) = 0$ and I satisfies the (PS)-condition. If*

(i) there exists an $F \subset E$ with $\dim F = n$ and $R > 0$ such that

$$\sup_{F \setminus B_R} I \leq 0,$$

(ii) there exists an $N \subset E$ with $\operatorname{codim} N = m$ and $\alpha > 0, \rho > 0$ such that

$$\inf_{N \cap \partial B_\rho} \geq \alpha,$$

then I has at least $n - m$ pairs of critical points with positive critical values.

The following $\mathbb{Z}_2$ version of the Mountain Pass Theorem will be get infinitively many solutions.

Theorem 1.9 ([21]). *Let E be an infinite-dimensional Banach space, and $I \in C(E, \mathbb{R})$ be even, satisfying the (PS) condition, and having $I(0) = 0$. Assume that $E = V \oplus X$, where V is finite dimensional. Suppose that the following hold.*

(a) There are constants $\nu, \rho > 0$ such that $\inf_{\partial B_\nu \cup X} I \geq \rho$.
(b) For each finite-dimensional subspace $\hat{E} \subset E$, there is an $\sigma = \sigma(\hat{E})$ such that $I \leq 0$ on $\hat{E} \setminus B_\sigma$.

Then I possesses an unbounded sequence of critical values.

If the space X is reflexive and separable, then there exists $\{e_i\} \subset X$ and $\{f_i\} \subset X^*$ such that

$$X = \overline{\langle e_i, i \in \mathbb{N} \rangle}, \quad X^* = \overline{\langle f_i, i \in \mathbb{N} \rangle}, \quad \langle e_i, f_j \rangle = \delta_{i,j},$$

where $\delta_{i,j}$ denotes the Kronecker symbol. Put

$$X_k = \operatorname{span}\{e_k\}, \quad Y_k = \bigoplus_{i=1}^{k} X_i, \quad Z_k = \overline{\bigoplus_{i=k}^{\infty} X_i}. \tag{1.5}$$

The following Fountain Theorem is established by Bartsch [27].

Theorem 1.10 ([27]). *Assume $\varphi \in C^1(X, \mathbb{R})$ satisfies the (PS) condition, $\varphi(-u) = \varphi(u)$. For every $k \in \mathbb{N}$, there exists $\rho_k > r_k > 0$, such that*

(i) $a_k := \max_{u \in Y_k, \|u\| = \rho_k} \varphi(u) \leq 0$;
(ii) $b_k := \inf_{u \in Z_k, \|u\| = r_k} \varphi(u) \to +\infty$ as $k \to \infty$,

where Y_k and Z_k are defined by (1.5). Then φ has a sequence of critical values tending to $+\infty$.

Here we recall Fountain Theorem under $(C)_c$ condition

Theorem 1.11. *Let X be a Banach space with the norm $\| \cdot \|$ and let X_j be a sequence of subspace of X with $\dim X_j < \infty$ for each $j \in \mathbb{N}$. Further, $X = \overline{\bigoplus_{j \in \mathbb{N}} X_j}$, the closure of the direct sum of all X_j. Set $W_k = \bigoplus_{j=0}^{k} X_j$, $Z_k = \overline{\bigoplus_{j=k}^{\infty} X_j}$. Consider an even functional $\Phi \in C^1(X, \mathbb{R})$ (i.e. $\Phi(-u) = \Phi(u)$ for all $u \in E$). If, for every $k \in \mathbb{N}$, there exist $\rho_k > r_k > 0$ such that*

(Φ1) $a_k := \max_{u \in W_k, \|u\| = \rho_k} \Phi(u) \leq 0$,
(Φ2) $b_k := \inf_{u \in Z_k, \|u\| = r_k} \Phi(u) \to +\infty$, as $k \to \infty$,
(Φ3) the $(C)_c$ holds at any level $c > 0$.

Then Φ has an unbounded sequence of critical values.

Remark 1.3. $(C)_c$ condition is weaker than the (PS) condition. However, it was shown in [28] that from $(C)_c$ condition a deformation lemma follows and, as a consequence, we can also get minimax theorems.

The following critical point theorem established by Kajikiya in [29].

Theorem 1.12 ([29]). *Let X be infinite dimensional Banach space and $I \in C^1(X, \mathbb{R})$ satisfy (Φ_1) and (Φ_2) below.*

(Φ_1) I *is even, bounded from below,* $I(0) = 0$ *and* I *satisfies the (PS) condition.*

(Φ_2) *For each* $k \in \mathbb{N}$, *there exists an* $A_k \in \Gamma_k$ *such that* $\sup_{u \in A_k} I(u) < 0$.

Then I *admits a sequence of critical points* $\{u_k\}$ *such that* $I(u_k) \leq 0$, $u_k \neq 0$ *and* $\lim_{k \to \infty} u_k = 0$.

Recently, Liu and Wang [30] obtained an extension of Clark's theorem as follows.

Theorem 1.13 ([30]). *Let* X *be a Banach space,* $\Phi \in C^1(X, \mathbb{R})$. *Assume* Φ *is even and satisfies the (PS) condition, bounded from below, and* $\Phi(0) = 0$. *If for any* $k \in \mathbb{N}$, *there exists a* k-*dimensional subspace* X^k *of* X *and* $\rho_k > 0$ *such that* $\sup_{X^k \cap S_{\rho_k}} \Phi < 0$, *where* $S_\rho = \{u \in X : \|u\| = \rho\}$, *then at least one of the following conclusions holds.*

(i) *There exists a sequence of critical points* $\{u_k\}$ *satisfying* $\Phi(u_k) < 0$ *for all* k *and* $\|u_k\| \to 0$ *as* $k \to \infty$.

(ii) *There exists* $r > 0$ *such that for any* $0 < a < r$ *there exists a critical point* u *such that* $\|u\| = a$ *and* $\Phi(u) = 0$.

1.5 Ricceri's Variational Results

The main part of this section is dedicated to Ricceri's recent multiplicity results. First, we discuss three critical points-type results with one or two parameters. In the next section a general variational principle of Ricceri is presented. Finally, a new kind of multiple result of Bonanno is given which guarantees the existence of two critical points for a functional.

1.5.1 *Three Critical Point Results*

For the reader's convenience, we recall the revised form of Ricceri's three critical points theorem.

Theorem 1.14 ([31]). *Let* X *be a reflexive real Banach space.* $\Phi \colon X \mapsto \mathbb{R}$ *is a continuously Gâteaux differentiable and sequentially weakly lower semicontinuous functional whose Gâteaux derivative admits a continuous inverse on* X^* *and* Φ *is bounded on each bounded subset of* X; $J \colon X \mapsto \mathbb{R}$ *is a continuously Gâteaux differentiable functional whose Gâteaux derivative* *is compact; $I \subseteq \mathbb{R}$ an interval. Assume that*

$$\lim_{\|u\| \to +\infty} (\Phi(u) + \lambda J(u)) = +\infty$$

for all $\lambda \in I$, and that there exists $\rho \in \mathbb{R}$ such that

$$\sup_{\lambda \in I} \inf_{u \in X} (\Phi(u) + \lambda(J(u) + \rho)) < \inf_{u \in X} \sup_{\lambda \in I} (\Phi(u) + \lambda(J(u) + \rho)). \qquad (1.6)$$

Then, there exists an open interval $\Lambda \subseteq I$ and a positive real number q with the following property: for every $\lambda \in \Lambda$ and every C^1 functional $\Psi \colon X \mapsto \mathbb{R}$ with compact derivative, there exists $\delta > 0$ such that, for each $\mu \in [0, \delta]$ the equation

$$\Phi'(u) + \lambda J'(u) + \mu \Psi'(u) = 0$$

has a least three solutions in X whose norms are less than q.

We will need the following result, which is Proposition 1.3 in [32] with J replaced by $-J$, to show the minimax inequality (1.6) in Theorem 1.14.

Proposition 1.13 ([32]). *Let X be a non-empty set, and $\Phi : X \to \mathbb{R}$, $J : X \to \mathbb{R}$ two real functions. Assume that $\Phi(u) \geq 0$ for every $u \in X$ and that there exists $u_0 \in X$ such that $\Phi(u_0) = J(u_0) = 0$. Further, assume that there exists $u_1 \in X$, $r > 0$ such that*

(i) $r < \Phi(u_1)$
(ii) $\sup_{\Phi(u)<r}(-J(u)) < r \dfrac{-J(u_1)}{\Phi(u_1)}$.

Then for every $h > 1$ and for every $\rho \in \mathbb{R}$ satisfying

$$\sup_{\Phi(u)<r} (-J(u)) + \frac{r \dfrac{-J(u_1)}{\Phi(u_1)} - \sup_{\Phi(u)<r}(-J(u))}{h} < \rho < r \frac{-J(u_1)}{\Phi(u_1)}$$

one has

$$\sup_{\lambda \in I} \inf_{u \in X} (\Phi(u) + \lambda(J(u) + \rho)) < \inf_{u \in X} \sup_{\lambda \in [0,a]} (\Phi(u) + \lambda(J(u) + \rho))$$

where

$$a = \frac{hr}{r \dfrac{-J(u_1)}{\Phi(u_1)} - \sup_{\Phi(u)<r}(-J(u))}.$$

Here is another proposition to show the minimax inequality (1.6) in Theorem 1.14.

Proposition 1.14 ([33]). *Let X be a non-empty set and Φ, Ψ two real functions on X. Assume that there are $r > 0$ and $x_0, x_1 \in X$ such that*

$$\Phi(x_0) = -\Psi(x_0) = 0, \quad \Phi(x_1) > r, \quad \sup_{x \in \Phi^{-1}(]-\infty, r])} -\Psi(x) < r\frac{-\Psi(x_1)}{\Phi(x_1)}.$$

Then, for each h satisfying

$$\sup_{x \in \Phi^{-1}(]-\infty, r])} -\Psi(x) < h < r\frac{-\Psi(x_1)}{\Phi(x_1)},$$

one has

$$\sup_{\lambda \geq 0} \inf_{x \in X} (\Phi(x) + \lambda(h + \Psi(x))) < \inf_{x \in X} \sup_{\lambda \geq 0} (\Phi(x) + \lambda(h + \Psi(x))).$$

The following result is proved in [31] that, on the basis of [34], can be equivalently stated as follows

Theorem 1.15 ([31]). *Let X be a separable and reflexive real Banach space; $\Phi : X \to \mathbb{R}$ a continuously Gâteaux differentiable and sequentially weakly lower semicontinuous functional whose Gâteaux derivative admits a continuous inverse on X^*, $\Psi : X \to \mathbb{R}$ a continuously Gâteaux differentiable functional whose Gâteaux derivative is compact. Assume that*

(i) $\lim_{\|u\| \to \infty} \Phi(u) + \lambda\Psi(u) = \infty$ *for all* $\lambda > 0$;
and there are $r \in \mathbb{R}$ *and* $u_0, u_1 \in X$ *such that*
(ii) $\Phi(u_0) < r < \Phi(u_1)$;
(iii) $\inf_{u \in \Phi^{-1}([-\infty, r])} \Psi(u) > \frac{(\Phi(u_1) - r)\Psi(u_0) + (r - \Phi(u_0))\Psi(u_1)}{\Phi(u_1) - \Phi(u_0)}$.

Then there exist an open interval $\Lambda \in (0, \infty)$ and a positive real number q such that for each $\lambda \in \Lambda$ and every continuously Gâteaux differentiable functional $J : X \to \mathbb{R}$ with compact derivative, there exists $\sigma > 0$ such that for each $\mu \in [0, \sigma]$, the equation

$$\Phi'(u) + \lambda\Psi'(u) + \mu J'(u) = 0$$

has at least three solutions in X whose norms are less than q.

We denote by $\mathcal{W}_X$ the class of all functionals $\Phi : X \to \mathbb{R}$ possessing the following property: if $\{u_n\}$ is a sequence in X converging weakly to $u \in X$ and $\liminf_{n \to \infty} \Phi(u_n) \leq \Phi(u)$, then $\{u_n\}$ has a subsequence converging strongly to u. For instance, if X is uniformly convex and $j : [0, +\infty[\to \mathbb{R}$ is a continuous, strictly increasing function, then, by a classical result, the functional $x \to j(\|x\|)$ belongs to the class $\mathcal{W}_X$. Also, given an operator $S : X \to X^*$, we say that S admits a continuous inverse on X^* if there exists a continuous operator $T : X^* \to X$ such that $T(S(x)) = x$ for all $x \in X$. Here, X^* denote the dual space X.

Theorem 1.16 ([35]). *Let X be a separable and reflexive real Banach space; $\Phi : X \to \mathbb{R}$ a coercive, sequentially weakly lower semicontinuous C^1 functional, belonging to $\mathcal{W}_X$, bounded on each bounded subset of X and whose derivative admits a continuous inverse on X^*; $\Psi : X \to \mathbb{R}$ a C^1 functional with compact derivative. Assume that Φ has a strict local minimum x_0 with $\Phi(x_0) = \Psi(x_0) = 0$. Finally, setting*

$$\alpha = \max \left\{ 0, \limsup_{\|x\| \to +\infty} \frac{\Psi(x)}{\Phi(x)}, \limsup_{x \to x_0} \frac{\Psi(x)}{\Phi(x)} \right\},$$

$$\beta = \sup_{x \in \Phi^{-1}(]0,+\infty[)} \frac{\Psi(x)}{\Phi(x)},$$

assume that $\alpha < \beta$.

Then, for each compact interval $[a,b] \subset]\frac{1}{\beta}, \frac{1}{\alpha}[$ (with the conventions $\frac{1}{0} = +\infty$, $\frac{1}{+\infty} = 0$), there exists $r > 0$ with the following property: for every $\lambda \in [a,b]$ and every C^1 functional $J : X \to \mathbb{R}$ with compact derivative, there exists $\delta > 0$ such that, for each $\mu \in [0,\delta]$, the equation

$$\Phi'(x) = \lambda \Psi'(x) + \mu J'(x)$$

has at least three weak solutions whose norms are less than r.

We point out that this type of critical point theorems was introduced by Ricceri in the very nice and seminal paper [36] (see also [37]). Moreover, we emphasize that, very recently, in [35] Ricceri had obtained a three critical points theorem which has been used for perturbed problems as can be seen, for instance, in [38].

The following result has been obtained in [39] and it is a more precise version of Theorem 3.2 of [40].

Theorem 1.17 ([39]). *Let X be a reflexive real Banach space, $\Phi : X \to \mathbb{R}$ be a coercive, continuously Gâteaux differentiable and sequentially weakly lower semicontinuous functional whose Gâteaux derivative admits a continuous inverse on X^*, and $\Psi : X \to \mathbb{R}$ be a continuously Gâteaux differentiable functional whose Gâteaux derivative is compact such that*

$$\Phi(0) = \Psi(0) = 0.$$

Assume that there exist $r > 0$ and $\bar{x} \in X$, with $r < \Phi(\bar{x})$, such that:

(a_1) $\dfrac{\sup_{\Phi(x) \le r} \Psi(x)}{r} < \dfrac{\Psi(\bar{x})}{\Phi(\bar{x})}$;

(a_2) *for each* $\lambda \in \Lambda_r :=]\frac{\Phi(\bar{x})}{\Psi(\bar{x})}, \frac{r}{\sup_{\Phi(x) \leq r} \Psi(x)}[$ *the functional* $\Phi - \lambda\Psi$ *is coercive.*

Then, for each $\lambda \in \Lambda_r$, *the functional* $\Phi - \lambda\Psi$ *has at least three distinct critical points in* X.

1.5.2 A General Variational Principle

A general critical points theorem due to Bonanno and Molica Bisci that is a generalization of a previous result of Ricceri [36] and that here we state in a smooth version for the reader's convenience.

Theorem 1.18 ([41]). *Let* X *be a reflexive real Banach space, let* $\Phi, \Psi : X \to \mathbb{R}$ *be two Gâteaux differentiable functionals such that* Φ *is sequentially weakly lower semicontinuous and coercive and* Ψ *is sequentially weakly upper semicontinuous. For every* $r > \inf_X \Phi$, *let us put*

$$\varphi(r) := \inf_{u \in \Phi^{-1}(]-\infty, r[)} \frac{(\sup_{v \in \Phi^{-1}(]-\infty, r[)} \Psi(v)) - \Psi(u)}{r - \Phi(u)}$$

and

$$\gamma := \liminf_{r \to +\infty} \varphi(r), \quad \delta := \liminf_{r \to (\inf_X \Phi)^+} \varphi(r).$$

Then, one has

(a) *for every* $r > \inf_X \Phi$ *and every* $\lambda \in]0, \frac{1}{\varphi(r)}[$, *the restriction of the functional* $I_\lambda = \Phi - \lambda\Psi$ *to* $\Phi^{-1}(] - \infty, r[)$ *admits a global minimum, which is a critical point (local minimum) of* I_λ *in* X.

(b) *If* $\gamma < +\infty$ *then, for each* $\lambda \in]0, \frac{1}{\gamma}[$, *the following alternative holds: either*

(b_1) I_λ *possesses a global minimum,*

or

(b_2) *there is a sequence* $\{u_n\}$ *of critical points (local minima) of* I_λ *such that* $\lim_{n \to +\infty} \Phi(u_n) = +\infty$.

(c) *If* $\delta < +\infty$ *then, for each* $\lambda \in]0, \frac{1}{\delta}[$, *the following alternative holds: either*

(c_1) *there is a global minimum of* Φ *which is a local minimum of* I_λ,

or

(c_2) *there is a sequence of pairwise distinct critical points (local minima) of* I_λ *which weakly converges to a global minimum of* Φ.

We recall Bonanno's critical point theorems. For a given non-empty set X, and two functionals $\Phi, \Psi : X \to \mathbb{R}$, we define the following two functions:

$$\beta(r_1, r_2) = \inf_{v \in \Phi^{-1}(]r_1, r_2[)} \frac{\sup_{u \in \Phi^{-1}(]r_1, r_2[)} \Psi(u) - \Psi(v)}{r_2 - \Phi(v)},$$

$$\rho(r_1, r_2) = \sup_{v \in \Phi^{-1}(]r_1, r_2[)} \frac{\Psi(v) - \sup_{u \in \Phi^{-1}(]-\infty, r_1[)} \Psi(u)}{\Phi(v) - r_1}$$

for all $r_1, r_2 \in \mathbb{R}$, $r_1 < r_2$.

Theorem 1.19 ([42]). *Let X be a reflexive real Banach space, $\Phi : X \to \mathbb{R}$ be a sequentially weakly lower semicontinuous, coercive and continuously Gâteaux differentiable functional whose Gâteaux derivative admits a continuous inverse on X^* and $\Psi : X \to \mathbb{R}$ be a continuously Gâteaux differentiable functional whose Gâteaux derivative is compact. Put $I_\lambda = \Phi - \lambda\Psi$ and assume that there are $r_1, r_2 \in \mathbb{R}$, $r_1 < r_2$, such that*

$$\beta(r_1, r_2) < \rho(r_1, r_2).$$

Then, for each $\lambda \in \left] \frac{1}{\rho(r_1, r_2)}, \frac{1}{\beta(r_1, r_2)} \right[$ there is $u_{0,\lambda} \in \Phi^{-1}(]r_1, r_2[)$ such that $I_\lambda(u_{0,\lambda}) \leq I_\lambda(u) \; \forall u \in \Phi^{-1}(]r_1, r_2[)$ and $I_\lambda'(u_{0,\lambda}) = 0$.

Here attach the theorem obtained by Bonanno [43].

Theorem 1.20 ([43]). *Let X be a real Banach space and let Φ, Ψ : $X \to \mathbb{R}$ be two continuously Gâteaux differentiable functionals such that Φ is bounded from below and $\Phi(0) = \Psi(0) = 0$. Fix $r > 0$ such that $\sup_{\Phi(u) < r} \Psi(u) < +\infty$ and assume that, for each $\lambda \in]0, \frac{r}{\sup_{\Phi(u) < r} \Psi(u)}[$, the functional $I_\lambda := \Phi - \lambda\Psi$ satisfies (PS)-condition and it is unbounded from below. Then, for each $\lambda \in]0, \frac{r}{\sup_{\Phi(u) < r} \Psi(u)}[$, the functional I_λ admits two distinct critical points.*

Chapter 2

Quasilinear Fourth-Order Problems

2.1 Introduction

The fourth-order equation of nonlinearity furnishes a model to study traveling waves in suspension bridges. In [44], Lazer and McKenna gave survey results in this direction. This fourth-order semilinear elliptic problem can be considered as an analogue of a class of second-order problems which have been studied by many authors. In particular, the deformations of an elastic beam in an equilibrium state, whose two ends are simply supported, can be described by fourth-order boundary value problems and, also for this reason, the existence and multiplicity of solutions for this kind of problems have been widely investigated (see, for instance, [45–61] and references therein). Nonlinear elliptic equations of p-biharmonic type have been studied by many authors, see for instance [44, 62–77] and the references cited therein.

2.2 Multiple Solutions of Quasilinear Fourth-Order Problems

Consider the following fourth order partial differential equation coupled with Navier boundary conditions

$$\begin{cases} \Delta\left(|\Delta u|^{p-2}\Delta u\right) - \mathrm{div}(|\nabla u|^{p-2}\nabla u) = \lambda f(x,u) + \mu g(x,u), & \text{in } \Omega, \\ u = \Delta u = 0, & \text{on } \partial\Omega, \end{cases}$$
(2.1)

where $\Omega \subset \mathbb{R}^N (N \geq 1)$ is a non-empty bounded open set with a sufficient smooth boundary $\partial\Omega$, $p > \max\{1, N/2\}$, $\lambda > 0$, $\mu > 0$ and $f, g \colon \Omega \times \mathbb{R} \to \mathbb{R}$ are two L^1-Carathéodory functions.

We recall that a function $f \colon \Omega \times \mathbb{R} \to \mathbb{R}$ is said to be L^1-Carathéodory if

- $x \to f(x,t)$ is measurable for every $t \in \mathbb{R}$;
- $t \to f(x,t)$ is continuous for a.e. $x \in \Omega$;
- for every $\varrho > 0$ there exists a function $l_\varrho \in L^1(\Omega)$ such that

$$\sup_{|t| \le \varrho} |f(x,t)| \le l_\varrho(x)$$

for a.e. $x \in \Omega$.

In this section, X will denote the Sobolev space $W^{2,p}(\Omega) \cap W_0^{1,p}(\Omega)$ equipped with the norm

$$\|u\| = \left(\int_\Omega |\Delta u(x)|^p + |\nabla u(x)|^p dx \right)^{1/p}.$$

Let

$$K := \sup_{u \in X \setminus \{0\}} \frac{\sup_{x \in \Omega} |u(x)|}{\|u\|}. \tag{2.2}$$

Since $p > \max\{1, N/2\}$, $W^{2,p}(\Omega) \cap W_0^{1,p}(\Omega) \hookrightarrow C^0(\overline{\Omega})$ is compact, and one has $K < +\infty$. As usual, a weak solution of the problem (2.1) is any $u \in X$ such that

$$\int_\Omega |\Delta u(x)|^{p-2} \Delta u(x) \Delta \xi(x) dx + \int_\Omega |\nabla u(x)|^{p-2} \nabla u(x) \nabla \xi(x) dx$$

$$= \lambda \int_\Omega f(x, u(x)) \xi(x) dx + \mu \int_\Omega g(x, u(x)) \xi(x) dx \tag{2.3}$$

for every $\xi \in X$.

The aim of this section is to establish the existence of a non-empty open interval $\Lambda \subseteq I$ and a positive real number q with the following property: for each $\lambda \in \Lambda$ and for each L^1-Carathéodory function $g : \Omega \times \mathbb{R} \to \mathbb{R}$, there is $\delta > 0$ such that, for each $\mu \in [0, \delta]$, the problem (2.1) admits at least three weak solutions whose norms in X are less than q.

Now, fix $x^0 \in \Omega$ and pick γ with $\gamma > 0$ such that $B(x^0, \gamma) \subset \Omega$ where $B(x^0, \gamma)$ denotes the ball with center at x^0 and radius of γ. Put

$$Q = \int_{B(x^0, \gamma) \setminus B(x^0, \gamma/2)} \left| \frac{12}{\gamma^3} |x - x^0| - \frac{24}{\gamma^2} l + \frac{9}{\gamma} \frac{l}{|x - x^0|} \right|^p dx,$$

$$R = \frac{\pi^{N/2}}{\Gamma(N/2)} \int_{(\gamma/2)^2}^{\gamma^2} \left| \frac{12(N+1)}{\gamma^3} \sqrt{t} + \frac{9(N-1)}{\gamma} \frac{1}{\sqrt{t}} - \frac{24N}{\gamma^2} \right|^p t^{N/2-1} dt$$

and

$$\theta = K(R+Q)^{1/p} \qquad (2.4)$$

where $l = (\sum_{i=1}^{N} x_i^2)^{1/2}$, $|x - x^0| = \sqrt{\sum_{i=1}^{N}(x_i - x_i^0)^2}$ and $m(\Omega)$ denotes the volume of Ω. We also let $F(x,t) = \int_0^t f(x,s)ds$ for all $(x,t) \in \Omega \times \mathbb{R}$. Our main result is formulated as follows:

Theorem 2.1. *Assume that there exist a positive constant r and a function $w \in X$ such that*

(A1) $\|w\|^p > pr$;

(A2) $\int_\Omega \sup_{s \in [-K \sqrt[p]{pr}, K \sqrt[p]{pr}]} F(x,s)dx < pr \dfrac{\int_\Omega F(x,w(x))dx}{\|w\|^p}$;

(A3) $pK^p m(\Omega) \limsup_{|s| \to +\infty} \dfrac{F(x,s)}{|s|^p} < \dfrac{1}{r\eta}$ *for almost every $x \in \Omega$ and for some η satisfying*

$$\eta > \frac{1}{pr\dfrac{\int_\Omega F(x,w(x))dx}{\|w\|^p} - \int_\Omega \sup_{s \in [-K \sqrt[p]{pr}, K \sqrt[p]{pr}]} F(x,s)dx}.$$

Then, there exist a non-empty open interval $\Lambda \subseteq [0, r\eta)$ and a positive real number q with the following property: for each $\lambda \in \Lambda$ and for an arbitrary L^1-Carathéodory function $g: \Omega \times \mathbb{R} \mapsto \mathbb{R}$, there exists $\delta > 0$ such that, for each $\mu \in [0, \delta]$, the problem (2.1) has at least three solutions whose norms in X are less than q.

Let us first present a consequence of Theorem 2.1 for a fixed test function w.

Corollary 2.1. *Assume that there exist two positive constants c and d with $c < \theta d$ such that*

(A4) $F(x,s) \geq 0$ *for a.e. $x \in \Omega\backslash B(x^0, \gamma/2)$ and all $s \in [0,d]$;*

(A5) $\int_\Omega \sup_{(x,s) \in \Omega \times [-c,c]} F(x,s)dx < \left(\dfrac{c}{\theta d}\right)^p \int_{B(x^0, \gamma/2)} F(x,d)dx$;

(A6) $c^p m(\Omega) \limsup_{|s| \to +\infty} \dfrac{F(x,s)}{|s|^p} < \dfrac{1}{\eta}$ *for almost every $x \in \Omega$ and for some η satisfying*

$$\eta > \frac{1}{\left(\dfrac{c}{\theta d}\right)^p \int_{B(x^0, \gamma/2)} F(x,d)dx - \int_\Omega \sup_{s \in [-c,c]} F(x,s)dx}.$$

Then, there exist a non-empty open interval $\Lambda \subseteq [0, \frac{1}{p}\left(\frac{c}{K}\right)^p \eta)$ and a positive real number q with the following property: for each $\lambda \in \Lambda$ and for an arbitrary L^1-Carathéodory function $g: \Omega \times \mathbb{R} \mapsto \mathbb{R}$, there exists $\delta > 0$ such that, for each $\mu \in [0, \delta]$, the problem (2.1) has at least three solutions whose norms in X are less than q.

Remark 2.1. We remark that the authors in [68] had already studied the problem (2.1) when $\mu = 0$. Under weaker assumptions as for Theorem 1 of [68], Corollary 2.1 ensures a more precise conclusion. In fact, our condition (A6) is weaker than the condition (iii) in Theorem 1 of [68]. For example, if F is autonomous, let $F(s) = \frac{s^p}{\ln(2+s^2)}$. Clearly, function F satisfies our condition (A6) but does not satisfy (iii) in Theorem 1 of [68].

The proof of Corollary 2.1 is based on the following technical lemma.

Lemma 2.1. *Assume that there exist two positive constants c and d with $c < \theta d$. Under Assumptions (A4) and (A5) of Corollary 2.1, there exist $r > 0$ and $w \in X$ such that $\|w\|^p > pr$ and*

$$\int_\Omega \sup_{s \in [-K\sqrt[p]{pr}, K\sqrt[p]{pr}]} F(x,s)dx < pr\frac{\int_\Omega F(x, w(x))dx}{\|w\|^p}.$$

Proof. Let

$$w(x) = \begin{cases} 0 & \text{for } x \in \Omega \backslash B(x^0, \gamma), \\ d\left(\dfrac{4}{\gamma^3}|x - x^0|^3 - \dfrac{12}{\gamma^2}|x - x^0|^2 \right. & \\ \left. +\dfrac{9}{\gamma}|x - x^0| - 1\right) & \text{for } x \in B(x^0, \gamma) \backslash B(x^0, \gamma/2), \\ d & \text{for } x \in B(x^0, \gamma/2), \end{cases}$$

$$(2.5)$$

where $|x - x^0| = \sqrt{\sum_{i=1}^N (x_i - x_i^0)^2}$ and $r = \frac{1}{p}\left(\frac{c}{K}\right)^p$. We have

$$\frac{\partial w(x)}{\partial x_i} = \begin{cases} 0 & \text{for } x \in \Omega \backslash B(x^0, \gamma) \cup B(x^0, \gamma/2), \\ d\left(\dfrac{12}{\gamma^3}|x - x^0|(x_i - x_i^0)\right. & \\ \left. -\dfrac{24}{\gamma^2}(x_i - x_i^0) + \dfrac{9(x_i - x_i^0)}{\gamma|x - x^0|}\right) & \text{for } x \in B(x^0, \gamma) \backslash B(x^0, \gamma/2) \end{cases}$$

and

$$\frac{\partial^2 w(x)}{\partial^2 x_i} = \begin{cases} 0 & \text{for } x \in \Omega \backslash B(x^0, \gamma) \cup B(x^0, \gamma/2), \\ d\left(\dfrac{12}{\gamma^3|x - x^0|}(x_i - x_i^0)^2\right. & \\ \left. -\dfrac{24}{\gamma^2} + \dfrac{9(|x - x^0|^2 - (x_i - x_i^0)^2)}{\gamma|x - x^0|^3}\right) & \text{for } x \in B(x^0, \gamma) \backslash B(x^0, \gamma/2). \end{cases}$$

It is easy to verify that $w \in W^{2,p}(\Omega) \cap W_0^{1,p}(\Omega)$, and in particular, one has

$$\|w\|^p = (R + Q)d^p,$$

and consequently from (2.4) we see that

$$\|w\| = \frac{\theta d}{K}.$$

Moreover, by the assumption $c < \theta d$, we can get that

$$\frac{\|w\|^p}{p} > \frac{1}{p}\left(\frac{d\theta}{K}\right)^p > \frac{1}{p}\left(\frac{c}{K}\right)^p = r.$$

Since, $0 \le w(x) \le d$, for each $x \in \Omega$, the condition (A4) ensures that

$$\int_{\Omega \setminus B(x^0,\gamma)} F(x,w(x))dx + \int_{B(x^0,\gamma)\setminus B(x^0,\gamma/2)} F(x,w(x))dx \ge 0.$$

Hence, from the condition (A5), $r = \frac{1}{p}\left(\frac{c}{K}\right)^p$ and the above inequality we have

$$\int_{\Omega} \sup_{s\in[-K\sqrt[p]{pr}, K\sqrt[p]{pr}]} F(x,s)dx < \left(\frac{c}{\theta d}\right)^p \int_{B(x^0,\gamma/2)} F(x,d)dx$$

$$\le pr\frac{\int_{\Omega} F(x,w(x))dx}{\|w\|^p}.$$

Thus

$$\int_{\Omega} \sup_{s\in[-K\sqrt[p]{pr}, K\sqrt[p]{pr}]} F(x,s)dx < pr\frac{\int_{\Omega} F(x,w(x))dx}{\|w\|^p},$$

so the proof is complete. □

Proof of Corollary 2.1. From Lemma 2.1 we see that Assumptions (A1) and (A2) of Theorem 2.1 are fulfilled for w given in (2.5). Also, from (A6), one has that (A3) is satisfied. Hence, the conclusion follows directly from Theorem 2.1. □

Remark 2.2. The statement of Corollary 2.1 mainly depends upon the choice of test function w assumed in Theorem 2.1. With the choice of w given in (2.5) we have the present structure of Corollary 2.1. So, other candidates for test function w in (2.5) can be considered to have other versions of the statement of Corollary 2.1.

We end this section by giving the following example to illustrate Corollary 2.1.

Example 2.1. Consider the problem

$$\begin{cases} u^{(iv)} - u'' = \lambda f(u) + \mu g(x,u) & \text{in }]0, 2\pi[, \\ u(0) = u(2\pi) = u''(0) = u''(2\pi) = 0 \end{cases} \tag{2.6}$$

where

$$f(s) = \begin{cases} s^2 & s \le 1 \\ \dfrac{1}{s^2} & s > 1 \end{cases}$$

and $g : [0, 2\pi] \times \mathbb{R} \to \mathbb{R}$ is a fixed L^1-Carathéodory function. Choose $p = 2$, $x^0 = \pi$ and $\gamma = \pi$, and noticing that $K = 1/2\pi$ (see Proposition 2.1 of [51]), one has $\theta = \frac{\sqrt{15(509\pi^2 - 720\pi^2 \ln 2 + 40)}}{5\pi^2}$. So, we see that all the assumptions of Corollary 2.1 are satisfied by choosing, for instance $c = 10^{-3}$ and $d = 1$. Thus, for each

$$\kappa > 20^{-6}\pi^2 \times \frac{1}{\frac{10^{-3} \times 25\pi^5}{90(509\pi^2 - 720\pi^2 \ln 2 + 40)} - \frac{20^{-9}\pi}{3}}$$

there exists an open interval $\Lambda \subset [0, \kappa]$ and a positive real number q such that, for each $\lambda \in \Lambda$ and for each L^1-Carathéodory function $g : [0, 2\pi] \times \mathbb{R} \to \mathbb{R}$, there is $\delta > 0$ such that, for each $\mu \in [0, \delta]$, the problem (2.6) admits at least three weak solutions whose norms in $W^{2,2}([0, 2\pi]) \cap W_0^{1,2}([0, 2\pi])$ are less than q.

Now we can give the proof of our main result.

Proof of Theorem 2.1. For each $u \in X$, let

$$\Phi(u) = \frac{\|u\|^p}{p}, \qquad J(u) = -\int_\Omega F(x, u(x))dx$$

and

$$\Psi(u) = -\int_\Omega \int_0^{u(x)} g(x, s)dsdx.$$

Under the assumptions of Theorem 2.1, Φ is a continuously Gâteaux differentiable and sequentially weakly lower semicontinuous functional. Moreover, the Gâteaux derivative of Φ admits a continuous inverse on X^*. Ψ and J are continuously Gâteaux differential functional whose Gâteaux derivative is compact. Obviously, Φ is bounded on each bounded subset of X. In particular, for each $u, \xi \in X$,

$$\Phi'(u)(\xi) = \int_\Omega |\Delta u(x)|\Delta u(x)\Delta \xi(x)dx + \int_\Omega |\nabla u(x)|^{p-2}\nabla u(x)\nabla \xi(x)dx,$$

$$J'(u)(\xi) = -\int_\Omega f(x, u(x))\xi(x)dx$$

and

$$\Psi'(u)(\xi) = -\int_\Omega g(x, u(x))\xi(x)dx.$$

Hence, it follows from (2.3) that the weak solutions of the problem (2.1) are exactly the solutions of the equation

$$\Phi'(u) + \lambda J'(u) + \mu \Psi'(u) = 0.$$

Furthermore, from (A3) there exist two constants ζ, $\tau \in \mathbb{R}$ with $0 < \zeta < 1/r\eta$ such that

$$pK^p m(\Omega) F(x, s) \leq \zeta |s|^p + \tau$$

for a.e. $x \in \Omega$ and all $s \in \mathbb{R}$. Fix $u \in X$. Then

$$F(x, u(x)) \leq \frac{1}{pK^p m(\Omega)} \left(\zeta |u(x)|^p + \tau \right)$$

for all $x \in \Omega$. Then, for any fixed $\lambda \in]0, r\eta]$, since $\sup_{x \in \Omega} |u(x)| \leq K\|u\|$, we get

$$\Phi(u) + \lambda J(u) = \frac{\|u\|^p}{p} - \lambda \int_\Omega F(x, u(x)) dx$$

$$\geq \frac{\|u\|^p}{p} - \frac{r\eta}{pK^p m(\Omega)} \left(\zeta \int_\Omega |u(x)|^p dx + \tau \right)$$

$$\geq \frac{1}{p}(1 - \zeta r\eta)\|u\|^p - \frac{r\eta}{pK^p m(\Omega)} \tau,$$

and so,

$$\lim_{\|u\| \to +\infty} (\Phi(u) + \lambda J(u)) = +\infty.$$

We claim that there exist $r > 0$ and $w \in X$ such that

$$\sup_{\Phi(u) < r} (-J(u)) < r \frac{-J(w)}{\Phi(w)}.$$

Note that

$$\sup_{x \in \Omega} |u(x)| \leq K\|u\|$$

for each $u \in X$, and so

$$\{u \in X : \Phi(u) < r\} = \{u \in X : \|u\|^p < pr\}$$

$$\subseteq \{u \in X : |u(x)| < K \sqrt[p]{pr} \text{ for all } x \in \Omega\},$$

and it follows that

$$\sup_{\Phi(u) < r} (-J(u)) < \int_\Omega \sup_{t \in [-K\sqrt[p]{pr}, K\sqrt[p]{pr}]} F(x, t) dx.$$

Now from (A2) we have

$$\int_\Omega \sup_{t \in [-K\sqrt[p]{pr}, K\sqrt[p]{pr}]} F(x, t) dx < pr \frac{\int_\Omega F(x, w(x)) dx}{\|w\|^p},$$

and so

$$\sup_{\Phi(u) < r} (-J(u)) < r \frac{-J(w)}{\Phi(w)}.$$

Also from (A1) we have $\Phi(w) > r$. Next recall from (A3) that

$$\eta > \frac{1}{r\frac{-J(w)}{\Phi(w)} - \sup_{\Phi(u)<r}(-J(u))}.$$

Choose

$$\nu > \eta \left(r\frac{-J(w)}{\Phi(w)} - \sup_{\Phi(u)<r}(-J(u)) \right)$$

and note $\nu > 1$. Also, since

$$\eta > \frac{1}{r\frac{-J(w)}{\Phi(w)} - \sup_{\Phi(u)<r}(-J(u))},$$

we have

$$\sup_{\Phi(u)<r}(-J(u)) + \frac{1}{\eta} < r\frac{-J(w)}{\Phi(w)}$$

and so with our choice of ν we have

$$\sup_{\Phi(u)<r}(-J(u)) + \frac{r\frac{-J(w)}{\Phi(w)} - \sup_{\Phi(u)<r}(-J(u))}{\nu} < r\frac{-J(w)}{\Phi(w)}.$$

Therefore, from Proposition 2.1 (with $u_0 = 0$ and $u_1 = w$) for every $\rho \in \mathbb{R}$ satisfying

$$\sup_{\Phi(u)<r}(-J(u)) + \frac{r\frac{-J(w)}{\Phi(w)} - \sup_{\Phi(u)<r}(-J(u))}{\nu} < \rho < r\frac{-J(w)}{\Phi(w)}$$

we have (note $\sigma = r\eta$)

$$\sup_{\lambda \in \mathbb{R}} \inf_{u \in X} (\Phi(u) + \lambda(J(u) + \rho)) < \inf_{u \in X} \sup_{\lambda \in [0,r\eta]} (\Phi(u) + \lambda(J(u) + \rho)).$$

Now, all assumptions of Theorem 1.14 are satisfied. Hence, the conclusion follows directly from Theorem 1.14. □

Notes and Comments

In recent years, Ricceri's three critical points theorem was widely used to solve differential equations, see [67, 78–86] and reference therein.

A nonlinear fourth-order equation furnishes a model to study traveling waves in suspension bridges, so it's important to Physics. Several results are known concerning the existence of multiple solutions for fourth-order boundary value problems, and we refer the reader to [50, 51, 87, 88] and the references cited therein. This section was obtained in [89].

2.3 Infinitely Many Solutions of Quasilinear Fourth-Order Problems

In this section we investigate the existence of infinitely many weak solutions for the problem (2.1) drop the term $\operatorname{div}(|\nabla u|^{p-2}\nabla u)$, that is to say we consider the following problem

$$\begin{cases} \Delta(|\Delta u|^{p-2}\Delta u) = \lambda f(x,u) + \mu g(x,u), & \text{in } \Omega, \\ u = \Delta u = 0, & \text{on } \partial\Omega, \end{cases} \tag{2.7}$$

where Ω is an open bounded subset of $\mathbb{R}^N$ with a smooth enough boundary $\partial\Omega$, $(N \geq 1)$, $p > \max\{1, N/2\}$, Δ is the usual Laplace operator, λ and μ are non-negative parameters and $f, g \in C^0(\bar{\Omega} \times \mathbb{R})$.

The aim of the this section is to localize precise intervals of parameters for λ and μ, namely Λ and I, such that for each $\lambda \in \Lambda$ and $\mu \in I$, problem (2.7) admits infinitely many weak solutions. To this end, we require a suitable oscillation behavior, either at infinity or at zero, for the antiderivative F of f and an appropriate growth condition on G, the antiderivative of g (Theorems 2.2 and 2.4). We emphasize that to get our conclusions we did not assumed conditions neither of symmetry nor of sign on the nonlinearities f and g. However, we emphasize that the case in which f and g satisfy suitable sign conditions is also considered in carrying out some consequences of the main results for the autonomous version of problem (2.7) (Theorem 2.3 and Corollary 2.2).

In this section X denotes the space $W^{2,p}(\Omega) \cap W_0^{1,p}(\Omega)$ endowed with the norm

$$\|u\| = \left(\int_\Omega |\Delta u(x)|^p dx \right)^{1/p} \quad \forall u \in X. \tag{2.8}$$

The Rellich Kondrachov Theorem assures that X is compactly embedded in $C^0(\bar{\Omega})$, being

$$k := \sup_{u \in X \setminus \{0\}} \frac{\|u\|_{C^0(\bar{\Omega})}}{\|u\|} < +\infty. \tag{2.9}$$

Moreover, if $N \geq 3$, $\partial\Omega$ is of class $C^{1,1}$ and $p \in]N/2, +\infty[$, owing to Theorem 2 and Remark 1 of [90], one has the following upper bound

$$k \leq |\Omega|^{2/N+1/p'-1} \frac{\Gamma(1+N/2)^{2/N}}{N(N-2)\pi} \left[\frac{\Gamma(1+p')\Gamma(N/(N-2)-p')}{\Gamma(N/(N-2))} \right]^{1/p'},$$

where Γ denotes the Gamma function, p' the conjugate exponent of p and $|\Omega|$ the Lebesgue measure of the open set Ω.

Let $f, g \in C^0(\bar{\Omega} \times \mathbb{R})$ and let us put

$$F(x,t) := \int_0^t f(x,\xi)d\xi, \qquad G(x,t) := \int_0^t g(x,\xi)d\xi \quad \forall(x,t) \in \bar{\Omega} \times \mathbb{R}.$$

Fixed $\lambda > 0$ and $\mu \geq 0$, let us define $\Phi, \psi_{\lambda_\mu} : X \to \mathbb{R}$ by putting

$$\Phi(u) := \frac{1}{p}\|u\|^p, \quad \psi_{\lambda_\mu}(u) := \int_\Omega \left[F(x,u(x)) + \frac{\mu}{\lambda}G(x,u(x))\right]dx \quad (2.10)$$

for every $u \in X$. It is simple to verify that Φ and ψ_{λ_μ} are well defined, as well as Gâteaux differentiable. Moreover, in view of the fact that Φ is continuous and convex, it turns out sequentially weakly lower semicontinuous, while, since Ψ has compact derivative, it results sequentially weakly continuous. In particular, one has

$$\Phi'(u)(v) = \int_\Omega |\Delta u(x)|^{p-2}\Delta u(x)\Delta v(x)dx,$$

$$\psi'_{\lambda_\mu}(u)(v) = \int_\Omega \left[f(x,u(x)) + \frac{\mu}{\lambda}g(x,u(x)))\right]v(x)dx$$

for every $u, v \in X$.

We explicitly observe that, in view of (2.9), one has that, for every $r > 0$

$$\Phi^{-1}(]-\infty, r[) := \{u \in X : \Phi(u) < r\} \subseteq \{u \in C^0(\bar{\Omega}) : \|u\|_{C^0} < k(pr)^{1/p}\}.$$
$$(2.11)$$

Finally, if we recall that a weak solution of problem (2.7) is a function $u \in X$ such that

$$\int_\Omega |\Delta u(x)|^{p-2}\Delta u(x)\Delta v(x)dx - \int_\Omega [\lambda f(x,u(x)) + \mu g(x,u(x))]\,v(x)dx = 0,$$

for all $v \in X$, it is obvious that our goal is to find critical points of the functional $\Phi - \lambda\psi_{\lambda,\mu}$.

Fixed $x^0 \in \Omega$, let us pick $0 < s_1 < s_2$ such that $B(x^0, s_2) \subseteq \Omega$ and put

$$L := L(x_0, s_1, s_2, N, k) := \frac{(s_2 - s_1)^{2p}\Gamma\left(1 + \frac{N}{2}\right)}{(4Nk)^p\pi^{N/2}(s_2^N - s_1^N)}, \quad (2.12)$$

where Γ denotes the Gamma function and k is defined in (2.9).

Theorem 2.2. *Assume that*

(A7) $F(x,t) \geq 0$ for every $(x,t) \in \Omega \times [0,+\infty[$;

(A8) *There exist* $x^0 \in X$, $0 < s_1 < s_2$ *as considered in (2.12) such that, if we put*

$$\alpha := \liminf_{t \to +\infty} \frac{\int_\Omega \max_{|\xi| \leq t} F(x, \xi) dx}{t^p}, \qquad \beta := \limsup_{t \to +\infty} \frac{\int_{B(x^0, s_1)} F(x, t) dx}{t^p},$$

one has

$$\alpha < L\beta. \tag{2.13}$$

Then, for every $\lambda \in \Lambda := \frac{1}{pk^p}]\frac{1}{L\beta}, \frac{1}{\alpha}[$ *and for every* $g \in C^0(\bar\Omega \times \mathbb{R})$ *such that*

(A9) $G(x, t) \geq 0$ *for every* $(x, t) \in \bar\Omega \times [0, +\infty[$,
(A10) $G_\infty := \limsup_{t \to +\infty} \frac{\int_\Omega \max_{|\xi| \leq t} G(x, \xi) dx}{t^p} < +\infty$,

for every $\mu \in I := [0, \mu^*[$, *where* $\mu^* := \frac{1}{pk^p G_\infty}(1 - \lambda \alpha pk^p)$, *problem (2.7) admits an unbounded sequence of weak solutions.*

Proof. Fix $\lambda \in \Lambda$, let $g \in C^0(\bar\Omega \times \mathbb{R})$ be satisfying (A9)-(A10) and pick $\mu \in [0, \mu^*[$. We want to apply Theorem 1.18 with $X = W^{2,p}(\Omega) \cap W_0^{1,p}(\Omega)$ endowed with the norm introduced in (2.8), and

$$\Phi(u) := \frac{1}{p}\|u\|^p, \qquad \Psi(u) := \psi_{\lambda_\mu}(u)$$

for every $u \in X$, where ψ_{λ_μ} is the functional defined in (2.10). For every $r > 0$, in view of (2.11), we can produce the following estimate from above of the number $\varphi(r)$ defined in Theorem 1.18

$$\varphi(r) \leq \frac{\sup_{\Phi^{-1}(]-\infty, r[)} \Psi}{r}$$

$$\leq \frac{\int_\Omega \max_{|\xi| \leq k(pr)^{1/p}} F(x, \xi) dx}{r} \tag{2.14}$$

$$+ \frac{\mu}{\lambda} \frac{\int_\Omega \max_{|\xi| \leq k(pr)^{1/p}} G(x, \xi) dx}{r}. \tag{2.15}$$

Let now a sequence $\{t_n\}$ of positive numbers such that $t_n \to +\infty$ and

$$\lim_{n \to +\infty} \frac{\int_\Omega \max_{|\xi| \leq t_n} F(x, \xi) dx}{t_n^p} = \alpha. \tag{2.16}$$

Let us define the sequence $\{r_n\}$ in $\mathbb{R}$ by putting $r_n = \frac{1}{p}\left(\frac{t_n}{k}\right)^p$ for every $n \in \mathbb{N}$. Exploiting (2.14), (2.16), (2.13) and assumption (g_2) one has

$$\gamma \leq \liminf_{n \to +\infty} \varphi(r_n) \leq pk^p \left(\alpha + \frac{\mu}{\lambda} G_\infty\right) < +\infty. \tag{2.17}$$

Hence, since $\mu \in [0, \mu^*[$, it is easy to verify that

$$\gamma \leq pk^p \left(\alpha + \frac{\mu^*}{\lambda} G_\infty \right) = \frac{1}{\lambda},$$

that is

$$\Lambda \subseteq \left] 0, \frac{1}{\gamma} \right[. \tag{2.18}$$

Our next step consists in verifying that

$$\Phi - \lambda \Psi \quad \text{is unbounded from below.} \tag{2.19}$$

Since $\frac{1}{\lambda} < pk^p L\beta$, there exists a sequence $\{\tau_n\}$ of positive numbers and $\eta > 0$ such that $\tau_n \to +\infty$ and

$$\frac{1}{\lambda} < \eta < pk^p L \frac{\int_{B(x^0, s_1)} F(x, \tau_n) dx}{\tau_n^p} \tag{2.20}$$

for every $n \in \mathbb{N}$ large enough. Let $\{w_n\}$ be a sequence in X defined by putting

$$w_n(x) = \tag{2.21}$$

$$\begin{cases} \tau_n & \text{if } x \in B(x^0, s_1) \\ \frac{\tau_n}{s_2 - s_1} [s_2 - s_1 - 2\frac{(|x-x^0|-s_1)^2}{s_2 - s_1}] & \text{if } x \in B(x^0, (s_1 + s_2)/2) \setminus B(x^0, s_1) \\ 2\frac{\tau_n}{s_2 - s_1} \frac{(|x-x^0|-s_2)^2}{s_2 - s_1}] & \text{if } x \in B(x^0, s_2) \setminus B(x^0, (s_1 + s_2)/2) \\ 0 & \text{if } x \in \Omega \setminus B(x^0, s_2). \end{cases}$$

Fixed $n \in \mathbb{N}$ and $i \in \{1, 2, \ldots, N\}$, a simple computation shows that

$$\frac{\partial w_n}{\partial x_i}(x) = \tag{2.22}$$

$$\begin{cases} 0 & \text{if } x \in B(x^0, s_1) \\ -4\frac{\tau_n}{(s_2 - s_1)^2}(|x - x^0| - s_1)\frac{x_i - x_i^0}{|x-x^0|} & \text{if } x \in B(x^0, (s_1 + s_2)/2) \setminus \bar{B}(x^0, s_1) \\ 4\frac{\tau_n}{(s_2 - s_1)^2}(|x - x^0| - s_2)\frac{x_i - x_i^0}{|x-x^0|} & \text{if } x \in B(x^0, s_2) \setminus \bar{B}(x^0, (s_1 + s_2)/2) \\ 0 & \text{if } x \in \Omega \setminus \bar{B}(x^0, s_2), \end{cases}$$

as well as

$$\frac{\partial^2 w_n}{\partial x_i^2}(x) = \tag{2.23}$$

$$\begin{cases} 0 & \text{if } x \in B(x^0, s_1) \\ -4\frac{\tau_n}{(s_2 - s_1)^2} \left[1 - s_1 \frac{|x-x^0|^2 - (x_i - x_i^0)^2}{|x-x^0|^3} \right] & \text{if } x \in B(x^0, (s_1 + s_2)/2) \setminus \bar{B}(x^0, s_1) \\ 4\frac{\tau_n}{(s_2 - s_1)^2} \left[1 - s_2 \frac{|x-x^0|^2 - (x_i - x_i^0)^2}{|x-x^0|^3} \right] & \text{if } x \in B(x^0, s_2) \setminus \bar{B}(x^0, (s_1 + s_2)/2) \\ 0 & \text{if } x \in \Omega \setminus \bar{B}(x^0, s_2). \end{cases}$$

Hence,

$$\Delta w_n(x) = \sum_{i=1}^{N} \frac{\partial^2 w_n}{\partial x_i^2}(x) \tag{2.24}$$

$$= \begin{cases} 0 & \text{if } x \in B(x^0, s_1) \\ -4\frac{\tau_n}{(s_2-s_1)^2}\left[N - \frac{s_1(N-1)}{|x-x^0|}\right] & \text{if } x \in B(x^0, (s_1+s_2)/2) \setminus \bar{B}(x^0, s_1) \\ 4\frac{\tau_n}{(s_2-s_1)^2}\left[N - \frac{s_2(N-1)}{|x-x^0|}\right] & \text{if } x \in B(x^0, s_2) \setminus \bar{B}(x^0, (s_1+s_2)/2) \\ 0 & \text{if } x \in \Omega \setminus \bar{B}(x^0, s_2). \end{cases}$$

Observed that

$$N - \frac{s_1(N-1)}{|x-x^0|} = 1 + \frac{|x-x^0| - s_1}{|x-x^0|}(N-1),$$

as well as

$$N - \frac{s_2(N-1)}{|x-x^0|} = 1 + \frac{|x-x^0| - s_2}{|x-x^0|}(N-1),$$

since,

$$0 \le \frac{|x-x^0| - s_1}{|x-x^0|} \le 1$$

for every $x \in B(x^0, (s_1+s_2)/2) \setminus \bar{B}(x^0, s_1)$, and

$$-1 < \frac{|x-x^0| - s_2}{|x-x^0|} < 0$$

for every $x \in B(x^0, s_2) \setminus \bar{B}(x^0, (s_1+s_2)/2)$, we obtain that

$$|\Delta w_n(x)| \le \frac{4N}{(s_2-s_1)^2}\tau_n,$$

for every $x \in B(x^0, s_2) \setminus \bar{B}(x^0, s_1)$. This implies that

$$\Phi(w_n) = \frac{1}{p}\int_{B(x^0,s_2)\setminus\bar{B}(x^0,s_1)} |\Delta w_n(x)|^p \, dx$$

$$\le \frac{1}{p}\frac{(4N)^p}{(s_2-s_1)^{2p}}\tau_n^p \frac{\pi^{N/2}}{\Gamma(1+N/2)}(s_2^N - s_1^N)$$

$$= \frac{\tau_n^p}{pk^pL}. \tag{2.25}$$

Putting together (A7) and (A9) one has

$$\Psi(w_n) = \int_{\Omega}\left[F(x, w_n(x)) + \frac{\mu}{\lambda}G(x, w_n(x))\right] dx \ge \int_{B(x^0,s_1)} F(x, \tau_n)dx. \tag{2.26}$$

At this point, from (2.25), (2.26) and (2.20) we achieve

$$\Phi(w_n) - \lambda\Psi(w_n) \leq \frac{\tau_n^p}{pk^p L} - \lambda \int_{B(x^0, s_1)} F(x, \tau_n)dx < \frac{\tau_n^p}{pk^p L}(1 - \lambda\eta)$$

for every $n \in \mathbb{N}$ large enough, which leads to (2.19), taking in mind that $\tau_n \to +\infty$ and $1 - \lambda\eta < 0$.

Finally we can apply Theorem 1.18 (case (b)) and obtain the existence of an unbounded sequence $\{u_n\}$ of critical points of the functional $\Phi - \lambda\Psi$. This completes the proof in view of the relation between the critical points of $\Phi - \lambda\Psi$ and the weak solutions of problem (2.7). □

Remark 2.3. We explicitly observe that, for $\mu = 0$, Theorem 2.2 gives back Theorem 3.1 of [65]. While, whenever $\alpha = 0$ and $\beta = +\infty$ the interval Λ becomes $]0, +\infty[$. Moreover, for every g satisfying (A9) and such that $G_\infty = 0$ the interval I becomes $[0, +\infty[$.

In the autonomous case it is possible to establish an estimate of the best constant L that is present in assumption (2.13). To this end, taking into account the Lipschitz continuous $s : \bar{\Omega} \to \mathbb{R}_0^+$ defined by

$$s(x) = d(x, \partial\Omega) \qquad \forall x \in \bar{\Omega}.$$

It is easy to prove that there exists $y_0 \in \Omega$ such that

$$\bar{s} = s(y^0) = \max_{x \in \Omega} s(x).$$

Moreover, putting

$$L' = \frac{\bar{s}^{2p}}{(4Nk)^p|\Omega|} \, \frac{\bar{\mu}^N(1 - \bar{\mu})^{2p}}{1 - \bar{\mu}^N}. \tag{2.27}$$

where $\bar{\mu} \in]0, 1[$ is the point where the function $\frac{\mu^N(1-\mu)^{2p}}{1-\mu^N}$ achieves its maximum in $]0, 1[$, the following autonomous version of Theorem 2.2 holds.

Theorem 2.3. *Let* $h : \mathbb{R} \to \mathbb{R}$ *be a continuous function such that:*

(A11) $H(t) = \int_0^t h(\xi)dx \geq 0$ *for every* $t \in [0, +\infty[$;
(A12) Put

$$\alpha' := \liminf_{t \to +\infty} \frac{\max_{|\xi| \leq t} H(t)}{t^p}, \qquad \beta' := \limsup_{t \to +\infty} \frac{H(t)}{t^p},$$

one has

$$\alpha' < L'\beta',$$

where L' *is defined (2.27).*

Then, for every $\lambda \in \frac{1}{pk^p|\Omega|} \left] \frac{1}{L'\beta'}, \frac{1}{\alpha'} \right[$ *and for every* $q \in C^0(\mathbb{R})$ *such that*

(A13) $Q(t) = \int_0^t q(\xi)d\xi \geq 0$ *for every* $t \in [0, +\infty[$,

(A14) $Q_\infty := \limsup_{t \to +\infty} \frac{\max_{|\xi| \leq t} Q(\xi)dx}{t^p} < +\infty$,

for every $\mu \in I := [0, \mu^*[$, *where* $\mu^* := \frac{1}{pk^p|\Omega|Q_\infty}(1 - \lambda\alpha pk^p)$, *the following problem*

$$\begin{cases} \Delta(|\Delta u|^{p-2}\Delta u) = \lambda h(u) + \mu q(u), & \text{in } \Omega, \\ u = \Delta u = 0, & \text{on } \partial\Omega, \end{cases} \tag{2.28}$$

admits an unbounded sequence of weak solutions.

Proof. Put $x^0 = y^0$, $s_2 = \bar{s}$, $s_1 = \bar{\mu}\bar{s}$, $f(x,t) = h(t)$ and $g(x,t) = q(t)$ for every $(x,t) \in \bar{\Omega} \times \mathbb{R}$. Obviously (A11) implies (A7). Moreover,

$$\alpha = |\Omega|\alpha', \quad \beta = \frac{\pi^{N/2}}{\Gamma(1 + N/2)}(\bar{s}\bar{\mu})^N\beta', \quad L = \frac{|\Omega|\Gamma(1 + N/2)}{\pi^{N/2}(\bar{\mu}\bar{s})^N}L'.$$

Hence, in view of (A12) one has

$$\alpha < |\Omega|L'\beta' = L\beta,$$

that is (A8) holds and the conclusion follows directly from Theorem 2.2 once observed that $G(x,t) = Q(t)$ for every $(x,t) \in \bar{\Omega} \times \mathbb{R}$ and $G_\infty = |\Omega|Q_\infty$ so that (A9) and (A10) are satisfied. $\qquad\square$

Remark 2.4. We explicitly point out that in view of the definitions of $\bar{s}$ and $\bar{\mu}$, according to the approach that we developed, in the case $\beta' < +\infty$, the constant L' introduced in (2.27) and involved in assumption (A14) is the best possible.

An immediate consequence of Theorem 2.3 is the following corollary.

Corollary 2.2. *Let* $h : \mathbb{R} \to \mathbb{R}$ *be a continuous and nonnegative function such that*

$$\liminf_{t \to +\infty} \frac{H(t)}{t^p} < L' \limsup_{t \to +\infty} \frac{H(t)}{t^p}, \tag{2.29}$$

being L' *defined in (2.27). Then, for every*

$$\lambda \in \Lambda' := \frac{1}{pk^p|\Omega|} \left] \frac{1}{L' \limsup_{t \to +\infty} \frac{H(t)}{t^p}}, \frac{1}{\liminf_{t \to +\infty} \frac{H(t)}{t^p}} \right[,$$

for every $q \in C^0(\mathbb{R})$ *such that*

(A15) $tq(t) \geq 0$ *for every* $t \in \mathbb{R}$,
(A16) $\lim_{|t| \to +\infty} \frac{q(t)}{|t|^{p-1}} = 0$

and for every $\mu \geq 0$ *problem (2.28) admits an unbounded sequence of weak solutions.*

Proof. It follows from Theorem 2.3 observing that, in view of the non-negativity of h, (A11) holds and $\alpha' = \liminf_{t \to +\infty} \frac{H(t)}{t^p}$, as well as (A15) implies (A13). Moreover, by (A15) one has

$$0 \leq \limsup_{t \to +\infty} \frac{\max_{|\xi| \leq t} Q(t)}{t^p} = \limsup_{t \to +\infty} \frac{\max\{Q(t), Q(-t)\}}{t^p}.$$

Exploiting, (A16) and owing to Hôpital rule we have

$$\lim_{t \to +\infty} \frac{Q(t)}{t^p} = \lim_{t \to +\infty} \frac{Q(-t)}{t^p} = \pm \lim_{t \to +\infty} \frac{q(\pm t)}{t^{p-1}} = 0.$$

Hence, $Q_\infty = 0$ and our conclusion follows. $\qquad\square$

Example 2.2. Let $\Omega =\,]0, 2l[$, with $l > 0$, $p = 2$ and $h : \mathbb{R} \to \mathbb{R}$ be a function defined by putting

$$h(t) := \begin{cases} 2t\left(1 + a\sin^2(\ln(a^2 + \ln^2 t)) + \sin(2\ln(a^2 + \ln^2 t))\dfrac{a\ln t}{a^2 + \ln^2 t}\right) \\ \quad \text{if } t \in]0, +\infty[, \\ 0 \quad \text{if } t \in]-\infty, 0], \end{cases}$$

where $a > \frac{2^{10}}{\pi^2} - 1$. For every $\lambda \in \frac{1}{l^4} \left]\frac{2^7}{1+a}, \frac{\pi^2}{8}\right[\subset \Lambda'$, for every $1 < s < p$ and for every $\mu \geq 0$ the following problem

$$\begin{cases} u^{iv} = \lambda h(u) + \mu|u|^{s-2}u \quad \text{in }]0, 2l[, \\ u(0) = u(2l) = 0, \\ u''(0) = u''(2l) = 0, \end{cases}$$

admits an unbounded sequence of weak solutions. In particular, if $2l = 1$ and $a > \frac{2^{10}}{\pi^2} - 1$, for every $\lambda \in \left]\frac{2^{11}}{1+a}, 2\pi^2\right[$, for every $1 < s < p$ and for every $\mu \geq 0$ the following problem

$$\begin{cases} u^{iv} = \lambda h(u) + \mu|u|^{s-2}u \quad \text{in }]0, 1[, \\ u(0) = u(1) = 0, \\ u''(0) = u''(1) = 0, \end{cases}$$

admits an unbounded sequence of weak solutions. To see this, we apply Corollary 2.2. In view of the choice of a, since

$$-1 \leq \sin(2\ln(a^2 + \ln^2 t)) \leq 1,$$

and

$$-\frac{1}{2} \leq \frac{a \ln t}{a^2 + \ln^2 t} \leq \frac{1}{2}$$

for every $t \in]0, +\infty[$, it is simple to verify that h is non-negative as well as continuous. Moreover,

$$H(t) = \int_0^t h(\xi) d\xi = \begin{cases} t^2 \left(1 + a \sin^2(\ln(a^2 + \ln^2 t))\right) & \text{if } t \in]0, +\infty[, \\ 0 & \text{if } t \in]-\infty, 0]. \end{cases}$$

Hence, if we put

$$a_n = \exp\left(\sqrt{\exp(n\pi) - a^2}\right) \quad \text{and} \quad b_n = \exp\left(\sqrt{\exp((2n+1)\pi/2) - a^2}\right)$$

for every $n \in \mathbb{N}$ such that $n > 2\frac{\ln a}{\pi}$, one has that

$$\liminf_{t \to +\infty} \frac{H(t)}{t^p} \leq \lim_{n \to +\infty} \frac{H(a_n)}{a_n^2} = 1 \tag{2.30}$$

and

$$\limsup_{t \to +\infty} \frac{H(t)}{t^p} \geq \lim_{n \to +\infty} \frac{H(b_n)}{b_n^2} = 1 + a. \tag{2.31}$$

We can observe that, fixed $u \in W^{2,2}([0, 2l]) \cap W_0^{1,2}([0, 2l])$, in view of the following Poincaré inequality (see [91])

$$\int_0^{2l} |u'(t)|^2 \, dt \leq \left(\frac{2l}{\pi}\right)^2 \int_0^{2l} |u''(t)|^2 \, dt,$$

and since

$$|u(t)| \leq \frac{\sqrt{2l}}{2} \left(\int_0^{2l} |u'(s)|^2 \, ds\right)^{\frac{1}{2}}$$

for every $t \in [0, 2l]$, one has that

$$\|u\|_{C^0([0,2l])} \leq \frac{l\sqrt{2l}}{\pi} \|u\|,$$

namely

$$k \leq \frac{l\sqrt{2l}}{\pi}. \tag{2.32}$$

Hence, from (2.27), (2.32) and the definition of $\bar{\mu}$, it follows that

$$L' = \frac{l^4}{(4 \cdot 1 \cdot k)^2 2l} \bar{\mu}(1 - \bar{\mu})^3 \geq \frac{l^4}{16 \cdot 2l} \frac{\pi^2}{2l^3} \bar{\mu}(1 - \bar{\mu})^3 \geq \frac{\pi^2}{64} \frac{1}{2^4} = \frac{\pi^2}{2^{10}}. \tag{2.33}$$

Putting together (2.30), (2.31) and the definition of L' one has

$$\liminf_{t \to +\infty} \frac{H(t)}{t^p} = 1 < \pi^2 \frac{1 + a}{2^{10}} \leq L' \limsup_{t \to +\infty} \frac{H(t)}{t^p}$$

so the assumptions of Corollary 2.2 are satisfied. Moreover, in view of the definition of $\bar{\mu}$ and L' one has

$$\frac{1}{pk^p|\Omega|L'\limsup_{t\to+\infty}\frac{H(t)}{t^p}} \leq \frac{2^7}{(1+a)l^4},$$

while, taking in mind (2.32),

$$\frac{1}{4lk^2} \geq \frac{1}{4l}\frac{\pi^2}{2l^3} = \frac{\pi^2}{8l^4}.$$

Arguing in a similar way, but applying case (c) of Theorem 1.18, it is possible to establish the existence of infinitely many weak solutions to problem (2.7) converging at zero. More precisely, the following result holds.

Theorem 2.4. *Assume that*

(A17) there exists $\tau > 0$ such that $F(x,t) \geq 0$ for every $(x,t) \in \bar{\Omega} \times [0,\tau]$,

(A18) there exist $x^0 \in X$, $0 < s_1 < s_2$ as considered in (2.12) such that, if we put

$$\alpha^0 := \liminf_{t\to 0^+} \frac{\int_\Omega \max_{|\xi|\leq t} F(x,\xi)dx}{t^p}, \beta^0 := \limsup_{t\to 0^+} \frac{\int_{B(x^0,s_1)} F(x,t)dx}{t^p},$$

one has

$$\alpha^0 < L\beta^0. \tag{2.34}$$

Then, for every $\lambda \in \Lambda := \frac{1}{pk^p}\left]\frac{1}{L\beta^0}, \frac{1}{\alpha^0}\right[$ and for every $g \in C^0(\bar{\Omega} \times \mathbb{R})$ such that

(A19) $G(x,t) \geq 0$ for every $(x,t) \in \bar{\Omega} \times [0,\tau]$,

(A20) $G_0 := \limsup_{t\to 0^+} \frac{\int_\Omega \max_{|\xi|\leq t} G(x,\xi)dx}{t^p} < +\infty$,

for every $\mu \in I := [0,\mu^*[$, where $\mu^* := \frac{1}{pk^p G_0}(1 - \lambda\alpha^0 pk^p)$, problem (2.7) admits a sequence $\{u_n\}$ of weak solutions such that $u_n \to 0$.

Proof. Fix $\lambda \in \Lambda$, let $g \in C^0(\bar{\Omega} \times \mathbb{R})$ be satisfying (A19)-(A20) and pick $\mu \in [0,\mu^*[$. We want to apply Theorem 1.18 case (c), with $X = W^{2,p}(\Omega) \cap W_0^{1,p}(\Omega)$ endowed with the norm introduced in (2.8), and

$$\Phi(u) := \frac{1}{p}\|u\|^p, \qquad \Psi(u) := \psi_{\lambda_\mu}(u)$$

for every $u \in X$, where ψ_{λ_μ} is the functional defined in (2.10).

Let $\{t_n\}$ be a sequence of positive numbers such that $t_n \to 0^+$ and

$$\lim_{n \to +\infty} \frac{\int_\Omega \max_{|\xi| \leq t_n} F(x,\xi)dx}{t_n^p} = \alpha^0 < +\infty. \tag{2.35}$$

Putting $r_n = \frac{1}{p}\left(\frac{t_n}{k}\right)^p$ for every $n \in \mathbb{N}$ and working as in the proof of Theorem 2.2, it follows that $\delta < +\infty$, where δ is as defined in Theorem 1.18, as well as $\Lambda \subseteq \left]0, \frac{1}{\delta}\right[$.

We claim that

$$\Phi - \lambda\Psi \text{ has not a local minimum at zero.} \tag{2.36}$$

Let $\{\tau_n\}$ be a sequence of positive numbers in $]0, \tau[$ and $\eta > 0$ such that $\tau_n \to 0^+$ and

$$\frac{1}{\lambda} < \eta < pk^p L \frac{\int_{B(x^0,s_1)} F(x,\tau_n)dx}{\tau_n^p} \tag{2.37}$$

for every $n \in \mathbb{N}$ large enough. Let $\{w_n\}$ be the sequence in X defined in (2.21). From (A17) and (A19) one has that (2.26) holds. Hence, putting together (2.25), (2.26) and (2.37) we achieve

$$\Phi(w_n) - \lambda\Psi(w_n) < \frac{\tau_n^p}{pk^p L}(1 - \lambda\eta) < 0 = \Phi(0) - \lambda\Psi(0)$$

for every $n \in \mathbb{N}$ large enough, that implies claim (2.36) in view of the fact that $\|w_n\| \to 0$.

Observed that $\min_X \Phi = \Phi(0)$, the conclusion follows from the alternative of Theorem 1.18 case (c). $\square$

Remark 2.5. By analogy with Theorem 2.3 and Corollary 2.2, if the "lim inf" and the "lim sup" are considered for $t \to 0^+$ and reasoning as in Theorem 2.4, it is possible to obtain further multiplicity results of arbitrarily small weak solutions of problem (2.28).

Notes and Comments

Nonlinear elliptic equations of p-biharmonic type had been studied by many Authors, see for instance [49, 53, 62, 65–67, 88, 91] and the references cited therein. In particular, in [64], via perturbation theory, it has been investigated problem where, also if the principal term f is odd, the perturbation g has not symmetry and therefore the Symmetric Mountain Pass Theorem cannot be applied directly to get infinitely many solutions. On this topic, see also [63], where the goal is achieved by using a variant of the Fountain Theorem. In [64], it deals with problem where the left-hand side of the equation involves an operator that is more general than the p-biharmonic. While in [44], a concrete example of application of such mathematical model to describe a physical phenomenon is also pointed out.

Here, starting by the multiplicity results obtained in [65] and [66], we are interested in looking for a class of perturbations for which (2.7) still preserves multiple solutions. In this direction, in [67], the existence of at least three solutions is obtained for (2.7). This section was obtained in [92].

2.4 Quasilinear Fourth-Order Problems with Singular Term

In this section, we want to investigate the following problem

$$\begin{cases} \Delta_p^2 u + \dfrac{|u|^{p-2}u}{|x|^{2p}} = \lambda f(x,u) & \text{in } \Omega, \\ u = \Delta u = 0 & \text{on } \partial\Omega, \end{cases} \tag{2.38}$$

where Ω is a bounded domain in $\mathbb{R}^N (N \geq 5)$ containing the origin and with smooth boundary $\partial\Omega$, $1 < p < N/2$, and $f : \Omega \times \mathbb{R} \to \mathbb{R}$ is a Carathéodory function such that

(A21) $|f(x,t)| \leq a_1 + a_2|t|^{q-1}$, $\forall (x,t) \in \Omega \times \mathbb{R}$,

for some non-negative constants a_1, a_2 and $q \in]p, p^*[$, where

$$p^* := \frac{pN}{N-2p}.$$

In this work, our goal is to obtain the existence of two distinct weak solutions for problem (2.38).

Now, we establish the main abstract result of this section. We recall that c_q is the constant of the embedding $W_0^{1,p}(\Omega) \cap W^{2,p}(\Omega) \hookrightarrow L^q(\Omega)$ for each $q \in [1, p^*[$, and c_1 stands for c_q with $q = 1$; see (2.42).

Theorem 2.5. *Let* $f : \Omega \times \mathbb{R} \to \mathbb{R}$ *be a Carathéodory function such that condition (A21) holds. Moreover, assume that*

(A22) there exist $\theta > p$ *and* $M > 0$ *such that*

$$0 < \theta F(x,t) \leq t f(x,t),$$

for each $x \in \Omega$ *and* $|t| \geq M$. *Then, for each* $\lambda \in]0, \lambda^*[$, *problem (2.38) admits at least two distinct weak solutions, where*

$$\lambda^* := \frac{q}{q a_1 c_1 p^{1/p} + a_2 c_q^q p^{q/p}}.$$

In conclusion we present a concrete example of application of Theorem 2.5 whose construction is motivated by [93, Example 4.1].

Example 2.3. We consider the function f defined by

$$f(x,t) := \begin{cases} c + dqt^{q-1}, & \text{if } x \in \Omega, t \geq 0, \\ c - dq(-t)^{q-1}, & \text{if } x \in \Omega, t < 0, \end{cases}$$

for each $(x,t) \in \Omega \times \mathbb{R}$, where $1 < p < q < p^*$ and c, d are two positive constants. Fixed $p < \theta < q$ and

$$r > \max \left\{ \left[\frac{(\theta - 1)c}{d(q - \theta)} \right]^h, \left(\frac{c}{d} \right)^h \right\},$$

with $h = \frac{1}{q-1}$, we prove that f verifies the assumptions requested in Theorem 2.5. Condition (A21) of Theorem 2.5 is easily verified. We observe that

$$F(x,t) = ct + d|t|^q,$$

for each $(x,t) \in \Omega \times \mathbb{R}$. Taking into account that, condition (A22) is verified (see Example 4.1 of [93]) and clearly $f(x,0) = 0$ in Ω, problem (2.38) has at least two weak solutions for every $\lambda \in]0, \lambda^*[$, where λ^* is the constant introduced in the statement of Theorem 2.5.

Remark 2.6. Thanks to Talenti's inequality, it is possible to obtain an estimate of the embeddings constants c_1, c_q. By the Sobolev embedding theorem there exists a positive constant c such that

$$\|u\|_{L^{p^*}(\Omega)} \leq c\|u\|, \quad (\forall u \in X) \tag{2.39}$$

see [90]. The best constant that appears in (2.39) is

$$c := \frac{1}{N^2 \pi} \left(\frac{\Gamma^2 \left(\frac{N}{2} \right)}{\Gamma(\frac{N}{2p^*})\Gamma((\frac{N}{2}) - (\frac{N}{2p^*}))} \right)^{2/N} \eta^{1-1/p}, \tag{2.40}$$

where

$$\eta := \frac{p-1}{p},$$

see, for instance [90].

Due to (2.40), as a simple consequence of Hölder's inequality, it follows that

$$c_q \leq \frac{\text{meas}(\Omega)^{\frac{p^* - q}{p^* q}}}{N^2 \pi} \left(\frac{\Gamma^2 \left(\frac{N}{2} \right)}{\Gamma(\frac{N}{2p^*})\Gamma((\frac{N}{2}) - (\frac{N}{2p^*}))} \right)^{2/N} \eta^{1-1/p},$$

where "meas(Ω)" denotes the Lebesgue measure of the set Ω.

For completeness, we recall that a careful and interesting analysis of singular elliptic problems was developed in the monographs [94] as well as the papers [65–67, 70, 72, 73, 85, 95] and references therein.

Let Ω be a bounded domain in $\mathbb{R}^N$ ($N \geq 5$) containing the origin and with smooth boundary $\partial\Omega$. Further, denote by X the space $W_0^{1,p}(\Omega) \cap W^{2,p}(\Omega)$ endowed with the norm

$$\|u\| := \left(\int_\Omega |\Delta u|^p dx \right)^{1/p}.$$

Let $1 < p < N/2$, we recall classical Hardy's inequality, which says that

$$\int_\Omega \frac{|u(x)|^p}{|x|^{2p}} dx \leq \frac{1}{H} \int_\Omega |\Delta u(x)|^p dx, \quad (\forall u \in X) \tag{2.41}$$

where $H := (\frac{N(p-1)(N-2p)}{p^2})^p$; see, for instance, the paper [96].

By the compact embedding $X \hookrightarrow L^q(\Omega)$ for each $q \in [1, p^*[$, there exists a positive constant c_q such that

$$\|u\|_{L^q(\Omega)} \leq c_q \|u\|, \quad (\forall u \in X) \tag{2.42}$$

where c_q is the best constant of the embedding.

Let us define $F(x, \xi) := \int_0^\xi f(x, t) dt$, for every (x, ξ) in $\Omega \times \mathbb{R}$. Moreover, we introduce the functional $I_\lambda : X \to \mathbb{R}$ associated with (2.38),

$$I_\lambda := \Phi(u) - \lambda\Psi(u), \quad (\forall u \in X)$$

where

$$\Phi(u) := \frac{1}{p} \left(\int_\Omega |\Delta u(x)|^p dx + \int_\Omega \frac{|u(x)|^p}{|x|^{2p}} dx \right), \quad \Psi(u) := \int_\Omega F(x, u) dx.$$

From Hardy's inequality (2.41), it follows that

$$\frac{\|u\|^p}{p} \leq \Phi(u) \leq \left(\frac{H+1}{pH} \right) \|u\|^p, \tag{2.43}$$

for every $u \in X$.

Fixing the real parameter λ, a function $u : \Omega \to \mathbb{R}$ is said to be a weak solution of (2.38) if $u \in X$ and

$$\int_\Omega |\Delta u|^{p-2} \Delta u \Delta v dx + \int_\Omega \frac{|u|^{p-2}}{|x|^{2p}} uv dx - \lambda \int_\Omega f(x, u) v dx = 0,$$

for every $v \in X$. Hence, the critical points of I_λ are exactly the weak solutions of (2.38).

Definition 2.1. Let X be a reflexive real Banach space. The operator $T : X \to X^*$ is said to satisfy the (S_+) condition if the assumptions $u_n \rightharpoonup u_0$ in X and $\limsup_{n \to +\infty} \langle T(u_n) - T(u_0), u_n - u_0 \rangle \leq 0$ imply $u_n \to u_0$ in X.

Proposition 2.1. *The operator* $T : X \to X^*$ *defined by*

$$T(u)(v) := \int_\Omega |\Delta u|^{p-2} \Delta u \Delta v \, dx + \int_\Omega \frac{|u|^{p-2}}{|x|^{2p}} uv \, dx,$$

for every u, $v \in X$, *is strictly monotone.*

Proof. Clearly T is coercive. Taking into account (2.2) of [97] for $p > 1$ there exists a positive constant C_p such that if $p \geq 2$, then

$$\langle |x|^{p-2}x - |y|^{p-2}y, x - y \rangle \geq C_p |x - y|^p,$$

if $1 < p < 2$, then

$$\langle |x|^{p-2}x - |y|^{p-2}y, x - y \rangle \geq C_p \frac{|x - y|^p}{(|x| + |y|)^{2-p}},$$

where $\langle \cdot, \cdot \rangle$ denotes the usual inner product in $\mathbb{R}^N$. Thus, it is easy to see that, if $p \geq 2$, then, for any u, $v \in X$, with $u \neq v$, we have

$$\langle T(u) - T(v), u - v \rangle \geq C_p \int_\Omega |\Delta u - \Delta v|^p dx = C_p \|u - v\|^p > 0,$$

and if $1 < p < 2$, then

$$\langle T(u) - T(v), u - v \rangle \geq C_p \int_\Omega \frac{|\Delta u - \Delta v|^2}{(|\Delta u| + |\Delta v|)^{2-p}} dx > 0,$$

for every u, $v \in X$, which means that T is strictly monotone. $\square$

Now, we proof Theorem 2.5.

Proof. Our aim is to apply Theorem 1.20 to problem (2.38) in the case $r = 1$ to the space $X := W_0^{1,p}(\Omega) \cap W^{2,p}(\Omega)$ with the norm

$$\|u\| := \left(\int_\Omega |\Delta u|^p dx \right)^{1/p},$$

and to the functionals Φ, $\Psi : X \to \mathbb{R}$ be defined by

$$\Phi(u) := \frac{1}{p} \left(\int_\Omega |\Delta u(x)|^p dx + \int_\Omega \frac{|u(x)|^p}{|x|^{2p}} dx \right)$$

and

$$\Psi(u) := \int_\Omega F(x, u) dx,$$

for all $u \in X$. The functional Φ is in $C^1(X, \mathbb{R})$ and $\Phi' : X \to X^*$ is strictly monotone (see Proposition 2.1). Now we prove that Φ' is a mapping of type (S_+). Let $u_n \rightharpoonup u$ in X and $\limsup_{n \to +\infty} \langle \Phi'(u_n) - \Phi'(u), u_n - u \rangle \leq 0$. Since Φ' is strictly monotone, then

$$\limsup_{n \to +\infty} \langle K'(u_n) - K'(u), u_n - u \rangle \leq 0,$$

where $K' : X \to X^*$ defined as

$$K(u) := \frac{1}{p} \int_\Omega |\Delta u|^p dx, \quad (\forall u \in X),$$

and

$$K'(u)(v) = \int_\Omega |\Delta u|^{p-2} \Delta u \Delta v dx,$$

for every $v \in X$. Then $u_n \to u$ in X (see Theorem 3.1 of [98]). So, Φ' is a mapping of type (S_+). By Theorem 3.1 from [98], we get that $\Phi' : X \to X^*$ is a homeomorphism. Moreover, thanks to condition (A21) and to the compact embedding $W_0^{1,p}(\Omega) \cap W^{2,p}(\Omega) \hookrightarrow L^q(\Omega)$, Ψ is $C^1(X, \mathbb{R})$ and has compact derivative and

$$\Psi'(u)(v) = \int_\Omega f(x, u)v dx,$$

for every $v \in X$. Now we prove that $I_\lambda = \Phi - \lambda \Psi$ satisfies (PS)-condition for every $\lambda > 0$. Namely, we will prove that any sequence $\{u_n\} \subset X$ satisfying

$$d := \sup_n I_\lambda(u_n) < +\infty, \quad \|I_\lambda'(u_n)\|_{X^*} \to 0, \tag{2.44}$$

contains a convergent subsequence. For n large enough, we have by (2.44)

$$d \geq I_\lambda(u_n) = \frac{1}{p} \left(\int_\Omega |\Delta u_n|^p dx + \int_\Omega \frac{|u_n|^p}{|x|^{2p}} dx \right) - \lambda \int_\Omega F(x, u_n) dx,$$

then

$$\begin{aligned}
I_\lambda(u_n) &\geq \frac{1}{p} \left(\int_\Omega |\Delta u_n|^p dx + \int_\Omega \frac{|u_n|^p}{|x|^{2p}} dx \right) - \frac{\lambda}{\theta} \int_\Omega f(x, u_n) u_n dx \\
&> \left(\frac{1}{p} - \frac{1}{\theta} \right) \left(\int_\Omega |\Delta u_n|^p dx \right) + \frac{1}{\theta} \left(\int_\Omega |\Delta u_n|^p dx + \int_\Omega \frac{|u_n|^p}{|x|^{2p}} dx \right. \\
&\quad \left. - \lambda \int_\Omega f(x, u_n) u_n dx \right) \\
&\geq \left(\frac{1}{p} - \frac{1}{\theta} \right) \|u_n\|^p + \frac{1}{\theta} \langle I'(u_n), u_n \rangle.
\end{aligned}$$

Due to (2.44), we can actually assume that $\left| \frac{1}{\theta} \langle I_\lambda'(u_n), u_n \rangle \right| \leq \|u_n\|$. Thus,

$$d + \|u_n\| \geq I_\lambda(u_n) - \frac{1}{\theta} \langle I_\lambda'(u_n), u_n \rangle \geq \left(\frac{1}{p} - \frac{1}{\theta} \right) \|u_n\|^p.$$

It follows from this quadratic inequality that $\{\|u_n\|\}$ is bounded. By the Eberlian–Smulyan theorem, passing to a subsequence if necessary, we can assume that $u_n \rightharpoonup u$. Then $\Psi'(u_n) \to \Psi'(u)$ because of compactness. Since $I_\lambda'(u_n) = \Phi'(u_n) - \lambda\Psi'(u_n) \to 0$, then $\Phi'(u_n) \to \lambda\Psi'(u)$. Since Φ' is a homeomorphism, then $u_n \to u$ and so I_λ satisfies (PS)-condition.

From (A22), by standard computations, there is a positive constant C such that

$$F(x,t) \geq C|t|^\theta \tag{2.45}$$

for all $x \in \Omega$ and $|t| > M$. In fact, setting $a(x) := \min_{|\xi|=M} F(x,\xi)$ and

$$\varphi_t(s) := F(x,st), \quad \forall s > 0, \tag{2.46}$$

by (A22), for every $x \in \Omega$ and $|t| > M$ one has

$$0 < \theta\varphi_t(s) = \theta F(x,st) \leq st f(x,st) = s\varphi_t'(s), \quad \forall s > \frac{M}{|t|}.$$

Therefore,

$$\int_{M/|t|}^1 \frac{\varphi_t'(s)}{\varphi_t(s)}ds \geq \int_{M/|t|}^1 \frac{\theta}{s}ds.$$

Then

$$\varphi_t(1) \geq \varphi_t\left(\frac{M}{|t|}\right)\frac{|t|^\theta}{M^\theta}.$$

Taking into account of (2.46), we obtain

$$F(x,t) \geq F\left(x, \frac{M}{|t|t}\right)\frac{|t|^\theta}{M^\theta} \geq a(x)\frac{|t|^\theta}{M^\theta} \geq C|t|^\theta,$$

where $C > 0$ is a constant. Thus (2.45) is proved.

Fixed $u_0 \in X \setminus \{0\}$, for each $t > 1$ one has

$$I_\lambda(tu_0) \leq \frac{1}{p}t^p\|u_0\|^p - \lambda Ct^\theta \int_\Omega |u_0|^\theta dx.$$

Since $\theta > p$, this condition guarantees that I_λ is unbounded from below. Fixed $\lambda \in]0, \lambda^*[$, from (2.43) it follows that

$$\|u\| < p^{1/p}, \tag{2.47}$$

for each $u \in X$ such that $u \in \Phi^{-1}(]-\infty, 1)$. Moreover, the compact embedding $X \hookrightarrow L^1(\Omega)$, (A21), (2.47) and the compact embedding $X \hookrightarrow L^q(\Omega)$ imply that, for each $u \in \Phi^{-1}(]-\infty, 1)$, we have

$$\Psi(u) \leq a_1\|u\|_{L^1(\Omega)} + \frac{a_2}{q}\|u\|_{L^q(\Omega)}^q$$

$$\leq a_1c_1\|u\| + \frac{a_2}{q}c_q^q\|u\|^q$$

$$< a_1c_qp^{1/p} + \frac{a_2}{q}c_q^qp^{q/p},$$

and so,

$$\sup_{\Phi(u)<1} \Psi(u) \le a_1 c_q p^{1/p} + \frac{a_2}{q} c_q^q p^{q/p} = \frac{1}{\lambda^*} < \frac{1}{\lambda}. \tag{2.48}$$

From (2.48) one has

$$\lambda \in]0, \lambda^*[\subseteq \left]0, \frac{1}{\sup_{\Phi(u)<1} \Psi(u)}\right[.$$

So all hypotheses of Theorem 1.20 are verified. Therefore, for each $\lambda \in]0, \lambda^*[$, the functional I_λ admits two distinct critical points that are weak solutions of problem (2.38). □

Notes and Comments

Singular elliptic problems had been intensively studied in the last decades. Among others, we mentioned the works [99–107]. Stationary problems involving singular nonlinearities, as well as the associated evolution equations, describe naturally several physical phenomena and applied economical models. For instance, nonlinear singular boundary value problems arise in the context of chemical heterogeneous catalysts and chemical catalyst kinetics, in the theory of heat conduction in electrically conducting materials, singular minimal surfaces, as well as in the study of non-Newtonian fluids and boundary layer phenomena for viscous fluids. Moreover, nonlinear singular elliptic equations are also encountered in glacial advance, in transport of coal slurries down conveyor belts and in several other geophysical and industrial contents.

Recently, motivated by this large interest, the problem

$$\begin{cases} \Delta_p^2 u = \dfrac{|u|^{p-2}u}{|x|^{2p}} + g(\lambda, x, u) & \text{in } \Omega \\ u, \Delta u|_{\partial\Omega} = 0, \end{cases} \tag{2.49}$$

where $g :]0, +\infty[\times\Omega \times \mathbb{R} \to \mathbb{R}$ is a suitable function, had been extensively investigated.

For instance, when $p = 2$, Wang and Shen [100] considered the problem (2.49), assuming that the nonlinearity has the form $g(\lambda, x, u) = f(x, u)$. In this setting, the existence of non-trivial solutions by using variational methods is established. Successively, Berchio et al. [101] considered the case $g(\lambda, x, u) = (1 + u)^q$, studied the behavior of extremal solutions to biharmonic Gelfand-type equations under Steklov boundary conditions. Also in [102,104,106] the author was interested in the existence and multiplicity solutions for this kind of singular elliptic problems. Precisely, the existence of multiple solutions was proved by Chung [102] through a variant of the three critical points theorem by Bonanno [32]. Pérez-Llanos and Primo [106] studied the optimal exponent q to have solvability of problem with $g(\lambda, x, u) = u^q + cf$. Sign-changing solutions was investigated by Pei and Zhang [104].

Also in presence of p-biharmonic operator, singular equations had been investigated. For instance, Xie and Wang, in [103], studied the problem (2.49) has infinitely many solutions with positive energy levels. Later, Huang and Liu [105] obtained the existence of sign-changing solutions of p-biharmonic equations with Hardy potential by using the method of invariant sets of descending flow. This section was obtained in [108].

2.5 Semilinear Fourth-Order Problems on $\mathbb{R}^N$

Consider a nonlinear fourth-order elliptic equation

$$\Delta^2 u - \Delta u + a(x)u = \lambda b(x)f(u) + \mu g(x, u), \quad x \in \mathbb{R}^N, u \in H^2(\mathbb{R}^N), \quad (2.50)$$

where λ is a positive parameter, $a(x)$ and $b(x)$ are positive functions and $N > 2$. $f : \mathbb{R} \to \mathbb{R}$ is a continuous function. $g : \mathbb{R}^N \times \mathbb{R} \to \mathbb{R}$ is a Carathéodory function satisfies the following condition,

$$g(x, s) \le C(1 + |s|^q),$$

where $0 < q < \frac{N+4}{N-4}$ and C is a positive constant.

We also assumes that the potentials $a(x)$ and $b(x)$ satisfy the following conditions:

(A23) $\inf_{x \in \mathbb{R}^N} a(x) \ge \delta > 0$ and for each $M > 0, \text{meas}\{x \in \mathbb{R}^N : a(x) \le M\} < +\infty$, where δ is a constant and meas denotes the Lebesgue measure in $\mathbb{R}^N$.

(A24) $b \in L^1(\mathbb{R}^N) \cap L^\infty(\mathbb{R}^N)$, $b \ge 0$ and

$$\sup_{\mathbb{R}>0} \text{essinf}_{|x|<\mathbb{R}} b(x) > 0.$$

The aim of the present section is to show under novel assumptions ensures the existence of a compact interval $[a, b] \subseteq [\gamma, +\infty)$ and a positive real number r such that, for each $\lambda \in [a, b]$, problem (2.50) admits at least three weak solutions whose norms in $H^2(\mathbb{R}^N)$ are less than r.

Following the suggestions of Bartsch and Wang [109], due to (A23), we can defines the Hilbert space

$$E = \left\{ u \in H^2(\mathbb{R}^N) : \int_{\mathbb{R}^N} \left((\Delta u)^2 + |\nabla u|^2 + a(x)u^2 \right) dx < \infty \right\}$$

endowed with the inner product

$$(u,v)_E = \int_{\mathbb{R}^N} (\Delta u \Delta v + \nabla u \cdot \nabla v + a(x)uv)\, dx \text{ for } u,v \in E$$

and consequently with the induced norm which we denote by $\|\cdot\|$. The condition (A23) implies that the space E can be continuously embedded into $L^r(\mathbb{R}^N)$ whenever $2 \leq r \leq 2^*$ and the embedding is compact when $2 \leq r < 2^*$. Here, 2^* denotes the critical Sobolev exponent, i.e., $2^* = 2N/(N-4)$ for $N \geq 5$ and $2^* = +\infty$ for $N = 1,2,3,4$. Actually, we have the following embedding result.

Lemma 2.2. $E \hookrightarrow L^r(\mathbb{R}^N)$ *is compact, when* $2 \leq r < 2^*$.

Proof. Let $\{u_n\} \subset E$ be a sequence of E such that $u_n \rightharpoonup u$ weakly in E. Then $\{\|u_n\|\}$ is bounded and $u_n \to u$ strongly in $L^r_{loc}(\mathbb{R}^N)$ for $2 \leq r < 2^*$. We first show that $u_n \to u$ strongly in $L^2(\mathbb{R}^N)$. It suffices to prove that $\delta_n := \|u_n\|_{L^2} \to \|u\|_{L^2}$. Assume, up to a subsequence, that $\delta_n \to \delta$. For any bounded domain Ω in $\mathbb{R}^N$,

$$\int_\Omega |u_n|^2 dx \leq \int_{\mathbb{R}^N} |u_n|^2 dx \to \delta^2,$$

hence $\delta > \|u\|_{L^2}$. Let

$$A(R,M) := \{x \in \mathbb{R}^N \setminus B_R : a(x) \geq M\},$$
$$B(R,M) := \{x \in \mathbb{R}^N \setminus B_R : a(x) < M\}.$$

Then

$$\int_{A(R,M)} u_n^2 dx \leq \int_{\mathbb{R}^N} \frac{a(x)}{M} u_n^2 dx \leq \frac{\|u_n\|^2}{M}.$$

Choose $t \in (1, N/(N-4))$ such that $1/t + 1/t' = 1$. Then,

$$\int_{B(R,M)} u_n^2 dx \leq \left(\int_{B(R,M)} |u_n|^{2t} dx \right)^{1/t} (\operatorname{meas} B(R,M))^{1/t'}$$
$$\leq \|u_n\|^2 (\operatorname{meas} B(R,M))^{1/t'}.$$

Note that $\|u_n\|$ is bounded and that condition (A23) holds. We may choose R, M large enough such that $\frac{\|u_n\|^2}{M}$ and $\operatorname{meas} B(R,M)$ are small enough. Hence, for an arbitrary $\epsilon > 0$, we have that

$$\int_{\mathbb{R}^N \setminus B_R} u_n^2 dx = \int_{A(R,M)} u_n^2 dx + \int_{B(R,M)} u_n^2 dx \leq \epsilon.$$

Therefore,

$$\|u\|_{L^2}^2 = \|u\|_{L^2(B_R)}^2 + \|u\|_{L^2(\mathbb{R}^N \setminus B_R)}^2 \geq \lim_{n \to \infty} \|u_n\|_{L^2(B_R)}^2 \geq \delta^2 - \epsilon.$$

It means that $\delta = \|u\|_{L^2}$. Finally, by Gagliardo-Nirenberg inequality and the norm $\|\Delta u\|_{L^2}$ is equivalent to the norm $\|(\nabla u)^2\|_{L^2}$ (for more details see [110, p. 164]), we have that $u_n \to u$ strongly in $L^r(\mathbb{R}^N)$ for $2 \leq r < 2^*$. This completes the proof of the lemma. □

Now, we state our main result.

Theorem 2.6. *Assume that (A23) and (A24), $f \in C^0(\mathbb{R})$ and the following conditions are satisfied:*

(A25) there exists $C > 0$ and $q <]0, 1[$ such that

$$|f(x)| < C|s|^q \qquad \text{for each } x \in \mathbb{R}.$$

(A26) $f(s) = o(|s|)$ as $s \to 0$.
(A27) $\sup_{s \in \mathbb{R}} F(s) > 0$ where $F(s) = \int_0^s f(t)dt$.

Then, if we set

$$\gamma = \frac{1}{2} \inf \left\{ \frac{\|u\|^2}{\int_{\mathbb{R}^N} F(u(x))b(x)dx} : u \in X, \int_{\mathbb{R}^N} F(u)b(x)dx > 0 \right\},$$

for each compact interval $\Lambda := [a, b] \subset]\gamma, +\infty[$, there exists $r > 0$ with the following property: for every $\lambda \in \Lambda$ and every $g(x, t)$ with subcritical growth, there exists $\delta > 0$ such that, for each $\mu \in [0, \delta]$, the problem (2.50) has at least three weak solutions whose norms in E are less than r.

Remark 2.7. In [111], Yin and Wu had already studied problem (2.50) when $\lambda = 1$, $\mu = 0$. When nonlinear term f satisfies Ambrosetti–Rabinowitz condition and f is odd, they got problem (2.50) has multiple solutions. Property (A25) is a sublinearity growth assumption at infinity on f which complements the Ambrosetti–Rabinowitz condition. At last the conclusion of Theorem 2.6 gives a much more precise information about the fourth-order equation (2.50), namely, one can see that (2.50) is stable with respect to small perturbations.

When $\mu = 0$, we study this problem for small values of the parameter λ. More precisely, we consider the following problem

$$\Delta^2 u - \Delta u + a(x)u = \lambda b(x)f(u); \qquad x \in \mathbb{R}^N; u \in H^2(\mathbb{R}^N). \qquad (2.51)$$

Theorem 2.7. *Assume that conditions (A23), (A24) are satisfies and $f \in C^0(\mathbb{R})$ be a function which satisfies (A25), (A26) and (A27) as the same in Theorem 2.6. Then there exist a number $c_f = \max_{s>0} \frac{f(s)}{s}$, for every $0 \le \lambda < c_f^{-1} \|b/a\|_{L^\infty}$, problem (2.51) has only the trivial solution.*

Now, we prove Theorem 2.6 and Theorem 2.7.

Proof of Theorem 2.6. We are going to apply Theorem 1.16, taking

$$\Phi(u) = \frac{1}{2}\|u\|^2, \Psi(u) = \int_{\mathbb{R}^N} b(x)F(u(x))dx \text{ and } J(u) = \int_{\mathbb{R}^N} G(x,u)dx$$

for all $u \in X$. Note that $\Phi(u)$ is a coercive, sequentially weakly lower semi-continuous C^1 functional with a continuous inverse on X^* which belongs to $\mathcal{W}_X$. The latter assertion is a classical result, since the space X is uniformly convex and $\Phi(u) = h(\|u\|)$ with $h(t) = \frac{1}{2}t^2 : [0, +\infty[\to \mathbb{R}$, which is a continuous and strictly increasing function. Because $\Phi(u)$ is continuous, it is bounded on each bounded subset of X. Clearly, 0 is the only global minimum of $\Phi(u)$. Standard arguments based on the hypothesis $(\tilde{a})$ and on the fact that E is continuously embedded in $L^r(\mathbb{R}^N)$ when $2 \le r \le 2^*$ show that the functional Ψ is well defined, it is of class C^1, and satisfies

$$\Psi'(u)(v) = \int_{\mathbb{R}^N} b(x)f(u(x))v(x)dx$$

for all $u, v \in E$.

Now, we fix a number $\epsilon > 0$; in view of (A25) and (A26) there exist ρ_1, ρ_2 with $0 < \rho_1 < \rho_2$ such that

$$b(x)F(s) < \epsilon a(x)|s|^2 \tag{2.52}$$

for a.e. $x \in \mathbb{R}^N$ and all $s \in \mathbb{R} \setminus ([-\rho_2, -\rho_1] \cup [\rho_1, \rho_2])$. Then, as F is bounded on $([-\rho_2, -\rho_1] \cup [\rho_1, \rho_2])$, we can choose $D > 0$ and $2 < q < 2^*$ in such a way that

$$b(x)F(s) < \epsilon a(x)|s|^2 + D|s|^q$$

for a.e. $x \in \mathbb{R}^N$ and all $s \in \mathbb{R}$. Thus, by continuous embedding,

$$\Psi(u) \le \epsilon\|u\|^2 + D\kappa_q^q\|u\|^q$$

for all $u \in E$. Hence,

$$\limsup_{u \to 0} \frac{2\Psi(u)}{\|u\|^2} \le 2\epsilon. \tag{2.53}$$

Further, by (2.52) again, for each $u \in E \setminus \{0\}$, we obtain

$$\frac{\Psi(u)}{\|u\|^2} = \frac{\int_{\mathbb{R}^N(|u| \leq \rho_2)} b(x)F(u(x))dx}{\|u\|^2} + \frac{\int_{\mathbb{R}^N(|u| > \rho_2)} b(x)F(u(x))dx}{\|u\|^2}$$

$$\leq \frac{\sup_{[-\rho_2, \rho_2]} F(u(x)) \int_{\mathbb{R}^N} b(x)dx}{\|u\|^2} + \epsilon.$$

So, we get

$$\limsup_{\|u\| \to +\infty} \frac{2\Psi(u)}{\|u\|^2} \leq 2\epsilon. \tag{2.54}$$

Since ϵ is arbitrary, from (2.53) and (2.54) it follows that

$$\max\left\{ \limsup_{\|u\| \to +\infty} \frac{2\Psi(u)}{\|u\|^2}, \limsup_{u \to 0} \frac{2\Psi(u)}{\|u\|^2} \right\} \leq 0.$$

Thus, by using the notation of Theorem 1.16, we have $\alpha = 0$ and by our assumption $0 < \beta \leq +\infty$. Therefore, for $\gamma = 1/\beta$, the conclusion follows from Theorem 1.16. $\square$

Example 2.4. Let k, h and ξ be arbitrary real positive numbers. We choose $f : \mathbb{R} \to \mathbb{R}$ defined by

$$f(s) = \begin{cases} s|s|[4m|s| + 3n], & |s| \leq \xi, \\ kse^{-h|s|} & |s| \geq \xi, \end{cases}$$

where

$$m = m(k, h, \xi) = -\frac{ke^{-h\xi}(h\xi + 1)}{4\xi^2}, \quad n = n(k, h, \xi) = \frac{ke^{-h\xi}(2 + h\xi)}{3\xi}.$$

Then, we take as potentials $a(x) = |x|^2 + l$ with l a positive constant and $b(x) = e^{-|x|^2}$, $x \in \mathbb{R}^N$. It follows that the assumptions (A23), (A24), (A25), (A26) and (A27) of Theorem 2.6 hold.

Proof of Theorem 2.7. Assume that $u \in H^2(\mathbb{R}^N)$ is a solution of (2.51). Multiplying (2.51) by the test function $u \in H^2(\mathbb{R}^N)$, we obtain

$$\|u\|^2 = \int_{\mathbb{R}^N} \left((\Delta u)^2 + |\nabla u|^2 + a(x)u^2\right) dx = \lambda \int_{\mathbb{R}^N} b(x)f(u)u\,dx$$

$$\leq \lambda \|b/a\|_{L^\infty} c_f \int_{\mathbb{R}^N} a(x)u^2 \leq \lambda \|b/a\|_{L^\infty} c_f \|u\|^2.$$

Now, fix $0 \leq \lambda < c_f^{-1}\|b/a\|_{L^\infty}$ arbitrarily, the above estimate implies $u = 0$, which concludes the proof. $\square$

Last, we give the following final remarks.

Remark 2.8. The reader can observe that we considered only a Bartsch-type (coercivity-based) compactness result. The compactness result can be replaced by symmetric-based compactness arguments together with the principle of symmetric criticality (see reference [112] for a closely related approach). In this situation, the proof of the result is all most the same as Theorem 2.6, so we omit the details.

Notes and Comments

The same variational methods as were used there, which are based on the seminal paper of [35], have been applied to obtain multiple solutions (see, for instance, [35, 38, 86, 110, 112–123]). This section was obtained in [124].

Chapter 3

Kirchhoff Problems

3.1 Introduction

This chapter was motivated by some works that have appeared in recent years concerning with the following the Kirchhoff-type problem

$$- \left(a + b \int_\Omega |\nabla u|^2 dx \right) \Delta u = g(x, u), \qquad \text{in } \Omega, \qquad (3.1)$$

where $\Omega \subseteq \mathbb{R}^N$, $a > 0, b \geq 0$ and u satisfies some initial or boundary condition.

The problem (3.1) is related to the stationary analogue of the Kirchhoff equation

$$u_{tt} - \left(a + b \int_\Omega |\nabla_x u|^2 dx \right) \Delta_x u = g(x, u) \qquad (3.2)$$

which was proposed by Kirchhoff in 1883 (see [125]) as an generalization of the well-known d'Alembert's equation

$$\rho \frac{\partial^2 u}{\partial t^2} - \left(\frac{P_0}{h} + \frac{E}{2L} \int_0^L \left| \frac{\partial u}{\partial x} \right|^2 dx \right) \frac{\partial^2 u}{\partial x^2} = g(x, u)$$

for free vibrations of elastic strings. Kirchhoff's model takes into account the changes in length of the string produced by transverse vibrations. It is pointed in [126] that the problem (3.2) model several physical and biological systems, where u describes a process which depends on the average of itself (for example, population density). Some early classical studies of Kirchhoff equations were those of Bernstein [127] and Pohožaev [128]. However, Eq. (3.2) received great attention only after Lions [129] proposed an abstract framework for the problem. Some interesting results for (3.2) can be found in [130–132] and the references therein.

There is a large number of references to the Kirchhoff equations with non-constant potential $V(x)$ which has a compactness property for the functional (3.4) corresponded to the problem (3.3). We refer the readers to see the following references [126, 133–162].

3.2   Radial Solutions of Kirchhoff Problems

In this section, we study the existence, non-existence, and multiplicity of solutions to the following Kirchhoff equation in $\mathbb{R}^N$ ($N \geq 3$):

$$\left(1 + b \int_{\mathbb{R}^N} (|\nabla u|^2 + V(x)u^2)dx\right) [-\Delta u + V(x)u] = f(u) \text{ in } \mathbb{R}^N, \quad (3.3)$$

where $b > 0$ is a parameter, $V(x)$ and $f(u)$ satisfy the following hypotheses:

(B1) $V(x) \in C(\mathbb{R}^N, \mathbb{R})$ and $V(x) \equiv V(|x|) \geq V_0 > 0$ for all $x \in \mathbb{R}^N$;

(B2) $\lim_{|x| \to \infty} V(x) = V(\infty) \in (0, +\infty)$;

(B3) $f \in C(\mathbb{R}, \mathbb{R})$ and $\lim_{s \to 0} \frac{f(s)}{s} = 0$;

(B4) there exists $l \in (\Lambda, V(\infty))$ such that $\lim_{|s| \to \infty} \frac{f(s)}{s} = l$, where Λ is the infimum of the spectrum of the Schrödinger operator $-\Delta + V$, i.e.,

$$\Lambda = \inf_{u \in H^1(\mathbb{R}^N) \backslash \{0\}} \frac{\int_{\mathbb{R}^N} (|\nabla u|^2 + V(x)u^2)dx}{\int_{\mathbb{R}^N} u^2 dx};$$

(B5) $f(-s) = f(s)$ for all $s \in \mathbb{R}$.

Let us define the functional $I_b(u) : H_r^1(\mathbb{R}^N) \to \mathbb{R}$ by

$$I_b(u) = \frac{1}{2} \int_{\mathbb{R}^N} (|\nabla u|^2 + V(x)u^2)dx + \frac{b}{4} \left(\int_{\mathbb{R}^N} (|\nabla u|^2 + V(x)u^2)dx\right)^2 \quad (3.4)$$
$$- \int_{\mathbb{R}^N} F(u)dx,$$

where $H_r^1(\mathbb{R}^N)$ denotes a radial function Sobolev space and $F(u) = \int_0^u f(t)dt$. Then, the critical points of $I_b \in C^1$ provide the solutions of (3.3).

Using the variation methods, we establish the following existence and non-existence results.

Theorem 3.1. *Suppose that $V(x)$ satisfies (B1), (B2) and (B3) (particularly, $f \in (\mathbb{R}, \mathbb{R}^+)$), (B4) hold, then there exists $b^* > 0$ such that problem (3.3) has a positive radial solution for $b \in (0, b^*)$ which possesses exponential decay at infinity. Moreover, assume $N \leq 4$, there exists $\bar{b} > 0$ such that problem (3.3) has no nontrivial solution for $b \geq \bar{b}$.*

Remark 3.1. The variational approach of this type functional was developed by Wang and Zhou [163] who applied it to nonlinear stationary Schrödinger–Poisson systems. The nonlocal term of problem (3.3) is different from the Schrödinger–Poisson systems. Using the critical point theory to the functional I_b related to Kirchhoff in $H_r^1(\mathbb{R}^N)$, we prove the boundedness of a (PS) sequence and show the existence of radial solutions.

The existence and multiplicity of solutions to problem (3.3) is as follows.

Theorem 3.2. *Assume that $V(x)$ satisfies (B1), (B2) and F satisfies (B3)–(B5). Let λ_k be the eigenvalue of the operator $-\Delta + V(x)$ (see (3.7) below). Suppose that $l > \lambda_k$, then there exists $\tilde{b}$ such that problem (3.3) has at least k pairs of nontrivial radial solutions for $0 < b \leq \tilde{b}$.*

Remark 3.2. We can observe that Li *et al.* [164] get the following result: assume that $N \geq 3$, and a, b are positive constants, $\lambda \geq 0$ is a parameter. If the conditions

(H1) $f \in C(\mathbb{R}^+, \mathbb{R}^+)$ and $|f(t)| \leq C(|t| + |t|^{p-1})$ for all $t \in \mathbb{R}^+ = [0, \infty)$ and some $p \in (2, 2^*)$, where $2^* = 2N/(N-2)$ for $N \geq 3$;

(H2) $\lim_{t \to 0} \frac{f(t)}{t} = 0$;

(H3) $\lim_{t \to \infty} \frac{f(t)}{t} = \infty$,

hold, then there exists $\lambda_0 > 0$ such that for any $\lambda \in [0, \lambda_0)$, (3.17) has at least one positive solution. In our Theorem 3.1, we study the asymptotically linear nonlinearities which different with (H3) called superlinear nonlinearities, generally. This result can be regarded as an extension of the result of [164]. We can also observe that we get a multiplicity result when nonlinearities has the odd property but in [164] only a solution obtained. We believe that even the nonlinearities has the odd property, it is hard to get multiplicity solutions by using the same method in [164].

Remark 3.3. The condition $\Lambda < l < V(\infty)$ implies that $V(x) \not\equiv$ constant. If we replace $f(u)$ by $f(|x|, u)$, and modify the conditions (B3), (B4) slightly (for example, we may just require that the limits hold uniformly in x), then the above theorems are still true.

Here we use standard notation. $H_r^1(\mathbb{R}^N)$ denotes a radial function Sobolev space equipped with the norm

$$\|u\|_{H^1} := \left(\int_{\mathbb{R}^N} (|\nabla u|^2 + |u|^2) dx \right)^{1/2},$$

which is equivalent to

$$\|u\| := \left(\int_{\mathbb{R}^N} (|\nabla u|^2 + V(x)|u|^2) dx \right)^{1/2},$$

by (B1) and (B2).

To apply the critical point theorem 1.3 to the functional, we recall some propositions related to the energy functional I_b. This implies that I_b is well defined in $H_r^1(\mathbb{R}^N)$. Indeed, we have the following proposition.

Proposition 3.1. *The functional I_b is continuously differentiable on $H_r^1(\mathbb{R}^N)$ and its critical point u is a weak solution of (3.3) (see [165, Lemma 1] for more details). Besides, for any $\psi \in H_r^1(\mathbb{R}^N)$*

$$\langle I_b'(u), \psi \rangle = \int_{\mathbb{R}^N} (\nabla u \cdot \nabla \psi + V(x)u\psi) dx$$

$$+ b \int_{\mathbb{R}^N} (|\nabla u|^2 + V(x)u^2) dx \int_{\mathbb{R}^N} (\nabla u \cdot \nabla \psi + V(x)u\psi) dx$$

$$- \int_{\mathbb{R}^N} f(u)\psi dx. \tag{3.5}$$

We recall the spectrum theory for the self-adjoint operator [166]. It is known that if $-\Delta + V(x)$ is self-adjoint in $H^2(\mathbb{R}^N)$, then the residual spectrum $\sigma_{\text{res}}(-\Delta + V(x))$ is empty, where

$$\sigma_{\text{res}}(-\Delta + V(x)) := \{\lambda \in \mathbb{C} : \text{Ker}(-\Delta + V(x)) = \{0\},$$

$$\text{Ran}(-\Delta + V(x)) \text{ is not dense in } H^2(\mathbb{R}^N)\},$$

where Ran refers to the range space of the operator $-\Delta + V(x)$. Since $V \in L_{loc}^2(\mathbb{R}^N)$ satisfies $V(x) \geq V_0 > 0$, the spectral analysis gives that $-\Delta + V(x)$ is bounded below by some positive constant. Thus, we have $0 < \lambda_1 \leq \lambda_2 \leq \cdots \leq \lambda_k \leq \cdots \leq \Sigma_{\text{res}} := \inf\{\sigma_{ess}(-\Delta + V(x))\}$, where the eigenvalues are given by

$$\lambda_1 = \inf_{u \in H^1(\mathbb{R}^N)\backslash\{0\}} \frac{\int_{\mathbb{R}^N} (|\nabla u|^2 + V(x)u^2) dx}{\int_{\mathbb{R}^N} u^2 dx} \tag{3.6}$$

and

$$\lambda_k = \inf_{u \in H^1(\mathbb{R}^N) \cap E_k^\perp} \frac{\int_{\mathbb{R}^N} (|\nabla u|^2 + V(x)u^2) dx}{\int_{\mathbb{R}^N} u^2 dx}, \tag{3.7}$$

for $k \geq 2$ with denoting the eigenspaces $E_k := \bigoplus_{1 \leq j \leq k} \text{Ker}(-\Delta + V(x) - \lambda_k)$. Moreover, let us denote the eigenfunction e_k corresponding to λ_k.

Now, we will proof the main results. Firstly, we show that the functional possesses the mountain pass geometry.

Lemma 3.1. *Assume that (B1)–(B3), and (B4) hold. Then*

(i) there exist $\rho > 0$, $\alpha > 0$ such that $I(u) \geq \alpha > 0$ for all $u \in H^1_r(\mathbb{R}^N)$ with $\|u\| = \rho$;

(ii) there exist $e \in H^1_r(\mathbb{R}^N)$ with $\|e\| > \rho$ and $b^* > 0$ such that $I(e) < 0$ for $0 < b < b^*$.

Proof. (i) Using the conditions (B3) and (B4), we have that for any $\varepsilon > 0$, there exist $p > 1$ and $A = A(\varepsilon, p) > 0$ such that for all $s > 0$,

$$F(s) \leq \frac{1}{2}\varepsilon s^2 + A s^{p+1}. \tag{3.8}$$

Observing that the term $\frac{b}{4}\left(\int_{\mathbb{R}^N}(|\nabla u|^2 + V(x)u^2)dx\right)^2$ is nonnegative and by the Sobolev inequality, we deduce that

$$I_b(u) \geq \frac{1}{2}(1 - C_1\varepsilon)\|u\|^2 - AC_2\|u\|^{p+1}.$$

Therefore, part (i) is proved if we choose $\varepsilon = \frac{1}{2}C_1$ and $\|u\| = \rho > 0$ sufficiently small.

(ii) Since $l > \Lambda$, (B1), and (B4), there exists a nonnegative function $u \in H^1_r(\mathbb{R}^N)$ such that

$$\int_{\mathbb{R}^N}(|\nabla u|^2 + V(x)u^2)dx < l\int_{\mathbb{R}^N}u^2 dx.$$

By Fatou's Lemma, we obtain that

$$\lim_{t\to+\infty}\frac{I_0(tu)}{t^2} = \frac{1}{2}\int_{\mathbb{R}^N}(|\nabla u|^2 + V(x)u^2)dx - \lim_{t\to+\infty}\frac{F(tu)}{t^2u^2}u^2 dx$$

$$\leq \frac{1}{2}\left(\int_{\mathbb{R}^N}(|\nabla u|^2 + V(x)u^2)dx - l\int_{\mathbb{R}^N}u^2 dx\right) < 0.$$

Consequently, there exists $e \in H^1_r(\mathbb{R}^N)$ with $\|e\| > \rho$ such that $I_0(e) < 0$. Since $I_b(e) \to I_0(e)$ as $b \to 0^+$, we deduce that there exists $b^* > 0$ such that $I_b(e) < 0$, for all $0 < b < b^*$. $\qquad\square$

For α and e given by Lemma 3.1, Theorem 1.3 provides that there exists a (PS) sequence $\{u_k\} \subset H^1_r(\mathbb{R}^N)$ such that

$$I_b(u_k) \to c > 0 \text{ and } I'_b(u_k) \to 0 \text{ in } H^*_r(\mathbb{R}^N) \text{ as } k \to \infty, \tag{3.9}$$

where $H^*_r(\mathbb{R}^N)$ denotes the dual space of $H^1_r(\mathbb{R}^N)$. Now we prove that the $(PS)_c$ sequence $\{u_k\}$ is bounded.

Lemma 3.2. *Suppose that (B1), (B2), (B3), and (B4) hold, then the (PS) sequence $\{u_k\}$ obtained by (3.9) is bounded in $H^1_r(\mathbb{R}^N)$.*

Proof. Since $I_b'(u_k) \to 0$ in $H_r^1(\mathbb{R}^N)$ as $k \to \infty$, for k sufficiently large, we have

$$\langle I_b'(u_k), u_k \rangle \leq \|I_b'(u_k)\|_{H_r^*(\mathbb{R}^N)} \|u_k\| \leq \|u_k\|,$$

that is,

$$\int_{\mathbb{R}^N} (|\nabla u_k|^2 + V(x)u_k^2)dx + b \left(\int_{\mathbb{R}^N} (|\nabla u_k|^2 + V(x)u_k^2)dx \right)^2$$
$$- \int_{\mathbb{R}^N} f(u_k)u_k dx \leq \|u_k\|. \tag{3.10}$$

Noting that $f(u_k(x))u_k(x) \leq Cu_k^2(x)$ for all $x \in \mathbb{R}^N$ by (B4), where C is a positive constant. By (3.10), one has

$$\|u_k\|^2 + b\|u_k\|^4 \leq \|u_k\| + \frac{l}{\Lambda}\|u_k\|^2,$$

which implies that $\|u_k\|$ is bounded. $\qquad\square$

Lemma 3.3. *If $\{u_k\}$ is a bounded sequence in $H_r^1(\mathbb{R}^N)$ such that $I_b'(u_k) \to 0$ in $H_r^*(\mathbb{R}^N)$ as $k \to \infty$, then there exists a convergent subsequence still denoted by $\{u_k\}$ in $H_r^1(\mathbb{R}^N)$.*

Proof. Since $\{u_k\}$ is bounded and the Sobolev embedding $H_r^1(\mathbb{R}^N) \hookrightarrow L^p(\mathbb{R}^N)$ ($p \in [2, 2^*)$, here $2^* = \frac{2N}{N-2}$) is compact (see [5, Corollary 1.26]), we might assume that, up to subsequence, there exists $u_k \in H_r^1(\mathbb{R}^N)$ such that

$$u_k \rightharpoonup u \text{ weakly in } H_r^1(\mathbb{R}^N), \tag{3.11}$$
$$u_k \to u \text{ strongly in } L^p(\mathbb{R}^N), p \in [2, 2^*),$$
$$u_k \to u \text{ a.e. in } \mathbb{R}^N.$$

Note that

$$\langle I_b'(u_k), u_k - u \rangle = \int_{\mathbb{R}^N} (\nabla u_k \cdot \nabla(u_k - u) + V(x)u_k(u_k - u))dx$$
$$+ b \int_{\mathbb{R}^N} (|\nabla u_k|^2 + V(x)u_k^2) \, dx \left(\int_{\mathbb{R}^N} (\nabla u_k \cdot \nabla(u_k - u) + V(x)u_k(u_k - u))dx \right)$$
$$- \int_{\mathbb{R}^N} f(u_k)(u_k - u)dx \to 0.$$

By using (3.11), we see that $\|u_k\|$ converges to $\|u\|$, which implies the strong convergence in $H_r^1(\mathbb{R}^N)$. $\qquad\square$

Proof of Theorem 3.1. By Lemma 3.1, Theorem 1.3, and Lemma 3.2, we can obtain the (PS) sequence $\{u_k\}$, which is bounded in $H_r^1(\mathbb{R}^N)$. By Lemma 3.3, we establish the existence of a nontrivial critical point in $H_r^1(\mathbb{R}^N)$. Moreover, by using $u \in H_r^1(\mathbb{R}^N)$, we know that $u \in W_{loc}^{2,p}(\mathbb{R}^N)$, hence $u \in C^{1,\alpha}(\mathbb{R}^N)$. Then the bootstrap argument concludes that $u \in C^{2,\alpha}(\mathbb{R}^N)$. Multiplying equation (3.3) by u^- and integrating over $\mathbb{R}^N$, we find $\|u^-\| \leq 0$. This implies that u is positive. Finally, we can deduce by $V(x) \geq V_0 > 0$ and the comparison principle that there exist C_1, C_2 such that $u \leq C_2 e^{-C_1|x|}$ for $|x| \geq R$.

Next, we prove the non-existence result. Suppose that $u \in H_r^1(\mathbb{R}^N)$ is a solution of (3.3). Then, multiplying equation (3.3) by u and integrating by parts, we have

$$\int_{\mathbb{R}^N}(|\nabla u|^2+V(x)u^2)dx+b\left(\int_{\mathbb{R}^N}(|\nabla u|^2+V(x)u^2)dx\right)^2-\int_{\mathbb{R}^N}f(u)udx = 0.$$
(3.12)

Since the Sobolev embedding $H_r^1(\mathbb{R}^N) \hookrightarrow L^4(\mathbb{R}^N)$ is continuous, so there exists a constant C_4, such that

$$\int_{\mathbb{R}^N}|u|^4dx \leq C_4^4\|u\|^4.$$
(3.13)

According to (B3) and (B4), there exists $C = C(V_0)$ such that

$$f(u)u \leq V_0u^2 + C|u|^4.$$
(3.14)

Substituting (3.13) and (3.14) into (3.12), we have

$$0 = \int_{\mathbb{R}^N}(|\nabla u|^2 + V(x)u^2)dx + b\left(\int_{\mathbb{R}^N}|\nabla u|^2 + V(x)u^2dx\right)^2 - \int_{\mathbb{R}^N}f(u)udx$$

$$\geq \int_{\mathbb{R}^N}(|\nabla u|^2 + (V(x) - V_0)u^2)dx + \frac{b}{C_4^4}\int_{\mathbb{R}^N}|u|^4dx - C\int_{\mathbb{R}^N}|u|^4dx$$

$$\geq \left(\frac{b}{C_4^4} - C\right)\int_{\mathbb{R}^N}|u|^4dx.$$

Therefore, if $b > C_4^4C$, then u must be zero. $\qquad\square$

Proof of Theorem 3.2. Similar to the proof of Lemma 3.1(i), we see that Theorem 1.8(ii) holds with $m = 0$. By the same method used in the proof of Lemma 3.2, we can obtain that (PS) sequence is bounded. Since we consider the space $H_r^1(\mathbb{R}^N)$, we have the compactness. Therefore, $I_b(u)$ satisfies condition (PS)$_c$ for $c > 0$. Let E_k, λ_k be given by (3.7) for any k.

By Theorem 1.8, to prove Theorem 3.2, we need only, for any $k \geq 1$, find a k-dimensional subspace E_k of $H_r^1(\mathbb{R}^N)$ and $R_k > 0$ such that

$$I_b(u) \leq 0 \text{ for all } u \in E_k \setminus B_{R_k}.$$
(3.15)

Since $l > \lambda_k$, there is $\varepsilon > 0$ such that $l - \varepsilon > \lambda_k$. By the definition of λ_k, we see that for all $u \in E_k \backslash \{0\}$,

$$\int_{\mathbb{R}^N} (|\nabla u|^2 + V(x)u^2)dx \le \lambda_k \int_{\mathbb{R}^N} |u|^2 dx < (l - \varepsilon) \int_{\mathbb{R}^N} |u|^2 dx. \qquad (3.16)$$

By (B3) and (B4), there exists $C > 0$ such that $|F(s)| \le C|s|^2$ for all $s \in \mathbb{R}$. Thus, it follows from (B4) and the dominated convergence theorem that

$$\lim_{t \to \infty} \int_{\mathbb{R}^N} \frac{F(tu)}{t^2} = \frac{l}{2} \int_{\mathbb{R}^N} |u|^2 dx.$$

Therefore, by (3.16), for $u \in E_k \backslash \{0\}$,

$$\lim_{t \to \infty} \frac{I_0(tu)}{t^2} = \frac{1}{2}\|u\|^2 - \lim_{t \to \infty} \int_{\mathbb{R}^N} \frac{F(tu)}{t^2} dx$$
$$< \frac{l - \varepsilon}{2} \int_{\mathbb{R}^N} |u|^2 dx - \frac{l}{2} \int_{\mathbb{R}^N} |u|^2 dx < 0,$$

which, jointly with $\dim E_k < \infty$, implies that there exists $R_k > 0$ such that

$$\sup_{E_k \backslash R_k} I_0(u) < 0.$$

Noting that $I_b(u) \to I_0(u)$ as $b \to 0^+$, we see that there exists $\tilde{b} > 0$ such that $I_b(u) < 0$ for all $u \in E_k \backslash R_k$ and $0 < b < \tilde{b}$. The proof is complete. $\square$

Notes and Comments

Recently, using variational method, Jin and Wu [167] obtained the existence of infinitely many radial solutions for problem (3.3) with $V(x) = 1$ in $\mathbb{R}^N$ by using the Fountain Theorem. Next, Azzollini *et al.* [168] got a multiplicity result concerning the critical points of a class of functionals involving local and nonlocal nonlinearities, then they applied their result to the nonlinear elliptic Kirchhoff equation (3.3) in $\mathbb{R}^N$ assuming on the local nonlinearity has the general hypotheses introduced by Berestycki and Lions [169]. In [147], Azzollini presented a very simple proof of the existence of at least one nontrivial solution for a Kirchhoff-type equation on $\mathbb{R}^N$, for $N > 3$. In particular, in the first part of the paper he was interested in studying the existence of a positive solution to the elliptic Kirchhoff equation under the effect of a nonlinearity satisfying the general Berestycki–Lions assumptions. In the second part, he looked for ground states using minimizing arguments on a suitable natural constraint. Very recently, Li *et al.* [164] used Azzollini's idea to study the existence of at least one positive radial solution to the following nonlinear Kirchhoff-type equation:

$$\left(a + \lambda \int_{\mathbb{R}^N} |\nabla u|^2 + \lambda b \int_{\mathbb{R}^N} u^2\right)[-\Delta u + bu] = f(u) \qquad (3.17)$$

in $\mathbb{R}^N$, where $N \geq 3$, a, b are positive constants, and $\lambda \geq 0$ is a parameter. The main result does not assume the usual compactness conditions. A cut-off functional associated with (3.17) is utilized to obtain bounded Palais–Smale sequences. Their result can be regarded as an extension of a classical result for the semilinear equation

$$-\Delta u + bu = f(u)$$

in $\mathbb{R}^N$ to the case of the nonlinear Kirchhoff-type equation (3.17). The choice of parameters λ for which (3.17) admits at least one solution depends on the nonlinearity f, the constants N, a and b, a Sobolev embedding constant and several test-functions used in the proof. Actually, in [164], their nonlinear term was superlinear at infinity and $V(x) = 1$. Motivated by [164], we want to prove that the existence and multiplicity of solutions of problem (3.3) when nonlinear term f is an asymptotically linear nonlinearities. We will also show the effect of the potential $V(x)$ and get a non-existence result. There still a large number of references to the Kirchhoff equations with non-constant potential $V(x)$ which had a compactness property for the functional (3.4) corresponded to the problem (3.3). We refer the readers to see the following references [133–135, 135–145]. This section was obtained in [170].

3.3   Multiple Solutions of Nonhomogeneous Problems

In the same spirit of [143, 167, 168], we study an nonhomogeneous Kirchhoff equation on all the space $\mathbb{R}^N$, namely we consider the problem

$$-\left(a + b \int_{\mathbb{R}^N} |\nabla u|^2 dx\right)\Delta u + V(x)u = f(x, u) + h(x), \qquad \text{in } \mathbb{R}^N. \quad (3.18)$$

In the sense, the problem turns out to be a generalization of the well-known nonhomogeneous Schrödinger equation:

$$-\Delta u + V(x)u = f(x, u) + h(x), \qquad \text{in } \mathbb{R}^N.$$

In this section, we are interested in looking for the existence of multiple solutions of the problem (3.18). We make the following assumptions.

(B6) $V \in C(\mathbb{R}^N, \mathbb{R})$ satisfies $\inf_{x \in \mathbb{R}^N} V(x) \geq a_1 > 0$, where $a_1 > 0$ is a constant. Moreover, for every $M > 0$, meas $(\{x \in \mathbb{R}^N : V(x) \leq M\}) < \infty$, where meas denotes the Lebesgue measure in $\mathbb{R}^N$.

(B7) $f \in C(\mathbb{R}^N \times \mathbb{R}, \mathbb{R})$ and, for some $2 < p < 2^* = \frac{N-2}{2N}$, $a_2 > 0$,

$$|f(x, z)| \leq a_2(1 + |z|^{p-1}),$$

for a.e. $x \in \mathbb{R}^N$ and all $z \in \mathbb{R}$.

(B8) There exists $\mu > 4$ such that

$$\mu F(x, z) := \mu \int_0^z f(x, y)dy \leq zf(x, z),$$

for every $x \in \mathbb{R}^N$ and all $z \in \mathbb{R}$.

(B9) $f(x, z)/z \to 0$, as $z \to 0$, uniformly for $x \in \mathbb{R}^N$.

(B10) $\inf_{x \in \mathbb{R}^N, \, |z|=1} F(x, z) > 0$.

Before stating our main results, we give several notations. Let

$$E := \left\{ u \in H^1(\mathbb{R}^N) : \int_{\mathbb{R}^N} \left(|\nabla u|^2 + V(x)u^2\right) dx < \infty \right\}.$$

Then E is a Hilbert space with the inner product

$$(u, v)_E = \int_{\mathbb{R}^N} \left(\nabla u \cdot \nabla v + V(x)uv\right) dx$$

and the norm $\|u\|_E = (u, u)_E^{1/2}$. Obviously, the embedding $E \hookrightarrow L^s(\mathbb{R}^N)$ is continuous, for any $s \in [2, 2^*]$.

We can now state the first result:

Theorem 3.3. *Suppose that $h \in L^2(\mathbb{R}^N)$ and $h \not\equiv 0$. Let (B6), (B7)–(B10) hold, then there exists a constant $m_0 > 0$ such that problem (3.18) admits at least two different solutions in E when $\|h\|_{L^2} < m_0$.*

Remark 3.4.

(a) The condition in (B6), which implies the compactness of embedding of the working space E (see Lemma 3.4) and contains the coercivity condition: $V(x) \to \infty$ as $|x| \to \infty$, was first introduced by T. Bartsch and Z.-Q. Wang in [109] to overcome the lack of compactness.

(b) It is not difficult to find out functions f satisfying (B7)–(B10), for example,

$$f(x, z) = |z|^{s-2}z, \quad 4 < s < 6.$$

(c) To the best of our knowledge, it seems that Theorem 3.3 is the first result about the existence of multiple solutions for the nonhomogeneous Kirchhoff equations in $\mathbb{R}^N$.

In the following, we consider the problem in (3.18) with $V(x) \equiv 1$ and $f(x, u) = |u|^{p-2}u$, that is

$$- \left(a + b \int_{\mathbb{R}^N} |\nabla u|^2 dx \right) \Delta u + u = |u|^{p-2}u + h(x), \qquad \text{in } \mathbb{R}^N. \quad (3.19)$$

We can now state the second result:

Theorem 3.4. *Suppose that $h \in C^1(\mathbb{R}^N) \cap L^2(\mathbb{R}^N)$ is a radial function and $h \not\equiv 0$. Let $4 < p < 6$, then there exists a constant $m_1 > 0$ such that problem (3.19) admits at least two different radial solutions in $H^1(\mathbb{R}^N)$ when $\|h\|_{L^2} < m_1$.*

In order to apply critical point theory, we first state a key property of E to our proof.

Lemma 3.4 ([17], Lemma 3.4). *Under assumption (B6), the embedding $E \hookrightarrow L^s(\mathbb{R}^N)$ is compact for any $s \in [2, 2^*)$.*

Problem (3.18) has a variational structure. Indeed we consider the functional

$$I : E \to \mathbb{R}$$

defined by

$$I(u) = \frac{a}{2} \int_{\mathbb{R}^N} |\nabla u|^2 dx + \frac{b}{4} \left(\int_{\mathbb{R}^N} |\nabla u|^2 dx \right)^2 + \frac{1}{2} \int_{\mathbb{R}^N} V(x)u^2 dx$$
$$- \int_{\mathbb{R}^N} F(x, u)dx - \int_{\mathbb{R}^N} h(x)u dx.$$

Evidently, the action functional I belongs to $C^1(E, \mathbb{R})$ with derivative given by

$$\langle I'(u), v \rangle = \left(a + b \int_{\mathbb{R}^N} |\nabla u|^2 dx \right) \int_{\mathbb{R}^N} \nabla u \cdot \nabla v dx + \int_{\mathbb{R}^N} V(x)uv dx$$
$$- \int_{\mathbb{R}^N} f(x, u)v dx - \int_{\mathbb{R}^N} h(x)v dx. \quad (3.20)$$

Now we give a sketch of how to look for two distinct critical points of the functional I. First, we consider a minimization of I constrained in a neighborhood of zero, by using the Ekeland's variational principle (see [18], or [171]), we can find a critical point of I which achieves the local minimum of I. Moreover, this local minimum is negative, see the Step 1 in the proof of Theorem 3.3. Next, around "zero", by the Mountain Pass Theorem (see [172], or [21]), we can also get a critical point of I and its level is positive (see Step 2 in the proof of Theorem 3.3). Because these two critical points are in different level, they must be distinct.

Before going to the proof of Theorem 3.3, we give some useful preliminary results.

Lemma 3.5. *Suppose that (B6), (B7) and (B9) hold. Let $h \in L^2(\mathbb{R}^N)$. Then there exist some constants $\rho, \alpha, m_0 > 0$ such that $I(u)\big|_{\|u\|_E = \rho} \geq \alpha$ for all h satisfying $\|h\|_{L^2} < m_0$.*

Proof. It follows from (B7), (B9) and the equality $F(x, z) = \int_0^1 f(x, tz)z\, dt$ that for every $\varepsilon > 0$ there exists a constant $C_\varepsilon > 0$ such that

$$F(x, z) \leq \varepsilon |z|^2 + C_\varepsilon |z|^p$$

for any $x \in \mathbb{R}^N$ and all $z \in \mathbb{R}$. Thus, using $b \geq 0$, the Hölder inequality and $H^1(\mathbb{R}^N) \hookrightarrow L^s(\mathbb{R}^N)$ for any $s \in [2, 2^*]$, we have

$$I(u) \geq \frac{a}{2} \int_{\mathbb{R}^N} |\nabla u|^2 dx + \frac{1}{2} \int_{\mathbb{R}^N} V(x) u^2 dx - \int_{\mathbb{R}^N} F(x, u) dx - \int_{\mathbb{R}^N} h(x) u\, dx$$

$$\geq \frac{\min\{a, 1\}}{2} \|u\|_E^2 - \varepsilon \|u\|_{L^2}^2 - C_\varepsilon \|u\|_{L^p}^p - \|h\|_{L^2} \|u\|_{L^2}$$

$$\geq \frac{\min\{a, 1\}}{2} \|u\|_E^2 - \frac{\varepsilon}{a_1} \|u\|_E^2 - a_2 \|u\|_E^p - \frac{1}{\sqrt{a_1}} \|h\|_{L^2} \|u\|_E$$

$$= \|u\|_E \left[\left(\frac{\min\{a, 1\}}{2} - \frac{\varepsilon}{a_1} \right) \|u\|_E - a_2 \|u\|_E^{p-1} - \frac{1}{\sqrt{a_1}} \|h\|_{L^2} \right],$$

where a_1 is a lower bound of the potential V from (B6) and $a_2 > 0$ is a constant. Taking $\varepsilon = a_1 \min\{a, 1\}/4$ and setting

$$g(t) = \frac{\min\{a, 1\}}{4} t - a_2 t^{p-1}$$

for $t \geq 0$, we see that there exists a constant $\rho > 0$ such that $\max_{t \geq 0} g(t) = g(\rho) > 0$. Taking $m_0 := \frac{1}{2}\sqrt{a_1}\, g(\rho)$, then it follows that there exists a constant $\alpha > 0$ such that $I(u)\big|_{\|u\|_E = \rho} \geq \alpha$ for all h satisfying $\|h\|_{L^2} < m_0$. $\square$

Lemma 3.6. *Suppose that (B6), (B8) and (B10) hold, then there exists a function $v \in E$ with $\|v\|_E > \rho$ such that $I(v) < 0$, where ρ is given by Lemma 3.5.*

Proof. Let

$$k(t) = t^{-\mu}F(x, tz) - F(x, z), \ t \geq 1,$$

then, we get

$$k'(t) = t^{-\mu-1}(f(x, tz)tz - \mu F(x, tz)) \geq 0$$

for all $t \geq 1$ by (B8). Hence, it follows that $k(t) \geq k(1) = 0$ for all $t \geq 1$, that is

$$F(x, tz) \geq t^{\mu}F(x, z)$$

for all $x \in \mathbb{R}^N, z \in \mathbb{R}$ and $t \geq 1$. Thus, by $\mu > 4$ and (B10) we have

$$
\begin{aligned}
I(tu) &= \frac{at^2}{2}\int_{\mathbb{R}^N}|\nabla u|^2 dx + \frac{bt^4}{4}\left(\int_{\mathbb{R}^N}|\nabla u|^2 dx\right)^2 + \frac{t^2}{2}\int_{\mathbb{R}^N}V(x)u^2 dx \\
&\quad - \int_{\mathbb{R}^N}F(x, tu)dx - t\int_{\mathbb{R}^N}h(x)u dx \\
&\leq \frac{at^2}{2}\int_{\mathbb{R}^N}|\nabla u|^2 dx + \frac{bt^4}{4}\left(\int_{\mathbb{R}^N}|\nabla u|^2 dx\right)^2 + \frac{t^2}{2}\int_{\mathbb{R}^N}V(x)u^2 dx \\
&\quad - t^{\mu}\int_{\mathbb{R}^N}F(x, u)dx - t\int_{\mathbb{R}^N}h(x)u dx \\
&\to -\infty,
\end{aligned}
$$

as $t \to +\infty$ for $u \in E, u \neq 0$. The lemma is proved by taking $v = t_0 u$ with $t_0 > 0$ large enough and $u \neq 0$. $\square$

Lemma 3.7. *Assume that (B6), (B7)–(B9) hold, and $\{u_n\} \subset E$ is a bounded Palais–Smale sequence of I, then $\{u_n\}$ has a strongly convergent subsequence in E.*

Proof. Consider a sequence $\{u_n\}$ in E which satisfies

$$I(u_n) \to c, \ I'(u_n) \to 0, \ \text{and} \ \sup_n \|u_n\|_E < +\infty.$$

Going if necessary to a subsequence, we can assume that $u_n \rightharpoonup u$ in E. In view of Lemma 3.4, $u_n \to u$ in $L^s(\mathbb{R}^N)$ for any $s \in [2, 2^*)$. By (3.20) and

$b \geq 0$, we easily get

$$\langle I'(u_n) - I'(u), u_n - u \rangle$$

$$= \left(a + b \int_{\mathbb{R}^N} |\nabla u_n|^2 dx \right) \int_{\mathbb{R}^N} \nabla u_n \cdot \nabla(u_n - u) dx + \int_{\mathbb{R}^N} V(x)|u_n - u|^2 dx$$

$$- \left(a + b \int_{\mathbb{R}^N} |\nabla u|^2 dx \right) \int_{\mathbb{R}^N} \nabla u \cdot \nabla(u_n - u) dx$$

$$- \int_{\mathbb{R}^N} (f(x, u_n) - f(x, u))(u_n - u) dx$$

$$= \left(a + b \int_{\mathbb{R}^N} |\nabla u_n|^2 dx \right) \int_{\mathbb{R}^N} |\nabla(u_n - u)|^2 dx$$

$$+ \int_{\mathbb{R}^N} V(x)|u_n - u|^2 dx$$

$$- b \left(\int_{\mathbb{R}^N} |\nabla u|^2 dx - \int_{\mathbb{R}^N} |\nabla u_n|^2 dx \right) \int_{\mathbb{R}^N} \nabla u \cdot \nabla(u_n - u) dx$$

$$- \int_{\mathbb{R}^N} (f(x, u_n) - f(x, u))(u_n - u) dx$$

$$\geq \min\{a, 1\} \|u_n - u\|_E^2 - b \left(\int_{\mathbb{R}^N} |\nabla u|^2 dx - \int_{\mathbb{R}^N} |\nabla u_n|^2 dx \right)$$

$$\times \int_{\mathbb{R}^N} \nabla u \cdot \nabla(u_n - u) dx - \int_{\mathbb{R}^N} (f(x, u_n) - f(x, u))(u_n - u) dx.$$

One has

$$\min\{a, 1\} \|u_n - u\|_E^2 \leq \langle I'(u_n) - I'(u), u_n - u \rangle$$

$$+ \int_{\mathbb{R}^3} (f(x, u_n) - f(x, u))(u_n - u) dx$$

$$+ b \left(\int_{\mathbb{R}^N} |\nabla u|^2 dx - \int_{\mathbb{R}^N} |\nabla u_n|^2 dx \right) \int_{\mathbb{R}^N} \nabla u \cdot \nabla(u_n - u) dx.$$

It is clear that

$$\langle I'(u_n) - I'(u), u_n - u \rangle \to 0, \text{ as } n \to \infty.$$

According to assumptions (B7) and (B9), there exists a constant $a_3 > 0$ such that

$$f(x, u) \leq \varepsilon|u| + a_3|u|^{p-1}$$

for a.e. $x \in \mathbb{R}^N$, and all $z \in \mathbb{R}$. Using the Hölder inequality, we obtain

$$\int_{\mathbb{R}^N} (f(x, u_n) - f(x, u))(u_n - u) dx$$

$$\leq \int_{\mathbb{R}^N} \left[\varepsilon(|u_n| + |u|) + a_3 \left(|u_n|^{p-1} + |u|^{p-1} \right) \right] |u_n - u| dx$$

$$\leq \varepsilon \left(\|u_n\|_{L^2} + \|u\|_{L^2} \right) \|u_n - u\|_{L^2} + a_3(\|u_n\|_{L^p}^{p-1} + \|u\|_{L^p}^{p-1}) \|u_n - u\|_{L^p}.$$

Since $u_n \to u$ in $L^s(\mathbb{R}^N)$ for any $s \in [2, 2^*)$, we have

$$\int_{\mathbb{R}^N} (f(x, u_n) - f(x, u))(u_n - u)dx \to 0, \text{ as } n \to \infty.$$

Define the linear functional $g : E \to \mathbb{R}$ by

$$g(w) = \int_{\mathbb{R}^N} \nabla u \cdot \nabla w dx.$$

Since $g(w) \leq \|u\|_E \|w\|_E$, we can deduce that g is continuous on E. Using $u_n \rightharpoonup u$ in E, we get

$$\int_{\mathbb{R}^N} \nabla u \cdot \nabla(u_n - u)dx \to 0, \text{ in } n \to \infty.$$

Thus, by the boundedness of $\{u_n\}$ in E, we have

$$b\left(\int_{\mathbb{R}^N} |\nabla u|^2 dx - \int_{\mathbb{R}^N} |\nabla u_n|^2 dx\right) \int_{\mathbb{R}^N} \nabla u \cdot \nabla(u_n - u)dx \to 0, \text{ as } n \to \infty.$$

so that $\|u_n - u\|_E \to 0$ as $n \to \infty$. $\qquad\square$

Proof of Theorem 3.3. The proof of this theorem is divided in two steps.

Step 1 There exists a function $u_0 \in E$ such that $I'(u_0) = 0$ and $I(u_0) < 0$.

Observe that assumptions (B7)–(B9) and the equality $F(x, z) = \int_0^1 f(x, tz)z dt$ imply that there exist two positive constants a_4, a_5 such that

$$F(x, z) \geq a_4 |z|^\mu - a_5 |z|^2$$

for all $x \in \mathbb{R}^N$, and all $z \in \mathbb{R}$.

Since $h \in L^2(\mathbb{R}^N)$ and $h \not\equiv 0$, we can choose a function $\psi \in E$ such that

$$\int_{\mathbb{R}^3} h(x)\psi(x)dx > 0.$$

Hence, we have

$$I(t\psi) \leq \frac{at^2}{2} \int_{\mathbb{R}^N} |\nabla \psi|^2 dx + \frac{bt^4}{4} \left(\int_{\mathbb{R}^N} |\nabla \psi|^2 dx\right)^2 + \frac{t^2}{2} \int_{\mathbb{R}^N} V(x)\psi^2 dx$$

$$-a_4 t^\mu \|\psi\|_{L^\mu}^\mu + a_5 t^2 \|\psi\|_{L^2}^2 - t \int_{\mathbb{R}^3} h(x)\psi dx < 0$$

for $t > 0$ small enough, thus, we obtain

$$c_0 = \inf\{I(u) : u \in \overline{B}_\rho\} < 0,$$

where $\rho > 0$ is given by Lemma 3.5, $B_\rho = \{u \in E : \|u\|_E < \rho\}$. By the Ekeland's variational principle, there exists a sequence $\{u_n\} \subset \overline{B}_\rho$ such that

$$c_0 \leq I(u_n) < c_0 + \frac{1}{n},$$

and

$$I(w) \geq I(u_n) - \frac{1}{n}\|w - u_n\|_E$$

for all $w \in \overline{B}_\rho$. Then by a standard procedure, we can show that $\{u_n\}$ is a bounded Palais-Smale sequence of I. Therefore Lemma 3.7 implies that there exists a function $u_0 \in E$ such that $I'(u_0) = 0$ and $I(u_0) = c_0 < 0$.

Step 2 There exists a function $\widetilde{u}_0 \in E$ such that $I'(\widetilde{u}_0) = 0$ and $I(\widetilde{u}_0) > 0$.

From Lemma 3.5, Lemma 3.6 and the Mountain Pass Theorem, there is a sequence $\{u_n\} \subset E$ such that

$$I(u_n) \to \widetilde{c}_0 > 0, \quad \text{and} \quad I'(u_n) \to 0.$$

In view of Lemma 3.7, we only need to check that $\{u_n\}$ is bounded in E. By (B8) we have

$$\widetilde{c}_0 + 1 + \|u_n\|_E$$
$$\geq I(u_n) - \frac{1}{\mu}\langle I'(u_n), u_n \rangle$$
$$= \left(\frac{a}{2} - \frac{a}{\mu}\right)\int_{\mathbb{R}^N} |\nabla u|^2 dx + \left(\frac{b}{4} - \frac{b}{\mu}\right)\left(\int_{\mathbb{R}^N} |\nabla u|^2 dx\right)^2$$
$$+ \left(\frac{1}{2} - \frac{1}{\mu}\right)\int_{\mathbb{R}^N} V(x)u^2 dx + \int_{\mathbb{R}^3}\left(\frac{1}{\mu}f(x, u_n)u_n - F(x, u_n)\right)dx$$
$$+ \left(\frac{1}{\mu} - 1\right)\int_{\mathbb{R}^3} h(x)u_n dx$$
$$\geq \left(\frac{1}{2} - \frac{1}{\mu}\right)\|u_n\|_E^2 + \left(\frac{1}{4} - \frac{1}{\mu}\right)\left(\int_{\mathbb{R}^N} |\nabla u|^2 dx\right)^2$$
$$+ \left(\frac{1}{\mu} - 1\right)\frac{1}{\sqrt{a_1}}\|h\|_{L^2}\|u_n\|_E$$

for n large enough. Since $\mu > 4$ and $\|h\|_{L^2} < m_0$, it follows that $\{u_n\}$ is bounded in E. $\qquad\square$

Problem (3.19) has a variational structure. Indeed we consider the functional

$$I : H^1(\mathbb{R}^N) \to \mathbb{R}$$

defined by

$$I(u) = \frac{a}{2} \int_{\mathbb{R}^N} |\nabla u|^2 dx + \frac{b}{4} \left(\int_{\mathbb{R}^N} |\nabla u|^2 dx \right)^2 + \frac{1}{2} \int_{\mathbb{R}^N} u^2 dx$$

$$- \int_{\mathbb{R}^N} |u|^p dx - \int_{\mathbb{R}^N} h(x) u dx.$$

Evidently, the action functional I belongs to $C^1(H^1(\mathbb{R}^N), \mathbb{R})$ with derivative given by

$$\langle I'(u), v \rangle = \left(a + b \int_{\mathbb{R}^N} |\nabla u|^2 dx \right) \int_{\mathbb{R}^N} \nabla u \cdot \nabla v dx + \int_{\mathbb{R}^N} uv dx$$

$$- \int_{\mathbb{R}^N} |u|^{p-2} uv dx - \int_{\mathbb{R}^N} h(x) v dx.$$

Since equation (3.19) is set on $\mathbb{R}^N$, it is well known that the Sobolev embedding $H^1(\mathbb{R}^N) \hookrightarrow L^s(\mathbb{R}^N)$ $(2 \le s \le 6)$ is not compact, and then it is usually difficult to prove that a minimizing sequence or a Palais-Smale sequence is strongly convergent if we seek solutions of problem (3.19) by variational methods. To overcome this difficulty we restrict ourselves to radial functions $u = u(r)$, $r = |x|$. More precisely we shall consider the functional I on the space of the radial functions

$$H_r^1(\mathbb{R}^N) := \{ u \in H^1(\mathbb{R}^N) : u = u(r), \ r = |x| \}.$$

$H_r^1(\mathbb{R}^N)$ is a natural constraint for I, namely any critical point $u \in H_r^1(\mathbb{R}^N)$ of $I_{|H_r^1(\mathbb{R}^N)}$ is also a critical point of I. Then we are reduced to look for critical points of $I_{|H_r^1(\mathbb{R}^N)}$. We recall (see [169]) that, for $2 < s < 6$, $H_r^1(\mathbb{R}^N)$ is compactly embedded into $L^s(\mathbb{R}^N)$. In the following, we still denote $I_{|H_r^1(\mathbb{R}^N)}$ by I.

The rest proof of Theorem 3.4 is similar to the proof of Theorem 3.3.

Notes and Comments

Some interesting studies by variational methods can be found in [104, 126, 173–177] for Kirchhoff-type problems (3.1). But the problems that they studied are ones in a bounded domain of $\Omega \subset \mathbb{R}^N$. In Jin and Wu [167], the authors obtained the existence of infinitely many radial solutions for problem (3.18) with $h = 0$ in $\mathbf{R}^N$ by using the Fountain Theorem. In Wu [135], the author got four new existence results for nontrivial solutions and a sequence of high energy solutions for problem (3.18) with $h = 0$ in $\mathbb{R}^N$ which was obtained by using a Symmetric Mountain Pass Theorem. In Azzollini, d'Avenia and Pomponio [168], the authors got a multiplicity result concerning the critical points of a class of functionals involving local and nonlocal nonlinearities and apply it to the nonlinear elliptic Kirchhoff equation (3.18) with $h = 0$ in $\mathbb{R}^N$ assuming on the local nonlinearity the general hypotheses introduced by Berestycki and Lions. This section was obtained in [136].

3.4   Multiple Solutions with Superlinear Nonlinearities

In this section, we consider the following Kirchhoff equation

$$-\left(1 + b \int_{\mathbb{R}^3} |\nabla u|^2 dx\right) \Delta u + V(x)u = f(x, u), \quad x \in \mathbb{R}^3, \qquad (3.21)$$

where $b > 0$ is a positive constant, the potential V and the nonlinearity f are allowed to be sign-changing.

Recently, Wu [135] study this type of equation with positive coercive potential V. Four new existence results for nontrivial solutions and a sequence of high energy solutions for problem (3.21) are obtained by using a Symmetric Mountain Pass Theorem. Actually, coercive potential V was introduced by Rabinowitz [178] to overcome the lack of compact Sobolev embedding. Later, many authors [133, 136–138, 140, 144, 149, 179–183] using this type of potential. Very recently, the potential V can be zero at some points had also been considered [139, 184–187].

In some of the aforementioned references, the potential V is always assumed to be positive or vanish at infinity and the following famous Ambrosetti–Rabinowitz condition ((AR) for short) is usually required.

(AR) There exists $\mu > 4$ such that

$$0 < \mu F(x, u) \leq u f(x, u), \quad u \neq 0.$$

It is well known that the role of (AR) is to ensure the boundedness of the Palais–Smale (PS) sequences of the energy functional, which is very crucial in applying the critical point theory.

Motivated by the above works, in this section, we consider another case of f being superlinear, that is, $f(x, u)/u \to +\infty$ as $u \to \infty$. Furthermore, the potential V and the primitive of f are also allowed to be sign-changing, which is quite different from the previous results. Before stating our main results, we give the following assumption on $V(x)$.

(B11) $V \in C(\mathbb{R}^3, \mathbb{R})$ and $\inf_{x \in \mathbb{R}^3} V(x) > -\infty$. Moreover, there exists a constant $d_0 > 0$ such that for any $M > 0$,

$$\lim_{|y| \to \infty} \text{meas}\{x \in \mathbb{R}^3 : |x - y| \le d_0, V(x) \le M\} = 0,$$

where $\text{meas}(\cdot)$ denotes the Lebesgue measure in $\mathbb{R}^3$.

Inspired by Zhang and Xu [188], we can find a constant $V_0 > 0$ such that $\tilde{V}(x) := V(x) + V_0 \ge 1$ for all $x \in \mathbb{R}^3$ and let $\tilde{f}(x, u) := f(x, u) + V_0 u$, $\forall (x, u) \in \mathbb{R}^3 \times \mathbb{R}$. Then it is easy to verify the following lemma.

Lemma 3.8. *Equation* (3.21) *is equivalent to the following problem*

$$-\left(1 + b \int_{\mathbb{R}^3} |\nabla u|^2 dx\right) \Delta u + \tilde{V}(x) u = \tilde{f}(x, u), \quad x \in \mathbb{R}^3. \tag{3.22}$$

In what follows, we let $\mu > 4$ and impose some assumptions on $\tilde{f}$ and its primitive $\tilde{F}$ as follows:

(B12) $\tilde{f} \in C(\mathbb{R}^3 \times \mathbb{R}, \mathbb{R})$, and there exist constants $c_1, c_2 > 0$ and $q \in [4, 6)$ such that

$$|\tilde{f}(x, u)| \le c_1 |u|^3 + c_2 |u|^{q-1}.$$

(B13) $\lim_{|u| \to \infty} \frac{|\tilde{F}(x,u)|}{|u|^4} = \infty$ a.e. $x \in \mathbb{R}^3$ and there exists constants $c_3 \ge 0$, $r_0 \ge 0$ and $\tau \in (0, 2)$ such that

$$\inf_{x \in \mathbb{R}^3} \tilde{F}(x, u) \ge c_3 |u|^\tau \ge 0, \quad \forall (x, u) \in \mathbb{R}^3 \times \mathbb{R}, \ |u| \ge r_0,$$

where and in the sequel, $\tilde{F}(x, u) = \int_0^u \tilde{f}(x, s) ds$.

(B14) $\hat{\mathcal{F}}(x, u) := \frac{1}{4} u \tilde{f}(x, u) - \tilde{F}(x, u) \ge 0$, and there exists $c_4 > 0$ and $\kappa > 1$ such that

$$|\tilde{F}(x, u)|^\kappa \le c_4 |u|^{2\kappa} \hat{\mathcal{F}}(x, u), \quad \forall (x, u) \in \mathbb{R}^3 \times \mathbb{R}, \ |u| \ge r_0.$$

Now, we state our main results as follows.

Theorem 3.5. *Suppose that conditions (B11), (B12), (B13) and (B14) are satisfied. Then problem* (3.21) *possesses at least two different solutions.*

Remark 3.5. There are some functions not satisfying the condition (AR) for any $\mu > 4$. For example, the superlinear function $f(x, u) = \sin x \ln(1 + |u|) u^2$ does not satisfy condition (AR).

Remark 3.6. To the best of our knowledge, the condition (B11) is first given in [189], but $\inf_{x \in \mathbb{R}^3} V(x) > 0$ is required. From (B11), one can see that the potential $V(x)$ is allowed to be sign-changing. Therefore, the condition (B11) is weaken than [133, 136–140, 144, 149, 179–187].

Remark 3.7. It is not difficult to find the functions V and f satisfying the above conditions. For example, let $V(x)$ be a zigzag function with respect to $|x|$ defined by

$$V(x) = \begin{cases} 2n|x| - 2n(n-1) + a_0, & n-1 \le |x| \le (2n-1)/2, \\ -2n|x| + 2n^2 + a_0, & (2n-1)/2 \le |x| \le n, \end{cases}$$

where $n \in \mathbb{N}$ and $a_0 \in \mathbb{R}$.

Remark 3.8. Zhang *et al.* [180] study (3.21) with sign-changing potential V. They got multiple solution with odd nonlinearity. Here we do not need the nonlinearity is odd and also got two solutions for problem (3.21). Bahrouni [148] got infinitely many solutions for (3.21) with the potential and nonlinearity are both sign-changing. But he study the sublinear case and with odd nonlinearity. Here our results can be regarded as an extension of the result of [148, 180].

In this section, we use the following notation. Denotes the space $E = \{u \in H^1(\mathbb{R}^3) : \int_{\mathbb{R}^3}(|\nabla u|^2 + \tilde{V}(x)|u|^2)dx < \infty\}$ with the norm

$$\|u\|_E^2 = \int_{\mathbb{R}^3}(|\nabla u|^2 + \tilde{V}(x)|u|^2)dx.$$

Next, we make the following assumption instead of (B11):

(B15) $\tilde{V} \in C(\mathbb{R}^3, \mathbb{R})$ and $\inf_{x \in \mathbb{R}^3} \tilde{V}(x) > 0$. Moreover, there exists a constant $d_0 > 0$ such that for any $M > 0$,

$$\lim_{|y| \to \infty} \text{meas}\{x \in \mathbb{R}^3 : |x - y| \le d_0, V(x) \le M\} = 0.$$

Remark 3.9. Under assumptions (B15), we know from Lemma 3.1 in [189] that the embedding $E \hookrightarrow L^s(\mathbb{R}^3)$ is compact for $s \in [2, 6)$.

Now, functional I on E by

$$I(u) = \frac{1}{2}\int_{\mathbb{R}^3}(|\nabla u|^2 + \tilde{V}(x)u^2)dx + \frac{b}{4}\left(\int_{\mathbb{R}^3}|\nabla u|^2 dx\right)^2 - \int_{\mathbb{R}^3}\tilde{F}(x,u)dx,$$

$$\tag{3.23}$$

for all $u \in E$. By condition (B12), we have

$$|\tilde{F}(x, u)| \leq \frac{c_1}{4}|u|^4 + \frac{c_2}{q}|u|^q, \quad \forall(x, u) \in \mathbb{R}^3 \times \mathbb{R}. \tag{3.24}$$

Consequently, similar to the discussion in [135], under assumptions (B15) and (3.24), it is easy to prove that the functional I is of class $C^1(E, \mathbb{R})$. Moreover,

$$\langle I'(u), v \rangle = \left(1 + b \int_{\mathbb{R}^3} |\nabla u|^2 dx\right) \int_{\mathbb{R}^3} \nabla u \cdot \nabla v dx + \int_{\mathbb{R}^3} \tilde{V}(x) uv dx$$
$$- \int_{\mathbb{R}^3} \tilde{f}(x, u) v dx. \tag{3.25}$$

Hence, if $u \in E$ is a critical point of I, then u is a solution of equation (3.22).

Here, we give the sketch of how to look for two distinct critical points of the functional I. First, we consider a minimization of I constrained in a neighborhood of zero via the Ekeland's variational principle (see [5, 18]) and we can find a critical point of I which achieves the local minimum of I and the level of this local minimum is negative (see the Step 1 in the proof of Theorem 3.5); And then, around "zero" point, by using Mountain Pass Theorem (see [22]) we can also obtain other critical point of I with its positive level (see the Step 2 in the proof of Theorem 3.5). Obviously, these two critical points are different because they are in different levels.

Lemma 3.9. *Assume that the conditions (B15) and (B12) hold. Then there exist ρ, $\eta > 0$ such that $\inf\{I(u) : u \in E \text{ with } \|u\|_E = \rho\} > \eta$.*

Proof. From (3.24) and the Sobolev inequality, we have

$$\left|\int_{\mathbb{R}^3} \tilde{F}(x, u) dx\right| \leq \int_{\mathbb{R}^3} \left|\frac{c_1}{4}|u|^4 + \frac{c_2}{q}|u|^q\right| dx$$
$$= \frac{c_1}{4}\|u\|_4^4 + \frac{c_2}{q}\|u\|_q^q$$
$$\leq S_4\frac{c_1}{4}\|u\|_E^4 + S_q\frac{c_2}{q}\|u\|_E^q, \tag{3.26}$$

for any $u \in E$. Combining (3.23) with (3.26), we have

$$I(u) = \frac{1}{2} \int_{\mathbb{R}^3} (|\nabla u|^2 + \tilde{V}(x)u^2)dx + \frac{b}{4} \left(\int_{\mathbb{R}^3} |\nabla u|^2 dx \right)^2 - \int_{\mathbb{R}^3} \tilde{F}(x,u)dx$$

$$\geq \frac{1}{2}\|u\|_E^2 - \int_{\mathbb{R}^3} |\tilde{F}(x,u)|dx$$

$$\geq \frac{1}{2}\|u\|_E^2 - S_4 \frac{c_1}{4}\|u\|_E^4 - S_q \frac{c_2}{q}\|u\|_E^q$$

$$= \frac{1}{2}\|u\|_E^2 - C_1 4\|u\|_E^4 - C - 2\|u\|_E^q. \tag{3.27}$$

Since $q \in [4,6)$, we can easily get that there exists $\eta > 0$ such that this lemma holds if we let $\|u\|_E = \rho > 0$ small enough. $\quad\square$

Lemma 3.10. *Assume that the conditions (B15) and (B13) hold. Then there exists $v \in E$ with $\|v\|_E = \rho$ such that $I(v) < 0$, where ρ is given in Lemma 3.9.*

Proof. From (3.23), we have

$$\frac{I(tu)}{t^4} = \frac{t^2}{2}\|u\|_E^2 + \frac{bt^4}{4} \left(\int_{\mathbb{R}^3} |\nabla u|^2 dx \right)^2 - \frac{1}{t^4} \int_{\mathbb{R}^3} \tilde{F}(x,tu)dx.$$

Then, by (B13) and Fatou's Lemma we deduce that

$$\lim_{t \to \infty} \frac{I(tu)}{t^4} = \lim_{t \to \infty} \left[\frac{t^2}{2}\|u\|_E^2 + \frac{b}{4} \left(\int_{\mathbb{R}^3} |\nabla u|^2 dx \right)^2 - \frac{1}{t^4} \int_{\mathbb{R}^3} \tilde{F}(x,tu)dx \right]$$

$$= \lim_{t \to \infty} \frac{b}{4} \left(\int_{\mathbb{R}^3} |\nabla u|^2 dx \right)^2 - \frac{1}{t^4} \int_{\mathbb{R}^3} \tilde{F}(x,tu)dx$$

$$\leq \frac{b}{4} \left(\int_{\mathbb{R}^3} |\nabla u|^2 dx \right)^2 - \int_{\mathbb{R}^3} \lim_{t \to \infty} \frac{\tilde{F}(x,tu)}{t^4 u^4} u^4 dx$$

$$\leq C_3 \|u\|_E^4 - \int_{\mathbb{R}^3} \lim_{t \to \infty} \frac{\tilde{F}(x,tu)}{t^4 u^4} u^4 dx$$

$$= -\infty \text{ as } n \to \infty.$$

Thus, the lemma is proved by taking $v = t_0 u$ with $t_0 > 0$ large enough. $\quad\square$

Based on Lemmas 3.9 and 3.10, Theorem 1.4 implies that there is a sequence $\{u_n\} \subset E$ such that

$$I(u_n) \to c > 0 \text{ and } (1 + \|u_n\|_E)\|I'(u_n)\|_{E^*} \to 0, \text{ as } n \to \infty. \tag{3.28}$$

Lemma 3.11. *Assume that the conditions (B15), (B12), (B13) and (B14) hold. Then the sequence $\{u_n\}$ defined in (3.28) is bounded in E.*

Proof. Arguing by contradiction, we can assume $\|u_n\|_E \to \infty$. Define $v_n := \frac{u_n}{\|u_n\|_E}$. Clearly, $\|v_n\|_E = 1$ and $\|v_n\|_s \leq S_s\|v_n\|_E = S_s$, for $2 \leq s < 6$. Observe that for n large enough, from (3.28) and (B14) we have

$$c + 1 \geq I(u_n) - \frac{1}{4}\langle I'(u_n), u_n \rangle$$

$$= \frac{1}{4}\|u_n\|_E^2 + \int_{\mathbb{R}^3} \left(\frac{1}{4}\tilde{f}(x, u_n)u_n - \tilde{F}(x, u_n)\right) dx$$

$$\geq \int_{\mathbb{R}^3} \tilde{\mathcal{F}}(x, u_n)dx. \tag{3.29}$$

In view of (3.23) and (3.28), we have

$$\frac{1}{2} = \frac{I(u_n)}{\|u_n\|_E^2} + \frac{1}{\|u_n\|_E^2}\int_{\mathbb{R}^3} \tilde{F}(x, u_n)dx - \frac{b}{4\|u_n\|_E^2}\left(\int_{\mathbb{R}^3} |\nabla u|^2 dx\right)^2$$

$$\leq \frac{I(u_n)}{\|u_n\|_E^2} + \frac{1}{\|u_n\|_E^2}\int_{\mathbb{R}^3} |\tilde{F}(x, u_n)|dx$$

$$\leq \limsup_{n\to\infty} \left[\frac{I(u_n)}{\|u_n\|_E^2} + \frac{1}{\|u_n\|_E^2}\int_{\mathbb{R}^3} |\tilde{F}(x, u_n)|dx\right]$$

$$\leq \limsup_{n\to\infty} \int_{\mathbb{R}^3} \frac{|\tilde{F}(x, u_n)|}{\|u_n\|_E^2}dx. \tag{3.30}$$

For $0 \leq a < b$, let $\Omega_n(a, b) := \{x \in \mathbb{R}^3 : a \leq |u_n(x)| < b\}$. Going if necessary to a subsequence, we may assume that $v_n \rightharpoonup v$ in E. Then by Remark 3.9, we have $v_n \to v$ in $L^s(\mathbb{R}^3)$ for $2 \leq s < 6$, and $v_n \to v$ a.e. on $\mathbb{R}^3$.

We now consider the following two possible cases about v.

Case 1: If $v = 0$, then $v_n \to 0$ in $L^s(\mathbb{R}^3)$ for $2 \leq s < 6$, and $v_n \to 0$ a.e. on $\mathbb{R}^3$. Hence, it follows from (3.24) and $v_n := \frac{u_n}{\|u_n\|_E^2}$ that

$$\int_{\Omega_n(0, r_0)} \frac{|\tilde{F}(x, u_n)|}{\|u_n\|_E^2}dx = \int_{\Omega_n(0, r_0)} \frac{|\tilde{F}(x, u_n)|}{|u_n|^2}|v_n|^2 dx$$

$$\leq \left(\frac{c_1}{4}r_0^2 + \frac{c_2}{q}r_0^{q-2}\right)\int_{\Omega_n(0, r_0)} |v_n|^2 dx$$

$$\leq C_4 \int_{\mathbb{R}^3} |v_n|^2 dx \to 0, \text{ as } n \to \infty. \tag{3.31}$$

From (B14), we know that $\kappa > 1$. Thus, if we set $\kappa' = \kappa/(\kappa 1)$, then $2\kappa' \in (2,6)$. Hence, it follows from (B14) and (3.29) that

$$\int_{\Omega_n(r_0,\infty)} \frac{|\tilde{F}(x,u_n)|}{\|u_n\|_E^2} dx = \int_{\Omega_n(r_0,\infty)} \frac{|\tilde{F}(x,u_n)|}{|u_n|^2} |v_n|^2 dx$$

$$\leq \left[\int_{\Omega_n(r_0,\infty)} \left(\frac{|\tilde{F}(x,u_n)|}{|u_n|^2}\right)^\kappa dx\right]^{1/\kappa} \left[\int_{\Omega_n(r_0,\infty)} |v_n|^{2\kappa'} dx\right]^{1/\kappa'}$$

$$\leq c_4^{1/\kappa} \left[\int_{\Omega_n(r_0,\infty)} \tilde{\mathcal{F}}(x,u_n) dx\right]^{1/\kappa} \left[\int_{\Omega_n(r_0,\infty)} |v_n|^{2\kappa'} dx\right]^{1/\kappa'}$$

$$\leq c_4^{1/\kappa}(c+1)^{1/\kappa} \left[\int_{\Omega_n(r_0,\infty)} |v_n|^{2\kappa'} dx\right]^{1/\kappa'}$$

$$\leq C_5 \left[\int_{\Omega_n(r_0,\infty)} |v_n|^{2\kappa'} dx\right]^{1/\kappa'} \to 0, \qquad (3.32)$$

as $n \to \infty$. Combining (3.31) with (3.32), we have

$$\int_{\mathbb{R}^3} \frac{|\tilde{F}(x,u_n)|}{\|u_n\|_E^2} dx = \int_{\Omega_n(0,r_0)} \frac{|\tilde{F}(x,u_n)|}{\|u_n\|_E^2} dx + \int_{\Omega_n(r_0,\infty)} \frac{|\tilde{F}(x,u_n)|}{\|u_n\|_E^2} dx \to 0$$

as $n \to \infty$, which contradicts (3.30).

Case 2: If $v \neq 0$, we set $A := \{x \in \mathbb{R}^3 : v(x) \neq 0\}$. Then $\mathrm{meas}(A) > 0$. For a.e. $x \in A$, we have $\lim_{n\to\infty} |u_n(x)| = \infty$. Hence $A \subset \Omega_n(r_0,\infty)$ for $n \in \mathbb{N}$ large enough. It follows from (3.23), (3.24), (3.28) and Fatou's Lemma that

$$0 = \lim_{n\to\infty} \frac{c + o(1)}{\|u_n\|_E^4} = \lim_{n\to\infty} \frac{I(u_n)}{\|u_n\|_E^4}$$

$$= \lim_{n\to\infty} \left[\frac{1}{2\|u_n\|_E^2} + \frac{b}{4\|u_n\|_E^4}\left(\int_{\mathbb{R}^3} |\nabla u|^2 dx\right)^2 - \int_{\mathbb{R}^3} \frac{\tilde{F}(x,u_n)}{\|u_n\|_E^4} dx\right]$$

$$= \left[\frac{b}{4\|u_n\|_E^4} \left(\int_{\mathbb{R}^3} |\nabla u|^2 dx \right)^2 - \int_{\Omega_n(0,r_0)} \frac{|\tilde{F}(x,u_n)|}{|u_n|^4} |v_n|^4 dx \right.$$

$$\left. - \int_{\Omega_n(r_0,\infty)} \frac{|\tilde{F}(x,u_n)|}{|u_n|^4} |v_n|^4 dx \right]$$

$$\leq \frac{b}{4} + \limsup_{n\to\infty} \int_{\Omega_n(0,r_0)} \left(\frac{c_1}{4} + \frac{c_2}{q} |u_n|^{q-4} \right) |v_n|^4 dx$$

$$- \liminf_{n\to\infty} \left[\int_{\Omega_n(r_0,\infty)} \frac{|\tilde{F}(x,u_n)|}{|u_n|^4} |v_n|^4 dx \right]$$

$$\leq \frac{b}{4} + \left(\frac{c_1}{4} + \frac{c_2}{q} |r_0|^{q-4} \right) \limsup_{n\to\infty} \int_{\Omega_n(0,r_0)} |v_n|^4 dx$$

$$- \liminf_{n\to\infty} \left[\int_{\Omega_n(r_0,\infty)} \frac{|\tilde{F}(x,u_n)|}{|u_n|^4} |v_n|^4 dx \right]$$

$$\leq \frac{b}{4} + C_8 - \liminf_{n\to\infty} \int_{\Omega_n(r_0,\infty)} \frac{|\tilde{F}(x,u_n)|}{|u_n|^4} |v_n|^4 dx$$

$$= \frac{b}{4} + C_8 - \liminf_{n\to\infty} \int_{\mathbb{R}^3} \frac{|\tilde{F}(x,u_n)|}{|u_n|^4} [\chi_{\Omega_n(r_0,\infty)}(x)] |v_n|^4 dx$$

$$= C_9 - \int_{RR^3} \liminf_{n\to\infty} \frac{|\tilde{F}(x,u_n)|}{|u_n|^4} [\chi_{\Omega_n(r_0,\infty)}(x)] |v_n|^4 dx \to -\infty, \qquad (3.33)$$

as $n \to \infty$, which is a contradiction. Thus $\{u_n\}$ is bounded in E. The proof is completed. □

To complete our proof, we have to cite a result in [190].

Lemma 3.12. *Assume that p_1, $p_2 > 1$, r, $q \geq 1$ and $\Omega \subseteq \mathbb{R}^3$. Let $g(x,t)$ be a Carathéodory function on $\Omega \times \mathbb{R}$ and satisfy*

$$|g(x,t)| \leq a_1 |t|^{(p_1-1)/r} + a_2 |t|^{(p_2-1)/r}, \quad \forall (x,t) \in \Omega \times \mathbb{R},$$

where a_1, $a_2 \geq 0$. If $u_n \to u$ in $L^{p_1}(\Omega) \cap L^{p_2}(\Omega)$, and $u_n \to u$ a.e. $x \in \Omega$, then for any $v \in L^{p_1 q}(\Omega) \cap L^{p_2 q}(\Omega)$,

$$\lim_{n\to\infty} \int_\Omega |g(x,u_n) - g(x,u)|^r |v|^q dx \to 0. \qquad (3.34)$$

Lemma 3.13. *If the conditions (B15) and (B12) hold. Then any bounded sequence $\{u_n\}$ satisfying (3.28) has a convergent subsequence in E.*

Proof. Going if necessary to a subsequence, we may assume that $u_n \rightharpoonup u$ in E. Then by Remark 3.9, we have $v_n \to v$ in $L^s(\mathbb{R}^3)$, for $2 \leq s < 6$. Let us take $r \equiv 1$ in Lemma 3.12 and combine with $u_n \to u$ in $L^s(\mathbb{R}^3)$ for $2 \leq s < 6$, one can get

$$\lim_{n \to \infty} |\tilde{f}(x, u_n) - \tilde{f}(x, u)||u_n - u| dx \to 0, \text{ as } n \to \infty. \tag{3.35}$$

We observe that

$$\langle I'(u_n) - I'(u), u_n - u \rangle \to 0, \text{ as } n \to \infty, \tag{3.36}$$

and we have

$$\langle I'(u_n) - I'(u), u_n - u \rangle$$

$$= \int_{\mathbb{R}^3} V(x)|u_n - u|^2 dx + \left(1 + b\int_{\mathbb{R}^3} |\nabla u_n|^2 dx\right) \int_{\mathbb{R}^3} \nabla u_n \cdot \nabla(u_n - u) dx$$

$$- \left(1 + b\int_{\mathbb{R}^3} |\nabla u|^2 dx\right) \int_{\mathbb{R}^3} \nabla u \cdot \nabla(u_n - u) dx$$

$$- \int_{\mathbb{R}^3} [f(x, u_n) - f(x, u)](u_n - u) dx$$

$$= \|u_n - u\|_E^2 + \left(1 + b\int_{\mathbb{R}^3} |\nabla u_n|^2 dx\right) \int_{\mathbb{R}^3} |\nabla(u_n - u)|^2 dx$$

$$- \left(\int_{\mathbb{R}^3} |\nabla u|^2 dx - \int_{\mathbb{R}^3} |\nabla u_n|^2 dx\right) \int_{\mathbb{R}^3} \nabla u \cdot \nabla(u_n - u) dx$$

$$- \int_{\mathbb{R}^3} [f(x, u_n) - f(x, u)](u_n - u) dx$$

$$\geq \|u_n - u\|_E^2 - b\left(\int_{\mathbb{R}^3} |\nabla u|^2 dx - \int_{\mathbb{R}^3} |\nabla u_n|^2 dx\right)$$

$$- \int_{\mathbb{R}^3} \nabla u \cdot \nabla(u_n - u) dx - \int_{\mathbb{R}^3} [f(x, u_n) - f(x, u)](u_n - u) dx. \tag{3.37}$$

Then, (3.37) implies that

$$\|u_n - u\|_E^2 \leq \langle I'(u_n) - I'(u), u_n - u \rangle$$

$$+ b\left(\int_{\mathbb{R}^3} |\nabla u|^2 dx - \int_{\mathbb{R}^3} |\nabla u_n|^2 dx\right) \int_{\mathbb{R}^3} \nabla u \cdot \nabla(u_n - u) dx$$

$$+ \int_{\mathbb{R}^3} [f(x, u_n) - f(x, u)](u_n - u) dx. \tag{3.38}$$

Define the functional $h_u : E \to \mathbb{R}$ by

$$h_u(v) = \int_{\mathbb{R}^3} \nabla u \cdot \nabla v dx, \quad \forall v \in E.$$

Obviously, h_u is a linear functional on E. Furthermore,

$$|h_u(v)| \le \int_{\mathbb{R}^3} |\nabla u \cdot \nabla v| dx \le \|u\|_E \|v\|_E,$$

which implies h_u is bounded on E. Hence $h_u \in E^*$. Since $u_n \rightharpoonup u$ in E, it has $\lim_{n\to\infty} h_u(u_n) = h_u(u)$, that is, $\int_{\mathbb{R}^3} \nabla u \cdot \nabla (u_n - u) dx \to 0$ as $n \to \infty$. Consequently, by $v_n \to v$ in $L^s(\mathbb{R}^3)$, for $2 \le s < 6$ and the boundedness of $\{u_n\}$, it has

$$b\left(\int_{\mathbb{R}^3} |\nabla u|^2 dx - \int_{\mathbb{R}^3} |\nabla u_n|^2 dx\right)\int_{\mathbb{R}^3} \nabla u \cdot \nabla(u_n - u) dx \to 0, \quad n \to +\infty. \tag{3.39}$$

Consequently, (3.35), (3.36), (3.38), (3.39) imply that

$$u_n \to u \text{ in } E \text{ as } n \to \infty.$$

This completes the proof. $\qquad\square$

Now, we prove the main results.

Proof of Theorem 3.5. To complete the proof of Theorem 3.5, we need to consider the following two steps.

Step 1. We first show that there exists a function $u_0 \in E$ such that $I'(u_0) = 0$ and $I(u_0) < 0$. Let $r_0 = 1$, for any $|u| \ge 1$, from (B13), we have

$$\tilde{F}(x, u_n) \ge c_3 |u|^\tau > 0. \tag{3.40}$$

By (B12), for a.e. $x \in \mathbb{R}^3$ and $0 \le |u| \le 1$, there exists $M > 0$ such that

$$\left|\frac{\tilde{f}(x, u)u}{u^2}\right| \le \left|\frac{(c_1|u|^3 + c_2|u|^{q-1})|u|}{|u|^2}\right| \le M,$$

which implies that

$$\tilde{f}(x, u)u \ge -M|u|^2.$$

Using the equality $\tilde{F}(x, u) = \int_0^1 \tilde{f}(x, tu) dt$, for a.e. $x \in \mathbb{R}^3$ and $0 \le |u| \le 1$, we obtain

$$\tilde{F}(x, u) > -\frac{1}{2}M|u|^2. \tag{3.41}$$

In view of (3.40) and (3.41), we have for a.e. $x \in \mathbb{R}^3$ and all $u \in \mathbb{R}$ that

$$\tilde{F}(x, u) \ge -\frac{1}{2}M|u|^2 + c_3|u|^\tau.$$

Then, we have

$$\tilde{F}(x, t\psi) \geq -\frac{1}{2} M t^2 |\psi|^2 + t^\tau c_3 |\psi|^\tau. \tag{3.42}$$

Combining (3.23) with (3.42), we have

$$I(tu) = \frac{t^2}{2} \|u\|_E^2 + \frac{bt^4}{4} \left(\int_{\mathbb{R}^3} |\nabla u|^2 dx \right)^2 - \int_{\mathbb{R}^3} \tilde{F}(x, tu) dx$$

$$\leq \frac{t^2}{2} \|u\|_E^2 + \frac{bt^4}{4} \left(\int_{\mathbb{R}^3} |\nabla u|^2 dx \right)^2 + \frac{t^2 M}{2} \int_{\mathbb{R}^3} |u|^2 dx - t^\tau c_3 \int_{\mathbb{R}^3} |u|^\tau dx.$$

Since $\tau \in (0, 2)$, for t small enough, we can get that $I(tu) < 0$. Thus, we obtain

$$c_0 = \inf\{I(u) : u \in \bar{B}_\rho\} < 0,$$

where $\rho > 0$ is given by Lemma 3.9, $B_\rho = \{u \in E : \|u\|_E < \rho\}$. By the Ekeland's variational principle, there exists a sequence $\{u_n\} \subset B_\rho$ such that

$$c_0 \leq I(u_n) \leq c_0 + \frac{1}{n},$$

and

$$I(w) \geq I(u_n) - \frac{1}{n} \|w - u_n\|_E,$$

for all $w \in B_\rho$. Then, following the idea of [5], we can show that $\{u_n\}$ is a bounded Cerami sequence of I. Therefore, Lemma 3.13 implies that there exists a function $u_0 \in E$ such that $I'(u_0) = 0$ and $I(u_0) = c_0 < 0$.

Step 2. We now show that there exists a function $\tilde{u}_0 \in E$ such that $I'(\tilde{u}_0) = 0$ and $I(\tilde{u}_0) = \tilde{c}_0 > 0$. By Lemmas 3.9, 3.10 and 1.4, there is a sequence $\{u_n\} \in E$ satisfies (3.28). Moreover, Lemma 3.11 and 3.13 shows that this sequence has a convergent subsequence and is bounded in E. So, we complete the Step 2.

Therefore, combining the above two Steps and Lemma 3.8, we complete the proof of Theorem 3.5. □

Notes and Comments

This section was obtained in [191].

3.5 Multiple Solutions with Asymptotically Linear Nonlinearities

In this section, we consider the following nonhomogeneous Kirchhoff equation

$$\begin{cases} -\left(1 + b\int_{\mathbb{R}^3} |\nabla u|^2 dx\right)\Delta u + u = k(x)f(u) + h(x), & x \in \mathbb{R}^3, \\ u \in H^1(\mathbb{R}^3), \quad u > 0, \quad x \in \mathbb{R}^3, \end{cases} \tag{3.43}$$

where constant $b > 0$, k is a positive bounded function, $f \in C(\mathbb{R}, \mathbb{R}^+)$, $f(t) \equiv 0$ if $t < 0$ and $h \in L^2(\mathbb{R}^3)$, $h \geq 0$. Note that problem (3.43) with $b = 0$ and $\mathbb{R}^3$ replaced by $\mathbb{R}^N$, problem (3.43) reduces to

$$-\Delta u + u = k(x)f(u) + h(x) \quad \text{in } \mathbb{R}^N,$$

which can been looked as a generalization of the well-known Schrödinger equation.

Before stating our main results, we give some notations. Recall that the Sobolev's inequality with the best constant is

$$\|v\|_6 \leq S\|v\|.$$

Moreover, problem (3.43) has a variational structure. Indeed the corresponding action functional $I : H^1(\mathbb{R}^3) \longrightarrow \mathbb{R}$ of equation (3.43) is defined by

$$I(u) = \frac{a}{2}\int_{\mathbb{R}^3} |\nabla u|^2 dx + \frac{b}{4}\left(\int_{\mathbb{R}^3} |\nabla u|^2 dx\right)^2 + \frac{1}{2}\int_{\mathbb{R}^3} |u|^2 dx - \int_{\mathbb{R}^3} k(x)F(u)dx$$
$$- \int_{\mathbb{R}^3} h(x)udx.$$

By Lemma 2.1 in [167] or Lemma 1 in [135], the functional I is $C^1(H^1(\mathbb{R}^3), \mathbb{R})$ with the derivative given by

$$\langle I'(u), u \rangle = \left(a + b\int_{\mathbb{R}^3} |\nabla u|^2 dx\right)\int_{\mathbb{R}^3} \nabla u \cdot \nabla v dx + \int_{\mathbb{R}^3} uvdx$$
$$- \int_{\mathbb{R}^3} k(x)f(u)udx - \int_{\mathbb{R}^3} h(x)vdx. \tag{3.44}$$

Hence, if $u \in H^1(\mathbb{R}^3)$ is a nonzero critical point of I, then it is also a nonnegative solution of (3.43). In fact, by $f(t) \equiv 0$ if $t < 0$ and $h \geq 0$, we have $\langle I'(u), u^- \rangle = -(a+b\|u\|^2)\int_{\mathbb{R}^3} |\nabla u^-|^2 dx - \int_{\mathbb{R}^3}(u^-)^2 dx - \int_{\mathbb{R}^3} k(x)f(u)u^- dx - \int_{\mathbb{R}^3} h(x)u^- dx = 0$, where $u^- = \max\{-u, 0\}$. This yields that $u^- = 0$, then $u = u^+ - u^- = u^+ \geq 0$, where $u^+ = \max\{u, 0\}$. By the maximum principle, the nonzero critical point of I is the positive solution for problem (3.43).

Here is the main result of this section.

Theorem 3.6. *Suppose that $h \in L^2(\mathbb{R}^3)$, $h \geq 0$ and the following conditions hold:*

(B16) $f \in C(\mathbb{R}, \mathbb{R}^+)$, $f(0) = 0$, and $f(t) \equiv 0$ for $t < 0$.

(B17) $\lim_{t\to 0} \frac{f(t)}{t} = 0$.

(B18) $\lim_{t\to +\infty} \frac{f(t)}{t} = l < +\infty$.

(B19) $k(x)$ is a positive continuous function and there exists $R_0 > 0$ such that

$$\sup\{f(t)/t : \ t > 0\} < \inf\{1/k(x) : \ |x| \geq R_0\}.$$

(B20) Let

$$l_0 > \mu^* := \inf\left\{ \|u\|^2 : \ u \in H^1(\mathbb{R}^3), \ \int_{\mathbb{R}^3} k(x)F(u)dx \geq \frac{b}{2}l^2 \right\}$$

holds, where $l_0 = \min\left\{l, \frac{b}{2}l^2\right\}$, $F(t) = \int_0^t f(s)ds$.

Then problem (3.43) *has at least two positive solutions* u_0, $u \in H^1(\mathbb{R}^3)$ *satisfying* $I(u_0) < 0$ *and* $I(u) > 0$ *if* $\|h\|_2 < m$ *for some small* $m > 0$.

Remark 3.10. It is not difficult to find some functions k, f satisfying conditions of Theorem 1.1. For example, for any $R_0 > 0$, let

$$f(t) = \begin{cases} R_0 t^2/(1+t), & \text{if } t > 0, \\ 0, & \text{if } t < 0. \end{cases}$$

Clearly, f satisfies (B16)–(B18) with $l = R_0$. Moreover, $F(t) = R_0(\frac{1}{2}t^2 - t + \ln(1+t))$ and $\sup\{f(t)/t : \ t > 0\} = R_0$. Taking a positive continuous function $k(x)$

$$k(x) = \begin{cases} C_0/(1+|x|), & \text{if } |x| \leq \dfrac{R_0}{2}, \\ 1/(1+|R_0|), & \text{if } |x| \geq R_0, \end{cases}$$

where $C_0 = \frac{3M^3(1+R_0/2)}{4\pi(\ln 2 - 1/2)}$ for some $M > 0$. Note that

$$\inf\{1/k(x) : \ |x| \geq R_0\} = 1 + R_0 > R_0 = \sup\{f(t)/t : \ t > 0\},$$

then (B19) holds. To verify the condition (B20), we have to choose some special $R_0 > 0$. For any $R > 0$, taking $\psi \in C_0^\infty(\mathbb{R}^3, [0,1])$ such that $\psi(x) = 1$ if $|x| \leq R$, $\psi(x) = 0$ if $|x| \geq 2R$ and $|\nabla\psi(x)| \leq \frac{C}{R}$ for all $x \in \mathbb{R}^3$, where $C > 0$ is an arbitrary constant independent of x. Then we have, for any $R_0 > 2R$, we have

$$\int_{\mathbb{R}^3} k(x)F(\psi)dx \geq \int_{|x|\leq R} k(x)F(\psi)dx$$

$$\geq \frac{R_0 C_0}{1+R}\left(\ln 2 - \frac{1}{2}\right)|B_R(0)|$$

$$\geq \frac{3M^3(1+R_0/2)}{4\pi(\ln 2 - 1/2)}\frac{4\pi R^3 R_0}{3(1+R_0/2)}\left(\ln 2 - \frac{1}{2}\right)$$

$$= M^3 R_0 R^3,$$

$$\int_{\mathbb{R}^3} (|\nabla\psi|^2 + |\psi|^2)dx \leq \int_{|x|\leq 2R} \frac{C^2}{R^2}dx + \int_{|x|\leq 2R} dx$$

$$\leq \left(1 + \frac{C^2}{R^2}\right)\frac{32\pi}{3}R^3$$

$$\leq \frac{32\pi}{3}R\left(C^2 + R^2\right). \tag{3.45}$$

Taking $R_0 = l = 1$, $R = \frac{1}{M}R_0\sqrt[3]{\frac{b}{2}} = \frac{1}{M}\sqrt[3]{\frac{b}{2}}$ where M large enough such that $2R < \frac{R_0}{4}$, $\frac{40\pi}{3M^3} < 1$ and $\frac{40\pi}{3M^3}\frac{b}{2} < 1$. Let $C = \frac{1}{4M}\sqrt[3]{\frac{b}{2}}$. Then, we obtain that $\int_{\mathbb{R}^3} k(x)F(\psi)dx \geq \frac{bR_0}{2} = \frac{bl^2}{2}$. Moreover, in view of definition of μ^* and (3.45), one has

$$\mu^* \leq \int_{\mathbb{R}^3} (|\nabla\psi|^2 + |\psi|^2)dx \leq \frac{32\pi}{3}R\left(C^2 + R^2\right) = \frac{40\pi}{3M^3}\frac{b}{2} < l_0.$$

So, condition (B20) holds. Especially, the condition (B19) and above examples can also be found in [192] in which the asymptotically linear term $k(x)f(u)$ satisfying (B19) appeared firstly.

Remark 3.11. If $h \equiv 0$, we know that problem (3.43) has a positive ground state solution by using the method in [193] and a trivial solution($u(x) \equiv 0$). If $h \not\equiv 0$, a trivial solution($u(x) \equiv 0$) is replaced by the local minimum solution by Theorem 1.1. Note that the local minimal solution exists due to the homogeneous term which is looked as a small perturbation because $\|h\|_2 < m$ for small m.

In order to obtain our results, we have to overcome various difficulties. Since the embedding of $H^1(\mathbb{R}^3)$ into $L^p(\mathbb{R}^3)$, $p \in [2,6]$, is not compact, condition (B19) and (B20) are crucial to obtain the boundedness of Cerami sequence. Furthermore, in order to recover the compactness, we establish a compactness result $\int_{|x|\geq R}(|\nabla u_n|^2 + |u_n|^2)dx \leq \varepsilon$ which is similar to [193] but different from the one in [135, 140, 141, 146, 156, 167, 168]. In fact, this difficulty can be avoided, when problems are considered, restricting I to the subspace of $H^1(\mathbb{R}^3)$ consisting of radially symmetric functions [136,141,167] and constraint potential functions [135, 146], or, when one is looking for semi-classical states [156], by using perturbation methods or a reduction to a finite dimension by the projections method. Third, it is not difficult to find that every (PS) sequence is bounded because a variant of Ambrosetti–Rabinowitz condition is satisfied (see [135, 136, 140]). However, for the asymptotically linear case, we have to find another method to verify the boundedness of (PS) sequence.

In the following discussion, we denote various positive constants as C or C_i $(i = 1, 2, 3, \ldots)$ for convenience.

Firstly, we prove that problem (3.43) has a mountain pass type solution and a local minimum solution with $h \not\equiv 0$. For this purpose, we use a variant version of Mountain Pass Theorem 1.4, which allows us to find a so-called Cerami type (PS) sequence (Cerami sequence, in short). The properties of this kind of Cerami sequence sequences are very helpful in showing its boundedness in the asymptotically cases. The following lemmas will show that I has the so-called mountain pass geometry.

Lemma 3.14. *Suppose that* $h \in L^2(\mathbb{R}^3)$, $h \geq 0$, *(B16)–(B18) and (B19) hold. Then there exist* ρ, α, $m > 0$ *such that* $I(u)|_{\|u\| = \rho} \geq \alpha > 0$ *for* $\|h\|_2 < m$.

Proof. For any $\varepsilon > 0$, it follows from (B16)–(B18) that there exists $C_\varepsilon > 0$ such that

$$|f(t)| \leq \varepsilon|t| + C_\varepsilon|t|^5 \text{ for all } t \in \mathbb{R}. \tag{3.46}$$

Therefore, we have

$$|F(t)| \leq \frac{1}{2}\varepsilon|t|^2 + \frac{C_\varepsilon}{6}|t|^6 \text{ for all } t \in \mathbb{R}. \tag{3.47}$$

Furthermore, by (B16)-(B18) and (B19), there exists $C_1 > 0$ such that

$$k(x) \leq C_1 \text{ for all } x \in \mathbb{R}^3. \tag{3.48}$$

According to (3.47), (3.48) and the Sobolev inequality, we deduce

$$\left| \int_{\mathbb{R}^3} k(x)F(u)dx \right| \leq \frac{\varepsilon C_1}{2} \int_{\mathbb{R}^3} |u|^2 dx + \frac{C_1 C_\varepsilon}{6} \int_{\mathbb{R}^3} |u|^6 dx$$

$$\leq \frac{\varepsilon C_1}{2}\|u\|^2 + C_2\|u\|^6,$$

where $C_2 = \frac{C_1 C_\varepsilon S^6}{6}$. By $b > 0$, $h \in L^2(\mathbb{R}^3)$ and the Hölder inequality, one has

$$I(u) = \frac{a}{2} \int_{\mathbb{R}^3} |\nabla u|^2 dx + \frac{b}{4} \left(\int_{\mathbb{R}^3} |\nabla u|^2 dx \right)^2 + \frac{1}{2} \int_{\mathbb{R}^3} |u|^2 dx - \int_{\mathbb{R}^3} k(x)F(u)dx$$

$$- \int_{\mathbb{R}^3} h(x)u dx$$

$$\geq \frac{a}{2} \int_{\mathbb{R}^3} |\nabla u|^2 dx + \frac{1}{2} \int_{\mathbb{R}^3} |u|^2 dx - \frac{\varepsilon C_1}{2}\|u\|^2 - C_2\|u\|^6 - \|h\|_2\|u\|_2$$

$$\geq \frac{1}{2}\|u\|^2 - \frac{\varepsilon C_1}{2}\|u\|^2 - C_2\|u\|^6 - \|h\|_2\|u\|$$

$$\geq \|u\| \left(\frac{1 - \varepsilon C_1}{2}\|u\| - C_2\|u\|^5 - \|h\|_2 \right). \tag{3.49}$$

Taking $\varepsilon = \frac{1}{2C_1}$ and setting $g(t) = \frac{1}{4}t - C_2 t^5$ for $t \geq 0$, we see there exists $\rho = \left(\frac{1}{20C_2}\right)^{\frac{1}{4}}$ such that $\max_{t \geq 0} g(t) = g(\rho) := m > 0$. Then it follows from (3.49) that there exists $\alpha > 0$ such that $I(u)|_{\|u\|=\rho} \geq \alpha > 0$ for $\|h\|_2 < m$. Of course, ρ, m can be chosen small enough, we can obtain the same result: there exists $\alpha > 0$ such that $I(u)|_{\|u\|=\rho} \geq \alpha > 0$ for $\|h\|_2 < m$. $\qquad \square$

Lemma 3.15. *Suppose that $h \in L^2(\mathbb{R}^3)$, $h \geq 0$, (B16)–(B18) and (B19)–(B20) hold. Then there exist $v \in H^1(\mathbb{R}^3)$ with $\|v\| > \rho$, ρ is given by Lemma 3.14, such that $I(v) < 0$.*

Proof. By (B20) and $h \geq 0$, in view of the definition of μ^* and $l_0 > \mu^*$ with $l_0 = \min\left\{l, \frac{b}{2}l^2\right\}$, there is a nonnegative function $v \in H^1(\mathbb{R}^3)$ such that

$$\int_{\mathbb{R}^3} k(x)F(v)dx \geq \frac{b}{2}l^2, \quad \int_{\mathbb{R}^3} h(x)vdx \geq 0,$$

and $\mu^* \leq \|v\|^2 < l_0$. Then, we have

$$I(v) = \frac{a}{2}\int_{\mathbb{R}^3}|\nabla v|^2 dx + \frac{b}{4}\left(\int_{\mathbb{R}^3}|\nabla v|^2 dx\right)^2 + \frac{1}{2}\int_{\mathbb{R}^3}|v|^2 dx - \int_{\mathbb{R}^3}k(x)F(v)dx$$
$$- \int_{\mathbb{R}^3}h(x)vdx$$
$$\leq \frac{1}{2}\|v\|^2 + \frac{b}{4}\|v\|^4 - \frac{b}{2}l^2$$
$$\leq \frac{1}{2}\|v\|^2 - \frac{b}{4}l^2$$
$$< 0.$$

Choosing $\rho > 0$ small enough in Lemma 3.14 such that $\|v\| > \rho$, then this lemma is proved. $\qquad \square$

From Lemmas 3.14, 3.15 and Mountain Pass Theorem 1.4, there is a Cerami sequence $\{u_n\} \subset H^1(\mathbb{R}^3)$ such that

$$\|I'(u_n)\|_{H^*}(1 + \|u_n\|) \to 0 \quad \text{and} \quad I(u_n) \to c \text{ as } n \to \infty, \quad (3.50)$$

where H^* denotes the dual space of $H^1(\mathbb{R}^3)$. In the following, we shall prove that I satisfies the Cerami condition, that is, the Cerami sequence $\{u_n\}$ has a convergence subsequence.

Lemma 3.16. *Suppose that $h \in L^2(\mathbb{R}^3)$, $h \geq 0$, (B16)–(B18) and (B19) hold. Then $\{u_n\}$ defined in (3.50) is bounded in $H^1(\mathbb{R}^3)$.*

Proof. By contradiction, let $\|u_n\| \to \infty$. Define $w_n = u_n \|u_n\|^{-1}$. Clearly, $\{w_n\}$ is bounded in $H^1(\mathbb{R}^3)$ and there is a $w \in H^1(\mathbb{R}^3)$ such that, up to a sequence,

$$w_n \rightharpoonup w \quad \text{weakly in } H^1(\mathbb{R}^3),$$
$$w_n \to w \quad \text{a.e. in } \mathbb{R}^3,$$
$$w_n \to w \quad \text{strongly in } L^2_{loc}(\mathbb{R}^3)$$

as $n \to \infty$.

Firstly, we claim that w is nontrivial, that is $w \not\equiv 0$. Otherwise, if $w \equiv 0$, the Sobolev embedding implies that $w_n \to 0$ strongly in $L^2(B_{R_0})$, R_0 is given by (B19). By (B16)–(B18), there exists $C_3 > 0$ such that

$$\frac{f(t)}{t} \leq C_3 \text{ for all } t \in \mathbb{R}. \tag{3.51}$$

Then, for all $n \in N$, we have

$$0 \leq \int_{|x| < R_0} k(x) \frac{f(u_n)}{u_n} w_n^2 dx \leq C_3 \|k\|_\infty \int_{|x| < R_0} w_n^2 dx \to 0.$$

This yields

$$\lim_{n \to \infty} \int_{|x| < R_0} k(x) \frac{f(u_n)}{u_n} w_n^2 dx = 0. \tag{3.52}$$

Furthermore, by (B19), there exists a constant $\theta \in (0, 1)$ such that

$$\sup\{f(t)/t : t > 0\} \leq \theta \inf\{1/k(x) : |x| \geq R_0\}. \tag{3.53}$$

Then, for all $n \in N$, we have

$$\int_{|x| \geq R_0} k(x) \frac{f(u_n)}{u_n} w_n^2 dx \leq \theta \int_{|x| \geq R_0} w_n^2 dx \leq \theta \|w\|^2 = \theta < 1. \tag{3.54}$$

Combining (3.52) and (3.54), we obtain

$$\limsup_{n \to \infty} \int_{\mathbb{R}^3} k(x) \frac{f(u_n)}{u_n} w_n^2 dx < 1. \tag{3.55}$$

By (3.50), we get

$$0 \leq |\langle I'(u_n), u_n \rangle| \leq \|I'(u_n)\|_{H^*} \|u_n\| \leq \|I'(u_n)\|_{H^*} (1 + \|u_n\|) \tag{3.56}$$
$$\to 0$$

as $n \to \infty$. Together with $\|u_n\| \to \infty$ as $n \to \infty$, it follows that

$$\frac{\langle I'(u_n), u_n \rangle}{\|u_n\|^2} = o(1).$$

Together with $b > 0$, we have

$$o(1) = \frac{1}{\|u_n\|^2} \left(a \int_{\mathbb{R}^3} |\nabla u_n|^2 dx + \int_{\mathbb{R}^3} |u_n|^2 dx + b \left(\int_{\mathbb{R}^3} |\nabla u_n|^2 dx \right)^2 \right.$$

$$\left. - \int_{\mathbb{R}^3} k(x) f(u_n) u_n dx - \int_{\mathbb{R}^3} h(x) u_n dx \right)$$

$$\geq \|w_n\|^2 - \int_{\mathbb{R}^3} k(x) \frac{f(u_n)}{u_n} w_n^2 dx$$

$$\geq 1 - \int_{\mathbb{R}^3} k(x) \frac{f(u_n)}{u_n} w_n^2 dx,$$

where, and in what follows, $o(1)$ denotes a quantity which goes to zero as $n \to \infty$. Therefore, we deduce that

$$\int_{\mathbb{R}^3} k(x) \frac{f(u_n)}{u_n} w_n^2 dx + o(1) \geq 1,$$

which contradicts (3.55). So, $w \not\equiv 0$.

Furthermore, because $\|u_n\| \to \infty$ as $n \to \infty$, it follows from (3.56) that

$$\frac{\langle I'(u_n), u_n \rangle}{\|u_n\|^4} = o(1),$$

that is,

$$o(1) = \frac{b \left(\int_{\mathbb{R}^3} |\nabla u_n|^2 dx \right)^2}{\|u_n\|^4} - \frac{1}{\|u_n\|^2} \int_{\mathbb{R}^3} k(x) \frac{f(u_n)}{u_n} w_n^2 dx.$$

Together with (3.48), (3.51) and $b > 0$, one has

$$\frac{\left(\int_{\mathbb{R}^3} |\nabla u_n|^2 dx \right)^2}{\|u_n\|^4} = \frac{\left(\int_{\mathbb{R}^3} |(\nabla u_n|^2 + |u_n|^2) dx - \int_{\mathbb{R}^3} |u_n|^2 dx \right)^2}{\|u_n\|^4} = o(1).$$

This yields

$$\frac{\left(\int_{\mathbb{R}^3} |(\nabla u_n|^2 + |u_n|^2) dx \right)^2 - 2 \int_{\mathbb{R}^3} |(\nabla u_n|^2 + |u_n|^2) dx \int_{\mathbb{R}^3} |u_n|^2 dx + \left(\int_{\mathbb{R}^3} |u_n|^2 dx \right)^2}{\|u_n\|^4}$$

$$= o(1).$$

This means

$$1 - 2 \int_{\mathbb{R}^3} |w_n|^2 dx + \left(\int_{\mathbb{R}^3} |w_n|^2 dx \right)^2 = \left(1 - \int_{\mathbb{R}^3} |w_n|^2 dx \right)^2 = o(1).$$

Therefore, we have

$$\int_{\mathbb{R}^3} |w_n|^2 dx \to 1 \quad \text{as } n \to \infty.$$

By $\|w_n\| = 1$, we get $\int_{\mathbb{R}^3} |\nabla w_n|^2 dx \to 0$ as $n \to \infty$, thus $w_n \to 0$ strongly in $D^{1,2}(\mathbb{R}^3)$, therefore, $w_n \rightharpoonup 0$ weakly in $D^{1,2}(\mathbb{R}^3)$. Since $w_n \rightharpoonup w$ weakly in $H^1(\mathbb{R}^3)$, we have $w_n \rightharpoonup w$ weakly in $D^{1,2}(\mathbb{R}^3)$. By the uniqueness of the weak limitation, we have $w = 0$ which contradicts $w \neq 0$. Therefore, $\{u_n\}$ is a bounded in $H^1(\mathbb{R}^3)$. $\square$

Lemma 3.17. *Suppose that* $h \in L^2(\mathbb{R}^3)$, $h \geq 0$, *(B16)-(B18), and (B19) hold. Then for any* $\varepsilon > 0$, *there exist* $R(\varepsilon) > R_0$ *and* $n(\varepsilon) > 0$ *such that* $\{u_n\}$ *defined in (3.50) satisfies* $\int_{|x|\geq R}(|\nabla u_n|^2 + |u_n|^2)dx \leq \varepsilon$ *for* $n > n(\varepsilon)$ *and* $R \geq R(\varepsilon)$.

Proof. Let $\xi_R : R^3 \to [0,1]$ be a smooth function such that

$$\xi_R(x) = \begin{cases} 0, & 0 \leq |x| \leq R/2, \\ 1, & |x| \geq R. \end{cases} \tag{3.57}$$

Moreover, there exists a constant C_4 independent of R such that

$$|\nabla \xi_R(x)| \leq \frac{C_4}{R} \text{ for all } x \in R^3. \tag{3.58}$$

Then, for all $n \in N$ and $R \geq R_0$, by (3.57), (3.58) and the Hölder inequality, we have

$$\int_{\mathbb{R}^3} |\nabla(u_n\xi_R)|^2 dx$$

$$\leq \int_{\mathbb{R}^3} |\nabla u_n|^2 |\xi_R|^2 dx + \int_{\mathbb{R}^3} |u_n|^2 |\nabla \xi_R|^2 dx + 2\int_{\mathbb{R}^3} |u_n||\xi_R||\nabla u_n||\nabla \xi_R|dx$$

$$\leq \int_{R/2<|x|<R} |\nabla u_n|^2 dx + \int_{|x|>R} |\nabla u_n|^2 dx + \frac{C_4^2}{R^2}\int_{\mathbb{R}^3} |u_n|^2 dx$$

$$+2\left(\int_{\mathbb{R}^3} |\nabla u_n|^2 |\xi_R^2| dx\right)^{\frac{1}{2}} \left(\int_{\mathbb{R}^3} |u_n|^2 |\nabla \xi_R|^2 dx\right)^{\frac{1}{2}}$$

$$\leq \int_{R/2<|x|<R} |\nabla u_n|^2 dx + \int_{|x|>R} |\nabla u_n|^2 dx + \frac{C_4^2}{R^2}\int_{\mathbb{R}^3} |u_n|^2 dx$$

$$+2\left(\int_{R/2<|x|<R} |\nabla u_n|^2 dx + \int_{|x|>R} |\nabla u_n|^2 dx\right)^{\frac{1}{2}} \left(\frac{C_4^2}{R^2}\int_{\mathbb{R}^3} |u_n|^2 dx\right)^{\frac{1}{2}}$$

$$\leq \left(2 + \frac{C_4^2}{R^2} + \frac{2\sqrt{2}C_4}{R}\right)\|u_n\|^2$$

$$\leq \left(2 + \frac{C_4^2}{R_0^2} + \frac{2\sqrt{2}C_4}{R_0}\right)\|u_n\|^2.$$

This implies that

$$\|u_n\xi_R\| \leq C_5\|u_n\| \tag{3.59}$$

for all $n \in N$ and $R \geq R_0$, where $C_5 = \left(3 + \frac{C_4^2}{R_0^2} + \frac{2\sqrt{2}C_4}{R_0}\right)^{\frac{1}{2}}$. From Lemma 2.3, we know that $\{u_n\}$ is bounded in $H^1(R^3)$. Together with (3.50), we obtain that $I'(u_n) \to 0$ in $H^*(R^3)$. Moreover, by (3.59), for $\varepsilon > 0$, there exists $n(\varepsilon) > 0$ such that

$$\langle I'(u_n), \xi_R u_n \rangle \leq C_5 \|I'(u_n)\|_{H^*(R^3)} \|u_n\| \leq \frac{\varepsilon}{4}$$

for $n > n(\varepsilon)$ and $R > R_0$. Note that

$$\langle I'(u_n), \xi_R u_n \rangle$$

$$= \left(a + b \int_{R^3} |\nabla u_n|^2 dx\right) \int_{R^3} |\nabla u_n|^2 \xi_R dx + \int_{R^3} |u_n|^2 \xi_R dx$$

$$+ \left(a + b \int_{R^3} |\nabla u_n|^2 dx\right) \int_{R^3} u_n \nabla u_n \cdot \nabla \xi_R dx - \int_{R^3} k(x) f(u_n) u_n \xi_R dx$$

$$- \int_{R^3} h(x) u_n \xi_R dx$$

$$\leq \frac{\varepsilon}{4}.$$

This yields

$$\left(a + b \int_{R^3} |\nabla u_n|^2 dx\right) \int_{R^3} |\nabla u_n|^2 \xi_R dx + \int_{R^3} |u_n|^2 \xi_R dx$$

$$+ \left(a + b \int_{R^3} |\nabla u_n|^2 dx\right) \int_{R^3} u_n \nabla u_n \cdot \nabla \xi_R dx$$

$$\leq \int_{R^3} k(x) f(u_n) u_n \xi_R dx + \int_{R^3} h(x) u_n \xi_R dx + \frac{\varepsilon}{4}. \tag{3.60}$$

By (3.53), we have

$$k(x) f(u_n) u_n \leq \theta u_n^2 \quad \text{for } \theta \in (0, \min\{1, a\}) \text{ and } |x| \geq R_0.$$

This yields

$$\int_{R^3} k(x) f(u_n) u_n \xi_R dx \leq \theta \int_{R^3} u_n^2 \xi_R dx \tag{3.61}$$

for all $n \in N$ and $|x| \geq R_0$. For any $\varepsilon > 0$, there exists $R(\varepsilon) \geq R_0$ such that

$$\frac{1}{R^2} \leq \frac{4\varepsilon^2}{C_4^2} \quad \text{for all } R > R(\varepsilon). \tag{3.62}$$

Because $h \in L^2(R^3)$, $h \geq (\not\equiv)0$, there exists $\overline{\rho} = \overline{\rho}(\varepsilon)$ such that

$$\|h\|_{2, R^3 \setminus B_\rho(0)} < \varepsilon, \quad \forall \rho \geq \overline{\rho}. \tag{3.63}$$

By the Hölder inequality, (3.57), (3.63) and $\{u_n\}$ is bounded in $H^1(R^3)$, we have

$$\int_{R^3} h(x) u_n \xi_R dx \leq \|h(x) \xi_R\|_2 \|u_n\|_2 \leq \|h(x)\|_{2, |x| > R/2} \|u_n\|_2 \leq \frac{\varepsilon}{4} \tag{3.64}$$

for all $R > R(\varepsilon)$. By the Young inequality, (3.58) and (3.62), for all $n \in N$ and $R > R(\varepsilon)$, we obtain

$$\int_{\mathbb{R}^3} |u_n \nabla u_n \cdot \nabla \xi_R| dx$$

$$\leq \int_{\mathbb{R}^3} \sqrt{2\varepsilon} |\nabla u_n| \frac{1}{\sqrt{2\varepsilon}} |u_n| |\nabla \xi_R| dx$$

$$\leq \varepsilon \int_{\mathbb{R}^3} |\nabla u_n|^2 dx + \frac{1}{4\varepsilon} \int_{|x| \leq R} |u_n|^2 \frac{C_4^2}{R^2} dx$$

$$\leq \varepsilon \int_{\mathbb{R}^3} |\nabla u_n|^2 dx + \varepsilon \int_{|x| \leq R} |u_n|^2 dx$$

$$\leq \varepsilon \|u_n\|^2. \tag{3.65}$$

Combining $b > 0$, (3.60), (3.61), (3.64), (3.65) and $\{u_n\}$ is bounded in $H^1(R^3)$, there exists $C_6 > 0$ such that

$$\min\{1 - \theta, a\} \int_{\mathbb{R}^3} (|\nabla u_n|^2 + |u_n|^2) \xi_R dx$$

$$\leq \frac{\varepsilon}{2} + \varepsilon \|u_n\|^2 \left(a + b \int_{\mathbb{R}^3} |\nabla u_n|^2 dx\right)$$

$$\leq C_6 \varepsilon \text{ for all } R > R(\varepsilon).$$

Noting that C_6 is independent of ε. So, for any $\varepsilon > 0$, we can choose $R(\varepsilon) > R_0$ and $n(\varepsilon) > 0$ such that $\int_{|x| \geq R}(|\nabla u_n|^2 + |u_n|^2) dx \leq \varepsilon$ holds. $\square$

Lemma 3.18. *Suppose that* $h \in L^2(\mathbb{R}^3)$, $h \geq 0$, *(B16)–(B18), and (B19)–(B20) hold. Then the sequence* $\{u_n\}$ *in (3.50) has a convergent subsequence. Moreover, I possesses a nonzero critical point u in* $H^1(\mathbb{R}^3)$ *and* $I(u) > 0$.

Proof. By Lemma 3.16, the sequence $\{u_n\}$ in (3.50) is bounded in $H^1(\mathbb{R}^3)$. We may assume that, up to a subsequence $u_n \rightharpoonup u$ weakly in $H^1(\mathbb{R}^3)$ for some $u \in H^1(\mathbb{R}^3)$. Now, we shall show that $\|u_n\| \to \|u\|$ as $n \to \infty$.

By (3.44), we have

$$\langle I'(u_n), u_n \rangle = \left(a + b \int_{\mathbb{R}^3} |\nabla u_n|^2 dx\right) \int_{\mathbb{R}^3} |\nabla u_n|^2 dx + \int_{\mathbb{R}^3} u_n^2 dx$$

$$- \int_{\mathbb{R}^3} k(x) f(u_n) u_n - \int_{\mathbb{R}^3} h(x) u_n dx, \tag{3.66}$$

and

$$\langle I'(u_n), u \rangle = \left(a + b \int_{\mathbb{R}^3} |\nabla u_n|^2 dx\right) \int_{\mathbb{R}^3} \nabla u_n \cdot \nabla u dx + \int_{\mathbb{R}^3} u_n u dx$$

$$- \int_{\mathbb{R}^3} k(x) f(u_n) u - \int_{\mathbb{R}^3} h(x) u dx. \tag{3.67}$$

By (3.66), (3.67), $b > 0$ and $\{u_n\}$ is bounded in $H^1(\mathbb{R}^3)$, we easily get

$$\langle I'(u_n), u_n - u \rangle$$

$$= \left(a + b \int_{\mathbb{R}^3} |\nabla u_n|^2 dx \right) \int_{\mathbb{R}^3} |\nabla u_n|^2 dx + \int_{\mathbb{R}^3} u_n^2 dx$$

$$- \int_{\mathbb{R}^3} k(x) f(u_n) u_n dx - \int_{\mathbb{R}^3} h(x) u_n dx$$

$$- \left(a + b \int_{\mathbb{R}^3} |\nabla u_n|^2 dx \right) \int_{\mathbb{R}^3} \nabla u \cdot \nabla u_n dx - \int_{\mathbb{R}^3} u u_n dx + \int_{\mathbb{R}^3} k(x) f(u_n) u dx$$

$$+ \int_{\mathbb{R}^3} h(x) u dx$$

$$= \left(a + b \int_{\mathbb{R}^3} |\nabla u_n|^2 dx \right) \int_{\mathbb{R}^3} |\nabla(u_n - u)|^2 dx + \int_{\mathbb{R}^3} |u_n - u|^2 dx$$

$$+ \left(a + b \int_{\mathbb{R}^3} |\nabla u_n|^2 dx \right) \int_{\mathbb{R}^3} \nabla u \nabla(u_n - u) dx + \int_{\mathbb{R}^3} u(u_n - u) dx$$

$$- \int_{\mathbb{R}^3} k(x) f(u_n)(u_n - u) dx - \int_{\mathbb{R}^3} h(x)(u_n - u) dx$$

$$\geq a \int_{\mathbb{R}^3} |\nabla(u_n - u)|^2 dx + \int_{\mathbb{R}^3} |u_n - u|^2 dx$$

$$+ \left(a + b \int_{\mathbb{R}^3} |\nabla u_n|^2 dx \right) \int_{\mathbb{R}^3} \nabla u \nabla(u_n - u) dx + \int_{\mathbb{R}^3} u(u_n - u) dx$$

$$- \int_{\mathbb{R}^3} k(x) f(u_n)(u_n - u) dx - \int_{\mathbb{R}^3} h(x)(u_n - u) dx$$

$$\geq \|u_n - u\|^2 + \left(a + b \int_{\mathbb{R}^3} |\nabla u_n|^2 dx \right) \int_{\mathbb{R}^3} \nabla u \nabla(u_n - u) dx$$

$$+ \int_{\mathbb{R}^3} u(u_n - u) dx$$

$$- \int_{\mathbb{R}^3} k(x) f(u_n)(u_n - u) dx - \int_{\mathbb{R}^3} h(x)(u_n - u) dx.$$

One has

$$\|u_n - u\|^2$$

$$\leq \langle I'(u_n), u_n - u \rangle - \left(a + b \int_{\mathbb{R}^3} |\nabla u_n|^2 dx \right) \int_{\mathbb{R}^3} \nabla u \nabla(u_n - u) dx$$

$$- \int_{\mathbb{R}^3} u(u_n - u) dx$$

$$+ \int_{\mathbb{R}^3} k(x) f(u_n)(u_n - u) dx + \int_{\mathbb{R}^3} h(x)(u_n - u) dx. \tag{3.68}$$

It is clear that

$$\langle I'(u_n), u_n - u \rangle \to 0, \quad \text{as} \ n \to \infty. \tag{3.69}$$

Since $u_n \rightharpoonup u$ weakly in $H^1(\mathbb{R}^3)$, we obtain

$$\int_{\mathbb{R}^3} (\nabla u_n \cdot \nabla u + u_n u) dx$$

$$= \int_{\mathbb{R}^3} (|\nabla u|^2 + |u|^2) dx + o(1), \quad \text{as} \ n \to \infty. \tag{3.70}$$

By the continuity of embedding $H^1(\mathbb{R}^3) \hookrightarrow L^2(\mathbb{R}^3)$, we have that $u_n \rightharpoonup u$ weakly in $L^2(\mathbb{R}^3)$, that is,

$$\int_{\mathbb{R}^3} u_n u \, dx = \int_{\mathbb{R}^3} u^2 dx + o(1), \quad \text{as} \ n \to \infty. \tag{3.71}$$

By (3.70) and (3.71), we deduce

$$\int_{\mathbb{R}^3} \nabla u_n \cdot \nabla u \, dx = \int_{\mathbb{R}^3} |\nabla u|^2 dx + o(1), \quad \text{as} \ n \to \infty. \tag{3.72}$$

Combining the boundedness of $\{u_n\}$ in $H^1(\mathbb{R}^3)$, (3.71) and (3.72), we obtain

$$\left(a + b \int_{\mathbb{R}^3} |\nabla u_n|^2 dx \right) \int_{\mathbb{R}^3} \nabla u \nabla (u_n - u) dx + \int_{\mathbb{R}^3} u(u_n - u) dx$$

$$= o(1), \tag{3.73}$$

as $n \to \infty$.

Moreover, by (3.53), Lemma 3.17 and $u_n \rightharpoonup u$ in $H^1(\mathbb{R}^3)$, for any $\varepsilon > 0$ and n large enough, one has

$$\int_{|x| \geq R(\varepsilon)} k(x) f(u_n) u_n dx - \int_{|x| \geq R(\varepsilon)} k(x) f(u_n) u \, dx$$

$$= \int_{|x| \geq R(\varepsilon)} k(x) f(u_n)(u_n - u) dx$$

$$\leq \int_{|x| \geq R(\varepsilon)} |k(x) f(u_n)| |u_n - u| dx$$

$$\leq \left(\int_{|x| \geq R(\varepsilon)} |k^2(x) f^2(u_n)| dx \right)^{\frac{1}{2}} \left(\int_{|x| \geq R(\varepsilon)} |u_n - u|^2 dx \right)^{\frac{1}{2}}$$

$$\leq \theta \left(\int_{|x| \geq R(\varepsilon)} |u_n^2| dx \right)^{\frac{1}{2}} \left(\int_{|x| \geq R(\varepsilon)} |u_n - u|^2 dx \right)^{\frac{1}{2}}$$

$$\leq \theta \left(\int_{|x| \geq R(\varepsilon)} (|\nabla u_n|^2 + |u_n^2|) dx \right)^{\frac{1}{2}} \left(\int_{|x| \geq R(\varepsilon)} |u_n - u|^2 dx \right)^{\frac{1}{2}}$$

$$\leq \theta \varepsilon.$$

This and the compactness of embedding $H^1(\mathbb{R}^3) \hookrightarrow L^2_{loc}(\mathbb{R}^3)$ imply

$$\int_{\mathbb{R}^3} k(x)f(u_n)u_n dx = \int_{\mathbb{R}^3} k(x)f(u_n)u\,dx + o(1). \tag{3.74}$$

Since u_n is bounded in $H^1(\mathbb{R}^3)$ and the continuity of the Sobolev embedding of $H^1(\mathbb{R}^3)$ embedding in $L^2(\mathbb{R}^3)$, for any choice of $\varepsilon > 0$ and $\rho > 0$, the relation

$$\|u_n - u\|_{2,B_\rho(0)} < \varepsilon \tag{3.75}$$

holds for large n. By $h \in L^2(\mathbb{R}^3)$, for any $\varepsilon > 0$ there exists $\overline{\rho} = \overline{\rho}(\varepsilon)$ such that

$$\|h\|_{2,\mathbb{R}^3 \setminus B_\rho(0)} < \varepsilon, \quad \forall \rho \geq \overline{\rho}. \tag{3.76}$$

By (3.76) and (3.75), we have

$$\int_{\mathbb{R}^3} h(x)u_n dx - \int_{\mathbb{R}^3} h(x)u\,dx$$

$$\leq \int_{\mathbb{R}^3 \setminus B_\rho(0)} |h(x)(u_n - u)|dx + \int_{B_\rho(0)} |h(x)(u_n - u)|dx$$

$$\leq \|h(x)\|_{2,\mathbb{R}^3 \setminus B_\rho(0)} \|u_n - u\|_{2,\mathbb{R}^3 \setminus B_\rho(0)} + \|h(x)\|_{2,B_\rho(0)} \|u_n - u\|_{2,B_\rho(0)}$$

$$\leq \varepsilon \|u_n - u\|_{2,\mathbb{R}^3 \setminus B_\rho(0)} + \varepsilon \|h(x)\|_{2,B_\rho(0)}$$

$$\leq C_7 \varepsilon.$$

This yields

$$\int_{\mathbb{R}^3} h(x)u_n dx = \int_{\mathbb{R}^3} h(x)u\,dx + o(1). \tag{3.77}$$

By (3.68), (3.69), (3.73), (3.74) and (3.77), we have

$$\|u_n - u\|^2 = o(1), \quad \text{as } n \to \infty,$$

by $a > 0$. This yields that $\|u_n\| \to \|u\|$ as $n \to \infty$ and u is a nonzero critical point of I in $H^1(\mathbb{R}^3)$ and $I(u) > 0$ by Mountain Pass Theorem 1.4. $\quad\square$

Now, we give local properties of the variational functional I, which is required by using Ekeland's variational principle.

Lemma 3.19. *Suppose that $h \in L^2(\mathbb{R}^3)$, $h \geq (\neq)0$, (B16)–(B18) and (B19) hold. If $\|h\|_2 < m$, then there exists $u_0 \in H^1(\mathbb{R}^3)$ such that*

$$I(u_0) = \inf\{I(u) : u \in \overline{B}_\rho\} < 0, \quad where \ B_\rho = \{u \in H^1(\mathbb{R}^3) : \|u\| < \rho\},$$

m, ρ are given by Lemma 3.14 and u_0 is a positive solution of equation (3.43).

Proof. Because $h \in L^2(\mathbb{R}^3)$, $h \geq (\neq)0$, we can choose a function $\varphi \in H^1(\mathbb{R}^3)$ such that

$$\int_{\mathbb{R}^3} h(x)\varphi dx > 0. \tag{3.78}$$

Together with (B16), (B19) and (3.78), for $t > 0$, we have

$$
\begin{aligned}
I(t\varphi) &= \frac{at^2}{2} \int_{\mathbb{R}^3} |\nabla \varphi|^2 dx + \frac{bt^4}{4} \left(\int_{\mathbb{R}^3} |\nabla \varphi|^2 dx \right)^2 + \frac{t^2}{2} \int_{\mathbb{R}^3} |\varphi|^2 dx \\
&\quad - \int_{\mathbb{R}^3} k(x) F(t\varphi) dx - \int_{\mathbb{R}^3} h(x) t\varphi dx \\
&\leq \frac{t^2}{2} \|\varphi\|^2 + \frac{bt^4}{4} \|\varphi\|^4 - t \int_{\mathbb{R}^3} h(x)\varphi dx \\
&\leq 0
\end{aligned}
$$

for $t > 0$ small enough. Thus there exists u small enough such that $I(u) < 0$. By Lemma 3.14, we deduce that

$$c_0 := \inf_{u \in \overline{B}_\rho} I(u) < 0 < \inf_{u \in \partial \overline{B}_\rho} I(u).$$

By applying Ekeland's variational principle [18] in $\overline{B}_\rho$, there is a minimizing sequence $\{u_n\} \subset \overline{B}_\rho$ such that

(i) $c_0 \leq I(u_n) < c_0 + \dfrac{1}{n}$, (ii) $I(w) \geq I(u_n) - \dfrac{1}{n}\|w - u_n\|$ for all $w \in \overline{B}_\rho$.

Then, by a standard procedure, we can show that $\{u_n\}$ is a bounded (PS) sequence of I. Lemmas 3.17 and 3.18 imply that there exists $u_0 \in H^1(\mathbb{R}^3)$ such that $I'(u_0) = 0$ and $I(u_0) = c_0 < 0$. So this lemma is proved. □

Proof of Theorem 3.6. By Lemmas 3.14–3.18, we obtain the existence of a mountain pass solution u for problem (3.43) and $I(u) > 0$. By Lemma 3.19, we know that problem (3.43) has a local minimum solution u_0 and $I(u_0) < 0$. Thus, $u \neq u_0$ and u, u_0 are positive. Thus this theorem is proved. □

Notes and Comments

Some interesting studies for Kirchhoff-type problem (3.1) in a bounded domain Ω of $\mathbb{R}^N$ by variational methods can be found in [126] and [173–177, 194–201]. In the same spirit of [135, 167, 168] and [140, 141, 146, 156], we study a nonhomogeneous Kirchhoff equation (3.43) on the whole space $\mathbb{R}^3$. Especially, inspired by the paper [193, 202], we consider the asymptotically linear nonlinearity at infinity of problem (3.43). For the nonhomogeneous Kirchhoff problem, Chen and Li in [136] study it under the condition of superlinear nonlinearity at infinity. In [202], Wang and Zhou study the existence of two positive solutions for a nonhomogeneous elliptic equation (equation (3.43) with $a = 1$ and $b = 0$). In [193], Sun, Chen and Nieto study the existence of a ground state solution for some nonautonomous Schrödinger–Poisson systems involving the asymptotically linear nonlinearity at infinity without the nonhomogeneous term. But we shall the existence of two positive solutions for Kirchhoff-type problem (3.43) with a, $b > 0$, the asymptotically linear nonlinearity at infinity and the nonhomogeneous term. So, we cannot obtain the existence of a ground state solution for Kirchhoff-type problem (3.43) and the compactness result as like in [193] because of the nonhomogeneous term, and we cannot easily obtain the compactness result as like in [202] due to the nonlocal term (or $b \neq 0$). To our best knowledge, little has been done for nonhomogeneous Kirchhoff problems with respect to the asymptotically linear nonlinearity at infinity. This section was obtained in [203].

3.6 Infinitely Many Solutions with Sublinear Nonlinearities

Consider the following nonlinear Kirchhoff equation

$$\left(1 + b \int_{\mathbb{R}^3} \left(|\nabla u|^2 + V(x)u^2\right) dx\right) [-\Delta u + V(x)u] = f(x, u) \text{ in } \mathbb{R}^3, \quad (3.79)$$

where $b > 0$ is a constant, $V(x)$ is a nonnegative potential function and $f : \mathbb{R}^3 \times \mathbb{R} \mapsto \mathbb{R}$.

In the present section, we are concerned with the existence of infinitely many solutions for (3.79) without any growth conditions imposed on $f(x, u)$ at infinity with respect to u. Now, we make some assumptions on the functions V and f. Specifically, we make the following assumptions:

(B21) $V \in C(\mathbb{R}^3)$ and $\inf_{x \in \mathbb{R}^3} V(x) > 0$;

(B22) $f \in C(\mathbb{R}^3 \times [-\delta, \delta], \mathbb{R})$ with $\delta > 0$, and there exist constants $\nu \in (1, 2)$, $\mu \in (3/2, 2/(2 - \nu)]$ and a nonnegative function $\xi \in L^\mu(\mathbb{R}^3)$ such that

$$|f(x, u)| \leq \nu \xi(x)|u|^{\nu - 1}, \quad \forall (x, u) \in \mathbb{R}^3 \times [-\delta, \delta];$$

(B23) There exist an $x_0 \in \mathbb{R}^3$ and a constant $r_0 > 0$ such that

$$\liminf_{u \to 0} \left(\inf_{x \in B_{r_0}(x_0)} u^{-2} F(x, u)\right) > -\infty,$$

and

$$\limsup_{u \to 0} \left(\inf_{x \in B_{r_0}(x_0)} u^{-2} F(x, u) \right) = +\infty,$$

where $F(x, u) = \int_0^u f(x, t)dt$;

(B24) f is odd with respect to u.

Now we define precisely what we mean by weak solutions.

Definition 3.1. We say that $u \in H^1(\mathbb{R}^3)$ is a weak solution for (3.79) if for all $\varphi \in C_0^\infty(\mathbb{R}^3)$, there holds

$$\int_{\mathbb{R}^3} (\nabla u \cdot \nabla \varphi + V(x)u\varphi)dx + b\|u\|^2 \int_{\mathbb{R}^3} (\nabla u \cdot \nabla \varphi + V(x)u\varphi)dx$$

$$= \int_{\mathbb{R}^3} f(x, u)\varphi dx.$$

Our main result can be stated as follows:

Theorem 3.7. *Assume that conditions (B21) and (B22)–(B24) hold. Then the problem (3.79) possesses a sequence of weak solutions $\{u_k\}$ in $H^1(\mathbb{R}^3)$ with $u_k \to 0$ in $L^\infty(\mathbb{R}^3)$ as $k \to \infty$.*

A few remarks are in order:

Remark 3.12. Our Theorem 3.7 improves some related results in the existing literature. Compared with [144], there is no need to impose an additional condition on $V(x)$ except (B21). Moreover, there are many functions f satisfying (B22)–(B24) while they do not satisfy the conditions in [144, 152]. For instance, let

$$F(x, u) = \begin{cases} x_1 e^{-|x|^2} |u|^\alpha \sin^2(|u|^{-\varepsilon}), & \forall x = (x_1, x_2, x_3) \in \mathbb{R}^3 \text{ and } 0 < |u| \le 1, \\ 0, & \forall x = (x_1, x_2, x_3) \in \mathbb{R}^3 \text{ and } u = 0 \end{cases}$$

be the primitive function of f, where $\varepsilon > 0$ small enough and $\alpha \in (1 + \varepsilon, 2)$. Then it is easy to check that f satisfies conditions (B22)–(B24) with $\delta = 1$, $\nu = \alpha - \varepsilon$ and $\xi(x) = (\alpha + \varepsilon)(\alpha - \varepsilon)^{-1}|x_1|e^{-|x|^2}$.

Remark 3.13. It should be noted here that the nonlinear term $f(x, u)$ is only locally defined for $|u|$ small. This is in sharp contrast to the aforementioned references. To the best of our knowledge, there is little literature concerning infinitely many solutions for (3.79) in this situation.

Remark 3.14. It should be mentioned that infinitely many solutions for Kirchhoff problems in bounded domain without the usual oddness assumption on the nonlinearities have been investigated in [204–206]. In these papers, the authors can obtain a sequence of unbounded solutions for the corresponding Kirchhoff problems involving oscillating but non-necessarily odd nonlinearities.

We now make some comments on the key ingredients of the analysis in this section. Following the idea of dealing with the elliptic problem in [29], we first extend the nonlinear term $f(x, u)$ and introduce a modified nonlinear Kirchhoff equation. Next, we show by variational methods that the modified nonlinear Kirchhoff equation possesses a sequence of weak solutions $\{u_k\}$ in $H^1(\mathbb{R}^3)$ with $\{u_k\}$ converging to 0 in $H^1(\mathbb{R}^3)$. Finally, using the Moser-type iteration technique, we prove that $\{u_k\}$ converges to 0 in $L^\infty(\mathbb{R}^3)$.

In what follows, we will always assume that (B21) is satisfied. By Sobolev embedding, we have the following inequality

$$\|u\|_{L^p} \leq C(p)\|u\|. \tag{3.80}$$

In order to prove our main result via variational methods, we need to modify and extend $f(x, u)$ for u outside a neighborhood of 0 to get $\tilde{f}(x, u)$ as follows. Define a cut-off function $\chi \in C_c(\mathbb{R})$ such that $0 \leq \chi(t) \leq 1$, $\chi(t) \equiv 1$ for $|t| \leq \delta/2$, and $\chi(t) \equiv 0$ for $|t| \geq \delta$. Let

$$\tilde{f}(x, u) := \chi(u)f(x, u), \qquad \forall(x, u) \in \mathbb{R}^3 \times \mathbb{R} \tag{3.81}$$

and

$$\tilde{F}(x, u) := \int_0^u \tilde{f}(x, t)dt, \qquad \forall(x, u) \in \mathbb{R}^3 \times \mathbb{R}. \tag{3.82}$$

Consider the following modified nonlinear Kirchhoff equation

$$-\left(1 + b\int_{\mathbb{R}^3} \left(|\nabla u|^2 + V(x)u^2\right) dx\right) [\Delta u + V(x)u] = \tilde{f}(x, u) \text{ in } \mathbb{R}^3. \tag{3.83}$$

For $u \in E$, we now define functionals Ψ and I by

$$\Psi(u) := \int_{\mathbb{R}^3} \tilde{F}(x, u)dx$$

and

$$I(u) := \frac{1}{2}\|u\|^2 + \frac{b}{4}\|u\|^4 - \Psi(u) = \frac{1}{2}\|u\|^2 + \frac{b}{4}\|u\|^4 - \int_{\mathbb{R}^3} \tilde{F}(x, u)dx. \tag{3.84}$$

Then we have

Lemma 3.20. *Let (B21) and (B22) be satisfied. Then $\Psi \in C^1(E)$ and $\Psi' : E \to E^*$ is completely continuous. Moreover, it holds that for all $u, v \in E$,*

$$\langle \Psi(u), v \rangle = \int_{\mathbb{R}^3} \tilde{f}(x, u) v \, dx, \tag{3.85}$$

$$\langle I(u), v \rangle = \int_{\mathbb{R}^3} (\nabla u \cdot \nabla v + V(x)uv) dx + b\|u\|^2 \int_{\mathbb{R}^3} (\nabla u \cdot \nabla v + V(x)uv) dx \tag{3.86}$$

$$- \int_{\mathbb{R}^3} \tilde{f}(x, u) v \, dx.$$

In addition, if $u \in E \subseteq H^1(\mathbb{R}^3)$ is a critical point of I on E, then u is a weak solution for (3.83).

Proof. Firstly, we verify (3.85) by definition and prove that $\Psi \in C^1(E)$ with $\Psi' : E \to E^*$ being completely continuous. It follows from (B22) and (3.81) that

$$|\tilde{f}(x, u)| \leq \nu \xi(x) |u|^{\nu - 1} \text{ for all } (x, u) \in \mathbb{R}^3 \times \mathbb{R}, \tag{3.87}$$

which combined with (3.82) leads to

$$|\tilde{F}(x, u)| \leq \xi(x) |u|^{\nu} \text{ for all } (x, u) \in \mathbb{R}^3 \times \mathbb{R}. \tag{3.88}$$

For simplicity, we set

$$\mu^* := \frac{\nu \mu}{\mu - 1}. \tag{3.89}$$

Since $\nu \in (1, 2)$ and $\mu \in (\frac{3}{2}, \frac{2}{2 - \nu}]$ in (B22), we get $\mu^* \in (2, 6)$. Thus, for any $u \in E$, we obtain from (3.88), Hölder's inequality, and (3.80) that

$$\int_{\mathbb{R}^3} |\tilde{F}(x, u)| dx \leq \int_{\mathbb{R}^3} \xi(x) |u|^{\nu} dx \leq \|\xi\|_{L^\mu} \|u\|_{L^{\mu^*}}^{\nu} \leq C(\mu^*, \nu) \|\xi\|_{L^\mu} \|u\|^{\nu} < \infty. \tag{3.90}$$

Hence, Ψ is well defined. For any given $u \in E$, define an associated linear operator $J(u) : E \to \mathbb{R}$ as follows:

$$\langle J(u), v \rangle = \int_{\mathbb{R}^3} \tilde{f}(x, u) v(x) dx.$$

By virtue of (3.87), (3.80), and Hölder's inequality, we arrive at

$$|\langle J(u), v\rangle| \le \int_{\mathbb{R}^3} \nu \xi(x)|u|^{\nu-1}|v|dx$$

$$\le \nu \|\xi\|_{L^\mu} \|u\|_{L^{\mu^*}}^{\nu-1} \|v\|_{L^{\mu^*}}$$

$$\le C(\mu^*, \nu) \|\xi\|_{L^\mu} \|u\|^{\nu-1} \|v\|, \tag{3.91}$$

which implies that $J(u)$ is well defined and bounded in E. It follows from (3.87) that

$$|\tilde{f}(x, u + \eta v)v| \le \nu 2^{\nu-1} \xi(x)(|u|^{\nu-1}|v| + |v|^\nu),$$

for all $x \in \mathbb{R}^3$, $\eta \in [0,1]$ and $u, v \in \mathbb{R}$. For any $u, v \in E$, combining (3.90) and (3.91), we get $\xi(x)(|u|^{\nu-1}|v| + |v|^\nu) \in L^1(\mathbb{R}^3)$. Therefore, by Mean Value Theorem and Lebesgue's Dominated Convergence Theorem, we have

$$\lim_{t \to 0} \frac{\Psi(u + tv) - \Psi(u)}{t} = \lim_{t \to 0} \int_{\mathbb{R}^3} \tilde{f}(x, u + \theta(x)tv)dx$$

$$= \int_{\mathbb{R}^3} \tilde{f}(x, u)vdx = \langle J(u), v\rangle, \tag{3.92}$$

where $\theta(x) \in [0,1]$ depends only on u, v, and t. This shows that Ψ is Gâteaux differentiable on E and the Gâteaux derivative of Ψ at u is $J(u)$. In order to prove that $\Psi \in C^1(E)$ and $\Psi' : E \to E^*$ is completely continuous, it suffices to prove that $J : E \to E^*$ is completely continuous. To this end, we claim that if $u_n \rightharpoonup u_0$ weakly in E, then for any $R > 0$,

$$\tilde{f}(x, u_n) \to \tilde{f}(x, u_0) \text{ strongly in } L^{p_0}(B_R), \tag{3.93}$$

where

$$p_0 = \begin{cases} \dfrac{3}{2}, & \text{if } \mu > 2, \\ \max\left\{\dfrac{3}{2}, \dfrac{\mu}{\mu(\nu-1)+1}\right\}, & \text{if } \mu \le 2. \end{cases} \tag{3.94}$$

Arguing by contradiction. By Sobolev's embedding theorem, we assume that there exist constants R_0, $\varepsilon_0 > 0$ and a subsequence $\{u_{n_k}\}_{k \in \mathbb{N}}$ such that

$$u_{n_k} \to u_0 \text{ strongly in } L^{p_0^*}(B_{R_0}) \text{ and } u_{n_k} \to u_0 \text{ a.e. in } B_{R_0} \text{ as } k \to \infty \tag{3.95}$$

but

$$\int_{B_{R_0}} [\tilde{f}(x, u_{n_k}) - \tilde{f}(x, u_0)]^{p_0} dx \ge \varepsilon_0, \quad \forall k \in \mathbb{N}, \tag{3.96}$$

where $p_0^* := \frac{p_0\mu(\nu-1)}{\mu-p_0} \in [1,6)$ due to (B22) and (3.94). By (3.95), passing to a subsequence if necessary, we can assume that

$$\sum_{k=1}^{\infty} \|u_{n_k} - u_0\|_{L^{p_0^*}(B_{R_0})} < +\infty,$$

which yields $w(x) := \sum_{k=1}^{\infty} |u_{n_k}(x) - u_0(x)| \in L^{p_0^*}(B_{R_0})$. On the other hand, we derive from (3.87) and Hölder's inequality that

$$\begin{aligned}
[\tilde{f}(x, u_{n_k}) - \tilde{f}(x, u_0)]^{p_0} &\le [\nu\xi(x)(|u_{n_k}|^{\nu-1} + |u_0|^{\nu-1})]^{p_0} \\
&\le (2\nu)^{p_0}\xi(x)^{p_0}(|u_{n_k}|^{p_0(\nu-1)} + |u_0|^{p_0(\nu-1)}) \\
&\le \nu^{p_0}2^{p_0\nu+1}\xi(x)^{p_0}(|u_{n_k} - u_0|^{p_0(\nu-1)} + |u_0|^{p_0(\nu-1)}) \\
&\le \nu^{p_0}2^{p_0\nu+1}\xi(x)^{p_0}(|w|^{p_0(\nu-1)} + |u_0|^{p_0(\nu-1)}),
\end{aligned}$$
(3.97)

for all $k \in \mathbb{N}$ and $x \in \mathbb{R}^3$ and

$$\begin{aligned}
\int_{B_{R_0}} &\xi(x)^{p_0}(|w|^{p_0(\nu-1)} + |u_0|^{p_0(\nu-1)}) \\
&\le \|\xi\|_\mu^{p_0}\left(\|w\|_{L^{p_0^*}(B_{R_0})}^{p_0(\nu-1)} + \|u_0\|_{L^{p_0^*}(B_{R_0})}^{p_0(\nu-1)}\right) \\
&< +\infty,
\end{aligned}$$
(3.98)

which together with (3.95), (3.97), and Lebesgue's Dominated Convergence Theorem leads to

$$\lim_{k\to\infty} \int_{B_{R_0}} [\tilde{f}(x, u_{n_k}) - \tilde{f}(x, u_0)]^{p_0} dx = 0.$$

This contradicts with (3.96). Thus the claim is true.

Now let $u_n \rightharpoonup u$ weakly in E as $n \to \infty$, then $\{u_n\}$ is bounded in E and hence there exists a constant $D_0 > 0$ such that for all $n \in \mathbb{N}$,

$$\|u_n\|^{\nu-1} + \|u\|^{\nu-1} \le D_0.$$
(3.99)

For any $\varepsilon > 0$, by (B23), there exists $R_\varepsilon > 0$ such that

$$\left(\int_{\mathbb{R}^3 \setminus B_{R_\varepsilon}} \xi(x)^\mu dx\right)^{1/\mu} < \frac{\varepsilon}{2\nu D_0\tau_{\mu^*}^\nu}.$$
(3.100)

By (3.87), (3.99), (3.100), and Hölder's inequality, we have

$$\int_{\mathbb{R}^3 \setminus B_{R_\varepsilon}} |\tilde{f}(x, u_n) - \tilde{f}(x, u)| |v| dx$$

$$\leq \int_{\mathbb{R}^3 \setminus R_{R_\varepsilon}} \nu \xi(x) (|u_n|^{\nu-1} + |u|^{\nu-1}) |v| dx$$

$$\leq \nu \int_{\mathbb{R}^3 \setminus B_{R_\varepsilon}} \xi(x) (|u_n|^{\nu-1} + |u|^{\nu-1}) |v| dx$$

$$\leq \nu \left(\int_{\mathbb{R}^3 \setminus B_{R_\varepsilon}} \xi(x)^\mu dx \right)^{1/\mu} \left(\|u_n\|_{L^{\mu^*}}^{\nu-1} + \|u\|_{L^{\mu^*}}^{\nu-1} \right) \|v\|_{L^{\mu^*}}$$

$$\leq \nu \tau_{\mu^*}^\nu \left(\int_{\mathbb{R}^3 \setminus B_{R_\varepsilon}} \xi(x)^\mu dx \right)^{1/\mu} \left(\|u_n\|_{L^{\mu^*}}^{\nu-1} + \|u\|_{L^{\mu^*}}^{\nu-1} \right)$$

$$< \frac{\varepsilon}{2}, \quad \forall n \in \mathbb{N} \text{ and } \|v\| = 1. \tag{3.101}$$

For the R_ε given in (3.100), by the above claim and the Hölder inequality, there exists $N_\varepsilon \in \mathbb{N}$ such that

$$\int_{B_{R_\varepsilon}} |\tilde{f}(x, u_n) - \tilde{f}(x, u)| |v| dx \tag{3.102}$$

$$\leq \left(\int_{B_{R_\varepsilon}} |\tilde{f}(x, u_n) - \tilde{f}(x, u)|^{p_0} dx \right)^{1/p_0} \|v\|_{L^{p_0}(R_\varepsilon)}$$

$$\leq \left(\int_{B_{R_\varepsilon}} |\tilde{f}(x, u_n) - \tilde{f}(x, u)|^{p_0} dx \right)^{1/p_0} \|v\|_{L^{p_0}}$$

$$\leq \tau_{\bar{p}_0} \left(\int_{B_{R_\varepsilon}} |\tilde{f}(x, u_n) - \tilde{f}(x, u)|^{p_0} dx \right)^{1/p_0}$$

$$< \frac{\varepsilon}{2}, \quad \forall n \geq N_\varepsilon \text{ and } \|v\| = 1, \tag{3.103}$$

where $\bar{p}_0 := \frac{p_0}{p_0 - 1} \in [2, 3]$ by (3.94) is the conjugate index of p_0, and $\tau_{\bar{p}_0}$ is the Sobolev embedding constant. Then for any $\varepsilon > 0$, combining (3.101) and (3.102), we get

$$\|J(u_n) - J(u)\|_{E^*} = \sup_{\|v\|=1} |\langle J(u_n) - J(u), v\rangle|$$

$$= \sup_{\|v\|=1} \left| \int_{\mathbb{R}^3} [\tilde{f}(x, u_n) - \tilde{f}(x, u)] v \, dx \right|$$

$$\leq \sup_{\|v\|=1} \int_{B_{R_\varepsilon}} |\tilde{f}(x, u_n) - \tilde{f}(x, u)||v| dx$$

$$+ \sup_{\|v\|=1} \int_{\mathbb{R}^3 \setminus B_{R_\varepsilon}} |\tilde{f}(x, u_n) - \tilde{f}(x, u)||v| dx$$

$$\leq \frac{\varepsilon}{2} + \frac{\varepsilon}{2} = \varepsilon, \quad \forall n \geq N_\varepsilon.$$

This shows that $J : E \to E^*$ is completely continuous.

Finally, combining (3.85) and the form of I in (3.84), we immediately derive that $I \in C^1(E)$ and (3.86) holds. □

In order to apply Theorem 1.12, we shall show that the functional I defined in (3.84) satisfies conditions (Φ_1) and (Φ_2) in Theorem 1.12.

Lemma 3.21. *Let (B21) and (B22) be satisfied. Then I is bounded from below and satisfies (PS) condition.*

Proof. We first prove that I is bounded from below. It follows from (3.90) that for $u \in E$,

$$I(u) \geq \frac{1}{2}\|u\|^2 - \int_{\mathbb{R}^3} |\tilde{F}(x, u)| dx \geq \frac{1}{2}\|u\|^2 - C(\mu^*, \nu)\|\xi\|_{L^\mu}\|u\|^\nu, \quad (3.104)$$

which combined with $\nu < 2$ yields that I is bounded from below.

Next, we show that I satisfies (PS) condition. Let $\{u_n\}_{n \in \mathbb{N}} \subset E$ be a (PS)-sequence, namely

$$|I(u_n)| \leq C \text{ and } I'(u_n) \to 0 \text{ as } n \to \infty \quad (3.105)$$

for some $C > 0$ independent of n. Then we obtain from (3.104) and (3.105) that for all $n \in \mathbb{N}$,

$$C \geq \frac{1}{2}\|u_n\|^2 - C(\mu^*, \nu)\|\xi\|_{L^\mu}\|u_n\|^\nu,$$

which implies that $\{u_n\}_{n \in \mathbb{N}}$ is bounded in E. Thus there exists a subsequence $\{u_{n_k}\}_{k \in \mathbb{N}}$ such that

$$u_{n_k} \rightharpoonup u_0 \text{ weakly in } E \text{ as } k \to \infty. \quad (3.106)$$

Then

$$(1 + \|u_{n_k}\|^2)\langle u_{n_k}, u_{n_k} - u_0\rangle = \langle I(u_0), u_{n_k} - u_0\rangle + \int_{\mathbb{R}^3} \tilde{f}(x, u_{n_k})dx. \quad (3.107)$$

By virtue of (3.105) and (3.106), we derive

$$\langle I(u_{n_k}), u_{n_k} - u_0\rangle \to 0 \text{ and } \langle I(u_0), u_{n_k} - u_0\rangle \to 0 \text{ as } k \to \infty.$$

Due to Lemma 3.20, $\Psi' : E \to E^*$ is completely continuous. Hence

$$\int_{\mathbb{R}^3} (f(x, u_{n_k}) - f(x, u_0))(u_{n_k} - u_0)dx \to 0 \text{ as } k \to \infty. \quad (3.108)$$

Therefore, (3.107) implies that

$$\|u_{n_k} - u_0\|^2 \to 0 \text{ as } k \to \infty,$$

which indicates that $u_{n_k} \to u_0$ strongly in E as $k \to \infty$. Thus I satisfies (PS) condition.

The proof of Lemma 3.21 is finished. $\quad\square$

Next, motivated by [29] where the author deals with the elliptic problem, we have the following lemma which indicates that I satisfies (Φ_2) in Theorem 1.12.

Lemma 3.22. *Let (B21), (B22), and (B23) be satisfied. Then for each $k \in \mathbb{N}$, there exists an $A_k \subseteq E$ with genus $\gamma(A_k) = k$ such that $\sup_{u \in A_k} I(u) < 0$.*

Proof. By coordinate translation, without loss of generality, we assume $x_0 = 0$ in (B23). Let $\mathcal{C}$ denote the cube

$$\mathcal{C} := \left\{ x = (x_1, x_2, x_3) \,\Big|\, -\frac{r_0}{2} \le x_i \le \frac{r_0}{2}, i = 1, 2, 3 \right\},$$

where r_0 is the positive constant given in (B23). Obviously, $\mathcal{C} \subseteq B_{r_0}$. By (B23), there exist a constant $\rho > 0$ and two sequences of positive numbers $\delta_n \to 0$, $M_n \to \infty$ as $n \to \infty$ such that

$$F(x, u) \ge -\rho u^2, \quad \forall x \in \mathcal{C} \text{ and } |u| \le \delta \quad (3.109)$$

and

$$F(x, \delta_n)/\delta_n \ge M_n, \quad \forall x \in \mathcal{C} \text{ and } n \in \mathbb{N}. \quad (3.110)$$

For any fixed $k \in \mathbb{N}$, let $q \in \mathbb{N}$ be the smallest positive integer satisfying $q \ge k$. We divide the cube $\mathcal{C}$ equally into q small cubes by planes parallel to each face of $\mathcal{C}$ and denote them by $\mathcal{C}_i$ with $1 \le i \le q^3$. Then the edge of each $\mathcal{C}_i$ has the length of $a := \frac{r_0}{q}$. For each $1 \le i \le k$, we make a cube $\mathcal{D}_i$

in C_i such that D_i has the same center as that of C_i, the faces of D_i and C_i are parallel and the edge of D_i has the length of $\frac{a}{2}$.

Choose a function $\psi \in C_0^\infty(\mathbb{R})$ such that $\psi(t) \equiv 1$ for $t \in [-\frac{a}{4}, \frac{a}{4}]$, $\psi(t) \equiv 0$ for $t \in \mathbb{R} \setminus [-\frac{a}{2}, \frac{a}{2}]$, and $0 \leq \psi(t) \leq 1$ for all $t \in \mathbb{R}$. Define

$$\varphi(x) := \psi(x_1)\psi(x_2)\psi(x_3), \quad \forall x = (x_1, x_2, x_3) \in \mathbb{R}^3.$$

For each $1 \leq i \leq k$, let $y_i \in \mathbb{R}^3$ be the center of both C_i and D_i, and define

$$\varphi_i(x) = \varphi(x - y_i), \quad \forall x \in \mathbb{R}^3.$$

Then for all $1 \leq i \leq k$, it is easy to see that

$$\operatorname{supp} \varphi_i \subseteq C_i \qquad (3.111)$$

and

$$\varphi_i(x) = 1, \quad \forall x \in D_i, \quad 0 \leq \varphi_i(x) \leq 1, \quad \forall x \in \mathbb{R}^3. \qquad (3.112)$$

Set

$$V_k := \left\{ (t_1, t_2, \ldots, t_k) \in \mathbb{R}^k : \max_{1 \leq i \leq k} |t_i| = 1 \right\}$$

and

$$W_k := \left\{ \sum_{i=1}^k t_i \varphi_i : (t_1, t_2, \ldots, t_k) \in V_k \right\}.$$

Evidently, V_k is homeomorphic to the unit sphere in $\mathbb{R}^k$ by an odd mapping. Thus $\gamma(V_k) = k$. If we define the mapping $\mathcal{H} : V_k \to W_k$ by

$$\mathcal{H}(t_1, t_2, \ldots, t_k) = \sum_{i=1}^k t_i \varphi_i, \quad \forall (t_1, t_2, \ldots, t_k) \in V_k,$$

then $\mathcal{H}$ is odd and homeomorphic. Therefore $\gamma(W_k) = \gamma(V_k) = k$. Moreover, it is evident that W_k is compact and hence there is a constant $C_k > 0$ such that

$$\|u\| \leq C_k, \quad \forall u \in W_k. \qquad (3.113)$$

For any $s \in (0, \frac{\delta}{2})$ and $u = \sum_{i=1}^k t_i \varphi_i \in W_k$, by (3.84), (3.111), and (3.112), we have

$$
\begin{aligned}
I(su) &= \frac{1}{2}\|su\|^2 + \frac{b}{4}\|su\|^4 - \int_{\mathbb{R}^3} \tilde{F}\left(x, s\sum_{i=1}^k t_i \varphi_i\right) dx \\
&= \frac{s^2}{2}\|u\|^2 + \frac{s^4 b}{4}\|u\|^4 - \sum_{i=1}^k \int_{C_i} \tilde{F}\left(x, st_i\varphi_i\right) dx \\
&= \frac{s^2}{2}\|u\|^2 + \frac{s^4 b}{4}\|u\|^4 - \sum_{i=1}^k \int_{C_i} F\left(x, st_i\varphi_i\right) dx, \qquad (3.114)
\end{aligned}
$$

where the last equality holds by the definition of $\tilde{F}$ in (3.82) and the fact that $|st_i\varphi_i(x)| < \frac{\delta}{2}$ for all $1 \le i \le k$. By the definition of V_k, there exists some integer $1 \le i_u \le k$ such that $|t_{i_u}| = 1$. Then it follows that

$$\sum_{i=1}^{k} \int_{\mathcal{C}_i} F\left(x, st_i\varphi_i\right) dx = \int_{\mathcal{D}_{i_u}} F\left(x, st_{i_u}\varphi_{i_u}\right) dx + \int_{\mathcal{C}_{i_u} \backslash \mathcal{D}_{i_u}} F\left(x, st_{i_u}\varphi_{i_u}\right) dx$$
$$+ \sum_{i \ne i_u} \int_{\mathcal{C}_i} F\left(x, st_i\varphi_i\right) dx. \tag{3.115}$$

By (3.109) and (3.112), there holds

$$\int_{\mathcal{C}_{i_u} \backslash \mathcal{D}_{i_u}} F\left(x, st_{i_u}\varphi_{i_u}\right) dx + \sum_{i \ne i_u} \int_{\mathcal{C}_i} F\left(x, st_i\varphi_i\right) dx \ge -\rho r_0^3 s^2. \tag{3.116}$$

Here we use the fact that $|\mathcal{C}| = r_0^3$. For each $\delta_n \in (0, \frac{\delta}{2})$ given in (3.110), combining (B24), (3.110), and (3.113)–(3.116), we deduce

$$I(\delta_n u) \le \frac{C_k^2 \delta_n^2}{2} + \frac{C_k^4 \delta_n^4 b}{4s^2} + \rho r_0^3 \delta_n^2 - \int_{\mathcal{D}_{i_u}} F\left(x, st_{i_u}\varphi_{i_u}\right) dx$$
$$\le \delta_n^2 \left(\frac{C_k^2}{2} + \frac{C_k^4 \delta_n^2 b}{4s^2} + \rho r_0^3 - \frac{a^3 M_n}{8} \right). \tag{3.117}$$

Here we use the fact that $|\delta_n t_{i_u}\varphi_{i_u}(x)| \equiv \delta_n$ for all $x \in \mathcal{D}_{i_u}$ and $|\mathcal{D}_{i_u}| = \frac{a^3}{8}$. Since $\delta_n \to 0$ and $M_n \to \infty$ as $n \to \infty$, we can choose $n_0 \in \mathbb{N}$ large enough such that for $n \ge n_0$, the right-hand side of (3.117) is negative. Define

$$A_k := \{\delta_{n_0} u : u \in W_k\}.$$

Then we have

$$\gamma(A_k) = \gamma(W_k) = k \quad \text{and} \quad \sup_{u \in A_k} I(u) < 0.$$

The proof of Lemma 3.22 is completed. $\qquad\qquad\qquad\qquad\qquad\square$

Lemma 3.23. *If $\{u_k\}$ is a critical point sequence of I satisfying $u_k \to 0$ in E as $k \to \infty$, then $u_k \to 0$ in $L^\infty(\mathbb{R}^3)$ as $k \to \infty$.*

Proof. First, we shall use Moser iteration technique to show $\{u_k\} \subset L^\infty(\mathbb{R}^3)$. For every critical point u of I on E and any $K > 0$, we define

$$u^K(x) := \begin{cases} u(x), & \text{if } |u(x)| \le K, \\ \pm K, & \text{if } \pm u(x) > K. \end{cases} \tag{3.118}$$

By (3.86), we have for all $v \in E$,

$$\int_{\mathbb{R}^3} (\nabla u \cdot \nabla v + V(x)uv)dx + b\|u\|^2 \int_{\mathbb{R}^3} (\nabla u \cdot \nabla v + V(x)uv)dx = \int_{\mathbb{R}^3} \tilde{f}(x, u)v dx.$$
(3.119)

For $\beta \geq 0$, it is easy to see that $v := |u^K|^{2\beta} u^K \in E$. Thus we get from (3.119) that

$$\int_{\mathbb{R}^3} \tilde{f}(x, u)|u^K|^{2\beta} u^K dx = \int_{\mathbb{R}^3} \left[(1 + 2\beta)|u^K|^{2\beta}\nabla u \cdot \nabla u^K + V(x)u|u^K|^{2\beta} u^K \right] dx$$

$$+ b\|u\|^2 \int_{\mathbb{R}^3} \left[(1 + 2\beta)|u^K|^{2\beta}\nabla u \cdot \nabla u^K + V(x)u|u^K|^{2\beta} u^K \right] dx,$$
(3.120)

which combined with (3.87), (B21), and (3.118) yields

$$\frac{1}{(\beta + 1)^2} \| |u^K|^{\beta+1} \|^2 \leq \nu \int_{\mathbb{R}^3} \xi(x)|u|^{2\beta+\nu} dx.$$
(3.121)

Then it follows from Sobolev's embedding theorem and the Hölder inequality that

$$\frac{S^2}{(\beta + 1)^2} \|u^K\|_{6\beta+6}^{2\beta+2} \leq \nu\|\xi\|_\mu \|u\|_{\mu(2\beta+\nu)/(\mu-1)}^{2\beta+\nu}.$$
(3.122)

Let $K \to +\infty$ in (3.122), we get

$$\|u\|_{6\beta+6} \leq [c_0(\beta + 1)]^{1/(\beta+1)} \|u\|_{\mu(2\beta+\nu)/(\mu-1)}^{(2\beta+\nu)/(2\beta+2)},$$
(3.123)

where $c_0 = \max\{1, \sqrt{\nu\|\xi\|_\mu}/S\}$. Set $\beta_0 = 0$ and $6\beta_{n-1} + 6 = \mu(2\beta_n + \nu)/(\mu - 1)$, for all $n \in \mathbb{N}$, that is

$$\beta_n = \frac{2\bar{\mu} - \nu}{2(\bar{\mu} - 1)}(\bar{\mu}^n - 1), \quad \forall n \in \mathbb{N},$$
(3.124)

where $\bar{\mu} := 3(\mu-1)/\mu > 1$ since $\mu > 3/2$ in (B22). For each $n \in \mathbb{N}$, let $\zeta_n := \sum_{i=0}^{n-1} \frac{\ln(c_0(\beta_i+1))}{\beta_i+1}$ and $\sigma_n := \prod_{i=0}^{n-1} \frac{2\beta_i+\nu}{2\beta_i+2}$. Then by (3.124), we know that both $\{\zeta_n\}$ and $\{\sigma_n\}$ are convergent sequences with $\zeta := \lim_{n\to\infty} \zeta_n > 0$ and $\sigma := \lim_{n\to\infty} \sigma_n \in (0, 1)$. Note that $[c_0(\beta_i + 1)]^{(2\beta_j+\nu)/(2\beta_j+2)} \leq c_0(\beta_i + 1)$, for any $i < j$. Doing iteration by (3.123), we have

$$\|u\|_{\mu(2\beta_n+\nu)/(\mu-1)} \leq e^{\zeta_n} \|u\|_{\mu^*}^{\sigma_n}, \quad \forall n \in \mathbb{N},$$
(3.125)

where μ^* is given in (3.89). Let $n \to \infty$ in (3.125), we obtain $\|u\|_\infty \leq e^\zeta \|u\|_{\mu^*}^\sigma$. This together with (3.80) shows that if $\{u_k\}$ is a critical point sequence of I satisfying $u_k \to 0$ in E as $k \to \infty$, then $u_k \to 0$ in $L^\infty(\mathbb{R}^3)$ as $k \to \infty$. The proof is completed. $\qquad\square$

With Lemmas 3.20–3.23 at hand, we are now in a position to prove our main result.

Proof of Theorem 3.7. By the definition of I in (3.84), we see that I is an even functional and $I(0) = 0$. Besides, Proposition 3.20 and Lemmas 3.21–3.22 show that $I \in C^1(E, \mathbb{R})$ and satisfies conditions (Φ_1) and (Φ_2) in Theorem 1.12. Thus, by Theorem 1.12, we get a sequence of nontrivial critical points $\{u_k\}$ for I satisfying $I(u_k) \leq 0$ for all $k \in \mathbb{N}$ and $u_k \to 0$ in E as $k \to \infty$. By virtue of Lemma 3.23, $\{u_k\}$ is a sequence of weak solutions for (3.83) with $u_k \to 0$ in $L^\infty(\mathbb{R}^3)$ as $k \to \infty$. Therefore, there exists a $k_0 \in \mathbb{N}$ such that u_k is a weak solution of (3.79) for each $k \geq k_0$. This completes the proof of Theorem 3.7. □

Notes and Comments

There were huge literatures on the studies of the existence and behavior of solutions of (3.79). When potential function V is positive and coercive, Wu [135] gave the first result on the existence of nontrivial weak solutions, and he also obtained a sequence of high energy solutions for problem (3.79) if the nonlinearity is superlinear at infinity, which was extended by Ye and Tang [144] and Li and Wu [138] to the case of f is 4-superlinear at infinity. Moreover, Ye and Tang [144] also investigated existence results when f is sublinear at infinity. It is worth noticing that the high energy solutions problem in [135] had been revisited by Liu and He [140]. Cheng [133] covered the case that nonlinearity f is asymptotically linear at infinity. When potential function V is only a positive continuous function, Duan and Huang [152] established weak solutions provided that f is sublinear at infinity and satisfies a control at infinity globally.

When the potential function V is coercive, it is not hard to show that in a weighted function space the Kirchhoff problem has weak solutions (see e.g. [137, 182]). Furthermore, when V is asymptotically to a constant and positive, He and Zou [155, 156] obtained existence and concentration behavior of positive solutions and ground states solutions, respectively. Recently, ground state solution and concentration of positive solutions were established by several authors [153, 154, 157, 161]. There are also some studies on existence results for other potential function of the Kirchhoff system, refer to [141, 143, 145, 146, 207–209] and references therein. This section was obtained in L. Li-X. Zhong [210].

3.7   Multiple Solutions with Combined Nonlinearities

In this section, we study the existence of positive solutions for the Kirchhoff equation

$$-\Big(a + b \int_\Omega |\nabla u|^2 dx\Big)\Delta u = h(x)u^q + f(x,u), \quad x \in \Omega,$$

$$u = 0, \quad x \in \partial\Omega,$$

(3.126)

where Ω is a bounded smooth domain in R^N ($N = 1, 2, 3$), $a > 0$, $b > 0$, $0 < q < 1$.

To state the assumptions, we recall some results about the following two eigenvalue problems:

$$-\Delta u = \lambda u \text{ in } \Omega, \quad u = 0 \text{ on } \Omega,$$

(3.127)

and

$$-\Big(\int_\Omega |\nabla u|^2 dx\Big)\Delta u = \mu u^3 \text{ in } \Omega, \quad u = 0 \text{ on } \Omega.$$

(3.128)

Let λ_1 be the principal eigenvalue of (3.127) and let $\phi_1 > 0$ be its associated eigenfunction. It is known that λ_1 can be characterized by

$$\lambda_1 = \inf\Big\{\int_\Omega |\nabla u|^2 dx : u \in H_0^1(\Omega), \int_\Omega |u|^2 dx = 1\Big\},$$

where $H_0^1(\Omega)$ is the usual Sobolev space defined as the completion of $C_0^\infty(\Omega)$ with respect to the norm $\|u\| = \big(\int_\Omega |\nabla u|^2 dx\big)^{1/2}$. Moreover, define

$$\mu_1 = \inf\Big\{\|u\|^4 : u \in H_0^1(\Omega), \int_\Omega |u|^4 dx = 1\Big\}.$$

As shown in [194], there exists $\mu_1 > 0$ which is the principle eigenvalue of (3.128) and there is a corresponding eigenfunction of $\varphi_1 > 0$ in Ω.

Now, we assume that h, f satisfy the following conditions:

(B25) $h \in L^\infty(\Omega)$ and $h(x) \not\equiv 0$;

(B26) $f \in C(\Omega \times \mathbb{R})$, $f(x, 0) = 0$ for all $x \in \Omega$, $f(x, s) \geq 0$ for all $x \in \Omega$ and $s \geq 0$;

(B27)

$$\lim_{s \to 0^+} \frac{f(x,s)}{a\lambda_1 s + b\mu_1 s^3} = \alpha \in [0, 1), \quad \lim_{s \to +\infty} \frac{f(x,s)}{a\lambda_1 s + b\mu_1 s^3} = \beta \in (1, +\infty)$$

uniformly for a.e. $x \in \Omega$.

It is obvious that the values of $f(x,s)$ for $s < 0$ are irrelevant for us to seek for positive solutions of (3.126), and we may define

$$f(x,s) = 0 \quad \text{for } x \in \Omega, \ s \le 0.$$

We say that $u \in H_0^1(\Omega)$ is a positive (nonnegative) weak solution to problem (3.126) if $u > 0$ ($u \ge 0$) a.e. Ω and satisfies

$$\left(a + b \int_\Omega |\nabla u|^2 dx \right) \int_\Omega \nabla u \cdot \nabla v dx = \int_\Omega h(x) u^q v dx + \int_\Omega f(x,u) v dx$$

for all $v \in H_0^1(\Omega)$. By assumption (B26), we know that to seek a nonnegative weak solution of (3.126) is equivalent to finding a nonzero critical point of the following functional on $H_0^1(\Omega)$:

$$I(u) = \frac{1}{2} \int_\Omega |\nabla u|^2 dx + \frac{b}{4} \left(\int_\Omega |\nabla u|^2 dx \right)^2 - \frac{1}{q+1} \int_\Omega h(x)(u^+)^{q+1} dx$$
$$- \int_\Omega F(x,u^+) dx,$$

where $u^+ = \max\{0,u\}$, $F(x,s) = \int_0^s f(x,\sigma) d\sigma$. By (B26) and (B27), I is a C^1 functional. By the strong maximum principle, the nonzero critical points of I are positive solutions to problem (3.126) if $h(x) \ge 0$.

Our results are as follows.

Theorem 3.8. *Suppose that $N = 1,2,3$, $a > 0$, $b > 0$, $0 < q < 1$, h and f satisfy (B25), (B26), (B27). Assume further that exists $v \in H_0^1(\Omega)$ such that*

(B28) $\int_\Omega h(x)(v^+)^{q+1} dx > 0$.

Then there exists a constant $m > 0$ such that if $\|h\|_\infty < m$, problem (3.126) has a solution $u_1 \in H_0^1(\Omega)$, $u_1 \ge 0$ and $I(u_1) < 0$. Moreover, if $h(x) \ge 0$, then $u_1 > 0$ a. e. in Ω.

Theorem 3.9. *Suppose that $N = 1,2,3$, $a > 0$, $b > 0$, $0 < q < 1$, h and f satisfy (B25), (B26), (B27). Assume further $\beta\mu_1$ is not an eigenvalue of (3.128). Then there exists a constant $m > 0$ such that if $\|h\|_\infty < m$, problem (3.126) has a nonnegative solution $u_2 \in H_0^1(\Omega)$ with $u_2 > 0$ and $I(u_2) > 0$ if $h(x) \ge 0$.*

Remark 3.15. Theorem 3.8 for problem (3.126) with $a, b > 0$ generalizes [211, Theorem 1.1] where (3.126) with $a = 1$ and $b = 0$.

Corollary 3.1. *Suppose that $N = 1,2,3$, $a > 0$, $b > 0$, $0 < q < 1$, h and f satisfy (B25), (B26), (B27). Assume further that $\beta\mu_1$ is not an eigenvalue* of (3.128) and $h(x) \geq (\not\equiv)0$. Then there exists a constant $m > 0$ such that for all $h \in L^\infty(\Omega)$ with $\|h\|_\infty < m$, problem (3.126) has at least two positive solutions $u_1, u_2 \in H_0^1(\Omega)$ such that $I(u_1) < 0 < I(u_2)$.

Remark 3.16. If $h(x) \geq (\not\equiv)0$, it is easy to see that (B28) is always satisfied. Therefore, Corollary 3.1 is a straightforward conclusion of Theorems 3.8 and 3.9 by applying the strong maximum principle [3].

Now, we prove Theorem 3.8 by Ekeland's variational principle. We need the following lemmas.

Lemma 3.24. *Suppose that $N = 1, 2, 3$, $a > 0$, $b > 0$, $0 < q < 1$, h and f satisfy (B25), (B26), (B27). Then there exists a constant $m > 0$ such that if $\|h\|_\infty < m$, we have*

(a) There exist $\rho, \eta > 0$ such that $I(u)|_{\|u\|=\rho} \geq \eta > 0$.

(b) There exists an $e \in \mathbb{R} \setminus B_\rho(0)$ such that $I(e) < 0$.

Proof. (a) By (B27), $\beta \in (1, +\infty)$ and noticing that $f(x,s)/s^{p-1} \to 0$ as $s \to +\infty$ uniformly in $x \in \Omega$ for any fixed $p \in (4,6)$ if $N = 3$; $p \in (4, +\infty)$ if $N = 1, 2$. Given $\varepsilon \in (0,1)$, there exist $\delta, M_\varepsilon > 0$ satisfying $0 < \delta < +\infty$ such that

$$f(x,s) < (\alpha + \varepsilon)(a\lambda_1 s + b\mu_1 s^3), \quad 0 < s < \delta,$$

and

$$f(x,s) < M_\varepsilon s^{p-1}, \quad \delta < s,$$

where $p \in (4,6)$ if $N = 3$; $p \in (4, +\infty)$ if $N = 1, 2$. Together with (B26) and $f(x,s) = 0$ for $x \in \Omega$, $s \leq 0$, we obtain

$$f(x,s) < a\lambda_1(\alpha + \varepsilon)|s| + b\mu_1(\alpha + \varepsilon)|s|^3 + M_\varepsilon s^{p-1}, \quad s \in R.$$

This yields

$$F(x,s) \leq \frac{a\lambda_1}{2}(\alpha + \varepsilon)|s|^2 + \frac{b\mu_1}{4}(\alpha + \varepsilon)|s|^4 + A|s|^p, \quad s \in R, \qquad (3.129)$$

where $A = M_\varepsilon/p$. Furthermore, by (B27), for the above ε, we have

$$f(x,s) > (\beta - \varepsilon)(a\lambda_1 s + b\mu_1 s^3), \quad s > \delta_\infty.$$

Thus, we obtain

$$F(x,s) > (\beta - \varepsilon)\left(\frac{a\lambda_1}{2}s^2 + \frac{b\mu_1}{4}s^4\right), \quad s > \delta_\infty.$$

Together with (B26) and $f(x,s) = 0$ for $x \in \Omega$, $s \le 0$, there exists a constant $B > 0$ such that

$$F(x,s) \ge \frac{a}{2}(\beta - \varepsilon)\lambda_1|s|^2 + \frac{b}{4}(\beta - \varepsilon)\mu_1|s|^4 - B, \quad s \in R. \qquad (3.130)$$

Since $\alpha < 1$, we can choose $\varepsilon > 0$ such that $\varepsilon < 1 - \alpha$. By (B25), (3.129), $\lambda_1\|u\|_2^2 \le \|u\|^2$, $\mu_1\|u\|_4^4 \le \|u\|^2$, the Sobolev's embedding theorem: $\|u\|_{q+1}^{q+1} \le K\|u\|^{q+1}$, $\|u\|_{p+1}^{p+1} \le M\|u\|^{p+1}$ and the Young inequality, we have

$$I(u) = \frac{a}{2}\int_\Omega |\nabla u|^2 dx + \frac{b}{4}\left(\int_\Omega |\nabla u|^2 dx\right)^2 - \frac{1}{q+1}\int_\Omega h(x)(u^+)^{q+1}dx$$

$$- \int_\Omega F(x,u^+)dx$$

$$\ge \frac{a}{2}\|u\|^2 + \frac{b}{4}\|u\|^4 - \frac{\|h\|_\infty}{q+1}\|u^+\|_{q+1}^{q+1} - \frac{a}{2}(\alpha + \varepsilon)\lambda_1\|u^+\|_2^2$$

$$- \frac{b}{4}(\alpha + \varepsilon)\mu_1\|u^+\|_4^4 - A\|u^+\|_p^p$$

$$\ge \frac{a}{2}\|u\|^2 + \frac{b}{4}\|u\|^4 - \frac{\|h\|_\infty}{q+1}\|u\|_{q+1}^{q+1} - \frac{a}{2}(\alpha + \varepsilon)\|u\|^2$$

$$- \frac{b}{4}(\alpha + \varepsilon)\|u\|^4 - A\|u\|_p^p$$

$$\ge \frac{a[1 - (\alpha + \varepsilon)]}{2}\|u\|^2 + \frac{b[1 - (\alpha + \varepsilon)]}{4}\|u\|^4 - \frac{\|h\|_\infty K}{q+1}\|u\|^{q+1} - AM\|u\|^p$$

$$\ge \|u\|^2\left(C_1 - C_2\|h\|_\infty\|u\|^{q-1} - C_3\|u\|^{p-2}\right),$$

$$(3.131)$$

where $C_1 = \frac{a[1-(\alpha+\varepsilon)]}{2}$, $C_2 = \frac{K}{q+1}$ and $C_3 = AM$. Let

$$g(t) = C_2\|h\|_\infty t^{q-1} + C_3 t^{p-2} \quad \text{for } t \ge 0.$$

Clearly,

$$g'(t) = C_2(q-1)\|h\|_\infty t^{q-2} + (p-2)C_3 t^{p-3}.$$

From $g'(t_0) = 0$, we have

$$t_0 = (C_4\|h\|_\infty)^{\frac{1}{p-q-1}}, \quad 0 < q < 1 < 4 < p,$$

where $C_4 = \frac{C_2(1-q)}{(p-2)C_3}$. Then

$$g(t_0) = C_2\|h\|_\infty(C_4\|h\|_\infty)^{\frac{q-1}{p-q-1}} + C_3(C_4\|h\|_\infty)^{\frac{p-2}{p-q-1}} = C_5\|h\|_\infty^{\frac{p-2}{p-q-1}},$$

where $C_5 = C_2 C_4^{\frac{q-1}{p-q-1}} + C_3 C_4^{\frac{p-2}{p-q-1}}$ and $\frac{p-2}{p-q-1} > 0$ because $0 < q < 1 < 4 < p$. Thus, for any $p > 4$, there exists $m > 0$ such that $g(t_0) < C_1$ if

$\|h\|_\infty < m$. Then, if $\|h\|_\infty < m$ and taking $\rho = t_0$, from (3.131), (a) is proved.

(b) For $t > 0$ large enough, by (3.130) and $0 < q < 1$, taking $\varepsilon > 0$ such that $\varepsilon < \min\{\beta - 1, 1 - \alpha\}$, we have

$$I(t\varphi_1) = \frac{at^2}{2}\int_\Omega |\nabla\varphi_1|^2 dx + \frac{bt^4}{4}\left(\int_\Omega |\nabla\varphi_1|^2 dx\right)^2 - \frac{t^{q+1}}{q+1}\int_\Omega h(x)\varphi_1^{q+1} dx$$

$$- \int_\Omega F(x, t\varphi_1) dx$$

$$\leq \frac{at^2}{2}\|\varphi_1\|^2 + \frac{bt^4}{4}\|\varphi_1\|^4 - \frac{t^{q+1}}{q+1}\int_\Omega h(x)\varphi_1^{q+1} dx - \frac{at^2}{2}(\beta - \varepsilon)\lambda_1\|\varphi_1\|_2^2$$

$$- \frac{bt^4}{4}(\beta - \varepsilon)\mu_1\|\varphi_1\|_4^4 + B|\Omega|$$

$$\leq \frac{at^2}{2}\|\varphi_1\|^2 + \frac{bt^4}{4}\|\varphi_1\|^4 - \frac{t^{q+1}}{q+1}\int_\Omega h(x)\varphi_1^{q+1} dx - \frac{bt^4}{4}(\beta - \varepsilon)\|\varphi_1\|^4 + B|\Omega|$$

$$= \frac{at^2}{2}\|\varphi_1\|^2 - \frac{bt^4}{4}(\beta - \varepsilon - 1)\|\varphi_1\|^4 - \frac{t^{q+1}}{q+1}\int_\Omega h(x)\varphi_1^{q+1} dx + B|\Omega|$$

$$\to -\infty$$

as $t \to \infty$. So we can choose $t^0 > 0$ large enough and $e = t\varphi_1$ so that $I(e) < 0$ and $\|e\| > \rho$. $\square$

Proof of Theorem 3.8. Set ρ as in Lemma 3.24(a), define

$$\overline{B}_\rho = \{u \in H_0^1(\Omega) : \|u\| \leq \rho\}, \quad \partial B_\rho = \{u \in H_0^1(\Omega) : \|u\| = \rho\}$$

and $\overline{B}_\rho$ is a complete metric space with the distance

$$\text{dist}(u, v) = \|u - v\| \text{ for } u, v \in \overline{B}_\rho.$$

By Lemma 3.24,

$$I(u)|_{\partial B_\rho} \geq \eta > 0. \tag{3.132}$$

Clearly, $I \in C^1(\overline{B}_\rho, \mathbb{R})$, hence I is lower semicontinuous and bounded from below on $\overline{B}_\rho$. Let

$$c_1 = \inf\{I(u) : u \in \overline{B}_\rho\}. \tag{3.133}$$

We claim that

$$c_1 < 0. \tag{3.134}$$

Indeed, let $v \in H_0^1(\Omega)$ be given by (B28), that is, $\int_\Omega h(x)(v^+)^{q+1}dx > 0$, then for $t > 0$ small enough such that for any $\varepsilon > 0$, we have $|tv| < \varepsilon$. Therefore, together (B27) and $\alpha > 1$ imply

$$
I(tv) = \frac{at^2}{2} \int_\Omega |\nabla v|^2 dx + \frac{bt^4}{4} \left(\int_\Omega |\nabla v|^2 dx \right)^2 - \frac{t^{q+1}}{q+1} \int_\Omega h(x)(v^+)^{q+1}dx
$$

$$
- \int_\Omega F(x, tv^+)dx
$$

$$
\leq \frac{at^2}{2}\|v\|^2 + \frac{bt^4}{4}\|v\|^4 - \frac{t^{q+1}}{q+1} \int_\Omega h(x)(v^+)^{q+1}dx
$$

$$
- \frac{at^2}{2}(\alpha + \varepsilon)\lambda_1\|v\|_2^2 - \frac{bt^4}{4}(\alpha + \varepsilon)\mu_1\|v\|_4^4 < 0,
$$

if $t > 0$ small enough, because $0 < q < 1$. So (3.134) is proved.

By the Ekeland's variational principle [212, Theorem 1.1] in $\overline{B}_\rho$ and (3.133), there is a minimizing sequence $\{u_n\} \subset \overline{B}_\rho$ such that

(i) $c_1 < I(u_n) < c_1 + \frac{1}{n}$,
(ii) $I(w) \geq I(u_n) - \frac{1}{n}\|w - u_n\|$ for all $w \in \overline{B}_\rho$.

So, $I'(u_n) \to 0$ in $H_0^{-1}(\Omega)$ as $n \to \infty$. Moreover, by (i) and (ii), we obtain $I(u_n) \to c_1 < 0$ as $n \to \infty$.

From the above discussion, we know that $\{u_n\}$ is a bounded (PS) sequence, there exist a subsequence (still denoted by $\{u_n\}$) and $u_1 \in H_0^1(\Omega)$ such that

$$u_n \rightharpoonup u_1 \quad \text{weakly in } H_0^1(\Omega),$$

$$u_n \to u_1 \quad \text{a.e. in } \Omega, \tag{3.135}$$

$$u_n \to u_1 \quad \text{strongly in } L^r(\Omega)$$

as $n \to \infty$, where $r \in [1,6]$ if $N = 3$ and $r \in (1, +\infty)$ if $N = 1, 2$. Thus, we have $\lim_{n\to\infty}\langle I'(u_n), v \rangle = \langle I'(u_1), v \rangle = 0$ for all $v \in H_0^1(\Omega)$ and $\lim_{n\to\infty} I(u_n) = c_1 < 0$. Moreover, it follows from $\langle I'(u_1), u_1^- \rangle = (a + b\|u_1\|^2)\|u_1^-\|^2 = 0$ that $u_1 = u_1^+ \geq 0$, where $u_1^- = \max\{-u_1, 0\}$. Therefore, u_1 is a nonnegative critical point of I. Furthermore, if $h(x) \geq 0$, the strong maximum principle [3] implies that u_1 is a positive solution of problem (3.126). $\quad\square$

Now, we use a variant version of Mountain Pass Theorem to obtain a nonzero critical point of functional I; this theorem is used also in [211] and its proof can be found in [22], let us recall first this theorem.

Proof of Theorem 3.9. Let ρ, η and e be given in Lemma 3.24, applying Mountain Pass Theorem 1.4 with $\mu = 0$, $E = H_0^1(\Omega)$, and for c defined as in Mountain Pass Theorem 1.4, then there exists a sequence $\{u_n\} \subset H_0^1(\Omega)$ such that

$$I(u_n) \to c \geq \eta \quad \text{and} \quad (1 + \|u_n\|)\|I'(u_n)\|_{E^{-1}} \to 0$$

as $n \to \infty$. This implies that

$$\frac{a}{2}\int_\Omega |\nabla u_n|^2 dx + \frac{b}{4}\left(\int_\Omega |\nabla u_n|^2 dx\right)^2 - \frac{1}{q+1}\int_\Omega h(x)(u_n^+)^{q+1}dx \quad (3.136)$$

$$- \int_\Omega F(x, u_n^+)dx = c + o(1),$$

$$a\int_\Omega \nabla u_n \cdot \nabla \varphi dx + b\int_\Omega |\nabla u_n|^2 dx \int_\Omega \nabla u_n \cdot \nabla \varphi dx - \frac{1}{q+1}\int_\Omega h(x)(u_n^+)^q \varphi$$

$$(3.137)$$

$$- \int_\Omega f(x, u_n^+)\varphi dx = o(1), \quad \text{for } \varphi \in H_0^1(\Omega),$$

$$a\int_\Omega |\nabla u_n|^2 dx + b\left(\int_\Omega |\nabla u_n|^2 dx\right)^2 - \int_\Omega h(x)(u_n^+)^{q+1}dx - \int_\Omega f(x, u_n^+)u_n^+ dx$$

$$(3.138)$$

$$= o(1).$$

By the compactness of Sobolev embedding and the standard procedures, we know that, if $\{u_n\}$ is bounded in $H_0^1(\Omega)$, there exists $u_2 \in H_0^1(\Omega)$ such that $I'(u_2) = 0$ and $I(u_2) = c > 0$ and u_2 is a nonnegative weak solution of problem (3.126), which is positive if $h(x) \geq 0$ by the strong maximum principle. Moreover, u_2 is different from the solution u_1 obtained in Theorem 3.8 since $I(u_1) = c_1 < 0$. So, to prove Theorem 3.9, we only need to prove that $\{u_n\}$ given by (3.136)–(3.138) is bounded in $H_0^1(\Omega)$.

Next, we shall show that $\{u_n\}$ is bounded in $H_0^1(\Omega)$. By contradiction, we suppose that $\|u_n\| \to \infty$ as $n \to \infty$, and set $w_n = \frac{u_n}{\|u_n\|}$. Clearly, $\{w_n\}$ is bounded in $H_0^1(\Omega)$. Thus, there exist a subsequence, still denoted by $\{w_n\}$, and $w \in H_0^1(\Omega)$, such that

$$w_n \rightharpoonup w \quad \text{weakly in } H_0^1(\Omega),$$

$$w_n \to w \quad \text{a.e. in } \Omega,$$

$$w_n \to w \quad \text{strongly in } L^r(\Omega)$$

as $n \to \infty$, where $r \in [1, 6]$ if $N = 3$ and $r \in (1, +\infty)$ if $N = 1, 2$.

Similarly, $w_n^+ = \frac{u_n^+}{\|u_n\|}$ also satisfies

$$w_n^+ \rightharpoonup w^+ \quad \text{weakly in } H_0^1(\Omega),$$
$$w_n^+ \to w^+ \quad \text{a.e. in } \Omega,$$
$$w_n^+ \to w^+ \quad \text{strongly in } L^r(\Omega)$$

as $n \to \infty$. We first claim that $w \not\equiv 0$. Indeed, if $w \equiv 0$, then by (B25), we have

$$\lim_{n\to\infty} \int_\Omega h(x)(w_n^+)^{q+1}dx = 0. \tag{3.139}$$

Moreover, by (B26)–(B27), for any $\varepsilon > 0$, if $s > 0$ large enough, we obtain

$$(\beta - \varepsilon)a\lambda_1 s + (\beta - \varepsilon)b\mu_1 s^3 < f(x,s) < (\beta + \varepsilon)a\lambda_1 s + (\beta + \varepsilon)b\mu_1 s^3.$$

Therefore, we deduce

$$(\beta - \varepsilon)a\lambda_1 s - \varepsilon b\mu_1 s^3 < f(x,s) - \beta b\mu_1 s^3 < (\beta + \varepsilon)a\lambda_1 s + \varepsilon b\mu_1 s^3.$$

It implies that

$$\frac{(\beta - \varepsilon)\lambda_1}{\|u_n\|^2} \int_\Omega w_n^+ \varphi dx - \varepsilon b\mu_1 \int_\Omega (w_n^+)^3 \varphi dx$$
$$< \int_\Omega \frac{f(x,u_n^+) - b\beta\mu_1(u_n^+)^3}{\|u_n\|^3}\varphi dx$$
$$< \frac{(\beta + \varepsilon)\lambda_1}{\|u_n\|^2} \int_\Omega w_n^+ \varphi dx - \varepsilon b\mu_1 \int_\Omega (w_n^+)^3 \varphi dx$$

for any $\varphi \in H_0^1(\Omega)$. By the arbitrariness of ε, we obtain

$$\lim_{n\to+\infty} \int_\Omega \frac{f(x,u_n^+) - b\beta\mu_1(u_n^+)^3}{\|u_n\|^3}\varphi dx = 0. \tag{3.140}$$

Multiplying (3.137) by $\frac{1}{\|u_n\|^3}$, we have

$$\frac{a}{\|u_n\|^2} \int_\Omega \nabla w_n \cdot \nabla\varphi dx + b \int_\Omega \nabla w_n \cdot \nabla\varphi dx - \frac{1}{\|u_n\|^{3-q}} \int_\Omega h(x)(w_n^+)^q \varphi dx$$
$$- b\beta\mu_1 \int_\Omega (w_n^+)^3 \varphi dx - \int_\Omega \frac{f(x,u_n^+) - b\beta\mu_1(u_n^+)^3}{\|u_n\|^3}\varphi dx$$
$$= o(1).$$

$$\tag{3.141}$$

Letting $n \to \infty$ in (3.141), according to $\|u_n\| \to \infty$ as $n \to \infty$, (3.139), (3.140) and $b \neq 0$, we have

$$\int_\Omega \nabla w \cdot \nabla\varphi dx = \beta\mu_1 \int_\Omega (w^+)^3 \varphi dx$$

and $w \neq 0$. Hence, $\beta\mu_1$ is an eigenvalue of (3.128), which contradicts with the assumption. The proof is complete. $\qquad \square$

Notes and Comments

Some interesting studies by variational methods can be found in [176, 177, 194, 196, 197, 199, 200, 213, 214] references therein and for Kirchhoff-type problem (3.1), they considered it in a bounded domain Ω. For example, Perera and Zhang [194] obtained nontrivial solutions of (3.1) with asymptotically 4-linear terms by using Yang index. In [177], they revisited problem (3.1) and established the existence of a positive, a negative and a sign-changing solution by means of invariant sets of descent flow. Similar results can also be found in Mao and Zhang [176] and in Yang and Zhang [196]. Yang and Zhang in [197] obtained the existence of nontrivial solutions for (3.1) by using the local linking theory. Sun and Tang [199] proved the existence of a mountain pass type positive solution for problem (3.1) with the nonlinearity which is asymptotically linear near zero and superlinear at infinity. Sun and Liu [200] obtained a nontrivial solution via Morse theory by computing the relevant critical groups for problem (3.1) with the nonlinearity which is superlinear near zero but asymptotically 4-linear at infinity and asymptotically near zero but 4-linear at infinity. In [213], the authors obtained the existence of positive solutions for problem (3.126) with $h \equiv 0$ and $f(x,t) = \nu h(x,t)$ by using the topological degree argument and variational method, where h is a continuous function which is asymptotically linear at zero and is asymptotically 3-linear at infinity. By the way, very recently, Cheng, Wu and Liu [214] applied variant Mountain Pass Theorem and Ekeland's variational principle to study the existence of multiple nontrivial solutions for a class of Kirchhoff type problems with concave nonlinearity similar to our problem. But in their paper, the nonlinear term was superlinear at infinity. This section was obtained in [215].

Chapter 4

Nonlinear Field Problems

4.1 Introduction

The Einstein field equations (EFE; also known as "Einstein's equations") are the set of 10 equations in Albert Einstein's general theory of relativity that describes the fundamental interaction of gravitation as a result of spacetime being curved by matter and energy. First published by Einstein in 1915 as a tensor equation, the EFE equate local spacetime curvature (expressed by the Einstein tensor) with the local energy and momentum within that spacetime (expressed by the stressenergy tensor).

Similar to the way that electromagnetic fields are determined using charges and currents via Maxwell's equations, the EFE are used to determine the spacetime geometry resulting from the presence of massenergy and linear momentum, that is, they determine the metric tensor of spacetime for a given arrangement of stressenergy in the spacetime. The relationship between the metric tensor and the Einstein tensor allows the EFE to be written as a set of nonlinear partial differential equations when used in this way. The solutions of the EFE are the components of the metric tensor. The inertial trajectories of particles and radiation (geodesics) in the resulting geometry are then calculated using the geodesic equation.

In this book we only consider the following two equations: Schrödinger–Poisson equations (also called Schrödinger–Maxwell equations) and Klein–Gordon–Maxwell equations (also called Klein–Gordon–Maxwell system). Because we use variational method, here we only consider the standing wave solutions and solitary waves solutions.

4.2 Schrödinger–Maxwell Equations

4.2.1 *Infinitely Many Solutions with Superlinear Nonlinearities*

In this section, we study the system of Schrödinger–Maxwell equations

$$-\Delta u + V(x)u + \phi u = f(x, u), \quad \text{in } \mathbb{R}^3,$$
$$-\Delta \phi = u^2, \quad \text{in } \mathbb{R}^3. \tag{4.1}$$

The problem of finding infinitely many large energy solutions is a very classical problem: there is an extensive literature concerning the existence of infinitely many large energy solutions of a plethora of problems via the Symmetric Mountain Pass Theorem and Fountain Theorem (cf. Ambrosetti and Rabinowitz [172], Rabinowitz [21], Bartsch [27], Bartsch and Willem [216], Struwe [217], Willem [5], etc.). The infinitely many large energy solutions for system (4.1) are obtained in [218] with the following variant "Ambrosetti–Rabinowitz" type condition (AR for short),

(AR) There exists $\mu > 4$ such that for all $s \in \mathbb{R}$ and $x \in \mathbb{R}^3$,

$$\mu F(x, s) := \mu \int_0^s f(x, t) \, dt \leq s f(x, s).$$

After that, Li *et al.* [219] study (4.1) without the (AR) condition. They use variant Fountain Theorem established by Zou [220]. Later, some authors also study this problem without the (AR) condition, see Alves *et al.* [221] and Yang and Han [222].

In this section, we use the Fountain Theorem to find infinitely many large energy solutions to system (4.1). We can see that (4.1) can be proved directly with the Fountain Theorem under Cerami condition. We assume the following assumptions:

(C1) $V \in C(\mathbb{R}^3, \mathbb{R})$ satisfies $\inf_{x \in \mathbb{R}^3} V(x) \geq a_1 > 0$, where $a_1 > 0$ is a constant. Moreover, for every $M > 0$, meas($\{x \in \mathbb{R}^3 : V(x) \leq M\}$) $< \infty$, where meas denote the Lebesgue measure in $\mathbb{R}^3$.

(C2) $f \in C(\mathbb{R}^3 \times \mathbb{R}, \mathbb{R})$ and for some $2 < p < 2^* = 6$, $a_2 > 0$,

$$|f(x, z)| \leq a_2(|z| + |z|^{p-1}),$$

for a.e. $x \in \mathbb{R}^3$ and all $z \in \mathbb{R}$.

$$\lim_{z \to 0} \frac{f(x, z)}{z} = 0$$

uniformly for $x \in \mathbb{R}^3$.

(C3) $\lim_{|z| \to \infty} \frac{F(x,z)}{|z|^4} = +\infty$, uniformly in $x \in \mathbf{R}^3$ and $F(x, 0) \equiv 0$, $F(x, z) \geq 0$ for all $(x, z) \in \mathbb{R}^3 \times \mathbb{R}$.

(C4) There exists a constant $\theta \geq 1$ such that

$$\theta H(x, z) \geq H(x, sz)$$

for all $x \in \mathbb{R}^3$, $z \in \mathbb{R}$ and $s \in [0, 1]$, where $H(x, z) = zf(x, z) - 4F(x, z)$.

(C5) $f(x, -z) = -f(x, z)$ for any $x \in \mathbb{R}^3$ and all $z \in \mathbb{R}$.

The main results of the present article are as follows.

Theorem 4.1. *Assume that* (C1), (C2)–(C5) *hold, then system* (4.1) *has infinitely many solutions* $\{(u_k, \phi_k)\}$ *in* $H^1(\mathbb{R}^3) \times D^{1,2}(\mathbb{R}^3)$ *satisfying*

$$\frac{1}{2} \int_{\mathbb{R}^3} (|\nabla u_k|^2 + V(x)u_k^2) \, dx - \frac{1}{4} \int_{\mathbb{R}^3} |\nabla \phi_k|^2 \, dx + \frac{1}{2} \int_{\mathbb{R}^3} \phi_k u_k^2 \, dx$$

$$- \int_{\mathbb{R}^3} F(x, u_k) \, dx \to +\infty.$$

Remark 4.1. Obviously, (C3) can be derived from (AR). Under (AR), any (PS) sequence of the corresponding energy functional is bounded, which plays an important role of the application of variational methods. Indeed, there are many superlinear functions which do not satisfy the (AR) condition. For instance the function

$$f(x, z) = z^3 \ln(1 + |z|) \tag{4.2}$$

does not satisfy the (AR) condition. But it is easy to see this function satisfies (C3) and (C4). There are many functions which satisfy (C4), but do not satisfy condition (AR) for any $\mu > 4$. However, we cannot deduce condition (C4) from condition (AR). For example, let

$$f(x, u) = 5|u|^4 \int_0^u |t|^{1+\sin t} t \, dt + |u|^{6+\sin u} u,$$

then

$$F(x, z) = |z|^5 \int_0^z |t|^{1+\sin t} t \, dt \, .$$

It is easy to see that $f(x, u)$ satisfies condition (AR) for $\mu = 5$, but it does not satisfy (C4). Thus, (C4) is also superlinear conditions and complement with (AR).

Remark 4.2. In [222], Yang and Han used

(C4′) $\frac{f(x,u)}{u^3}$ is increasing for $u > 0$ and decreasing for $u < 0$, for all $x \in \mathbb{R}^3$.

to obtain a bounded Cerami sequence. Li *et al.* [219], used

(C4″) $H(x,s) \leq H(x,t)$ for all $(s,t) \in \mathbb{R}^+ \times \mathbb{R}^+$, $s \leq t$ and a.e. $x \in \mathbb{R}^3$

to solve the problem (4.1). (C4′) implies that (C4″), as we can see in [223, Lemma 2.2]. We see that our condition (C4) is more general than (C4″). If $\theta = 1$ we can get that $H(x,z)$ is increasing in $\mathbb{R}^+$ with respect to z. Moreover, (C4) gives some general sense of monotony when $\theta > 1$ and we can find some examples that satisfy (C4) but do not satisfy (C4″). For example, let

$$f(x,z) = 4z^3 \ln(1 + z^4) + 2 \sin z,$$

it follows that

$$H(x,z) = 4z^4 - 4\ln(1 + z^4) + 2z \sin z + 8 \cos z.$$

Let $\theta = 100$, we can prove by some simple computation that f satisfies (C4) but does not satisfy (C4″) any more.

It is clear that system (4.1) is the Euler–Lagrange equations of the functional $J : E \times D^{1,2}(\mathbb{R}^3) \to \mathbb{R}$ defined by

$$J(u,\phi) = \frac{1}{2}\|u\|_E^2 - \frac{1}{4}\int_{\mathbb{R}^3} |\nabla \phi|^2 \, dx + \frac{1}{2}\int_{\mathbb{R}^3} \phi u^2 \, dx - \int_{\mathbb{R}^3} F(x,u) \, dx.$$

Evidently, the action functional J belongs to $C^1(E \times D^{1,2}(\mathbb{R}^3), \mathbb{R})$ and its critical points are the solutions of (4.1). It is easy to know that J exhibits a strong indefiniteness, namely it is unbounded both from below and from above on infinitely dimensional subspaces. This indefiniteness can be removed using the reduction method described in [224], by which we are led to study a one variable functional that does not present such a strongly indefinite nature.

Now, we recall this method.

For any $u \in E$, the Lax–Milgram theorem (see [3]) implies there exists a unique $\phi_u \in D^{1,2}(\mathbb{R}^3)$ such that

$$-\Delta \phi_u = u^2$$

in a weak sense. We can write an integral expression for ϕ_u in the form:

$$\phi_u = \frac{1}{4\pi} \int_{\mathbb{R}^3} \frac{u^2(y)}{|x - y|} \, dy, \tag{4.3}$$

for any $u \in E$ (for details, see Section 2 of [218]). The functions ϕ_u possess the following properties:

Lemma 4.1 ([218, Lemma 2.2]). *For any $u \in E$, we have:*

(1) $\|\phi_u\|_{D^{1,2}} \leq a_3 \|u\|_{L^{12/5}}^2$, *where $a_3 > 0$ does not depend on u. As a consequence there exists $a_4 > 0$ such that*

$$\int_{\mathbb{R}^3} \phi_u u^2 \, dx \leq a_4 \|u\|_E^4;$$

(2) $\phi_u \geq 0$.

So, we can consider the functional $I : E \to \mathbb{R}$ defined by $I(u) = J(u, \phi_u)$. After multiplying $-\Delta\phi_u = u^2$ by ϕ_u and integration by parts, we obtain

$$\int_{\mathbb{R}^3} |\nabla\phi_u|^2 \, dx = \int_{\mathbb{R}^3} \phi_u u^2 \, dx.$$

Therefore, the reduced functional takes the form

$$I(u) = \frac{1}{2}\|u\|_E^2 + \frac{1}{4}\int_{\mathbb{R}^3} \phi_u u^2 \, dx - \int_{\mathbb{R}^3} F(x, u) \, dx.$$

From Lemma 2.2, I is well defined. Furthermore, it is well known that I is C^1 functional with derivative given by

$$\langle I'(u), v \rangle = \int_{\mathbb{R}^3} (\nabla u \cdot \nabla v + V(x)uv + \phi_u uv - f(x, u)v) \, dx. \qquad (4.4)$$

Now, we can apply Theorem 2.3 of [224] to our functional J and obtain:

Proposition 4.1. *The following statements are equivalent:*

(1) $(u, \phi) \in E \times D^{1,2}(\mathbb{R}^3)$ *is a critical point of J (i.e. (u, ϕ) is a solution of (4.1));*
(2) u *is a critical point of I and $\phi = \phi_u$.*

We choose an orthogonal basis $\{e_j\}$ of $X := E$ and define $W_k := \text{span}\{e_1, \ldots, e_k\}$, $Z_k := W_{k-1}^\perp$. To complete the proof of our theorems, we need the following lemma.

Lemma 4.2 ([218, Lemma 2.5]). *For any $2 \leq p < 2^*$, we have that*

$$\beta_k := \sup_{u \in Z_k, \|u\|_E = 1} \|u\|_{L^p} \to 0, \quad k \to \infty.$$

Now, we show that the functional I satisfies the Cerami condition.

Lemma 4.3. *Under the assumptions* (C2)–(C4), *the functional $I(u)$ satisfies the Cerami condition at any positive level.*

Proof. We suppose that $\{u_n\}$ is the Cerami sequence, that is for some $c \in \mathbb{R}^+$

$$I(u_n) = \frac{1}{2}\|u_n\|_E^2 + \frac{1}{4}\int_{\mathbb{R}^3} \phi_{u_n} u_n^2 \, dx - \int_{\mathbb{R}^3} F(x, u_n) \, dx \to c \quad (n \to \infty) \quad (4.5)$$

and

$$(1 + \|u_n\|_E)I'(u_n) \to 0 \quad (n \to \infty). \tag{4.6}$$

From (4.5) and (4.6), for n large enough, we have

$$1 + c \geq I(u_n) - \frac{1}{4}\langle I'(u_n), u_n \rangle$$

$$= \frac{1}{4}\|u_n\|_E^2 + \frac{1}{4}\int_{\mathbb{R}^3} f(x, u_n)u_n \, dx - \int_{\mathbb{R}^3} F(x, u_n) \, dx. \tag{4.7}$$

We claim that $\{u_n\}$ is bounded. Otherwise there should exist a subsequence of $\{u_n\}$ satisfying $\|u_n\|_E \to \infty$ as $n \to \infty$. Denote $w_n = \frac{u_n}{\|u_n\|_E}$, then $\{w_n\}$ is bounded. Up to a subsequence, for some $w \in E$, we obtain

$$w_n \rightharpoonup w \quad \text{in } E,$$
$$w_n \to w \quad \text{in } L^t(\mathbb{R}^3), \ 2 \leq t < 2^*, \tag{4.8}$$
$$w_n(x) \to w(x) \quad \text{a.e. in } \mathbb{R}^3.$$

Suppose, $w \neq 0$ in E. Dividing by $\|u_n\|_E^4$ in both sides of (4.5), by (1) of lemma 4.1 we obtain

$$\int_{\mathbb{R}^3} \frac{F(x, u_n)}{\|u_n\|_E^4} \, dx = \frac{1}{2\|u_n\|_E^2} + \frac{\int_{\mathbb{R}^3} \phi_{u_n} u_n^2 \, dx - c}{4\|u_n\|_E^4} + o(\|u_n\|_E^{-4}) \leq a_5 < \infty, \tag{4.9}$$

where a_5 is a positive constant. We consider this situation, $\Omega := \{x \in \mathbb{R}^3 : w(x) \neq 0\}$, by $(C3)$, for all $x \in \Omega$,

$$\frac{F(x, u_n)}{\|u_n\|_E^4} = \frac{F(x, u_n)}{|u_n|^4}w_n^4(x) \to +\infty \quad (n \to \infty).$$

Since $|\Omega| > 0$, using Fatou's Lemma, we obtain

$$\int_{\mathbb{R}^3} \frac{F(x, u(x)_n)}{\|u(x)_n\|_E^4} \, dx \to +\infty \quad (n \to \infty).$$

This contradicts (4.9).

On the another hand, if $w(x) = 0$, we can define a sequence $\{t_n\} \subset \mathbb{R}$:

$$I(t_n u_n) = \max_{t \in [0,1]} I(t u_n).$$

Fix any $m > 0$, let $\overline{w}_n = \sqrt{4m}\frac{u_n}{\|u_n\|_E} = \sqrt{4m}w_n$. By $(C2)$,

$$|f(x, z)| \leq a_2|z| + a_2|z|^{p-1},$$

for a.e. $x \in \mathbb{R}^3$ and all $z \in \mathbb{R}$. By the equality $F(x, z) = \int_0^1 f(x, tz) z \, dt$ we obtain

$$F(x, z) \le \frac{a_2}{2} |z|^2 + a_6 |z|^p \tag{4.10}$$

for any $x \in \mathbb{R}^3$ and all $z \in \mathbb{R}$, where $a_6 = \frac{a_2}{p}$. Due to (3.5), we obtain

$$\lim_{n \to \infty} \int_{\mathbb{R}^3} F(x, \overline{w}_n) \, dx \le \lim_{n \to \infty} \left(\frac{a_2}{2} \int_{\mathbb{R}^3} |\overline{w}_n|^2 \, dx + a_6 \int_{\mathbb{R}^3} |\overline{w}_n|^p \, dx \right) = 0.$$

Then for n large enough,

$$\begin{aligned}
I(t_n u_n) &\ge I(\overline{w}_n) \\
&= 2m + \frac{1}{4} \int_{\mathbb{R}^3} \phi_{\overline{w}_n} \overline{w}_n^2 \, dx - \int_{\mathbb{R}^3} F(x, \overline{w}_n) \, dx \ge m.
\end{aligned} \tag{4.11}$$

Due to (4.11), $\lim_{n \to \infty} I(t_n u_n) = +\infty$. Since $I(0) = 0$, and $I(u_n) \to c$, then $0 < t_n < 1$ and we have

$$\int_{\mathbb{R}^3} \left(\nabla t_n u_n \nabla t_n u_n + V(x) t_n u_n t_n u_n + \phi_{t_n u_n} t_n u_n t_n u_n - f(x, t_n u_n) t_n u_n \right) dx$$

$$= \langle I'(t_n u_n), t_n u_n \rangle$$

$$= t_n \frac{d}{dt} \Big|_{t=t_n} I(t u_n) = 0,$$

if n large enough. Thus, by (C4) we obtain

$$\begin{aligned}
I(u_n) &- \frac{1}{4} \langle I'(u_n), u_n \rangle \\
&= \frac{1}{4} \|u_n\|_E^2 + \int_{\mathbb{R}^3} \left[\frac{1}{4} f(x, u_n) u_n - F(x, u_n) \right] dx \\
&= \frac{1}{4} \|u_n\|_E^2 + \frac{1}{4} \int_{\mathbb{R}^3} H(x, u_n) \, dx \\
&\ge \frac{1}{4\theta} \|t_n u_n\|_E^2 + \frac{1}{4\theta} \int_{\mathbb{R}^3} H(x, t_n u_n) \, dx \\
&= \frac{1}{4\theta} \|t_n u_n\|_E^2 + \frac{1}{\theta} \int_{\mathbb{R}^3} \left[\frac{1}{4} f(x, t_n u_n) t_n u_n - F(x, t_n u_n) \right] dx \\
&= \frac{1}{\theta} I(t_n u_n) - \frac{1}{4\theta} \langle I'(t_n u_n), t_n u_n \rangle \to +\infty.
\end{aligned}$$

This contradicts (4.7). So $\{u_n\}$ is bounded. Going if necessary to a subsequence, we can assume that $u_n \rightharpoonup u$ in E. In view of Sobolev compact embedding, $u_n \to u$ in $L^s(\mathbb{R}^3)$ for any $s \in [2, 2^*)$. By (4.4), we easily get

$$\begin{aligned}
\|u_n - u\|_E^2 &= \langle I'(u_n) - I'(u), u_n - u \rangle + \int_{\mathbb{R}^3} (f(x, u_n) - f(x, u))(u_n - u) \, dx \\
&\quad - \int_{\mathbb{R}^3} (\phi_{u_n} u_n - \phi_u u)(u_n - u) \, dx.
\end{aligned}$$

It is clear that

$$\langle I'(u_n) - I'(u), u_n - u \rangle \to 0.$$

According to assumptions $(C2)$, there exists $a_6 > 0$ such that

$$f(x, u) \le \frac{a_2}{2}|u| + a_6|u|^{p-1}$$

for a.e. $x \in \mathbb{R}^3$, and all $z \in \mathbb{R}$. Using the Hölder inequality, we obtain

$$\int_{\mathbb{R}^3} (f(x, u_n) - f(x, u))(u_n - u)\,dx$$

$$\le \int_{\mathbb{R}^3} \left[\frac{a_2}{2}(|u_n| + |u|) + a_6\left(|u_n|^{p-1} + |u|^{p-1}\right) \right] |u_n - u|\,dx$$

$$\le \frac{a_2}{2}\left(\|u_n\|_{L^2}^2 + \|u\|_{L^2}^2\right)\|u_n - u\|_{L^2}^2 + a_6\left(\|u_n\|_{L^p}^{p-1} + \|u\|_{L^p}^{p-1}\right)\|u_n - u\|_{L^p}.$$

Since $u_n \to u$ in $L^s(\mathbb{R}^3)$ for any $s \in [2, 2^*)$, we have

$$\int_{\mathbb{R}^3} (f(x, u_n) - f(x, u))(u_n - u)\,dx \to 0, \quad \text{as } n \to \infty.$$

By the Hölder inequality, Sobolev inequality and Lemma 4.1, we have

$$\left| \int_{\mathbb{R}^3} \phi_{u_n} u_n(u_n - u)\,dx \right| \le \|\phi_{u_n} u_n\|_{L^2}\|u_n - u\|_{L^2}$$

$$\le \|\phi_{u_n}\|_{L^6}\|u_n\|_{L^3}\|u_n - u\|_{L^2}$$

$$\le a_8\|\phi_{u_n}\|_{D^{1,2}}\|u_n\|_{L^3}\|u_n - u\|_{L^2}$$

$$\le a_4 a_8\|u_n\|_{L^{12/5}}^2\|u_n\|_{L^3}\|u_n - u\|_{L^2},$$

where $a_8 > 0$ is a constant. Again using $u_n \to u$ in $L^s(\mathbb{R}^3)$ for any $s \in [2, 2^*)$, we have

$$\int_{\mathbb{R}^3} \phi_{u_n} u_n(u_n - u)\,dx \to 0, \quad \text{as } n \to \infty.$$

Similarly, we obtain

$$\int_{\mathbb{R}^3} \phi_u u(u_n - u)\,dx \to 0, \quad \text{as } n \to \infty.$$

Thus,

$$\int_{\mathbb{R}^3} (\phi_{u_n} u_n - \phi_u u)(u_n - u)\,dx \to 0, \quad \text{as } n \to \infty,$$

so that $\|u_n - u\|_E \to 0$. We get that $I(u)$ satisfies Cerami condition. $\qquad \square$

Proof of Theorem 4.1. Due to Lemma 4.3, $I(u)$ satisfies Cerami condition. Next, we verify that $I(u)$ satisfies the rest conditions of Theorem 1.11.

First, we verify that $I(u)$ satisfies (Φ1). It follows from (C3) that for any $M > 0$, there exists $\delta(M) > 0$, such that for all $x \in \mathbb{R}^3$, $|z| \geq \delta$, we have

$$F(x, z) \geq \frac{1}{4}M|z|^4. \tag{4.12}$$

Taking $\widetilde{M} := \sup_{|z| < \delta}\left(\frac{1}{4}M|z|^4 - \frac{F(x,z)}{|z|^2}\right)$, then by (4.12) we obtain

$$F(x, z) \geq \frac{1}{4}M|z|^4 - \widetilde{M}|z|^2$$

for a.e. $x \in \mathbb{R}^3$, and all $z \in \mathbb{R}$. Hence we have

$$I(u) \leq \frac{1}{2}\|u\|_E^2 + \frac{a_4}{4}\|u\|_E^4 - \frac{1}{4}M\|u\|_{L^4}^4 + \widetilde{M}\|u\|_{L^2}^2.$$

Since, on the finitely dimensional space W_k all norms are equivalent, we have that

$$I(u) \leq \frac{1}{2}\|u\|_E^2 + \frac{a_4}{4}\|u\|_E^4 - \frac{1}{4}Ma_{10}\|u\|_E^4 + \widetilde{M}a_{10}\|u\|_E^2,$$

where a_{10} is a constant. Now since $\frac{a_4}{4} - \frac{1}{4}Ma_{10} < 0$, when M is large enough, it follows that

$$a_k := \max_{u \in W_k, \|u\|_E = \rho_k} I(u) \leq 0$$

for some $\rho_k > 0$ large enough.

Secondly, we prove that $I(u)$ satisfies (Φ2). Due to (4.10), we have

$$I(u) \geq \frac{1}{2}\|u\|_E^2 - \varepsilon\|u\|_{L^2}^2 - a_6\|u\|_{L^p}^p$$

$$\geq \left(\frac{1}{2} - \frac{\varepsilon}{a_1}\right)\|u\|_E^2 - a_6\beta_k{}^p\|u\|_E^p,$$

where a_1 is a lower bound of $V(x)$ from (C1) and β_k are defined in Lemma 4.2. Choosing $r_k := (a_6 p \beta_k^p)^{1/(2-p)}$, we obtain

$$b_k = \inf_{u \in Z_k, \|u\|_E = r_k} I(u)$$

$$\geq \inf_{u \in Z_k, \|u\|_E = r_k}\left[\left(\frac{1}{2} - \frac{\varepsilon}{a_1}\right)\|u\|_E^2 - a_6\beta_k{}^p\|u\|_E^p\right]$$

$$\geq \left(\frac{1}{2} - \frac{\varepsilon}{a_1} - \frac{1}{p}\right)(a_6 p \beta_k^p)^{\frac{2}{2-p}}.$$

Because $\beta_k \to 0$ as $k \to 0$ and $p > 2$, we have

$$b_k \geq \left(\frac{1}{2} - \frac{\varepsilon}{a_1} - \frac{1}{p}\right)(a_6 p \beta_k^p)^{\frac{2}{2-p}} \to +\infty$$

for enough small ε. This proves (Φ2). Now, we apply Theorem 1.11 to complete the proof of Theorem 4.1. $\square$

Notes and Comments

In recent years, system (4.1) with $V(x) \equiv 1$ or being radially symmetric, had been widely studied under various conditions on f, see for example [225–232]. Specially, in [226, 227] it was proved the existence of a sequence of radial solutions for system (4.1) by using the Symmetric Mountain Pass Theorem in [28]. The case of nonradial potential $V(x)$ had been considered in [163], when f was asymptotically linear at infinity, and in [225, 233], when f was superlinear at infinity. Moreover, in [233], the authors considered system (4.1) with periodic potential $V(x)$, and the existence of infinitely many geometrically distinct solutions had been proved by the nonlinear superposition principle established in [234]. By the way, we would like to point out that nonexistence results for (4.1) can be found in [163, 226, 229, 230, 235]. This section was obtained in [236].

4.2.2   Multiple Solutions with Asymptotically Linear Non-linearities

In this section, we are concerned with the existence of two positive solutions for the following nonhomogeneous Schrödinger–Poisson system

$$\begin{cases} -\Delta u + u + K(x)\phi(x)u = a(x)f(u) + h(x), & x \in \mathbb{R}^3, \\ -\Delta \phi = K(x)u^2, & x \in \mathbb{R}^3, \\ u > 0, & x \in \mathbb{R}^3, \end{cases} \qquad (4.13)$$

where K, $h \in L^2(\mathbb{R}^3)$, K, $h \geq (\not\equiv)0$, a is a nonnegative function, the function $f \in C(\mathbb{R}, \mathbb{R}^+)$ and F is the primitive function of f with $F(t) = \int_0^t f(s)ds$.

Here, we state our main results as follows.

Theorem 4.2. *Suppose that K, $h \in L^2(\mathbb{R}^3)$, K, $h \geq (\not\equiv)0$, and the following conditions hold:*

(C6) $f \in C(\mathbb{R}, \mathbb{R}^+)$, $f(0) = 0$, and $f(t) \equiv 0$ for $t < 0$.
(C7) $\lim_{t \to 0} \frac{f(t)}{t} = 0$.
(C8) $\lim_{t \to +\infty} \frac{f(t)}{t} = l$ with $0 < l < +\infty$.
(C9) $a(x)$ *is a positive continuous function and there exists $R_0 > 0$ such that*

$$\sup\{f(t)/t : \ t > 0\} < \inf\{1/a(x) : \ |x| \geq R_0\}.$$

(C10) *There exists a constant $\beta \in (0, 1)$ such that*

$$(1 - \beta)l > \mu^* := \inf\left\{ \int_{\mathbb{R}^3} (|\nabla u|^2 + u^2)dx : \ u \in H^1(\mathbb{R}^3, \mathbb{R}^+), \right.$$

$$\left. \int_{\mathbb{R}^3} a(x)F(u)dx \geq \frac{l}{2}, \int_{\mathbb{R}^3} K(x)\phi_u(x)u^2 dx < 2\beta l \right\}.$$

Then there exists $m > 0$ such that system (4.13) has at least two positive solutions u_0, $u_1 \in H^1(\mathbb{R}^3)$ satisfying $I(u_0) < 0$ and $I(u_1) > 0$ if $\|h\|_2 < m$.

Remark 4.3. In this section, $K \not\equiv 0$ and $h(x) \not\equiv 0$, system (4.13) is nonhomogeneous and the existence of two positive solutions for system (4.13) has been proved in our Theorem 4.2. Note that the first local minimum solution exists due to the homogeneous term which is looked a small perturbation because $\|h\|_2 < m$. Moreover, the second solution u_1 is the mountain pass solution with the positive energy. Furthermore, functions K, a, f which satisfy the above conditions of Theorem 4.2 exist. For example, for any $R_0 > 0$ and $r > 0$, taking $\psi \in C_0^\infty(\mathbb{R}^3, [0, 1])$ such that $\psi(x) = 1$ if $|x| \leq r$, $\psi(x) = 0$ if $|x| \geq 2r$ and $|\nabla \psi(x)| \leq \frac{C}{r}$ for all $x \in \mathbb{R}^3$, where $C > 0$ is an arbitrary constant independent of x, and $K \in L^2(\mathbb{R}^3)$ such that $K(x) \geq 0$ for all $x \in \mathbb{R}^3$, $K(x) \not\equiv 0$ and $\|K\|_2^2 \leq \frac{9}{2 \times 32^2 \pi^2} S^{-2} \overline{S}^{-4} R_0^{-2} \left(C^2 + R_0^2\right)^{-2}$, where S, $\overline{S}$ are also seen in Section 2. Let

$$a(x) = \begin{cases} 1000/(1 + |x|), & \text{if } |x| \leq \frac{R_0}{2}, \\ 1/(1 + R_0), & \text{if } |x| \geq R_0, \end{cases}$$

and

$$f(t) = \begin{cases} R_0 t^2/(1 + t), & \text{if } t > 0, \\ 0, & \text{if } t \leq 0. \end{cases}$$

These functions are also seen in Remark 1.1 of [193].

Theorem 4.3. *Suppose that K, $h \in L^2(\mathbb{R}^3)$, $a \in L^3(\mathbb{R}^3)$. Let K, h and $a \geq (\not\equiv)0$. Assume (C6), (C7) and the following conditions hold:*

(C11) $\lim_{t \to +\infty} \frac{f(t)}{t^3} = l$ *with $0 < l < +\infty$.*

(C12) $\frac{F(t)}{t^4}$ *is nondecreasing for $t > 0$.*

(C13) *There exists a constant $\beta \in (0, 1)$ such that*

$$(1 - \beta)l > \mu^* := \inf \left\{ \int_{\mathbb{R}^3} (|\nabla u|^2 + u^2)dx : u \in H^1(\mathbb{R}^3, \mathbb{R}^+), \right.$$

$$\left. \int_{\mathbb{R}^3} a(x)u^4 dx \geq 1, \int_{\mathbb{R}^3} K(x)\phi_u(x)u^2 dx < \beta l \right\}.$$

Then there exists $m > 0$ such that system (4.13) has at least two positive solutions u_0, $u_1 \in H^1(\mathbb{R}^3)$ satisfying $I(u_0) < 0$ and $I(u_1) > 0$ if $\|h\|_2 < m$.

Remark 4.4. In Theorem 4.2, f is superlinear at zero and asymptotical 3-linear at infinity, of course, is also superlinear at infinity. We usual need the (AR) condition as in (4.64) with $\theta \in \left(0, \frac{1}{4}\right)$, to deal with this superlinear case (seen [218] and the references herein). But Here, (C12) only satisfies condition (4.64) with $\theta = \frac{1}{4} \notin \left(0, \frac{1}{4}\right)$. Furthermore, Since f is asymptotical 3-linear at infinity, (C9) is not meaning. We try to replaced (C9) by (C14) as follows:

(C14) $a(x)$ is a positive continuous function and there exists $R_0 > 0$ such that

$$\sup\{f(t)/t^3 : t > 0\} < \inf\{1/a(x) : |x| \geq R_0\}.$$

We find that other conditions are needed to prove our result. So, we cannot consider (C14). In order to obtain the compact result: $\int_{|x| \geq R}(|\nabla u_n|^2 + |u_n|^2)dx \leq \varepsilon$, we assume that $a \in L^3(\mathbb{R}^3)$ and $a \geq (\neq)0$.

Remark 4.5. It is not difficult to find some functions K, a, f satisfying conditions of Theorem 4.3. For example, for any $r > 0$, taking $\psi \in C_0^\infty(\mathbb{R}^3, [0,1])$ such that $\psi(x) = 1$ if $|x| \leq r$, $\psi(x) = 0$ if $|x| \geq 2r$ and $|\nabla\psi(x)| \leq \frac{C}{r}$ for all $x \in \mathbb{R}^3$, where $C > 0$ is an arbitrary constant independent of x, and $K \in L^2(\mathbb{R}^3)$ such that $K(x) \geq 0$ for all $x \in \mathbb{R}^3$, $K(x) \neq 0$ and $\|K\|_2^2 \leq \frac{9}{2 \times 32^2 \pi^2} S^{-2} \overline{S}^{-4} R_0^{-2} \left(C^2 + R_0^2\right)^{-2}$ Setting $a(x) = \frac{3\sqrt[3]{1+R_0}}{4\pi R_0^3} \frac{1}{\sqrt[3]{1+|x|}}$ if $|x| \leq r$, $a(x) = 0$ if $|x| \geq r$. Let $f(t) = t^3$ if $t \geq 0$ and $f(t) = 0$ if $t < 0$. Clearly, f satisfies (C6), (C7), (C11) and (C12) and $l = 1 \in (0, +\infty)$. Taking $\beta = \frac{1}{2}$. Furthermore, for any $r < R_0$, we have

$$\int_{\mathbb{R}^3} a(x)\psi(x)^4 dx \geq \frac{3\sqrt[3]{1 + R_0}}{4\pi R_0^3} \int_{|x| \leq r} \frac{1}{\sqrt[3]{1 + |x|}} dx$$

$$\geq \frac{3\sqrt[3]{1 + R_0}}{4\pi R_0^3} \frac{1}{\sqrt[3]{1 + R_0}} \int_{|x| \leq R_0} dx$$

$$= \frac{3\sqrt[3]{1 + R_0}}{4\pi R_0^3} \frac{1}{\sqrt[3]{1 + R_0}} \frac{4\pi}{3} R_0^3 = 1,$$

$$\int_{\mathbb{R}^3} (|\nabla\psi|^2 + |\psi|^2)dx \leq \int_{|x| \leq 2r} \frac{C^2}{r^2} dx + \int_{|x| \leq 2r} dx$$

$$\leq \left(1 + \frac{C^2}{r^2}\right) \frac{32\pi}{3} r^3$$

$$= \frac{32\pi}{3} r \left(C^2 + r^2\right). \tag{4.14}$$

This and (4.21) in Section 2 yield

$$\int_{\mathbb{R}^3} K(x)\phi_\psi \psi^2 dx \leq S^2 \overline{S}^4 \|K\|_2^2 \|\psi\|^4$$

$$\leq S^2 \overline{S}^4 \|K\|_2^2 \frac{32^2 \pi^2}{9} r^2 \left(C^2 + r^2\right)^2$$

$$\leq S^2 \overline{S}^4 \|K\|_2^2 \frac{32^2 \pi^2}{9} R_0^2 \left(C^2 + R_0^2\right)^2$$

$$\leq \beta l = \frac{1}{2}.$$

Taking $R_0 = 1$, $r = \frac{1}{8} R_0 = \frac{1}{8}$ and $C = \frac{r}{4} = \frac{1}{32}$. Moreover, in view of the definition of μ^* and (4.14), one has

$$\mu^* \leq \int_{\mathbb{R}^3} (|\nabla \psi|^2 + |\psi|^2) dx \leq \frac{32\pi}{3} r \left(C^2 + r^2\right) < \frac{R_0}{2} = (1 - \beta)l.$$

So, condition (C13) holds.

As we know, in order to obtain two different solutions, for the asymptotically linear case, the method is standard. Precisely, similar to [237], by the Ekeland's variational principle [22], it is not difficult to get a weak solution u_0 for $\|h\|_2$ suitably small. Moreover, u_0 is the local minimizer of I and $I(u_0) < 0$. However, under our assumptions, it seems difficult to get the Mountain Pass solution (different from the local minimum solution) of (4.13) by applying the Mountain Pass Theorem as the mentioned references because $h(x) \geq (\not\equiv)0$, the nonlinearity is asymptotically and the working space is $H^1(\mathbb{R}^3)$. We have to find new ways to show that a Cerami sequence is bounded in $H^1(\mathbb{R}^3)$. Once a Cerami sequence is bounded in $H^1(\mathbb{R}^3)$, the usual strategy is try to show this sequence converges to a solution different from u_0, but this seems not so easy because the embedding of $H^1(\mathbb{R}^3) \hookrightarrow L^p(\mathbb{R}^3)$ ($p \in (2, 6)$) is not compact. In fact, firstly, this difficulty can be avoided by restricting I to the subspace of $H^1(\mathbb{R}^3)$ such as radially functions subspace usually denoted by $H_r^1(\mathbb{R}^3)$, see [226, 228, 230, 231, 238, 239]. Especially, many authors avoid the lack of the compactness by the external potential $V(x)$, some conditions are assumed on $V(x)$ to make the working space which is a subspace of $H^1(\mathbb{R}^3)$ have compactness embeddings, see [218, 230, 240, 241]. However, for the asymptotically case, we have to find another method to verify Cerami condition. Motivated by [193, 202], we consider system (4.13) with the following two asymptotically cases at infinity: asymptotically linear and asymptotically 3-linear in this section, respectively. Secondly, in order to recover the compactness, we establish the equi-absolutely-continuity at infinity: $\int_{|x| \geq R}(|\nabla u_n|^2 + |u_n|^2)dx \leq \varepsilon$ which is also called the compactness result in this section.

System (4.13) has a variational structure. Indeed we consider the functional

$$J : \ H^1(\mathbb{R}^3) \times D^{1,2}(\mathbb{R}^3) \to \mathbb{R}$$

defined by

$$J(u, \phi) = \frac{1}{2}\|u\|^2 - \frac{1}{4}\int_{\mathbb{R}^3} |\nabla \phi|^2 dx + \frac{1}{2}\int_{\mathbb{R}^3} K(x)\phi u^2 dx - \int_{\mathbb{R}^3} a(x)F(u)dx$$
$$- \int_{\mathbb{R}^3} h(x)u dx.$$

Evidently, the action functional J belongs to $C^1(H^1(\mathbb{R}^3) \times D^{1,2}(\mathbb{R}^3), \mathbb{R})$ and the partial derivatives in (u, ϕ) are given, for $\xi \in H^1(\mathbb{R}^3)$ and $\eta \in D^{1,2}(\mathbb{R}^3)$, by

$$\left\langle \frac{\partial J}{\partial u}(u, \phi), \xi \right\rangle = \int_{\mathbb{R}^3} (\nabla u \cdot \nabla \xi + u\xi + K(x)\phi u\xi - a(x)f(u)\xi - h(x)\xi)dx,$$

$$\left\langle \frac{\partial J}{\partial \phi}(u, \phi), \eta \right\rangle = \frac{1}{2}\int_{\mathbb{R}^3} (-\nabla \phi \cdot \nabla \eta + K(x)u^2\eta)dx.$$

Thus, the pair (u, ϕ) is a weak solution of system (4.13) if and only if it is a critical point of J in $H^1(\mathbb{R}^3) \times D^{1,2}(\mathbb{R}^3)$.

For all u in $H^1(\mathbb{R}^3)$, the Lax–Milgram theorem implies that there exists a unique $\phi_u \in D^{1,2}(\mathbb{R}^3)$ such that $-\Delta\phi_u = K(x)u^2$ in a weak sense. Then, insert ϕ_u into the first equation of (4.13), we have

$$-\Delta u + u + K(x)\phi_u(x)u = a(x)f(u) + h(x). \tag{4.15}$$

That is, system (4.13) can be easily transformed to a nonlinear Schrödinger equation (4.15) with a non-local term. Moreover, we can write an integral expression for ϕ_u in the explicit form:

$$\phi_u(x) = \int_{\mathbb{R}^3} \frac{K(y)u(y)^2}{|x - y|}dy \tag{4.16}$$

for any $u \in H^1(\mathbb{R}^3)$. So, we can consider the functional $I : H^1(\mathbb{R}^3) \to \mathbb{R}$ defined by $I(u) = J(u, \phi_u)$. After multiplying $-\Delta\phi_u = K(x)u^2$ by ϕ_u and integration by parts, we obtain

$$\int_{\mathbb{R}^3} |\nabla\phi_u|^2 dx = \int_{\mathbb{R}^3} K(x)\phi_u(x)u^2 dx. \tag{4.17}$$

Therefore, the reduced functional takes the form

$$\tilde{I}(u) = \frac{1}{2}\|u\|^2 + \frac{1}{4}\int_{\mathbb{R}^3} K(x)\phi_u(x)u^2 dx - \int_{\mathbb{R}^3} a(x)F(u)dx - \int_{\mathbb{R}^3} h(x)u dx,$$

$u \in H^1(\mathbb{R}^3)$.

Recall the Sobolev's inequalities with the best constant S and $\overline{S}$

$$\|v\|_6 \leq S\|v\|_{D^{1,2}(\mathbb{R}^3)}, \quad \|v\|_6 \leq \overline{S}\|v\|, \tag{4.18}$$

together with (4.17) and the Hölder's inequality, we have

$$\begin{aligned}
\|\phi_u\|^2_{D^{1,2}(\mathbb{R}^3)} &= \int_{\mathbb{R}^3} K(x)\phi_u(x)u^2 dx \\
&\leq \|K\|_2\|u^2\|_3\|\phi_u\|_6 \\
&= \|K\|_2\|u\|_6^2\|\phi_u\|_6 \\
&\leq S\overline{S}^2\|K\|_2\|u\|^2\|\phi_u\|_{D^{1,2}(R^3)}.
\end{aligned}$$

This yields

$$\|\phi_u\|_{D^{1,2}(\mathbb{R}^3)} \leq S\overline{S}^2\|K\|_2\|u\|^2, \tag{4.19}$$

$$\|\phi_u\|_6 \leq S\|\phi_u\|_{D^{1,2}(R^3)} \leq S^2\overline{S}^2\|K\|_2\|u\|^2. \tag{4.20}$$

Therefore, by the Hölder's inequality, (4.20) and (4.18) we have

$$\begin{aligned}
\int_{\mathbb{R}^3} K(x)\phi_u(x)u(x)^2 dx &\leq \|K\|_2\|u\|_6^2\|\phi_u\|_6 \\
&\leq S^2\overline{S}^4\|K\|_2^2\|u\|^4 \\
&:= C_0\|u\|^4. \tag{4.21}
\end{aligned}$$

Now, we shall look for the positive solution of problem (4.13). By assumption (C6), we know that to seek a nonnegative weak solution of problem (4.13) is equivalent to finding a nonzero critical point of the following functional I on $H^1(\mathbb{R}^3)$ defined by

$$I(u) = \frac{1}{2}\|u\|^2 + \frac{1}{4}\int_{\mathbb{R}^3} K(x)\phi_u(x)(u^+)^2 dx - \int_{\mathbb{R}^3} a(x)F(u^+)dx \tag{4.22}$$

$$- \int_{\mathbb{R}^3} h(x)u dx, \tag{4.23}$$

where $u^+ = \max\{u, 0\}$. Combining (4.20), (4.21), (C6)–(C8), and Lemma 3.3 in [193], I is well defined. Furthermore, I is C^1 and we have

$$\langle I'(u), v\rangle = \int_{\mathbb{R}^3} (\nabla u \cdot \nabla v + uv + K(x)\phi_u(x)u^+v - a(x)f(u^+)v - h(x)v)dx.$$

Hence, if $u \in H^1(\mathbb{R}^3)$ is a nonzero critical point of I, then (u, ϕ_u) with ϕ_u as in (4.16), is a nonnegative solution of (4.13). In fact, by (C6) and $h \geq 0$, we have $\langle I'(u), u^- \rangle = -\|u^-\|^2 - \int_{\mathbb{R}^3} h(x)u^- dx = 0$, where $u^- = \max\{-u, 0\}$. This yields that $u^- = 0$, then $u = u^+ - u^- = u^+ \geq 0$. By the strong maximum principle, the nonzero critical point of I is the positive solution for problem (4.13).

Next, we shall discuss system (4.13) with the two cases: asymptotically linear case and asymptotically cubic case at infinity, respectively.

Firstly, we prove that system (4.13) has a mountain pass type solution and a local minimum solution. For this purpose, we use a variant version of Mountain Pass Theorem, which allows us to find a so-called Cerami type (PS) sequence (Cerami sequence, in short). The properties of this kind of Cerami sequence are very helpful in showing its boundedness in the asymptotically case. The following lemmas will show that I defined in (4.23) has the so-called mountain pass geometry.

Lemma 4.4. *Suppose that K, $h \in L^2(\mathbb{R}^3)$, $K \geq (\not\equiv)0$, (C6)–(C8) and (C9) hold. Then there exist ρ, α, $m > 0$ such that $I(u)|_{\|u\|=\rho} \geq \alpha > 0$ for $\|h\|_2 < m$.*

Proof. For any $\varepsilon > 0$, it follows from (C6)–(C8) that there exists $C_\varepsilon > 0$ such that

$$|f(t)| \leq \varepsilon|t| + C_\varepsilon|t|^5 \text{ for all } t \in \mathbb{R}. \tag{4.24}$$

Therefore, we have

$$|F(t)| \leq \frac{1}{2}\varepsilon|t|^2 + \frac{C_\varepsilon}{6}|t|^6 \text{ for all } t \in \mathbb{R}. \tag{4.25}$$

Furthermore, by (C6)–(C8) and (C9), there exists $C_1 > 0$ such that

$$a(x) \leq C_1 \text{ for all } x \in \mathbb{R}^3. \tag{4.26}$$

According to (4.25), (4.26) and (4.18), we deduce

$$\left| \int_{\mathbb{R}^3} a(x)F(u^+)dx \right| \leq \frac{\varepsilon C_1}{2} \int_{\mathbb{R}^3} |u^+|^2 dx + \frac{C_1 C_\varepsilon}{6} \int_{\mathbb{R}^3} |u^+|^6 dx$$

$$\leq \frac{\varepsilon C_1}{2}\|u^+\|^2 + C_2\|u^+\|^6$$

$$\leq \frac{\varepsilon C_1}{2}\|u\|^2 + C_2\|u\|^6,$$

where $C_2 = \frac{C_1 C_\varepsilon \overline{S}^6}{6}$. Together with (4.16), $K \geq (\neq)0$, $h \in L^2(\mathbb{R}^3)$ and the Hölder inequality, one has

$$I(u) \geq \frac{1}{2}\|u\|^2 - \frac{\varepsilon C_1}{2}\|u\|^2 - C_2\|u\|^6 - \|h\|_2\|u\|_2$$

$$\geq \frac{1}{2}\|u\|^2 - \frac{\varepsilon C_1}{2}\|u\|^2 - C_2\|u\|^6 - \|h\|_2\|u\|$$

$$\geq \|u\| \left(\frac{1 - \varepsilon C_1}{2}\|u\| - C_2\|u\|^5 - \|h\|_2 \right). \tag{4.27}$$

Taking $\varepsilon = \frac{1}{2C_1}$ and setting $g(t) = \frac{1}{4}t - C_2 t^5$ for $t \geq 0$, we see that there exists $\rho = (\frac{1}{20C_2})^{\frac{1}{4}}$ such that $\max_{t \geq 0} g(t) = g(\rho) := m > 0$. Then it follows from (4.27) that there exists $\alpha > 0$ such that $I(u)|_{\|u\|=\rho} \geq \alpha > 0$ for $\|h\|_2 < m$. Of course, ρ can be chosen small enough, we can obtain the same result: there exist $\alpha > 0$, $m > 0$ such that $I(u)|_{\|u\|=\rho} \geq \alpha > 0$ for $\|h\|_2 < m$. $\square$

Lemma 4.5. *Suppose that K, $h \in L^2(\mathbb{R}^3)$, K, $h \geq (\neq)0$, (C6)–(C8) and (C9)–(C10) hold. Then there exists $v \in H^1(\mathbb{R}^3)$ with $\|v\| > \rho$, ρ is given by Lemma 4.4, such that $I(v) < 0$.*

Proof. By (C10) and $h \geq (\neq)0$, in view of the definition of μ^* and $(1 - \beta)l > \mu^*$, there is a nonnegative function $v \in H^1(\mathbb{R}^3)$ such that

$$\int_{\mathbb{R}^3} a(x)F(v)dx > \frac{l}{2}, \quad \int_{\mathbb{R}^3} K(x)\phi_v v^2 dx < 2\beta l, \quad \int_{\mathbb{R}^3} h(x)v dx \geq 0,$$

and $\mu^* \leq \|v\|^2 < (1 - \beta)l$. Then, we have

$$I(v) = \frac{1}{2}\|v\|^2 + \frac{1}{4}\int_{\mathbb{R}^3} K(x)\phi_v(x)v^2 dx - \int_{\mathbb{R}^3} a(x)F(v)dx - \int_{\mathbb{R}^3} h(x)v dx$$

$$\leq \frac{1}{2}\|v\|^2 + \frac{1}{4} \times 2\beta l - \frac{l}{2}$$

$$= \frac{1}{2}(\|v\|^2 - (1 - \beta)l) < 0.$$

Choosing $\rho > 0$ small enough in Lemma 4.4 such that $\|v\| > \rho$, then this Lemma is proved. $\square$

From Lemmas 4.4, 4.5 and Mountain Pass Theorem in [22], taking α as in Lemma 4.4 and v as in Lemma 4.5, there is a Cerami sequence $\{u_n\} \subset H^1(\mathbb{R}^3)$ such that

$$\|I'(u_n)\|_{H^{-1}}(1 + \|u_n\|) \to 0 \quad \text{and} \quad I(u_n) \to c \geq \alpha > 0 \text{ as } n \to \infty,$$

$$\tag{4.28}$$

Here, H^{-1} denotes the dual space of $H^1(\mathbb{R}^3)$ and c denotes by

$$c = \inf_{\gamma \in \tau} \max_{t \in [0,1]} I(\gamma(t)),$$

where

$$\tau = \{\gamma \in ([0,1], H^1(\mathbb{R}^3)) : \gamma(0) = 0, \gamma(1) = v\}.$$

In the following, we shall prove that I satisfies the Cerami condition, that is, the Cerami sequence $\{u_n\}$ has a convergence subsequence.

Lemma 4.6. *Suppose that* K, $h \in L^2(\mathbb{R}^3)$, K, $h \geq (\not\equiv)0$, (C6)–(C8) *and* (C9) *hold. Then* $\{u_n\}$ *defined in* (4.28) *is bounded in* $H^1(\mathbb{R}^3)$.

Proof. By contradiction, let $\|u_n\| \to \infty$. Define $w_n = u_n\|u_n\|^{-1}$. Clearly, $\{w_n\}$ is bounded in $H^1(\mathbb{R}^3)$ and there is a $w \in H^1(\mathbb{R}^3)$ such that, up to a subsequence,

$$\begin{cases} w_n \rightharpoonup w \quad \text{weakly in } H^1(\mathbb{R}^3), \\ w_n \to w \quad \text{a.e. in } \mathbb{R}^3, \\ w_n \to w \quad \text{strongly in } L^2_{loc}(\mathbb{R}^3) \end{cases}$$

as $n \to \infty$. Therefore, we obtain that $w_n^{\pm} = u_n^{\pm}\|u_n\|^{-1}$ and

$$\begin{cases} w_n^{\pm} \rightharpoonup w^{\pm} \quad \text{weakly in } H^1(\mathbb{R}^3), \\ w_n^{\pm} \to w^{\pm} \quad \text{a.e. in } \mathbb{R}^3, \\ w_n^{\pm} \to w^{\pm} \quad \text{strongly in } L^2_{loc}(\mathbb{R}^3) \end{cases}$$

as $n \to \infty$.

Firstly, we claim that w is nontrivial, that is $w \not\equiv 0$. Otherwise, if $w \equiv 0$, the Sobolev embedding implies that $w_n \to 0$ strongly in $L^2(B_{R_0})$, R_0 is given by (C9). By (C6)–(C8), there exists $C_3 > 0$ such that

$$\frac{f(t)}{t} \leq C_3 \text{ for all } t \in R. \tag{4.29}$$

Then, by (4.26) and (4.29), for all $n \in \mathbb{N}$, we have

$$0 \leq \int_{|x|<R_0} a(x)\frac{f(u_n^+)}{u_n^+}(w_n^+)^2 dx \leq C_1 C_3 \int_{|x|<R_0} (w_n^+)^2 dx$$

$$\leq C_1 C_3 \int_{|x|<R_0} w_n^2 dx \to 0.$$

This yields

$$\lim_{n \to \infty} \int_{|x|<R_0} a(x)\frac{f(u_n^+)}{u_n^+}(w_n^+)^2 dx = 0. \tag{4.30}$$

Furthermore, by (C9), there exists a constant $\theta \in (0,1)$ such that

$$\sup\{f(t)/t : t > 0\} \leq \theta \inf\{1/a(x) : |x| \geq R_0\}. \tag{4.31}$$

Then, for all $n \in \mathbb{N}$, we have

$$\int_{|x| \geq R_0} a(x) \frac{f(u_n^+)}{u_n^+}(w_n^+)^2 dx \leq \theta \int_{|x| \geq R_0} (w_n^+)^2 dx \leq \theta < 1. \tag{4.32}$$

Combining (4.30) and (4.32), we obtain

$$\limsup_{n \to \infty} \int_{\mathbb{R}^3} a(x) \frac{f(u_n^+)}{u_n^+}(w_n^+)^2 dx < 1. \tag{4.33}$$

By (4.28), we get

$$0 \leq |\langle I'(u_n), u_n \rangle| \leq \|I'(u_n)\|_{H^{-1}}(1 + \|u_n\|) \to 0 \tag{4.34}$$

as $n \to \infty$. Together with $\|u_n\| \to \infty$ as $n \to \infty$, it follows that

$$\frac{\langle I'(u_n), u_n \rangle}{\|u_n\|^2} = o(1),$$

that is,

$$o(1) = \|w_n\|^2 + \int_{\mathbb{R}^3} K(x)\phi_{w_n}(u_n^+)^2 dx - \int_{\mathbb{R}^3} a(x) \frac{f(u_n^+)}{u_n^+}(w_n^+)^2 dx$$

$$\geq 1 - \int_{\mathbb{R}^3} a(x) \frac{f(u_n^+)}{u_n^+}(w_n^+)^2 dx,$$

where, and in what follows, $o(1)$ denotes a quantity which goes to zero as $n \to \infty$. Therefore, we deduce that

$$\int_{\mathbb{R}^3} a(x) \frac{f(u_n^+)}{u_n^+}(w_n^+)^2 dx + o(1) \geq 1,$$

which contradicts (4.33). So, $w \not\equiv 0$.

Furthermore, because $\|u_n\| \to \infty$ as $n \to \infty$, it follows from (4.34) that

$$\frac{\langle I'(u_n), u_n \rangle}{\|u_n\|^4} = o(1),$$

that is,

$$o(1) = \frac{1}{\|u_n\|^2} + \int_{\mathbb{R}^3} K(x)\phi_{w_n}(w_n^+)^2 dx - \frac{1}{\|u_n\|^2} \int_{\mathbb{R}^3} a(x) \frac{f(u_n^+)}{u_n^+}(w_n^+)^2 dx.$$

Together with (4.26) and (4.29), one has

$$\int_{\mathbb{R}^3} K(x)\phi_{w_n}(w_n^+)^2 dx = o(1). \tag{4.35}$$

By the same method of Lemma 3.3 in [193], we can prove

$$\int_{\mathbb{R}^3} K(x)\phi_{w_n}(w_n^+)^2 dx = \int_{\mathbb{R}^3} K(x)\phi_w(w^+)^2 dx + o(1). \qquad (4.36)$$

Here we omit its proof. Equations (4.35) and (4.36) show that

$$\int_{\mathbb{R}^3} K(x)\phi_w(w^+)^2 dx = 0,$$

which implies $w^+ \equiv 0$. By (4.34), (C6), $h \in L^2(\mathbb{R}^3)$ and $\|u_n\| \to \infty$ as $n \to \infty$, we obtain

$$0 = \lim_{n\to\infty} \frac{\langle I'(u_n), u_n^-\rangle}{\|u_n\|^2} = -\lim_{n\to\infty} \|w_n^-\|^2.$$

This and $w_n^- \rightharpoonup w^-$ weakly in $H^1(\mathbb{R}^3)$ imply that $w^- = 0$. Thus $w = w^+ - w^- = 0$. That is a contradiction. Therefore, $\{u_n\}$ is a bounded in $H^1(\mathbb{R}^3)$. $\qquad \square$

From the idea of Lemma 3.4 in [193] or Lemma 2.1 in [202], we have the following lemma. The proof of this lemma follows from Lemma 3.4 in [193] (also seen Lemma 2.1 in [202]). Here we write it for the completeness because this lemma plays a key role to prove our Theorem 4.2.

Lemma 4.7. *Suppose that K, $h \in L^2(\mathbb{R}^3)$, K, $h \geq (\not\equiv)0$, (C6)–(C8), and (C9) hold. Then for any $\varepsilon > 0$, there exist $R(\varepsilon) > R_0$ and $n(\varepsilon) > 0$ such that $\{u_n\}$ defined in (4.28) satisfies $\int_{|x|\geq R}(|\nabla u_n|^2 + |u_n|^2)dx \leq \varepsilon$ for $n > n(\varepsilon)$ and $R \geq R(\varepsilon)$.*

Proof. Let $\xi_R : \mathbb{R}^3 \to [0,1]$ be a smooth function such that

$$\xi_R(x) = \begin{cases} 0, & 0 \leq |x| \leq R/2, \\ 1, & |x| \geq R. \end{cases} \qquad (4.37)$$

Moreover, there exists a constant C_4 independent of R such that

$$|\nabla\xi_R(x)| \leq \frac{C_4}{R} \text{ for all } x \in \mathbb{R}^3. \qquad (4.38)$$

Then, for all $n \in \mathbb{N}$ and $R \geq R_0$, by (4.37), (4.38) and the Hölder inequality, we have

$$\int_{\mathbb{R}^3} |\nabla(u_n\xi_R)|^2 dx$$

$$\leq \int_{\mathbb{R}^3} |\nabla u_n|^2 |\xi_R|^2 dx + \int_{\mathbb{R}^3} |u_n|^2 |\nabla\xi_R|^2 dx + 2\int_{\mathbb{R}^3} |u_n||\xi_R||\nabla u_n||\nabla\xi_R| dx$$

$$\leq \int_{R/2<|x|<R} |\nabla u_n|^2 dx + \int_{|x|>R} |\nabla u_n|^2 dx + \frac{C_4^2}{R^2}\int_{\mathbb{R}^3} |u_n|^2 dx$$

$$+2\left(\int_{\mathbb{R}^3} |\nabla u_n|^2 |\xi_R^2| dx\right)^{\frac{1}{2}}\left(\int_{\mathbb{R}^3} |u_n|^2 |\nabla\xi_R|^2 dx\right)^{\frac{1}{2}}$$

$$\leq \int_{R/2<|x|<R} |\nabla u_n|^2 dx + \int_{|x|>R} |\nabla u_n|^2 dx + \frac{C_4^2}{R^2}\int_{\mathbb{R}^3} |u_n|^2 dx$$

$$+2\left(\int_{R/2<|x|<R} |\nabla u_n|^2 dx + \int_{|x|>R} |\nabla u_n|^2 dx\right)^{\frac{1}{2}}\left(\frac{C_4^2}{R^2}\int_{\mathbb{R}^3} |u_n|^2 dx\right)^{\frac{1}{2}}$$

$$\leq \left(2 + \frac{C_4^2}{R^2} + \frac{2\sqrt{2}C_4}{R}\right)\|u_n\|^2$$

$$\leq \left(2 + \frac{C_4^2}{R_0^2} + \frac{2\sqrt{2}C_4}{R_0}\right)\|u_n\|^2.$$

This implies that

$$\|u_n\xi_R\| \leq C_5\|u_n\| \tag{4.39}$$

for all $n \in \mathbb{N}$ and $R \geq R_0$, where $C_5 = (3 + \frac{C_4^2}{R_0^2} + \frac{2\sqrt{2}C_4}{R_0})^{\frac{1}{2}}$. From Lemma 4.6, we know that $\{u_n\}$ is bounded in $H^1(\mathbb{R}^3)$. Together with (4.28), we obtain that $I'(u_n) \to 0$ in $H^{-1}(\mathbb{R}^3)$. Moreover, for $\varepsilon > 0$, there exists $n(\varepsilon) > 0$ such that

$$\langle I'(u_n), \xi_R u_n\rangle \leq C_5\|I'(u_n)\|_{H^{-1}(R^3)}\|u_n\| \leq \frac{\varepsilon}{4}$$

for $n > n(\varepsilon)$ and $R > R_0$. Note that

$$\langle I'(u_n), \xi_R u_n\rangle = \int_{\mathbb{R}^3} (|\nabla u_n|^2 + |u_n|^2)\xi_R dx + \int_{\mathbb{R}^3} u_n\nabla u_n \cdot \nabla\xi_R dx$$

$$+ \int_{\mathbb{R}^3} K(x)\phi_{u_n}(x)(u_n^+)^2\xi_R dx$$

$$- \int_{\mathbb{R}^3} a(x)f(u_n^+)u_n^+\xi_R dx - \int_{\mathbb{R}^3} h(x)u_n\xi_R dx$$

$$\leq \frac{\varepsilon}{4}.$$

This yields

$$\int_{\mathbb{R}^3} [(|\nabla u_n|^2 + |u_n|^2)\xi_R + u_n \nabla u_n \cdot \nabla \xi_R]dx$$

$$\leq \int_{\mathbb{R}^3} a(x)f(u_n^+)u_n^+\xi_R dx + \int_{\mathbb{R}^3} h(x)u_n\xi_R dx \qquad (4.40)$$

$$- \int_{\mathbb{R}^3} K(x)\phi_{u_n}(x)(u_n^+)^2\xi_R dx + \frac{\varepsilon}{4}$$

$$\leq \int_{\mathbb{R}^3} a(x)f(u_n^+)u_n^+\xi_R dx + \int_{\mathbb{R}^3} h(x)u_n\xi_R dx + \frac{\varepsilon}{4}. \qquad (4.41)$$

By (4.31), we have

$$a(x)f(u_n^+)u_n^+ \leq \theta(u_n^+)^2 \quad \text{for } \theta \in (0,1) \text{ and } |x| \geq R_0.$$

This yields

$$\int_{\mathbb{R}^3} a(x)f(u_n^+)u_n^+\xi_R dx \leq \theta \int_{\mathbb{R}^3} (u_n^+)^2\xi_R dx \leq \theta \int_{\mathbb{R}^3} u_n^2\xi_R dx \qquad (4.42)$$

for all $n \in \mathbb{N}$ and $|x| \geq R_0$. For any $\varepsilon > 0$, there exists $R(\varepsilon) \geq R_0$ such that

$$\frac{1}{R^2} \leq \frac{4\varepsilon^2}{C_4^2} \quad \text{for all } R > R(\varepsilon). \qquad (4.43)$$

Because $h \in L^2(\mathbb{R}^3)$, $h \geq 0$, there exists $\overline{\rho} = \overline{\rho}(\varepsilon)$ such that

$$\|h\|_{2,\mathbb{R}^3 \setminus B_\rho(0)} < \varepsilon, \quad \forall \rho \geq \overline{\rho}. \qquad (4.44)$$

By the Hölder inequality, (4.44), (4.37) and $\{u_n\}$ is bounded in $H^1(\mathbb{R}^3)$, we have

$$\int_{\mathbb{R}^3} h(x)u_n\xi_R dx \leq \|h(x)\xi_R\|_2\|u_n\|_2 \leq \|h(x)\|_{2,|x|>R/2}\|u_n\|_2 \leq \frac{\varepsilon}{4} (4.45)$$

for all $R > R(\varepsilon)$. By the Young inequality, (4.38) and (4.43), for all $n \in \mathbb{N}$ and $R > R(\varepsilon)$, we obtain

$$\int_{\mathbb{R}^3} |u_n \nabla u_n \cdot \nabla \xi_R|dx$$

$$= \int_{\mathbb{R}^3} \sqrt{2\varepsilon}|\nabla u_n|\frac{1}{\sqrt{2\varepsilon}}|u_n||\nabla \xi_R|dx$$

$$\leq \varepsilon \int_{\mathbb{R}^3} |\nabla u_n|^2 dx + \frac{1}{4\varepsilon} \int_{|x| \leq R} |u_n|^2 \frac{C_4^2}{R^2}dx$$

$$\leq \varepsilon \int_{\mathbb{R}^3} |\nabla u_n|^2 dx + \varepsilon \int_{|x| \leq R} |u_n|^2 dx$$

$$\leq \varepsilon \|u_n\|^2. \qquad (4.46)$$

Combining (4.41), (4.42), (4.45) and (4.46), there exists $C_6 > 0$ such that

$$\int_{\mathbb{R}^3} (|\nabla u_n|^2 + (1 - \theta)|u_n|^2)\xi_R dx \leq \frac{\varepsilon}{2} + \varepsilon \|u_n\|^2 \leq C_6 \varepsilon \quad \text{for all} \quad R > R(\varepsilon).$$

Noting that C_6 is independent of ε. So, for any $\varepsilon > 0$, we can choose $R(\varepsilon) > R_0$ and $n(\varepsilon) > 0$ such that $\int_{|x| \geq R}(|\nabla u_n|^2 + |u_n|^2)dx \leq \varepsilon$ holds. $\quad\square$

Lemma 4.8. *Suppose that K, $h \in L^2(\mathbb{R}^3)$, K, $h \geq (\not\equiv)0$, (C6)–(C8), and (C9)–(C10) hold. Then the sequence $\{u_n\}$ in (4.28) has a convergent subsequence. Moreover, u is a positive solution of problem (4.13) and $I(u) > 0$.*

Proof. By Lemma 4.6, the sequence $\{u_n\}$ in (4.28) is bounded in $H^1(\mathbb{R}^3)$. We may assume that, up to a subsequence, such that $u_n \rightharpoonup u$ weakly in $H^1(\mathbb{R}^3)$, $u_n \to u$ a.e. in $\mathbb{R}^3$ and $u_n \to u$ strongly in $L^2_{loc}(\mathbb{R}^3)$ for some $u \in H^1(\mathbb{R}^3)$. Now, we shall show that $\|u_n\| \to \|u\|$ as $n \to \infty$.
By (4.28), we have

$$\langle I'(u_n), u_n \rangle = \int_{\mathbb{R}^3} (|\nabla u_n|^2 + u_n^2 + K(x)\phi_{u_n}(x)(u_n^+)^2 - a(x)f(u_n^+)u_n^+$$

$$(4.47)$$

$$- h(x)u_n)dx = o(1), \qquad\qquad (4.48)$$

and

$$\langle I'(u_n), u \rangle = \int_{\mathbb{R}^3} (\nabla u_n \cdot \nabla u + u_n u + K(x)\phi_{u_n}(x)u_n^+ u \qquad (4.49)$$

$$- a(x)f(u_n^+)u^+ - h(x)u)dx = o(1).$$

Since $u_n \rightharpoonup u$ weakly in $H^1(\mathbb{R}^3)$, we obtain

$$\int_{\mathbb{R}^3} (\nabla u_n \cdot \nabla u + u_n u)dx = \int_{\mathbb{R}^3} (|\nabla u|^2 + |u|^2)dx + o(1). \qquad (4.50)$$

By the same argument of proof of Theorem 3.1 in [193], we have the following equalities:

$$\int_{\mathbb{R}^3} a(x)f(u_n^+)u_n^+ dx = \int_{\mathbb{R}^3} a(x)f(u_n^+)u^+ dx + o(1), \qquad (4.51)$$

and

$$\int_{\mathbb{R}^3} K(x)\phi_{u_n}(x)(u_n^+)^2 dx = \int_{\mathbb{R}^3} K(x)\phi_{u_n}(x)u_n^+ u dx + o(1). \qquad (4.52)$$

Moreover, $h \in L^2(\mathbb{R}^3)$ imply that for any $\varepsilon > 0$ there exists $\overline{\rho} = \overline{\rho}(\varepsilon)$ such that

$$\|h\|_{2, \mathbb{R}^3 \setminus B_\rho(0)} < \varepsilon, \quad \forall \rho \geq \overline{\rho}. \qquad (4.53)$$

Since $h \in L^2(\mathbb{R}^3)$, the Hölder inequality, $u_n \to u$ strongly in $L^2_{loc}(\mathbb{R}^3)$ and (4.53), we have

$$\int_{\mathbb{R}^3} h(x)u_n dx - \int_{\mathbb{R}^3} h(x)u dx$$

$$\leq \int_{\mathbb{R}^3 \backslash B_\rho(0)} |h(x)(u_n - u)| dx + \int_{B_\rho(0)} |h(x)(u_n - u)| dx$$

$$\leq \|h(x)\|_{2,\mathbb{R}^3 \backslash B_\rho(0)} \|u_n - u\|_{2,\mathbb{R}^3 \backslash B_\rho(0)} + \|h(x)\|_{2,B_\rho(0)} \|u_n - u\|_{2,B_\rho(0)}$$

$$\leq \varepsilon \|u_n - u\|_{2,\mathbb{R}^3 \backslash B_\rho(0)} + \varepsilon \|u_n - u\|_{2,B_\rho(0)}$$

$$\leq C_6 \varepsilon.$$

This yields

$$\int_{\mathbb{R}^3} h(x)u_n dx = \int_{\mathbb{R}^3} h(x)u dx + o(1). \tag{4.54}$$

By (4.47)–(4.52) and (4.54), we obtain

$$\int_{\mathbb{R}^3} (|\nabla u_n|^2 + u_n^2) dx - \int_{\mathbb{R}^3} (|\nabla u|^2 + |u|^2) dx = o(1).$$

This yields that $\|u_n\| \to \|u\|$ as $n \to \infty$ and u is a nonzero critical point of I in $H^1(\mathbb{R}^3)$ and $I(u) = c > 0$ by Mountain Pass Theorem in [22]. Therefore, u is a positive solution of problem (4.13). □

Now, we give local properties of the variational functional I, which is required by using Ekeland's variational principle.

Lemma 4.9. *Suppose that K, $h \in L^2(\mathbb{R}^3)$, K, a, $h \geq (\not\equiv)0$, (C6)–(C8) and (C9) hold. If $\|h\|_2 < m$, then there exists $u_0 \in H^1(\mathbb{R}^3)$ such that*

$$I(u_0) = \inf\{I(u) : u \in \overline{B}_\rho\} < 0, \quad \text{where } B_\rho = \{u \in H^1(\mathbb{R}^3) : \|u\| < \rho\},$$

m, ρ are given by Lemma 4.4 and u_0 is a positive solution of system (4.13).

Proof. Because $h \in L^2(\mathbb{R}^3)$, $h \geq (\not\equiv)0$, we can choose a nonnegative function $\varphi \in H^1(\mathbb{R}^3)$ such that

$$\int_{\mathbb{R}^3} h(x)\varphi dx > 0. \tag{4.55}$$

Together with (4.21), (C6), $a \geq (\not\equiv)0$ and (4.55), for $t > 0$, we have

$$I(t\varphi) = \frac{t^2}{2}\|\varphi\|^2 + \frac{1}{4}\int_{\mathbb{R}^3} K(x)\phi_{t\varphi}(x)(t\varphi)^2 dx - \int_{\mathbb{R}^3} a(x)F(t\varphi) dx$$

$$- \int_{\mathbb{R}^3} h(x)t\varphi dx$$

$$\leq \frac{t^2}{2}\|\varphi\|^2 + \frac{t^4}{4}C_0\|\varphi\|^4 - t\int_{\mathbb{R}^3} h(x)\varphi dx$$

$$\leq 0$$

for $t > 0$ small enough. Thus there exists u small enough such that $I(u) < 0$. By Lemma 4.4, we deduce that

$$c_0 := \inf_{u \in \overline{B}_\rho} I(u) < 0 < \inf_{u \in \partial \overline{B}_\rho} I(u).$$

By applying Ekeland's variational principle (Theorem 4.1 in [171]) in $\overline{B}_\rho$, there is a minimizing sequence $\{u_n\} \subset \overline{B}_\rho$ such that

(i) $c_0 \leq I(u_n) < c_0 + \dfrac{1}{n}$, (ii) $I(w) \geq I(u_n) - \dfrac{1}{n}\|w - u_n\|$ for all $w \in \overline{B}_\rho$.

Clearly, $\{u_n\}$ is a bounded (PS) sequence of I. Then, by a standard procedure, Lemmas 3.4 and 3.5 imply that there exists $u_0 \in H^1(\mathbb{R}^3)$ such that $I'(u_0) = 0$, $I(u_0) = c_0 < 0$. Moreover, $I(u_0) = c_0 < 0$ implies that $u_0 \neq 0$. Therefore, u_0 is a nonzero critical point of I, thus u_0 is a positive solution of problem (4.13). So this lemma is proved. □

Proof of Theorem 4.2. By Lemmas 4.4–4.9, we know that system (4.13) has two different positive solutions u_0 and u. Moreover, $I(u_0) = c_0 < 0$ and $I(u) > 0$. □

To obtain two positive solutions of system (4.13) with asymptotically 3-linear at infinity, we also use the same method as Theorem 4.2. We can obtain corresponding results by suitably modifying the proofs of Lemmas 4.4–4.9 as follows. Here, some proofs of the following lemmas which are the same as ones of Lemmas 4.4–4.9 are omitted.

Lemma 4.10. *Suppose that K, $h \in L^2(\mathbb{R}^3)$, $a \in L^3(\mathbb{R}^3)$, K, $a \geq (\not\equiv)0$, $h \geq 0$. Assume that* (C6), (C7) *and* (C11) *hold. Then there exist $\widetilde{\rho}$, $\widetilde{\alpha}$, $\widetilde{m} > 0$ such that $I(u)|_{\|u\|=\widetilde{\rho}} \geq \widetilde{\alpha} > 0$ for $\|h\|_2 < \widetilde{m}$.*

Proof. For any $\varepsilon > 0$, it follows from (C6), (C7) and (C11) that there exists $C'_\varepsilon > 0$ such that

$$|f(t)| \leq \varepsilon|t| + C'_\varepsilon|t|^3 \text{ for all } t \in \mathbb{R}. \tag{4.56}$$

Therefore, we have

$$|F(t)| \leq \frac{1}{2}\varepsilon|t|^2 + \frac{C'_\varepsilon}{4}|t|^4 \text{ for all } t \in \mathbb{R}. \tag{4.57}$$

According to (4.57), $a \in L^3(\mathbb{R}^3)$, $a(x) \geq (\not\equiv)0$, the Hölder inequality, and Sobolev embedding theorem, we deduce

$$
\left| \int_{\mathbb{R}^3} a(x)F(u^+)dx \right| \leq \frac{\varepsilon}{2} \int_{\mathbb{R}^3} a(x)|u^+|^2 dx + \frac{C'_\varepsilon}{4} \int_{\mathbb{R}^3} a(x)|u^+|^4 dx
$$

$$
\leq \frac{\varepsilon}{2} \int_{\mathbb{R}^3} a(x)|u|^2 dx + \frac{C'_\varepsilon}{4} \int_{\mathbb{R}^3} a(x)|u|^4 dx
$$

$$
\leq \frac{\varepsilon}{2} \|a(x)\|_3 \|u\|_3^2 + \frac{C'_\varepsilon}{4} \|a(x)\|_3 \|u\|_6^4
$$

$$
\leq \frac{\varepsilon C_7}{2} \|u\|^2 + C_8 \|u\|^4
$$

for some C_7, $C_8 > 0$. Together with (4.16), $h \in L^2(\mathbb{R}^3)$ and the Hölder inequality, one has

$$
I(u) = \frac{1}{2} \|u\|^2 + \frac{1}{4} \int_{\mathbb{R}^3} K(x)\phi_u(x)(u^+)^2 dx - \int_{\mathbb{R}^3} a(x)F(u^+)dx - \int_{\mathbb{R}^3} h(x)u dx
$$

$$
\geq \frac{1}{2} \|u\|^2 - \frac{\varepsilon C_7}{2} \|u\|^2 - C_8 \|u\|^4 - \|h\|_2 \|u\|
$$

$$
\geq \|u\| \left(\frac{1 - \varepsilon C_7}{2} \|u\| - C_8 \|u\|^3 - \|h\|_2 \right). \tag{4.58}
$$

Taking $\varepsilon = \frac{1}{2C_7}$ and setting $\tilde{g}(t) = \frac{1}{4}t - C_8 t^3$ for $t \geq 0$, we see there exists $\rho = (\frac{1}{12C_8})^{\frac{1}{2}}$ such that $\max_{t \geq 0} \tilde{g}(t) = \tilde{g}(\rho) := \tilde{m}$. Then it follows from (4.58) that there exists $\tilde{\alpha} > 0$ such that $I(u)|_{\|u\|=\tilde{\rho}} \geq \tilde{\alpha} > 0$ for $\|h\|_2 < \tilde{m}$. We also choose $\tilde{\rho}$ small enough to obtain the same result. $\square$

Lemma 4.11. *Suppose that* K, $h \in L^2(\mathbb{R}^3)$, $a \in L^3(\mathbb{R}^3)$, K, $a \geq (\not\equiv)0$, $h \geq 0$, (C6), (C7), (C11) *and* (C13) *hold. Then there exists* $\tilde{v} \in H^1(\mathbb{R}^3)$ *with* $\|\tilde{v}\| > \tilde{\rho}$, $\tilde{\rho}$ *is given by Lemma 4.10, such that* $I(\tilde{v}) < 0$.

Proof. By (C13), in view of the definition of μ^* and $(1 - \beta)l > \mu^*$, there is a nonnegative function $\tilde{v} \in H^1(\mathbb{R}^3)$ such that

$$
\int_{\mathbb{R}^3} a(x)\tilde{v}^4 dx \geq 1, \quad \int_{\mathbb{R}^3} K(x)\phi_{\tilde{v}}\tilde{v}^2 dx < \beta l, \quad \int_{\mathbb{R}^3} h(x)\tilde{v} dx > 0,
$$

and $\mu^* \leq \|\widetilde{v}\|^2 < (1-\beta)l$. Together with (C11), we have

$$\lim_{t \to +\infty} \frac{I(t\widetilde{v})}{t^4}$$

$$= \lim_{t \to +\infty} \left(\frac{1}{2t^2} \|\widetilde{v}\|^2 + \frac{1}{4} \int_{\mathbb{R}^3} K(x)\phi_{\widetilde{v}}(x)\widetilde{v}^2 dx - \int_{\mathbb{R}^3} a(x)\widetilde{v}^4 \frac{F(t\widetilde{v})}{(t\widetilde{v})^4} dx \right.$$

$$\left. - \frac{1}{t^3} \int_{\mathbb{R}^3} h(x)\widetilde{v} dx \right)$$

$$= \frac{1}{4} \int_{\mathbb{R}^3} K(x)\phi_{\widetilde{v}}(x)\widetilde{v}^2 dx - \frac{l}{4}$$

$$\leq \frac{\beta l - l}{4}$$

$$< 0.$$

Choosing $\widetilde{\rho} > 0$ small enough in Lemma 4.10 such that $\|v\| > \widetilde{\rho}$, then this Lemma is proved. $\square$

From Lemmas 4.10, 4.10 and Mountain Pass Theorem in [22], there is a sequence $\{u_n\} \subset H^1(\mathbb{R}^3)$ such that

$$\|I'(u_n)\|_{H^{-1}}(1 + \|u_n\|) \to 0 \quad \text{and} \quad I(u_n) \to \widetilde{c} \geq \widetilde{\alpha} > 0 \text{ as } n \to \infty, \quad (4.59)$$

where $\widetilde{c}$ denotes by

$$\widetilde{c} = \inf_{\gamma \in \tau} \max_{t \in [0,1]} I(\gamma(t)),$$

where

$$\tau = \{\gamma \in ([0,1], H^1(\mathbb{R}^3)) : \gamma(0) = 0, \gamma(1) = \widetilde{v}\}.$$

In the following, we shall prove that sequence $\{u_n\}$ has a convergence subsequence.

Lemma 4.12. *Suppose that K, $h \in L^2(\mathbb{R}^3)$, $a \in L^3(\mathbb{R}^3)$, K, $a \geq (\not\equiv)0$, $h \geq 0$, (C6), (C7) and (C12) hold. Then $\{u_n\}$ defined in (4.59) is bounded in $H^1(\mathbb{R}^3)$.*

Proof. By (4.59), we have

$$|\langle I'(u_n), u_n \rangle| \leq \|I'(u_n)\| \|u_n\| \leq (1 + \|u_n\|) \|I'(u_n)\|_{H^{-1}} \to 0$$

as $n \to \infty$. From (C6) and (C12), we obtain

$$f(t)t - 4F(t) \geq 0 \quad \text{for all } t \in \mathbb{R}.$$

Thus, we deduce

$$1 + \widetilde{c} \geq I(u_n) - \frac{1}{4}\langle I'(u_n), u_n \rangle$$

$$= \frac{1}{4}\|u_n\|^2 + \int_{\mathbb{R}^3} a(x)\left[\frac{1}{4}f(u_n^+)u_n^+ - F(u_n^+)\right]dx - \frac{3}{4}\int_{\mathbb{R}^3} h(x)u_n dx$$

$$\geq \frac{1}{4}\|u_n\|^2 - \frac{3}{4}\|h(x)\|_2\|u_n\|_2$$

$$\geq \frac{1}{4}\|u_n\|^2 - \frac{3}{4}\|h(x)\|_2\|u_n\| \tag{4.60}$$

for n large enough. This yields that $\{u_n\}$ is bounded in $H^1(\mathbb{R}^3)$, since $\|h\|_2 < \widetilde{m}$. $\qquad\square$

Lemma 4.13. *Suppose that K, $h \in L^2(\mathbb{R}^3)$, $a \in L^3(\mathbb{R}^3)$, K, $a \geq (\not\equiv)0$, $h \geq 0$, (C6), (C7), (C11) and (C12) hold. Then for any $\varepsilon > 0$, there exist $R(\varepsilon) > R_0$ and $n(\varepsilon) > 0$ such that $\{u_n\}$ defined in (4.59) satisfies $\int_{|x| \geq R}(|\nabla u_n|^2 + |u_n|^2)dx \leq \varepsilon$ for $n > \widetilde{n}(\varepsilon)$ and $R \geq \widetilde{R}(\varepsilon)$.*

Proof. Since $a \in L^3(\mathbb{R}^3)$ and $a(x) \geq (\not\equiv)0$, there exists $\overline{r} = \overline{r}(\varepsilon) > 0$ such that

$$\|a(x)\|_{3,\mathbb{R}^3 \setminus B_r(0)} < \varepsilon \text{ for all } \forall r > \overline{r}. \tag{4.61}$$

Let $\xi_R : \mathbb{R}^3 \to [0,1]$ be a smooth function defined by (4.37) and (4.38). By the same method of Lemma 4.7, we also obtain

$$\|u_n\xi_R\| \leq C_9\|u_n\|$$

for all $n \in \mathbb{N}$ and $R \geq \widetilde{R}_0(\varepsilon) > 2\overline{r}$. Moreover, for $\varepsilon > 0$, there exists $\widetilde{n}(\varepsilon) > 0$ such that

$$\langle I'(u_n), \xi_R u_n \rangle \leq C_9\|I'(u_n)\|_{H^{-1}(R^3)}\|u_n\| \leq \frac{\varepsilon}{4}$$

for $n > \widetilde{n}(\varepsilon)$ and $R > \widetilde{R}_0(\varepsilon) > 2\overline{r}$. By (4.56), the Hölder inequality, Sobolev embedding inequalities, (4.61) and the boundedness of u_n, we have

$$\int_{\mathbb{R}^3} a(x)f(u_n^+)u_n^+\xi_R dx \leq \varepsilon \int_{\mathbb{R}^3} a(x)(u_n^+)^2\xi_R dx + C'_\varepsilon \int_{\mathbb{R}^3} a(x)(u_n^+)^4\xi_R dx$$

$$\leq \varepsilon \int_{|x|>R/2} a(x)u_n^2 dx + C'_\varepsilon \int_{|x|>R/2} a(x)u_n^4 dx$$

$$\leq \varepsilon\|a(x)\|_{3,|x|>R/2}\|u_n\|_3^2 + C'_\varepsilon\|a(x)\|_{3,|x|>R/2}\|u_n\|_6^4$$

$$\leq \varepsilon\|a(x)\|_{3,\mathbb{R}^3 \setminus B_r(0)}\|u_n\|^2 + C'_\varepsilon\|a(x)\|_{3,\mathbb{R}^3 \setminus B_r(0)}\|u_n\|^4$$

$$\leq \varepsilon \tag{4.62}$$

for all $n \in \mathbb{N}$ and $|x| \geq \widetilde{R}_0(\varepsilon) > \bar{r}$.

Combining (4.41), (4.62), (4.45) and (4.46), there exists $C_{14} > 0$ such that

$$\int_{\mathbb{R}^3} (|\nabla u_n|^2 + |u_n|^2)\xi_R dx \leq \frac{3\varepsilon}{4} + \varepsilon\|u_n\|^2 \leq C_{14}\varepsilon \quad \text{for all } R > R(\varepsilon).$$

Noting that C_{14} is independent of ε. So, for any $\varepsilon > 0$, we can choose $R(\varepsilon) > \widetilde{R}_0$ and $\tilde{n}(\varepsilon) > 0$ such that $\int_{|x| \geq R}(|\nabla u_n|^2 + |u_n|^2)dx \leq \varepsilon$ holds. $\square$

Lemma 4.14. *Suppose that K, $h \in L^2(\mathbb{R}^3)$, $a(x) \in L^3(\mathbb{R}^3)$, K, $a \geq (\not\equiv)0$, $h \geq 0$, (C6), (C7), (C11), (C12) and (C13) hold. Then the sequence $\{u_n\}$ in (4.59) has a convergent subsequence. Moreover, I possesses a nonzero critical point $\tilde{u}$ in $H^1(\mathbb{R}^3)$, $I(\tilde{u}) > 0$ and $\tilde{u}$ is a positive solution of problem (4.13).*

Proof. By Lemma 4.6, the sequence $\{u_n\}$ in (4.59) is bounded in $H^1(\mathbb{R}^3)$. We may assume that, up to a subsequence $u_n \rightharpoonup \tilde{u}$ weakly in $H^1(\mathbb{R}^3)$ for some $\tilde{u} \in H^1(\mathbb{R}^3)$. Now, we shall show that $\|u_n\| \to \|\tilde{u}\|$ as $n \to \infty$. Under the conditions of this lemma, (4.50), (4.52) and (4.54) of Lemma 4.8 still hold. Now, we only need to prove that (4.51) still holds under conditions of Lemma 4.14. By Lemma 4.12, we know that u_n is bounded and weakly converge to $\tilde{u}$ in $H^1(\mathbb{R}^3)$. Together with Hölder inequality, (4.56) and Sobolev inequalities, we obtain

$$\int_{|x| \geq \widetilde{R}(\varepsilon)} a(x)f(u_n^+)u_n^+ dx - \int_{|x| \geq \widetilde{R}(\varepsilon)} a(x)f(u_n^+)\tilde{u}^+ dx$$

$$= \int_{|x| \geq \widetilde{R}(\varepsilon)} a(x)f(u_n^+)(u_n^+ - \tilde{u}^+)dx$$

$$\leq \int_{|x| \geq \widetilde{R}(\varepsilon)} |a(x)f(u_n^+)||u_n - \tilde{u}|dx$$

$$\leq \left(\int_{|x| \geq R(\varepsilon)} |f(u_n^+)|^2 dx\right)^{\frac{1}{2}} \left(\int_{|x| \geq R(\varepsilon)} a(x)^2|u_n - \tilde{u}|^2 dx\right)^{\frac{1}{2}}$$

$$\leq \left(\int_{|x| \geq R(\varepsilon)} (\varepsilon|u_n|^2 + 2\varepsilon C'_\varepsilon|u_n|^4 + {C'_\varepsilon}^2|u_n|^6)dx\right)^{\frac{1}{2}} \left(\int_{|x| \geq R(\varepsilon)} |u_n - \tilde{u}|^6 dx\right)$$

$$\|a(x)\|_{3,\mathbb{R}^3 \setminus B_r(0)}$$

$$\leq C_{15}\varepsilon.$$

This and the compactness of embedding $H^1(\mathbb{R}^3) \hookrightarrow L^2_{loc}(\mathbb{R}^3)$ imply (4.51). Combining (4.47), (4.50), (4.50), (4.51), (4.52) and (4.54), we have

$$\int_{\mathbb{R}^3} (|\nabla u_n|^2 + u_n^2)dx - \int_{\mathbb{R}^3} (|\nabla \tilde{u}|^2 + |\tilde{u}|^2)dx = o(1).$$

This yields that $\|u_n\| \to \|\widetilde{u}\|$ as $n \to \infty$ and $\widetilde{u}$ is a nonzero critical point of I in $H^1(\mathbb{R}^3)$ and $I(\widetilde{u}) = \widetilde{c} > 0$ by Mountain Pass Theorem in [22]. Therefore, $\widetilde{u}$ is a positive solution of problem (4.13). $\qquad\square$

Proof of Theorem 4.3. By the same method of Lemma 4.9, we can obtain system (4.13) has a local minimum positive solution $\widetilde{u}_0$ and $I(\widetilde{u}_0) < 0$. By Lemmas 4.10–4.14, we know that system (4.13) has a mountain pass solution and $I(\widetilde{u}) > 0$. Thus this theorem is proved. $\qquad\square$

Notes and Comments

When $K \equiv 0$, system (4.13) becomes into a single equation

$$-\Delta u + u = a(x)f(u) + h(x). \tag{4.63}$$

Problem (4.63) with $h(x) \equiv 0$ (homogeneous) had been studied extensively in the last decade, see [242–246] and so on. In these mentioned papers, the condition: $f(t)/t$ is nondecreasing in $t \geq 0$ is usually assumed to prove that the (PS) sequence is bounded. In the case of $h(x) \not\equiv 0$ (nonhomogeneous), Zhu in [247] proved that problem (4.63) has at least two positive solutions in $\mathbb{R}^N$ with $a(x) = 1$ and $f(t) = t^p (p \in (1, 2^* - 1)$ if $h(x)$ is small in some sense. After [247], there had been quite a lot of interesting existence results of positive solutions to problem (4.63) in $\mathbb{R}^N$, see [237, 248–250] and the references herein, the results in these papers were on the base of assuming that $f(t)$ satisfies usual Ambrosetti–Rabinowitz (AR) condition in [172]:

$$(AR) \quad 0 < F(t) = \int_0^t f(s)ds \leq \theta t f(t), \quad \text{for } t > 0 \tag{4.64}$$

and some $\theta \in \left(0, \frac{1}{2}\right)$. Wang and Zhou in [202] obtained the existence of two positive solutions for problem (4.63) in $\mathbb{R}^N (N \geq 3)$, for suitable a and h under the conditions (C6)–(C8) (seen in Theorem 4.2). From all above papers, we find that methods used in the homogeneous case are difficult to apply to the nonhomogeneous case of $h(x) \not\equiv 0$. However, in this section, we shall obtain solutions for nonhomogeneous Schrödinger-Poisson systems. Moreover, these systems have the asymptotical nonlinearity: the asymptotically linear or the asymptotically 3-linear at infinity. Clearly, the nonlinearity assumed in the following main results satisfies the (AR) condition as in (4.64) with $\theta = \frac{1}{4}$.

When $K \not\equiv 0$, Cerami and Vaira in [239] studied system (4.13) with $f(t) = |t|^{p-1}u(p \in (3,5))$ and $h(x) \equiv 0$ (homogeneous) and obtained the existence of positive ground state solutions by minimizing I restricted to the

Nehari manifold when K and a satisfy different assumptions, respectively. Sun, Chen and Nieto in [193] also studied system (4.13) with general f which is asymptotically linear at infinity ($\lim_{t\to+\infty} \frac{f(t)}{t} = l < +\infty$) and obtain the existence of positive ground state solution under suitable K and a by Mountain Pass Theorem. Wang and Zhou in [163] studied Schrödinger–Poisson systems with external potential, parameter λ and $f(x,t)$ which is asymptotically linear with respect to t at infinity

$$\begin{cases} -\Delta u + V(x)u + \lambda\phi(x)u = f(x,u), & x \in \mathbb{R}^3, \\ -\Delta\phi = u^2, \quad \lim_{|x|\to+\infty} \phi(x) = 0 & x \in \mathbb{R}^3, \end{cases} \quad (4.65)$$

and obtained a positive solution for small λ and not obtained any nontrivial solution for λ large. Zhu in [241] generalized system (4.65) with autonomous nonlinearity $f(t)$ to system (4.65) with non-autonomous nonlinearity $K(x)f(t)$ and obtained the same results as in [163] with the vanishing potential at infinity. Later, Zhu in [240] studied system (4.65) with $V(x) = \beta$ and asymptotically linear nonlinearity $f(x,t)t$ where $f(x,t)$ tends to $p(x)$ and $q(x) \in L^\infty(\mathbb{R}^3)$, respectively, as $t \to 0$ and $t \to +\infty$ and obtained existence and nonexistence results depending on the parameters β and λ. Furthermore, there are abundant results with respect to Schrödinger-Poisson systems, see [226–228, 230, 231, 238, 251] and so on. This section was obtained in [252].

4.2.3 *Quasilinear Schrödinger–Maxwell Equations*

Consider the following quasilinear Schrödinger–Poisson systems

$$\begin{cases} -\Delta u + u + K(x)\phi(x)u = a(x)f(u), & x \in \mathbb{R}^3, \\ -\operatorname{div}[(1 + \varepsilon^4|\nabla\phi|^2)\nabla\phi] = K(x)u^2, & x \in \mathbb{R}^3, \end{cases} \quad (4.66)$$

where $K \in L^2(\mathbb{R}^3)$, $K \geq (\not\equiv)0$, a is a positive bounded function, and $f \in C(\mathbb{R}, \mathbb{R}^+)$.

Here are the main results of this section.

Theorem 4.4. *Suppose that the following conditions hold:*

(C15) $f \in C(\mathbb{R}, \mathbb{R}^+)$, $f(0) = 0$, and $f(t) \equiv 0$ for $t < 0$.

(C16) $\lim_{t\to 0} \frac{f(t)}{t} = 0$.

(C17) $\lim_{t\to+\infty} \frac{f(t)}{t} = l < +\infty$.

(C18) $a(x)$ *is a positive continuous function and there exists $R_0 > 0$ such that*

$$\sup\{f(t)/t : t > 0\} < \inf\{1/a(x) : |x| \geq R_0\}.$$

(C19) *There exists a constant* $\beta \in (0,1)$ *such that*

$$(1-\beta)l > \mu^* := \inf\left\{\int_{\mathbb{R}^3}(|\nabla u|^2 + u^2)dx : u \in H^1(\mathbb{R}^3, \mathbb{R}^+),\right.$$

$$\left.\int_{\mathbb{R}^3} a(x)F(u)dx \geq \frac{l}{2}, \int_{\mathbb{R}^3} K(x)\phi_{\varepsilon,K}u^2 dx < \frac{4}{3}\beta l\right\},$$

where $F(t) = \int_0^t f(s)ds$.
(C20) $K \in L^2(\mathbb{R}^3)$, $K \geq (\not\equiv)0$ *for all* $x \in \mathbb{R}^3$.

Then system (4.66) *possesses a ground state solution* $(u_\varepsilon, \phi_{\varepsilon,K}(u_\varepsilon))$ *in* $H^1(\mathbb{R}^3)$ *for all* $\varepsilon > 0$.

Remark 4.6. When $\varepsilon \equiv 0$, system (4.66) has been studied in [193] and possesses a ground state solution in $H^1(\mathbb{R}^3)$ under the same conditions of Theorem 4.4. In this case, solvability of such Schrödinger–Poisson systems begins the unique positive solution of the linear Poisson equation in $D^{1,2}(\mathbb{R}^3)$ denoted by $\phi_{0,K}(u)$ which is the Newtonian potential of $K(x)u^2$ and has the explicit formula

$$\phi_{0,K}(u(x)) = \frac{1}{4\pi}\int_{\mathbb{R}^3}\frac{K(y)u^2(y)}{|x-y|}dy.$$

Clearly, this solution has some good properties. But when $\varepsilon > 0$, we will solve a quasilinear Poisson equation

$$-\text{div}[(1 + \varepsilon^4|\nabla\phi|^2)\nabla\phi] = K(x)|u|^2$$

which has a unique weak nonnegative solution $\phi_{\varepsilon,K}(u)$ in the space $D^{1,2} \cap D^{1,4}(\mathbb{R}^3)$ in the following Lemma 4.15. Moreover, Theorem 4.4 generalizes Theorem 1.1 in [193] where the author only studied the special situation, that is, $\varepsilon = 0$. Functions K, a, f satisfying Theorem 4.4 can be constructed by the same method as Remark 1.1 in [193].

Furthermore, we want to know the asymptotic behavior u_ε when $\varepsilon \to 0$. We have the following result.

Theorem 4.5. *If* $(u_\varepsilon, \phi_{\varepsilon,K}(u_\varepsilon))$ *denotes the ground state solution of system* (4.66) *obtained by Theorem 4.4, then* u_ε *is bounded in* $H^1(\mathbb{R}^3)$ *and any limit point of* $(u_\varepsilon, \phi_{\varepsilon,K}(u_\varepsilon))$ *when* $\varepsilon \to 0$ *is a solution* $(u_0, \phi_{0,K}(u_0))$ *of system* (4.66) *with* $\varepsilon \equiv 0$.

In order to obtain our results, we have to overcome various difficulties. First, the competing effect of the quasilinear non-local term in the functional of system (4.66) gives rise to some difficulties. Second, since the embedding of $H^1(\mathbb{R}^3)$ into $L^p(\mathbb{R}^3)$, $p \in [2,6]$, is not compact, condition (C18) is crucial to obtain the boundedness of Cerami sequence. Furthermore, in order to recover the compactness, we establish a compactness result $\int_{|x| \geq R}(|\nabla u_n|^2 + |u_n|^2)dx \leq \varepsilon'$ similar the one in [193] but different from the one in [239]. In fact, this difficulty can be avoided, when autonomous problems are considered, restricting the corresponding functional to the subspace of $H^1(\mathbb{R}^3)$ consisting of radially symmetric functions, or, when one is looking for semi-classical states, by using perturbation methods or a reduction to a finite dimension by the projections method. Third, it is not difficult to find that every (PS) sequence is bounded when $3 < p < 5$ in [239] because a variant of global Ambrosetti–Rabinowitz condition is satisfied when $3 < p < 5$ (see [228]). However, for the asymptotically linear case, we have to find another method to verify the boundedness of (PS) sequence.

System (4.66) has a variational structure. Its corresponding functional

$$J_\varepsilon : H^1(\mathbb{R}^3) \times (D^{1,2} \cap D^{1,4}(\mathbb{R}^3)) \to \mathbb{R}$$

defined by

$$J_\varepsilon(u, \phi) = \frac{1}{2}\|u\|^2 + \frac{1}{2}\int_{\mathbb{R}^3} K(x)\phi u^2 dx - \frac{1}{4}\int_{\mathbb{R}^3} |\nabla\phi|^2 dx - \frac{\varepsilon^4}{8}\int_{\mathbb{R}^3} |\nabla\phi|^4 dx$$
$$- \int_{\mathbb{R}^3} a(x)F(u)dx.$$

Evidently, from conditions of Theorem 4.4, the action functional $J_\varepsilon \in C^1(H^1(\mathbb{R}^3) \times (D^{1,2} \cap D^{1,4}(\mathbb{R}^3)), \mathbb{R})$ and the partial derivatives in (u, ϕ) are given, for $\zeta \in H^1(\mathbb{R}^3)$ and $\eta \in D^{1,2} \cap D^{1,4}(\mathbb{R}^3)$, we have

$$\left\langle \frac{\partial J_\varepsilon}{\partial u}(u, \phi), \zeta \right\rangle = \int_{\mathbb{R}^3}(\nabla u \cdot \nabla\zeta + u\zeta + K(x)\phi u\zeta - a(x)f(u)\zeta)dx,$$

$$\left\langle \frac{\partial J_\varepsilon}{\partial \phi}(u, \phi), \eta \right\rangle = -\frac{1}{2}\int_{\mathbb{R}^3}(\nabla\phi \cdot \nabla\eta + \varepsilon^4|\nabla\phi|^2\nabla\phi \cdot \nabla\eta - K(x)u^2\eta)dx.$$

Thus, we have the following result:

Proposition 4.2. *The pair (u, ϕ) is a weak solution of the system (4.66) if and only if it is a critical point of J_ε in $H^1(\mathbb{R}^3) \times (D^{1,2} \cap D^{1,4}(\mathbb{R}^3))$.*

Lemma 4.15. *Assume that (C20) holds. For any $u \in H^1(\mathbb{R}^3)$ and all $\varepsilon > 0$, there is a unique nonnegative weak solution $\phi_{\varepsilon,K}(u) \in D^{1,2} \cap D^{1,4}(\mathbb{R}^3)$ for*

$$-\text{div}[(1 + \varepsilon^4|\nabla\phi|^2)\nabla\phi] = K(x)u^2, \quad x \in \mathbb{R}^3. \tag{4.67}$$

Furthermore, for any $\psi \in D^{1,2} \cap D^{1,4}(\mathbb{R}^3)$ we have

$$\int_{\mathbb{R}^3} (1 + \varepsilon^4 |\nabla \phi_{\varepsilon,K}(u)|^2) \nabla \phi_{\varepsilon,K} \cdot \nabla \psi dx = \int_{\mathbb{R}^3} K(x) u^2 \psi dx.$$

Proof. Equation (4.67) is the special case of one of Lemma 3.1 in [253], so we write its proof for completeness.

For any $u \in H^1(\mathbb{R}^3) \setminus \{0\}$ and $K \in L^2(\mathbb{R}^3)$, by the Hölder inequality and Sobolev inequality, we have

$$\int_{\mathbb{R}^3} (K(x) u^2)^{6/5} dx \leq \|K(x)\|_2^{6/5} \|u\|_6^{12/5} \leq (S^*)^{6/5} \|K(x)\|_2^{6/5} \|u\|^{12/5}.$$

Therefore, $K(x) u^2 \in L^{6/5}$. The corresponding functional of (4.67) is

$$\widetilde{J}_\varepsilon(\phi) = \frac{1}{2} \int_{\mathbb{R}^3} |\nabla \phi|^2 dx + \frac{\varepsilon^4}{4} \int_{\mathbb{R}^3} |\nabla \phi|^4 dx - \int_{\mathbb{R}^3} K(x) u^2 \phi dx$$

for $\phi \in D^{1,2} \cap D^{1,4}(\mathbb{R}^3)$. Therefore, by the Hölder inequality and Sobolev inequality, we get

$$\widetilde{J}_\varepsilon(\phi) \geq \frac{1}{2} \int_{\mathbb{R}^3} |\nabla \phi|^2 dx + \frac{\varepsilon^4}{4} \int_{\mathbb{R}^3} |\nabla \phi|^4 dx - \|K(x) u^2\|_{6/5} \|\phi\|_6$$

$$\geq \frac{1}{2} \|\nabla \phi\|_2^2 + \frac{\varepsilon^4}{4} \|\nabla \phi\|_4^4 - S^{1/2} \|K(x) u^2\|_{6/5} \|\nabla \phi\|_2$$

$$\to +\infty$$

as $\|\phi\|_{D^{1,2} \cap D^{1,4}} \to +\infty$. That is, the functional $\widetilde{J}_\varepsilon(\phi)$ is coercive. So, $\widetilde{J}_\varepsilon$ has a bounded minimizing sequence $\{\phi_n\}$ such that

$$\widetilde{J}_\varepsilon(\phi_n) \to \inf_{D^{1,2} \cap D^{1,4}(\mathbb{R}^3)} \widetilde{J}_\varepsilon(\phi)$$

whenever $n \to \infty$.

Let $G_\varepsilon(\phi_n) = \frac{1}{2} \int_{\mathbb{R}^3} |\nabla \phi_n|^2 dx + \frac{\varepsilon^4}{4} \int_{\mathbb{R}^3} |\nabla \phi_n|^4 dx$ and $L(\phi_n) = \int_{\mathbb{R}^3} K(x) u^2 \phi_n dx$. Clearly, G_ε is a strictly convex functional and L is a linear functional. So, $\widetilde{J}_\varepsilon(\phi_n) = G_\varepsilon(\phi_n) - L(\phi_n)$ is a strictly convex functional. Furthermore, $\widetilde{J}_\varepsilon$ is C^1. So, by Mazuri's theorem (see, e.g., Theorem V.1.2 in [254]), $\widetilde{J}_\varepsilon$ is weakly lower semi-continuous on $D^{1,2} \cap D^{1,4}(\mathbb{R}^3)$. It follows from the least action principle (see, e.g., Theorem 1.1 in [171]) that $\widetilde{J}_\varepsilon$ has a minimum on $D^{1,2} \cap D^{1,4}(\mathbb{R}^3)$.

We claim that the minimum point of $\widetilde{J}_\varepsilon$ is unique. Otherwise, suppose that $\phi_{\varepsilon,K,1}(u)$ and $\phi_{\varepsilon,K,2}(u)$ are both minimum points of $\widetilde{J}_\varepsilon$. That is,

$$\phi_{\varepsilon,K,1}(u) = \phi_{\varepsilon,K,2}(u) = \inf_{D^{1,2} \cap D^{1,4}(\mathbb{R}^3)} \widetilde{J}_\varepsilon(\phi).$$

Because $\tilde{J}_\varepsilon$ is strictly convex, for each $\alpha \in (0,1)$, we obtain

$$\tilde{J}_\varepsilon(\alpha\phi_{\varepsilon,K,1}(u) + (1-\alpha)\phi_{\varepsilon,K,2}(u)) < \alpha\tilde{J}_\varepsilon(\phi_{\varepsilon,K,1}(u)) + (1-\alpha)\tilde{J}_\varepsilon(\phi_{\varepsilon,K,2}(u))$$

$$= \inf_{D^{1,2}\cap D^{1,4}(\mathbb{R}^3)} \tilde{J}_\varepsilon(\phi).$$

This is a contradiction. So, $\tilde{J}_\varepsilon$ has a unique minimum, then equation (4.67) has a unique weak solution $\phi_{\varepsilon,K}(u)$.

Next, we shall prove that the solution $\phi_{\varepsilon,K}(u)$ of equation (4.67) is non-negative. Denote by $\phi_{\varepsilon,K}^{\pm}(u) := \max\{\pm\phi_{\varepsilon,K}(u),0\}$ the positive (negative) part of $\phi_{\varepsilon,K}(u)$. Since $K(x)u^2 \geq 0$ and $\phi_{\varepsilon,K}(u)$ is a solution of equation (4.67), we deduce

$$-\mathrm{div}[(1 + \varepsilon^4|\nabla\phi_{\varepsilon,K}(u)|^2)\nabla\phi_{\varepsilon,K}(u)] \geq 0, \quad x \in \mathbb{R}^3.$$

Multiplying this equation by $\phi_{\varepsilon,K}^-(u)$ with $\phi_{\varepsilon,K}^-(u) = \max\{-\phi_{\varepsilon,K}(u),0\}$ and integrating on $\mathbb{R}^3$ by parts, we obtain

$$-\int_{\mathbb{R}^3}(|\nabla\phi_{\varepsilon,K}^-(u)|^2 + \varepsilon^4|\nabla\phi_{\varepsilon,K}^-(u)|^4) \geq 0.$$

This yields that $\|\phi_{\varepsilon,K}^-(u)\|_{D^{1,2}\cap D^{1,4}(\mathbb{R}^3)} = 0$, so, $\phi_{\varepsilon,K}^-(u) = 0$. Therefore, we obtain $\phi_{\varepsilon,K}(u) = \phi_{\varepsilon,K}^+(u) - \phi_{\varepsilon,K}^-(u) = \phi_{\varepsilon,K}^+(u) \geq 0$. Thus, $\phi_{\varepsilon,K}(u)$ is a nonnegative weak solution of (4.67).

From the above discussion, $\tilde{J}_\varepsilon$ achieves its minimum at a unique non-negative $\phi_{\varepsilon,K}(u) \in D^{1,2} \cap D^{1,4}(\mathbb{R}^3)$ and therefore

$$\langle\tilde{J}_\varepsilon'(\phi_{\varepsilon,K}(u)),\psi\rangle = 0, \quad \forall\psi \in D^{1,2} \cap D^{1,4}(\mathbb{R}^3).$$

$\square$

By Lemma 4.15, there exists a unique function $0 \leq \phi_{\varepsilon,K}(u) \in D^{1,2} \cap D^{1,4}(\mathbb{R}^3)$ such that

$$-\mathrm{div}[(1 + \varepsilon^4|\nabla\phi_{\varepsilon,K}(u)|^2)\nabla\phi_{\varepsilon,K}(u)] = K(x)u^2. \tag{4.68}$$

Substitute the solution $\phi_{\varepsilon,K}(u)$ in the first(Schrödinger) equation of the system (4.66), then get the corresponding functional $E_\varepsilon : H^1(\mathbb{R}^3) \to \mathbb{R}$ defined by $E_\varepsilon(u) = J_\varepsilon(u,\phi_{\varepsilon,K}(u))$. After multiplying (4.68) by $\phi_{\varepsilon,K}(u)$ and integration by parts, we obtain

$$\int_{\mathbb{R}^3}(1 + \varepsilon^4|\nabla\phi_{\varepsilon,K}(u)|^2)|\nabla\phi_{\varepsilon,K}(u)|^2 dx = \int_{\mathbb{R}^3} K(x)u^2\phi_{\varepsilon,K}(u)dx. \tag{4.69}$$

Therefore, the reduced functional takes the form

$$E_\varepsilon(u) = \frac{1}{2}\|u\|^2 + I_\varepsilon(u) - \int_{\mathbb{R}^3} a(x)F(u)dx,$$

where

$$I_\varepsilon(u) = \frac{1}{4}\int_{\mathbb{R}^3}|\nabla\phi_{\varepsilon,K}(u)|^2 dx + \frac{3\varepsilon^4}{8}\int_{\mathbb{R}^3}|\nabla\phi_{\varepsilon,K}(u)|^4 dx.$$

Clearly, $I_\varepsilon(u) \geq 0$ and $\int_{\mathbb{R}^3}K(x)u^2\phi_{\varepsilon,K}(u)dx \geq 0$. Now, we give the following definition:

Definition 4.1. $(u,\phi_{\varepsilon,K}(u))$ with $u \in H^1(\mathbb{R}^3)$ is a ground state solution of system (4.66), we mean that $(u,\phi_{\varepsilon,K}(u))$ is a solution of system (4.66) which has the least energy among all solutions of system (4.66), that is, $E'_\varepsilon(u) = 0$ and $E_\varepsilon(u) = \inf\{E_\varepsilon(v) : v \in H^1(\mathbb{R}^3) \setminus \{0\}$ and $E'_\varepsilon(v) = 0\}$.

Moreover, we have

Lemma 4.16. *For any $\varepsilon > 0$ the functional $u \mapsto I_\varepsilon(u)$ is C^1 on $H^1(\mathbb{R}^3)$ and its Fréchet-derivative satisfies*

$$\langle I'_\varepsilon(u),\psi\rangle = \int_{\mathbb{R}^3}K(x)\phi_{\varepsilon,K}(u)u\psi dx, \quad \forall u,\ \psi \in H^1(\mathbb{R}^3).$$

By suitably modifying the proof of Proposition 4.1 in [253], this lemma can be proved. Here we omit its proof.

By (4.69), the Hölder's inequality and Sobolev's inequalities, we have

$$\int_{\mathbb{R}^3}|\nabla\phi_{\varepsilon,K}(u)|^2 dx + \varepsilon^4\int_{\mathbb{R}^3}|\nabla\phi_{\varepsilon,K}(u)|^4 dx$$

$$= \int_{\mathbb{R}^3}K(x)u^2\phi_{\varepsilon,K}(u)dx$$

$$\leq \|\phi_{\varepsilon,K}(u)\|_6\|K\|_2\|u\|_6^2$$

$$\leq S^{1/2}S^*\|K\|_2\|\nabla\phi_{\varepsilon,K}(u)\|_2\|u\|^2.$$

This yields

$$\|\nabla\phi_{\varepsilon,K}(u)\|_2 \leq S^{1/2}S^*\|K\|_2\|u\|^2 := C_0\|u\|^2 \qquad (4.70)$$

and

$$\|\nabla\phi_{\varepsilon,K}(u)\|_2^2 + \varepsilon^4\|\nabla\phi_{\varepsilon,K}(u)\|_4^4$$
$$\leq S^{1/2}S^*\|K\|_2\|\nabla\phi_{\varepsilon,K}(u)\|_2\|u\|^2$$
$$\leq S(S^*)^2\|K\|_2^2\|u\|^4. \qquad (4.71)$$

From (4.71), we obtain

$$\varepsilon^4\|\nabla\phi_{\varepsilon,K}(u)\|_4^4 \leq S(S^*)^2\|K\|_2^2\|u\|^4 := C_1\|u\|^4. \qquad (4.72)$$

Combining (4.70), (4.72) and (C15), E_ε is well defined. Furthermore, together with Lemma 4.16, E_ε is a C^1 functional with derivative given by

$$\langle E'_\varepsilon(u), v \rangle = \int_{\mathbb{R}^3} (\nabla u \cdot \nabla v + uv + K(x)\phi_{\varepsilon,K}(u)uv - a(x)f(u)v)dx, \quad (4.73)$$

for all $v \in H^1(\mathbb{R}^3)$.

Now, we can apply Theorem 2.3 of [226] to the functional E_ε and obtain

Proposition 4.3. *The following statements are equivalent:*

(1) $(u, \phi) \in H^1(\mathbb{R}^3) \times (D^{1,2} \cap D^{1,4}(\mathbb{R}^3))$ *is a critical point of J_ε (i.e.* (u, ϕ) *is a solution of the system (4.66));*
(2) u *is a critical point of E_ε and $\phi = \phi_{\varepsilon,K}(u)$.*

Furthermore, in order to obtain our results, we also need the following Lemma.

Lemma 4.17. ([253], Lemma 3.2) *For all $\varepsilon > 0$ and f_ε, $f \in L^{6/5}(\mathbb{R}^3)$, let $\phi_\varepsilon(f_\varepsilon) \in D^{1,2} \cap D^{1,4}(\mathbb{R}^3)$ be a unique solution of $-\mathrm{div}[(1+\varepsilon^4|\nabla\phi|^2)\nabla\phi] = f_\varepsilon$ in $\mathbb{R}^3$ and $\phi_0(f) \in D^{1,2}(\mathbb{R}^3)$ be a unique solution of $-\Delta\phi = f$ in $\mathbb{R}^3$. Then:*

(i) if $f_\varepsilon \rightharpoonup f$ weakly in $L^{6/5}(\mathbb{R}^3)$ then $\phi_\varepsilon(f_\varepsilon) \rightharpoonup \phi_0(f)$ in $D^{1,2}(\mathbb{R}^3)$ as $\varepsilon \to 0$.
(ii) If $f_\varepsilon \to f$ strongly in $L^{6/5}(\mathbb{R}^3)$, then:

$$\phi_\varepsilon(f_\varepsilon) \to \phi_0(f) \quad \text{strongly in } D^{1,2}(\mathbb{R}^3),$$

$$\varepsilon\phi_\varepsilon(f_\varepsilon) \to 0 \quad \text{strongly in } D^{1,4}(\mathbb{R}^3)$$

as $\varepsilon \to 0$.

Now we prove that system (4.66) has a mountain pass type solution. For this purpose, we use a variant version of Mountain Pass Theorem in [22], which allows us to find a so-called Cerami type (PS) sequence (Cerami sequence, in short). The properties of this kind of (PS) sequence are very helpful in showing its boundedness in the asymptotically linear case. The following lemmas show that E_ε has the so-called mountain pass geometry.

Lemma 4.18. *Suppose that (C15)–(C17), (C18) and (C20) hold. Then there exist $\rho > 0$ and $\alpha > 0$ such that $E_\varepsilon(u)|_{\|u\|=\rho} \geq \alpha > 0$.*

Proof. For any $\widetilde{\varepsilon} > 0$, it follows from (C15)–(C17) that there exists $C_{\widetilde{\varepsilon}} > 0$ such that

$$|f(t)| \leq \widetilde{\varepsilon}|t| + C_{\widetilde{\varepsilon}}|t|^5 \text{ for all } t \in \mathbb{R}. \quad (4.74)$$

Therefore, we have

$$|F(t)| \leq \frac{1}{2}\widetilde{\varepsilon}|t|^2 + \frac{C_{\widetilde{\varepsilon}}}{6}|t|^6 \text{ for all } t \in \mathbb{R}. \tag{4.75}$$

Furthermore, by (C15)–(C17) and (C18), there exists $C_2 > 0$ such that

$$a(x) \leq C_2 \text{ for all } x \in \mathbb{R}^3. \tag{4.76}$$

According to (4.75), (4.76) and the Sobolev inequality, we deduce

$$\left| \int_{\mathbb{R}^3} a(x)F(u)dx \right| \leq \frac{\widetilde{\varepsilon}C_2}{2} \int_{\mathbb{R}^3} |u|^2 dx + \frac{C_2 C_{\widetilde{\varepsilon}}}{6} \int_{\mathbb{R}^3} |u|^6 dx$$

$$\leq \frac{\widetilde{\varepsilon}C_2}{2} \|u\|^2 + C_3\|u\|^6$$

for some $C_3 > 0$. Together with $I_\varepsilon(u) \geq 0$, one has

$$E_\varepsilon(u) = \frac{1}{2}\|u\|^2 + I_\varepsilon(u) - \int_{\mathbb{R}^3} a(x)F(u)dx$$

$$\geq \frac{1}{2}\|u\|^2 - \frac{\widetilde{\varepsilon}C_2}{2}\|u\|^2 - C_3\|u\|^6$$

$$\geq \|u\|\left(\frac{1 - \widetilde{\varepsilon}C_2}{2}\|u\| - C_3\|u\|^5 \right). \tag{4.77}$$

Taking $\widetilde{\varepsilon} \in (0, \frac{1}{2C_2})$. From (4.77), letting $\|u\| = \rho > 0$ small enough, there exists $\alpha > 0$ such that $I(u)|_{\|u\|=\rho} \geq \alpha > 0$. □

Lemma 4.19. *Suppose that* (C15)–(C17), (C18)–(C19) *and* (C20) *hold. Then there exists* $v \in H^1(\mathbb{R}^3)$ *with* $\|v\| > \rho$, ρ *is given by Lemma 4.18, such that* $E_\varepsilon(v) < 0$ *for all* $\varepsilon > 0$.

Proof. By (C19), in view of the definition of μ^* and $(1-\beta)l > \mu^*$, there is a nonnegative function $v \in H^1(\mathbb{R}^3)$ such that

$$\int_{\mathbb{R}^3} a(x)F(v)dx \geq \frac{l}{2}, \quad \int_{\mathbb{R}^3} K(x)\phi_{\varepsilon,K}(v)v^2 dx < \frac{4}{3}\beta l,$$

and $\mu^* \leq \|v\|^2 < (1-\beta)l$. From (4.69) and the definition of I_ε, we obtain

$$I_\varepsilon(v) \leq \frac{3}{8} \int_{\mathbb{R}^3} K(x)\phi_{\varepsilon,K}(v)v^2 dx < \frac{1}{2}\beta l.$$

Therefore, we have

$$E_\varepsilon(v) = \frac{1}{2}\|v\|^2 + I_\varepsilon(v) - \int_{\mathbb{R}^3} a(x)F(v)dx$$

$$\leq \frac{1}{2}\|v\|^2 + \frac{\beta l}{2} - \frac{l}{2}$$

$$= \frac{1}{2}(\|v\|^2 - (1-\beta)l) < 0.$$

Choosing $\rho > 0$ small enough in Lemma 4.18 such that $\|v\| > \rho$, then this lemma is proved. □

From Lemmas 4.18 and 4.19 and Mountain Pass Theorem in [22], there is a sequence $\{u_n\} \subset H^1(\mathbb{R}^3)$ such that

$$\|E'_\varepsilon(u_n)\|_{H^{-1}}(1 + \|u_n\|) \to 0 \quad \text{and} \quad E_\varepsilon(u_n) \to c \text{ as } n \to \infty, \quad (4.78)$$

where H^{-1} denotes the dual space of $H^1(\mathbb{R}^3)$. In the following, we shall prove that sequence $\{u_n\}$ has a convergence subsequence.

Lemma 4.20. *Suppose that* (C15)–(C17), (C18) *and* (C20) *hold. Then* $\{u_n\}$ *defined in* (4.78) *is bounded in* $H^1(\mathbb{R}^3)$.

Proof. By contradiction, let $\|u_n\| := \alpha_n \to \infty$. Define $w_n = u_n\|u_n\|^{-1} = \alpha_n^{-1}u_n$. Clearly, $\|w_n\| = 1$ and $\{w_n\}$ is bounded in $H^1(\mathbb{R}^3)$ and there is a $w \in H^1(\mathbb{R}^3)$ such that, up to a sequence(still denoted by $\{w_n\}$),

$$\begin{cases} w_n \rightharpoonup w & \text{weakly in } H^1(\mathbb{R}^3), \\ w_n \to w & \text{a.e. in } \mathbb{R}^3, \\ w_n \to w & \text{strongly in } L^2_{loc}(\mathbb{R}^3) \end{cases}$$

as $n \to \infty$.

Firstly, we claim that w is nontrivial, that is $w \not\equiv 0$. Otherwise, if $w \equiv 0$, the Sobolev embedding implies that $w_n \to 0$ strongly in $L^2_{loc}(B_{R_0})$, R_0 is given by (C18). Define

$$\widetilde{f}(t) = \begin{cases} \dfrac{f(t)}{t}, & t \neq 0, \\ 0, & t = 0. \end{cases}$$

Together with (C15)–(C17) with $l < +\infty$, there exists $C_4 > 0$ such that

$$\widetilde{f}(t) \leq C_4 \text{ for all } t \in \mathbb{R}. \quad (4.79)$$

Then, for all $n \in \mathbb{N}$, we have

$$0 \leq \int_{|x|<R_0} a(x)\widetilde{f}(u_n)w_n^2 dx \leq C_4|a|_\infty \int_{|x|<R_0} w_n^2 dx \to 0.$$

This yields

$$\lim_{n\to\infty} \int_{|x|<R_0} a(x)\widetilde{f}(u_n)w_n^2 dx = 0. \quad (4.80)$$

Furthermore, by (C18), there exists a constant $\theta \in (0,1)$ such that

$$\sup\{f(t)/t : t > 0\} \leq \theta \inf\{1/a(x) : |x| \geq R_0\}. \quad (4.81)$$

Then, for all $n \in \mathbb{N}$, we have

$$\int_{|x|\geq R_0} a(x)\widetilde{f}(u_n)w_n^2 dx \leq \theta \int_{|x|\geq R_0} w_n^2 dx < \theta < 1. \quad (4.82)$$

Combining (4.80) and (4.82), we obtain

$$\limsup_{n\to\infty} \int_{\mathbb{R}^3} a(x)\widetilde{f}(u_n)w_n^2 dx < 1. \tag{4.83}$$

By (4.78), we get

$$0 \le |\langle E_\varepsilon'(u_n), u_n\rangle| \le \|E_\varepsilon'(u_n)\|_{H^{-1}}\|u_n\| \le \|E_\varepsilon'(u_n)\|_{H^{-1}}(1 + \|u_n\|) \to 0$$

as $n \to \infty$. Together with $\alpha_n \to \infty$ as $n \to \infty$, it follows that

$$\alpha_n^{-2}\langle E_\varepsilon'(u_n), u_n\rangle = o(1).$$

So, by (4.73) and (4.69), we have

$$o(1) = \|w_n\|^2 + \alpha_n^{-2}\int_{\mathbb{R}^3} K(x)\phi_{\varepsilon,K}(u_n)u_n^2 dx - \int_{\mathbb{R}^3} a(x)\frac{f(u_n)}{u_n}w_n^2 dx$$

$$= \|w_n\|^2 + \alpha_n^{-2}\int_{\mathbb{R}^3} (1 + \varepsilon^4|\nabla\phi_{\varepsilon,K}(u_n)|^2)|\nabla\phi_{\varepsilon,K}(u_n)|^2 dx$$

$$- \int_{\mathbb{R}^3} a(x)\frac{f(u_n)}{u_n}w_n^2 dx$$

$$\ge 1 - \int_{\mathbb{R}^3} a(x)\frac{f(u_n)}{u_n}w_n^2 dx,$$

where, and in what follows, $o(1)$ denotes a quantity which goes to zero as $n \to \infty$. Therefore, we deduce that

$$\int_{\mathbb{R}^3} a(x)\frac{f(u_n)}{u_n}w_n^2 dx + o(1) \ge 1,$$

which contradicts (4.83). So, $w \not\equiv 0$.

By (4.78) and the definition of E_ε, we get

$$o(1) + \alpha_n^{-2}c = \alpha_n^{-2}E_\varepsilon(u_n)$$

$$= \frac{1}{2}\|w_n\|^2 + \alpha_n^{-2}I_\varepsilon(u_n) - \int_{\mathbb{R}^3}\frac{F(u_n)}{u_n^2}w_n^2 dx$$

$$= \frac{1}{2} + \alpha_n^{-2}I_\varepsilon(u_n) - \int_{\mathbb{R}^3}\frac{F(u_n)}{u_n^2}w_n^2 dx. \tag{4.84}$$

Define

$$\widetilde{F}(t) = \begin{cases} \dfrac{F(t)}{t^2}, & t \ne 0, \\ 0, & t = 0. \end{cases}$$

By (C15)–(C17) with $l < +\infty$ and (4.84), there exists $C_5 > 1/2$ such that

$$\widetilde{F}(t) \le C_5 \text{ for all } t \in \mathbb{R}.$$

So, we obtain

$$\int_{\mathbb{R}^3} \widetilde{F}(u_n) w_n^2 dx \le C_5 \int_{\mathbb{R}^3} w_n^2 dx \le C_5 \|w_n\|^2 = C_5. \tag{4.85}$$

Combining (4.84) and (4.85), we deduce

$$\alpha_n^{-2} I_\varepsilon(u_n) \le C_5 - \frac{1}{2} + o(1). \tag{4.86}$$

From Lemma 4.15, we know that $\phi_{\varepsilon,K}(u_n)$ is the unique solution of the equation

$$-\text{div}[(1 + \varepsilon^4 |\nabla \phi|^2) \nabla \phi] = K(x) u_n^2.$$

That is, the following equality holds:

$$-\text{div}[(1 + \varepsilon^4 |\nabla \phi_{\varepsilon,K}(u_n)|^2) \nabla \phi_{\varepsilon,K}(u_n)] = K(x) u_n^2. \tag{4.87}$$

Multiplying (4.87) by $\phi_{\varepsilon,K}(u_n)$ and integrating by parts, we find that

$$\widetilde{J}_\varepsilon(\phi_{\varepsilon,K}(u_n)) = -\frac{1}{2} \int_{\mathbb{R}^3} |\nabla \phi_{\varepsilon,K}(u_n)|^2 dx - \frac{3\varepsilon^4}{4} \int_{\mathbb{R}^3} |\nabla \phi_{\varepsilon,K}(u_n)|^4 dx$$

$$= -2 I_\varepsilon(u_n).$$

Together with (4.86), we get

$$-\frac{1}{2} \alpha_n^{-2} \widetilde{J}_\varepsilon(\phi_{\varepsilon,K}(u_n)) = \alpha_n^{-2} I_\varepsilon(u_n) \le C_5 - \frac{1}{2} + o(1). \tag{4.88}$$

From Lemma 4.15, $\phi_{\varepsilon,K}(w_n)$ satisfies

$$\int_{\mathbb{R}^3} K(x) \phi_{\varepsilon,K}(w_n) w_n^2 dx = \int_{\mathbb{R}^3} |\nabla \phi_{\varepsilon,K}(w_n)|^2 dx$$

$$+ \varepsilon^4 \int_{\mathbb{R}^3} |\nabla \phi_{\varepsilon,K}(w_n)|^4 dx. \tag{4.89}$$

Since $\phi_{\varepsilon,K}(u_n)$ is the minimizer of $\widetilde{J}_\varepsilon$ on $D^{1,2} \cap D^{1,4}(\mathbb{R}^3)$ and (4.89), we may write

$$\widetilde{J}_\varepsilon(\phi_{\varepsilon,K}(u_n)) \le \widetilde{J}_\varepsilon(\alpha_n^{2/3} \phi_{\varepsilon,K}(w_n))$$

$$= \frac{\alpha_n^{4/3}}{2} \int_{\mathbb{R}^3} |\nabla \phi_{\varepsilon,K}(w_n)|^2 dx + \frac{\alpha_n^{8/3} \varepsilon^4}{4} \int_{\mathbb{R}^3} |\nabla \phi_{\varepsilon,K}(w_n)|^4 dx$$

$$- \alpha_n^{2/3} \int_{\mathbb{R}^3} K(x) u_n^2 \phi_{\varepsilon,K}(w_n) dx$$

$$= \frac{\alpha_n^{4/3}}{2} \int_{\mathbb{R}^3} |\nabla \phi_{\varepsilon,K}(w_n)|^2 dx + \frac{\alpha_n^{8/3} \varepsilon^4}{4} \int_{\mathbb{R}^3} |\nabla \phi_{\varepsilon,K}(w_n)|^4 dx$$

$$- \alpha_n^{8/3} \int_{\mathbb{R}^3} K(x) w_n^2 \phi_{\varepsilon,K}(w_n) dx$$

$$= \left(\frac{\alpha_n^{4/3}}{2} - \alpha_n^{8/3} \right) \int_{\mathbb{R}^3} |\nabla \phi_{\varepsilon,K}(w_n)|^2 dx - \frac{3\alpha_n^{8/3} \varepsilon^4}{4} \int_{\mathbb{R}^3} |\nabla \phi_{\varepsilon,K}(w_n)|^4 dx$$

$$\le -\frac{3\alpha_n^{8/3}}{4} \left(\int_{\mathbb{R}^3} |\nabla \phi_{\varepsilon,K}(w_n)|^2 dx + \varepsilon^4 \int_{\mathbb{R}^3} |\nabla \phi_{\varepsilon,K}(w_n)|^4 dx \right),$$

because $\frac{\alpha_n^{4/3}}{2} - \alpha_n^{8/3} \leq -\frac{3\alpha_n^{8/3}}{4}$ for n large enough since $\alpha_n \to +\infty$ as $n \to \infty$. Together with (4.88), we obtain

$$\int_{\mathbb{R}^3} |\nabla \phi_{\varepsilon,K}(w_n)|^2 dx + \varepsilon^4 \int_{\mathbb{R}^3} |\nabla \phi_{\varepsilon,K}(w_n)|^4 dx$$

$$\leq -\frac{4}{3}\alpha_n^{-8/3}\tilde{J}_\varepsilon(\phi_{\varepsilon,K}(u_n))$$

$$\leq \frac{8\alpha_n^{-2/3}}{3}\left(C_5 - \frac{1}{2} + o(1)\right).$$

So, we have

$$\int_{\mathbb{R}^3} |\nabla \phi_{\varepsilon,K}(w_n)|^2 dx + \varepsilon^4 \int_{\mathbb{R}^3} |\nabla \phi_{\varepsilon,K}(w_n)|^4 dx \to 0 \quad \text{as} \quad n \to \infty. \quad (4.90)$$

By (4.89) and (4.90), we have

$$\int_{\mathbb{R}^3} K(x)\phi_{\varepsilon,K}(w_n)w_n^2 dx \to 0 \quad \text{as} \quad n \to \infty. \quad (4.91)$$

We can easily verify that

$$\int_{\mathbb{R}^3} K(x)\phi_{\varepsilon,K}(w_n)w_n^2 dx \to \int_{\mathbb{R}^3} K(x)\phi_{\varepsilon,K}(w)w^2 dx \quad \text{as} \quad n \to \infty. \quad (4.92)$$

Indeed, in view of the Sobolev embedding theorems and (3) of Lemma 2.1 in [239], $w_n \rightharpoonup w$ weakly in $H^1(\mathbb{R}^3)$, we obtain

$$\begin{array}{l} (a) \ w_n \rightharpoonup w \text{ weakly in } L^6(\mathbb{R}^3), \\ (b) \ w_n^2 \to w^2 \text{ strongly in } L^3_{loc}(\mathbb{R}^3), \\ (c) \ \phi_{\varepsilon,K}(w_n) \rightharpoonup \phi_{\varepsilon,K}(w) \text{ weakly in } D^{1,2}(\mathbb{R}^3), \\ (d) \ \phi_{\varepsilon,K}(w_n) \to \phi_{\varepsilon,K}(w) \text{ strongly in } L^6_{loc}(\mathbb{R}^3). \end{array} \quad (4.93)$$

For any choice of $\bar{\varepsilon} > 0$ and $\rho > 0$, the relation

$$\|w_n - w\|_{6,B_\rho(0)} < \bar{\varepsilon} \quad (4.94)$$

holds for large n. Using (c) of (4.93), for large n, we have

$$\left|\int_{\mathbb{R}^3} K(x)(\phi_{\varepsilon,K}(w_n) - \phi_{\varepsilon,K}(w))w^2 dx\right| = o(1). \quad (4.95)$$

Because w_n is bounded in $H^1(\mathbb{R}^3)$ and the continuity of the Sobolev embedding of $D^{1,2}(R^3)$ in $L^6(\mathbb{R}^3)$, then $\phi_{\varepsilon,K}(w_n)$ is bounded in $D^{1,2}(\mathbb{R}^3)$ and in $L^6(\mathbb{R}^3)$. Moreover, $K \in L^2(\mathbb{R}^3)$ implies that Kw_n^2 and Kw^2 belong to $L^{\frac{6}{5}}(\mathbb{R}^3)$ and that to any $\bar{\varepsilon} > 0$ there exists $\bar{\rho} = \bar{\rho}(\bar{\varepsilon})$ such that

$$\|K\|_{2,R^3 \setminus B_\rho(0)} < \bar{\varepsilon}, \quad \forall \rho \geq \bar{\rho}. \quad (4.96)$$

By (4.95), (4.94) and (4.96), we obtain

$$\left| \int_{\mathbb{R}^3} K(x)\phi_{\varepsilon,K}(w_n)w_n^2 dx - \int_{\mathbb{R}^3} K(x)\phi_{\varepsilon,K}(w)w^2 dx \right|$$

$$= \left| \int_{\mathbb{R}^3} K(x)\phi_{\varepsilon,K}(w_n)(w_n^2 - w^2)dx - \int_{\mathbb{R}^3} K(x)(\phi_{\varepsilon,K}(w) - \phi_{\varepsilon,K}(w_n))w^2 dx \right|$$

$$\leq \int_{\mathbb{R}^3} |K(x)\phi_{\varepsilon,K}(w_n)(w_n^2 - w^2)|dx + \int_{\mathbb{R}^3} |K(x)(\phi_{\varepsilon,K}(w_n) - \phi_{\varepsilon,K}(w))w^2|dx$$

$$\leq \|\phi_{\varepsilon,K}(w_n)\|_6 \left(\int_{\mathbb{R}^3} |K(x)(w_n^2 - w^2)|^{\frac{6}{5}} dx \right)^{\frac{5}{6}} + \overline{\varepsilon}$$

$$\leq C_6 \left(\int_{\mathbb{R}^3 \backslash B_\rho(0)} |K(x)(w_n^2 - w^2)|^{\frac{6}{5}} dx + \int_{B_\rho(0)} |K(x)(w_n^2 - w^2)|^{\frac{6}{5}} dx \right)^{\frac{5}{6}} + \overline{\varepsilon}$$

$$\leq C_6 \left(\|K\|_{2,\mathbb{R}^3 \backslash B_\rho(0)}^{\frac{6}{5}} \|w_n^2 - w^2\|_3^{\frac{6}{5}} + \|K\|_2^{\frac{6}{5}} \|w_n^2 - w^2\|_{3,B_\rho(0)}^{\frac{6}{5}} \right)^{\frac{5}{6}} + \overline{\varepsilon}$$

$$\leq C_7 \overline{\varepsilon}.$$

This proves (4.92). So, by (4.91) and (4.92), we obtain

$$\int_{\mathbb{R}^3} K(x)\phi_{\varepsilon,K}(w)w^2 dx = 0,$$

which implies that $w \equiv 0$. That is a contradiction. Therefore, $\{u_n\}$ is bounded in $H^1(\mathbb{R}^3)$. □

Lemma 4.21. *Suppose that* (C15)–(C17), (C18) *and* (C20) *hold. Then for any* $\varepsilon' > 0$, *there exist* $R(\varepsilon') > R_0$ *and* $n(\varepsilon') > 0$ *such that* $\{u_n\}$ *defined in* (4.78) *satisfies*

$$\int_{|x| \geq R} (|\nabla u_n|^2 + |u_n|^2)dx \leq \varepsilon'$$

for $n > n(\varepsilon')$ *and* $R \geq R(\varepsilon')$.

Proof. Let $\xi_R : \mathbb{R}^3 \to [0,1]$ be a smooth function such that

$$\xi_R(x) = \begin{cases} 0, & 0 \leq |x| \leq R/2, \\ 1, & |x| \geq R. \end{cases} \tag{4.97}$$

Moreover, there exists a constant C_8 independent of R such that

$$|\nabla \xi_R(x)| \leq \frac{C_8}{R} \text{ for all } x \in \mathbb{R}^3. \tag{4.98}$$

Then, for all $n \in \mathbb{N}$ and $R \geq R_0$, by (4.97), (4.98) and the Hölder inequality, we have

$$
\int_{\mathbb{R}^3} |\nabla(u_n \xi_R)|^2 dx \leq \int_{\mathbb{R}^3} |\nabla u_n|^2 |\xi_R|^2 dx + \int_{\mathbb{R}^3} |u_n|^2 |\nabla \xi_R|^2 dx
$$

$$
+ 2 \int_{\mathbb{R}^3} |u_n| |\xi_R| |\nabla u_n| |\nabla \xi_R| dx
$$

$$
\leq \int_{R/2 < |x| < R} |\nabla u_n|^2 dx + \int_{|x| \geq R} |\nabla u_n|^2 dx + \frac{C_8^2}{R^2} \int_{\mathbb{R}^3} |u_n|^2 dx
$$

$$
+ 2 \left(\int_{\mathbb{R}^3} |\nabla u_n|^2 |\xi_R^2| dx \right)^{\frac{1}{2}} \left(\int_{\mathbb{R}^3} |u_n|^2 |\nabla \xi_R|^2 dx \right)^{\frac{1}{2}}
$$

$$
\leq \int_{R/2 < |x| < R} |\nabla u_n|^2 dx + \int_{|x| \geq R} |\nabla u_n|^2 dx + \frac{C_8^2}{R^2} \int_{\mathbb{R}^3} |u_n|^2 dx
$$

$$
+ 2 \left(\int_{R/2 < |x| < R} |\nabla u_n|^2 dx + \int_{|x| \geq R} |\nabla u_n|^2 dx \right)^{\frac{1}{2}}
$$

$$
\left(\frac{C_8^2}{R^2} \int_{\mathbb{R}^3} |u_n|^2 dx \right)^{\frac{1}{2}}
$$

$$
\leq \left(2 + \frac{C_8^2}{R^2} + \frac{2\sqrt{2} C_8}{R} \right) \|u_n\|^2
$$

$$
\leq \left(2 + \frac{C_8^2}{R_0^2} + \frac{2\sqrt{2} C_8}{R_0} \right) \|u_n\|^2.
$$

This implies that

$$
\|u_n \xi_R\| \leq C_9 \|u_n\| \tag{4.99}
$$

for all $n \in \mathbb{N}$ and $R \geq R_0$, where $C_9 = (3 + \frac{C_8^2}{R_0^2} + \frac{2\sqrt{2} C_8}{R_0})^{\frac{1}{2}}$. From Lemma 4.20, we know that $\{u_n\}$ is bounded in $H^1(\mathbb{R}^3)$. Together with (4.78), we obtain that $E'_\varepsilon(u_n) \to 0$ in $H^{-1}(\mathbb{R}^3)$. Moreover, by (4.99), for $\varepsilon' > 0$, there exists $n(\varepsilon') > 0$ such that

$$
\langle E'_\varepsilon(u_n), \xi_R u_n \rangle \leq C_9 \|E'_\varepsilon(u_n)\|_{H^{-1}(\mathbb{R}^3)} \|u_n\| \leq \frac{\varepsilon'}{4}
$$

for $n > n(\varepsilon')$ and $R > R_0$. Note that

$$
\langle E'_\varepsilon(u_n), \xi_R u_n \rangle = \int_{\mathbb{R}^3} (|\nabla u_n|^2 + |u_n|^2) \xi_R dx + \int_{\mathbb{R}^3} u_n \nabla u_n \cdot \nabla \xi_R dx
$$

$$
+ \int_{\mathbb{R}^3} K(x) \phi_{\varepsilon, K}(u_n) u_n^2 \xi_R dx - \int_{\mathbb{R}^3} a(x) f(u_n) u_n \xi_R dx
$$

$$
\leq \frac{\varepsilon'}{4}.
$$

Together with Lemma 4.15, (C20) and the definition of ξ_R, we have

$$\int_{\mathbb{R}^3} [(|\nabla u_n|^2 + |u_n|^2)\xi_R + u_n \nabla u_n \cdot \nabla \xi_R]dx$$

$$\leq \int_{\mathbb{R}^3} a(x)f(u_n)u_n\xi_R dx - \int_{\mathbb{R}^3} K(x)\phi_{\varepsilon,K}(u_n)u_n^2\xi_R dx + \frac{\varepsilon'}{4}$$

$$\leq \int_{\mathbb{R}^3} a(x)f(u_n)u_n\xi_R dx + \frac{\varepsilon'}{4}. \tag{4.100}$$

By (4.81) and (C18), we have

$$a(x)f(u_n)u_n \leq \theta u_n^2 \text{ for } \theta \in (0,1) \text{ and } |x| \geq R_0.$$

This yields

$$\int_{\mathbb{R}^3} a(x)f(u_n)u_n\xi_R dx \leq \theta \int_{\mathbb{R}^3} u_n^2\xi_R dx \tag{4.101}$$

for all $n \in \mathbb{N}$ and $|x| \geq R_0$. For any $\varepsilon' > 0$, there exists $R(\varepsilon') \geq R_0$ such that

$$\frac{1}{R^2} \leq \frac{4\varepsilon'^2}{C_8^2} \text{ for all } R > R(\varepsilon'). \tag{4.102}$$

By the Young inequality, (4.98) and (4.102), for all $n \in \mathbb{N}$ and $R > R(\varepsilon')$, we obtain

$$\int_{\mathbb{R}^3} |u_n \nabla u_n \cdot \nabla \xi_R|dx$$

$$= \int_{\mathbb{R}^3} \sqrt{2\varepsilon'}|\nabla u_n|\frac{1}{\sqrt{2\varepsilon'}}|u_n||\nabla \xi_R|dx$$

$$\leq \varepsilon' \int_{\mathbb{R}^3} |\nabla u_n|^2 dx + \frac{1}{4\varepsilon'} \int_{|x| \leq R} |u_n|^2 \frac{C_8^2}{R^2}dx$$

$$\leq \varepsilon' \int_{\mathbb{R}^3} |\nabla u_n|^2 dx + \varepsilon' \int_{|x| \leq R} |u_n|^2 dx$$

$$\leq \varepsilon'\|u_n\|^2. \tag{4.103}$$

Combining (4.100), (4.101) and (4.103), there exists $C_6 > 0$ such that

$$(1-\theta) \int_{\mathbb{R}^3} (|\nabla u_n|^2 + |u_n|^2)\xi_R dx \leq \int_{\mathbb{R}^3} (|\nabla u_n|^2 + (1-\theta)|u_n|^2)\xi_R dx$$

$$\leq \frac{\varepsilon'}{4} + \varepsilon'\|u_n\|^2 \leq C_{10}\varepsilon'$$

for all $R > R(\varepsilon')$. Noting that C_{10} is independent of ε'. So, for any $\varepsilon > 0$, we can choose $R(\varepsilon') > R_0$ and $n(\varepsilon') > 0$ such that $\int_{|x| \geq R}(|\nabla u_n|^2 + |u_n|^2)dx \leq \varepsilon'$ holds. $\square$

Lemma 4.22. *Suppose that* (C15)–(C17), (C18), (C19) *and* (C20) *hold.*
Then the sequence $\{u_n\}$ *in* (4.78) *has a convergent subsequence. Moreover,*
E_ε *possesses a nonzero critical point* u *in* $H^1(R^3)$ *and* $E_\varepsilon(u) > 0$.

Proof. By Lemma 4.20, the sequence $\{u_n\}$ in (4.78) is bounded in $H^1(\mathbb{R}^3)$.
We may assume that, up to a subsequence $u_n \rightharpoonup u$ weakly in $H^1(\mathbb{R}^3)$ for
some $u \in H^1(\mathbb{R}^3)$. Now, we shall show that $\|u_n\| \to \|u\|$ as $n \to \infty$.

By (4.78), we have

$$\langle E'_\varepsilon(u_n), u_n \rangle = \int_{\mathbb{R}^3} (|\nabla u_n|^2 + u_n^2 + K(x)\phi_{\varepsilon,K}(u_n)u_n^2 - a(x)f(u_n)u_n)dx = o(1),$$

$$(4.104)$$

and

$$\langle E'_\varepsilon(u_n), u \rangle = \int_{\mathbb{R}^3} (\nabla u_n \cdot \nabla u + u_n u + K(x)\phi_{\varepsilon,K}(u_n)u_n u - a(x)f(u_n)u)dx = o(1).$$

$$(4.105)$$

Since $u_n \rightharpoonup u$ weakly in $H^1(\mathbb{R}^3)$, we obtain

$$\int_{\mathbb{R}^3} (\nabla u_n \cdot \nabla u + u_n u)dx = \int_{\mathbb{R}^3} (|\nabla u|^2 + |u|^2)dx + o(1). \qquad (4.106)$$

Moreover, by the Hölder inequality, (4.81), Lemma 4.21 and $u_n \to u$
strongly in $L^2_{loc}(\mathbb{R}^3)$, for any $\varepsilon' > 0$ and n large enough, one has

$$\int_{|x| \geq R(\varepsilon')} a(x)f(u_n)u_n dx - \int_{|x| \geq R(\varepsilon')} a(x)f(u_n)u dx$$

$$= \int_{|x| \geq R(\varepsilon')} a(x)f(u_n)(u_n - u)dx$$

$$\leq \int_{|x| \geq R(\varepsilon')} |a(x)f(u_n)||u_n - u|dx$$

$$\leq \left(\int_{|x| \geq R(\varepsilon')} |a^2(x)f^2(u_n)|dx\right)^{\frac{1}{2}} \left(\int_{|x| \geq R(\varepsilon')} |u_n - u|^2 dx\right)^{\frac{1}{2}}$$

$$\leq \theta \left(\int_{|x| \geq R(\varepsilon')} |u_n^2|dx\right)^{\frac{1}{2}} \left(\int_{|x| \geq R(\varepsilon')} |u_n - u|^2 dx\right)^{\frac{1}{2}}$$

$$\leq \theta \left(\int_{|x| \geq R(\varepsilon')} (|\nabla u_n|^2 + |u_n^2|)dx\right)^{\frac{1}{2}} \left(\int_{|x| \geq R(\varepsilon')} |u_n - u|^2 dx\right)^{\frac{1}{2}}$$

$$\leq C_{11}\varepsilon'.$$

This and the compactness of embedding $H^1(\mathbb{R}^3) \hookrightarrow L^2_{loc}(\mathbb{R}^3)$ imply

$$\int_{\mathbb{R}^3} a(x)f(u_n)u_n dx = \int_{\mathbb{R}^3} a(x)f(u_n)u dx + o(1). \qquad (4.107)$$

Furthermore, because $u_n \rightharpoonup u$ weakly in $H^1(\mathbb{R}^3)$, we obtain

(a) $u_n \rightharpoonup u$ weakly in $L^6(\mathbb{R}^3)$,

(b) $u_n^2 \to u^2$ strongly in $L^3_{loc}(\mathbb{R}^3)$,

(c) $\phi_{\varepsilon,K}(u_n) \rightharpoonup \phi_{\varepsilon,K}(u)$ weakly in $D^{1,2}(\mathbb{R}^3)$,

(d) $\phi_{\varepsilon,K}(u_n) \to \phi_{\varepsilon,K}(u)$ strongly in $L^6_{loc}(\mathbb{R}^3)$.

For any choice of $\varepsilon' > 0$ and $\rho > 0$, the relation

$$\|u_n - u\|_{6,B_\rho(0)} < \varepsilon' \tag{4.108}$$

holds for large n. Because u_n is bounded in $H^1(\mathbb{R}^3)$ and the continuity of the Sobolev embedding of $D^{1,2}(\mathbb{R}^3)$ in $L^6(\mathbb{R}^3)$, then $\phi_{\varepsilon,K}(u_n)$ is bounded in $D^{1,2}(\mathbb{R}^3)$ and in $L^6(\mathbb{R}^3)$. Moreover, $K \in L^2(\mathbb{R}^3)$ implies that Ku_n^2 and Ku^2 belong to $L^{\frac{6}{5}}(\mathbb{R}^3)$ and that to any $\varepsilon' > 0$ there exists $\widetilde{\rho} = \widetilde{\rho}(\varepsilon')$ such that

$$\|K\|_{2,\mathbb{R}^3 \setminus B_\rho(0)} < \varepsilon', \quad \forall \rho \ge \widetilde{\rho}. \tag{4.109}$$

By the Hölder inequality, (4.108) and (4.109), we obtain

$$\int_{\mathbb{R}^3} K(x)\phi_{\varepsilon,K}(u_n)u_n^2 dx - \int_{\mathbb{R}^3} K(x)\phi_{\varepsilon,K}(u_n)u_n u\, dx$$

$$\le \int_{\mathbb{R}^3} |K(x)\phi_{\varepsilon,K}(u_n)u_n(u_n - u)|dx$$

$$\le \|\phi_{\varepsilon,K}(u_n)\|_6 \left(\int_{\mathbb{R}^3} |K(x)u_n(u_n - u)|^{\frac{6}{5}}dx \right)^{\frac{5}{6}}$$

$$= \|\phi_{\varepsilon,K}(u_n)\|_6 \left(\int_{\mathbb{R}^3 \setminus B_\rho(0)} |K(x)u_n(u_n - u)|^{\frac{6}{5}}dx \right.$$

$$\left. + \int_{B_\rho(0)} |K(x)u_n(u_n - u)|^{\frac{6}{5}}dx \right)^{\frac{5}{6}}$$

$$\le C_{12}\left(\|K\|_{2,\mathbb{R}^3 \setminus B_\rho(0)}^{\frac{6}{5}}\|u_n(u_n - u)\|_3^{\frac{6}{5}} \right.$$

$$\left. + \|K\|_2^{\frac{6}{5}} \left[\int_{B_\rho(0)} |u_n|^6 dx \right]^{\frac{1}{5}} \left[\int_{B_\rho(0)} |u_n - u|^6 dx \right]^{\frac{1}{5}} \right)^{\frac{5}{6}}$$

$$\le C_{12}\left(\varepsilon'^{\frac{6}{5}}\|u_n(u_n - u)\|_3^{\frac{6}{5}} + \|K\|_2^{\frac{6}{5}}\|u_n\|_{6,B_\rho(0)}^{\frac{6}{5}}\|u_n - u\|_{6,B_\rho(0)}^{\frac{6}{5}} \right)^{\frac{5}{6}}$$

$$\le C_{12}\left(\varepsilon'^{\frac{6}{5}}\|u_n(u_n - u)\|_3^{\frac{6}{5}} + \varepsilon'^{\frac{6}{5}}\|K\|_2^{\frac{6}{5}}\|u_n\|_{6,B_\rho(0)}^{\frac{6}{5}} \right)^{\frac{5}{6}}$$

$$\le C_{13}\varepsilon'.$$

This yields

$$\int_{\mathbb{R}^3} K(x)\phi_{\varepsilon,K}(u_n)u_n^2 dx = \int_{\mathbb{R}^3} K(x)\phi_{\varepsilon,K}(u_n)u_n u dx + o(1). \quad (4.110)$$

By (4.104), (4.105), (4.106), (4.107) and (4.110), we have

$$\langle E_\varepsilon'(u_n), u_n - u \rangle = \int_{\mathbb{R}^3} (|\nabla u_n|^2 + u_n^2) dx - \int_{\mathbb{R}^3} (|\nabla u|^2 + |u|^2) dx = o(1).$$

This yields that $\|u_n\| \to \|u\|$ as $n \to \infty$ and u is a nonzero critical point of E_ε in $H^1(\mathbb{R}^3)$ and $E_\varepsilon(u) > 0$ by Mountain Pass Theorem in [22]. $\quad\square$

Proof of Theorem 4.4. Set the Nehari manifold

$$N_\varepsilon = \{u_\varepsilon \in H^1(\mathbb{R}^3) \setminus \{0\} : \langle E_\varepsilon'(u_\varepsilon), u_\varepsilon \rangle = 0\}.$$

By Lemma 4.22, we know that N_ε is not empty. For any $u_\varepsilon \in N_\varepsilon$, by Lemma 4.15 and (C20), we have

$$o(1) = \langle E_\varepsilon'(u_\varepsilon), u_\varepsilon \rangle = \|u_\varepsilon\|^2 + \int_{\mathbb{R}^3} K(x)\phi_{\varepsilon,K}(u_\varepsilon)u_\varepsilon^2 dx - \int_{\mathbb{R}^3} a(x)f(u_\varepsilon)u_\varepsilon dx$$

$$\geq \|u_\varepsilon\|^2 - \int_{\mathbb{R}^3} a(x)f(u_\varepsilon)u_\varepsilon dx.$$

Now, choose $\tilde{\varepsilon}$ such that $0 < \tilde{\varepsilon} < \min\{1, C_2^{-1}\}$ where C_2 is as in (4.76). By (4.74), (4.76) and the Sobolev inequality, we deduce

$$\left| \int_{\mathbb{R}^3} a(x)f(u_\varepsilon)u_\varepsilon dx \right| \leq \tilde{\varepsilon}C_2 \int_{\mathbb{R}^3} |u_\varepsilon|^2 dx + C_2 C_\varepsilon \int_{\mathbb{R}^3} |u_\varepsilon|^6 dx$$

$$\leq \tilde{\varepsilon}C_2 \|u_\varepsilon\|^2 + C_2 C_\varepsilon (S^*)^3 \|u_\varepsilon\|^6.$$

Therefore, for every $u_\varepsilon \in N$, we have

$$o(1) \geq \|u_\varepsilon\|^2 - \tilde{\varepsilon}C_2 \|u_\varepsilon\|^2 - C_2 C_\varepsilon (S^*)^3 \|u_\varepsilon\|^6. \quad (4.111)$$

We recall that $u_\varepsilon \neq 0$ whenever $u_\varepsilon \in N_\varepsilon$ and (4.111) implies

$$\|u_\varepsilon\| \geq \sqrt[4]{\frac{1 - \tilde{\varepsilon}C_2}{C_2 C_\varepsilon (S^*)^3}} > 0, \quad \text{for all } u_\varepsilon \in N_\varepsilon.$$

Hence any limit point of a sequence in the Nehari manifold is different from zero.

Now, we shall prove that E_ε is bounded from below on N_ε, that is, there exists $M > 0$ such that $E_\varepsilon(u_\varepsilon) \geq -M$ for all $u_\varepsilon \in N_\varepsilon$. Otherwise, there exists $\{u_n\} \subset N_\varepsilon$ such that

$$E_\varepsilon(u_n) < -n \text{ for all } n \in \mathbb{N}. \quad (4.112)$$

From (4.77), we have

$$E_\varepsilon(u_n) \geq \frac{1}{4}\|u_n\|^2 - C_3\|u_n\|^6.$$

This and (4.112) imply that $\|u_n\| \to +\infty$. Let $w_n = u_n\|u_n\|^{-1}$, there is $w \in H^1(\mathbb{R}^3)$ such that

$$\begin{cases} w_n \rightharpoonup w \text{ weakly in } H^1(\mathbb{R}^3), \\ w_n \to w \text{ a.e. in } \mathbb{R}^3, \\ w_n \to w \text{ strongly in } L^2_{loc}(\mathbb{R}^3) \end{cases}$$

as $n \to \infty$. Note that $E'_\varepsilon(u_n) = 0$ for $u_n \in N_\varepsilon$, as in the proof of Lemma 4.20, we obtain that $\|u_n\| \to +\infty$ is impossible. Then, E_ε is bounded from below on N_ε. So, we may define

$$\bar{c} = \inf\{E_\varepsilon(u_\varepsilon), \ u_\varepsilon \in N_\varepsilon\},$$

and $\bar{c} \geq -M$. Let $\{\bar{u}_n\} \subset N_\varepsilon$ be such that $E_\varepsilon(\bar{u}_n) \to \bar{c}$ as $n \to \infty$. Following almost the same procedures as proofs of Lemmas 4.20, 4.21, 4.22, we can show that $\{\bar{u}_n\}$ is bounded in $H^1(\mathbb{R}^3)$ and it has a convergence subsequence which strongly converges to $u_\varepsilon \in H^1(\mathbb{R}^3)\backslash\{0\}$. Then $E_\varepsilon(u_\varepsilon) = \bar{c}$ and $E'_\varepsilon(u_\varepsilon) = 0$. Therefore, $(u_\varepsilon, \phi_{\varepsilon,K}(u_\varepsilon))$ is a ground state solution of system (4.66). $\qquad\square$

Here we shall study the behavior of the solution $(u_\varepsilon, \phi_{\varepsilon,K}(u_\varepsilon))$ obtained via Theorem 4.4.

Proof of Theorem 4.5. From Lemma 2.3, with $f_\varepsilon = f = K(x)u^2$ for all $\varepsilon > 0$, we can easily check that when $\varepsilon \to 0$ we have

$$E_\varepsilon(u) \to E_0(u) \quad \text{for all } u \in H^1(\mathbb{R}^3). \tag{4.113}$$

From Lemma 4.19, we know that there exists $v \in H^1(\mathbb{R}^3)$ such that $E_0(v) < 0 = E_0(0)$. By (4.113), we deduce that there exists $\varepsilon^* > 0$ small enough such that $E_\varepsilon(v) < E_0(0)$ for $\varepsilon \in (0, \varepsilon^*)$. Since E_ε attains its minimum on $H^1(\mathbb{R}^3)$ at u_ε, we obtain that

$$E_\varepsilon(u_\varepsilon) \leq E_\varepsilon(v) < E_0(0) \quad \text{for all } u \in H^1(\mathbb{R}^3). \tag{4.114}$$

First of all, we claim that $\{u_\varepsilon\}$ is bounded in $H^1(\mathbb{R}^3)$. Indeed, by (4.114), there exists $\tilde{c} \in \mathbb{R}$ independent of ε such that $E_\varepsilon(u_\varepsilon) \leq \tilde{c}$. Now to prove $\{u_\varepsilon\}$ is bounded in $H^1(\mathbb{R}^3)$, assume by contradiction that there exists a subsequence, denoted by $\{u_\varepsilon\}$, satisfying

$$E_\varepsilon(u_\varepsilon) \leq \tilde{c} \quad \text{and} \quad \|u_\varepsilon\| \to \infty.$$

Set $\|u_\varepsilon\| := \alpha_\varepsilon \to \infty$. Define $w_\varepsilon = u_\varepsilon \|u_\varepsilon\|^{-1} = \alpha_\varepsilon^{-1} u_\varepsilon$. Clearly, $\{w_\varepsilon\}$ is bounded and $\|w_\varepsilon\| = 1$ in $H^1(\mathbb{R}^3)$. An argument similar to the one used in the proof of Lemma 4.20, we can obtain contradiction, so $\{u_\varepsilon\}$ is bounded in $H^1(\mathbb{R}^3)$. Therefore there exists u_0 such that, up to a subsequence, we have

$$u_\varepsilon \rightharpoonup u_0 \text{ weakly in } H^1(\mathbb{R}^3),$$
$$u_\varepsilon \to u_0 \text{ a.e. in } \mathbb{R}^3,$$
$$u_\varepsilon \to u_0 \text{ strongly in } L^p_{loc}(\mathbb{R}^3)(p \in [2,6])$$

as $\varepsilon \to 0$. Since $(u_\varepsilon, \phi_{\varepsilon,K}(u_\varepsilon))$ is the ground state solution of system (4.66), by Proposition 4.3 and (4.73), for $\psi \in C_c^\infty(\Omega)$ and a compact Ω such that supp$\psi \subset \Omega$, we obtain

$$\int_{\mathbb{R}^3} \nabla u_\varepsilon \cdot \nabla \psi dx + \int_{\mathbb{R}^3} u_\varepsilon \psi dx + \int_\Omega K(x)\phi_{\varepsilon,K}(u_\varepsilon)u_\varepsilon \psi dx - \int_\Omega a(x)f(u_\varepsilon)\psi dx = 0. \tag{4.115}$$

Since $u_\varepsilon \rightharpoonup u_0$ weakly in $H^1(\mathbb{R}^3)$, then

$$\int_{\mathbb{R}^3} \nabla u_\epsilon \cdot \nabla \psi dx + \int_{\mathbb{R}^3} u_\varepsilon \psi dx \to \int_{\mathbb{R}^3} \nabla u_0 \cdot \nabla \psi dx + \int_{\mathbb{R}^3} u_0 \psi dx \text{ as } \varepsilon \to 0^+. \tag{4.116}$$

Next, we shall prove that $K(x)u_\varepsilon^2 \rightharpoonup K(x)u_0^2$ weakly in $L^{6/5}(\mathbb{R}^3)$. In fact, let $\xi \in L^6(\mathbb{R}^3) = (L^{6/5}(\mathbb{R}^3))'$, then $K(x)\xi \in L^{3/2}(\mathbb{R}^3)$. Consider the subset of $\mathbb{R}^3$, $A_\lambda := \{x : |K(x)\xi| > \lambda\}$ and a compact subset Ω_0 of A_λ suitably chosen later. By the Hölder inequality, embedding theorem and $u_\varepsilon \rightharpoonup u_0$ in $H^1(\mathbb{R}^3)$, we write

$$\int_{\mathbb{R}^3} K(x)(u_\varepsilon - u_0)^2 \xi dx$$
$$= \int_{\mathbb{R}^3 - A_\lambda} K(x)(u_\varepsilon - u_0)^2 \xi dx + \int_{A_\lambda - \Omega_0} K(x)(u_\varepsilon - u_0)^2 \xi dx$$
$$+ \int_{\Omega_0} K(x)(u_\varepsilon - u_0)^2 \xi dx$$
$$\leq \lambda \|u_\varepsilon - u_0\|_2^2 + \|K(x)\xi\|_{L^{3/2}(A_\lambda - \Omega_0)}\|u_\varepsilon - u_0\|_6^2$$
$$+ \|K(x)\xi\|_{L^{3/2}(\Omega_0)}\|u_\varepsilon - u_0\|_{L^6(\Omega_0)}^2$$
$$\leq \lambda C_{14} + C_{15}\|K(x)\xi\|_{L^{3/2}(A_\lambda - \Omega_0)}$$
$$+ \|K(x)\xi\|_{L^{3/2}(\Omega_0)}\|u_\varepsilon - u_0\|_{L^6(\Omega_0)}^2. \tag{4.117}$$

For a given arbitrary $\delta > 0$, we fix first λ such that $\lambda C_{14} < \frac{\delta}{3}$. Next we choose a compact subset $\Omega_0 \subset A_\lambda$ such that $C_{15}\|K(x)\xi\|_{L^{3/2}(A_\lambda - \Omega_0)} <$

$\frac{\delta}{3}$ and subsequence of $\{u_\varepsilon\}$, still denoted by $\{u_\varepsilon\}$, $\|K(x)\xi\|_{L^{3/2}(\Omega_0)}\|u_\varepsilon - u_0\|^2_{L^6(\Omega_0)} < \frac{\delta}{3}$. Together with (4.117), we obtain

$$\int_{\mathbb{R}^3} K(x)(u_\varepsilon - u_0)^2 \xi dx \to 0 \qquad (4.118)$$

as $\varepsilon \to 0$. Since $u_0^2\xi^2 \in L^{3/2}(\mathbb{R}^3)$, by the same method, we can prove

$$\int_{\mathbb{R}^3} u_0^2\xi^2(u_\varepsilon - u_0)^2 dx \to 0 \qquad (4.119)$$

as $\varepsilon \to 0$. By the Hölder inequality, (4.118) and (4.119), we deduce

$$\int_{\mathbb{R}^3} K(x)u_\varepsilon^2\xi dx - \int_{\mathbb{R}^3} K(x)u_0^2\xi dx$$

$$= \int_{\mathbb{R}^3} K(x)(u_\varepsilon^2 - u_0^2)\xi dx$$

$$= \int_{\mathbb{R}^3} K(x)(u_\varepsilon - u_0)^2\xi dx + 2\int_{\mathbb{R}^3} K(x)(u_\varepsilon - u_0)u_0\xi dx$$

$$\leq \int_{\mathbb{R}^3} K(x)(u_\varepsilon - u_0)^2\xi dx + 2\|K(x)\|_2\left(\int_{\mathbb{R}^3} u_0^2\xi^2(u_\varepsilon - u_0)^2 dx\right)^{\frac{1}{2}}$$

$$\to 0$$

as $\varepsilon \to 0$. We infer that

$$\int_{\mathbb{R}^3} K(x)u_\varepsilon^2\xi dx \to \int_{\mathbb{R}^3} K(x)u_0^2\xi dx$$

as $\varepsilon \to 0$. Therefore $K(x)u_\varepsilon^2 \rightharpoonup K(x)u_0^2$ weakly in $L^{6/5}(\mathbb{R}^3)$.

Since $K(x)u_\varepsilon^2 \rightharpoonup K(x)u_0^2$ weakly in $L^{6/5}(\mathbb{R}^3)$, by Lemma 4.17, we obtain

$$\phi_{\varepsilon,K}(u_\varepsilon) \rightharpoonup \phi_{0,K}(u_0) \text{ weakly in } L^6(\mathbb{R}^3).$$

So, for all $\varrho \in L^{6/5}(\mathbb{R}^3)$, we have

$$\int_{\mathbb{R}^3} \phi_{\varepsilon,K}(u_\varepsilon)\varrho dx \to \int_{\mathbb{R}^3} \phi_{0,K}(u_0)\varrho dx. \qquad (4.120)$$

Furthermore, since $u_\varepsilon \rightharpoonup u_0$ weakly in $H^1(\mathbb{R}^3)$, by the Sobolev embedding theorem, we have

$$u_\varepsilon \to u_0 \text{ strongly in } L^p(\Omega)(p \in [2,6]). \qquad (4.121)$$

Clearly, $u_0\psi \in L^{6/5}(\mathbb{R}^3)$. Together with $K \in L^2(\mathbb{R}^3)$, by the Hölder inequality, (4.121) and (4.120), we infer that

$$\int_\Omega K(x)\phi_{\varepsilon,K}(u_\varepsilon)u_\varepsilon\psi dx - \int_\Omega K(x)\phi_{0,K}(u_0)u_0\psi dx$$

$$\leq \int_\Omega |K(x)||\phi_{\varepsilon,K}(u_\varepsilon)||u_\varepsilon - u_0||\psi|dx \qquad (4.122)$$

$$+ \int_\Omega |K(x)||\phi_{\varepsilon,K}(u_\varepsilon) - \phi_{0,K}(u_0)||u_0\psi|dx$$

$$\leq C_{16}\|K(x)\|_2\|\phi_{\varepsilon,K}(u_\varepsilon)\|_6\|u_\varepsilon - u_0\|_{6,\Omega}$$

$$+ C_{17}\int_{\mathbb{R}^3} |\phi_{\varepsilon,K}(u_\varepsilon) - \phi_{0,K}(u_0)||u_0\psi|dx$$

$$\to 0. \qquad (4.123)$$

By (C15)–(C17), (4.79) and (4.76), it is easy to prove that

$$\lim_{\varepsilon\to 0} \int_\Omega a(x)f(u_\varepsilon)\psi dx = \int_\Omega a(x)f(u_0)\psi dx. \qquad (4.124)$$

If not, there exists $\delta_0 > 0$, for all $\varepsilon_0 > 0$, exists $\varepsilon \in (0, \varepsilon_0)$ such that

$$\left|\int_\Omega a(x)f(u_\varepsilon)\psi dx - \int_\Omega a(x)f(u_0)\psi dx\right| \geq \delta_0.$$

That is, there exists $\delta_0 > 0$ and $\varepsilon_n \in (0, \varepsilon_0)$ for $\varepsilon_n \to 0$ such that

$$\left|\int_\Omega a(x)f(u_{\varepsilon_n})\psi dx - \int_\Omega a(x)f(u_0)\psi dx\right| \geq \delta_0 \qquad (4.125)$$

and subsequence of u_{ε_n}, still denoted by u_{ε_n}, is convergent in $L^p(\Omega)(p \in [2,6])$. By (4.121) and the uniqueness of limit, $u_{\varepsilon_n} \to u_0$ strongly in $L^p(\Omega)(p \in [2,6])$. Thus, we have

$$\lim_{n\to\infty} \int_{\mathbb{R}^3} a(x)(f(u_{\varepsilon_n}) - f(u_0))\psi dx = 0.$$

This contradicts (4.125). So, (4.124) holds.

Consequently, letting $\varepsilon \to 0$ in (4.115) and according to (4.116), (4.123) and (4.124), u_0 satisfies

$$\int_{\mathbb{R}^3} \nabla u_0 \cdot \nabla\psi dx + \int_{\mathbb{R}^3} u_0\psi dx + \int_{\mathbb{R}^3} K(x)\phi_{0,K}(u_0)u_0\psi dx$$

$$- \int_{\mathbb{R}^3} a(x)f(u_0)\psi dx = 0$$

for all $\psi \in C_c^\infty(\Omega)$. We conclude that u_ε converges weakly in $H^1(\mathbb{R}^3)$ to u_0. Thus, $\phi_{\varepsilon,K}(u_\varepsilon) \rightharpoonup \phi0, K(u_0)$ in $D^{1,2}(\mathbb{R}^3)$ by Lemma 4.17, where $(u_0, \phi0, K(u_0))$ is a solution of system (4.66) with $\varepsilon = 0$. $\square$

Notes and Comments

When $\varepsilon = 0$, this Schrödinger–Poisson system arises in an interesting physical model which describes the interaction of a charged particle with electromagnetic field (see [255] and the references therein). When $\varepsilon \neq 0$, system (4.66) firstly arises this form like

$$\begin{cases} i\partial_t u = -\frac{1}{2}\Delta u + (V + \phi(x))u, & x \in \mathbb{R}^3, \\ -\text{div}[\varepsilon(\nabla\phi)\nabla\phi] = |u|^2 - n^*, & x \in \mathbb{R}^3, \\ u(x,0) = u(x), & x \in \mathbb{R}^3, \end{cases}$$

which corresponds to a quantum mechanical model where the quantum effects are important, as in the case of microstructures (see for example Markowich, Ringhofer and Schmeiser [256]). The charge density $n(x,t)$ derives from the Schrödinger wave function $u(x,t)$ by $n(x,t) = |u(x,t)|^2$, while n^* and V represent respectively a dopant-density and a real effective potential which are time-independent. More details dealing with the phenomenon may be found in [257, 258] and references therein. After that the field dependent dielectric constant in Poisson equation has the form

$$\varepsilon(\nabla\phi) = \varepsilon^0 + \varepsilon^1 |\nabla\varphi|^2, \ \varepsilon^0, \ \varepsilon^1 > 0.$$

Existence and uniqueness of global strong solutions and existence results of solutions of the form $u(x,t) = e^{i\omega t}u(x)(\omega, \ u(x) \in \mathbb{R})$ are obtained under suitable conditions, respectively. Moreover, in [253], authors obtained that the existence of standing waves (actually ground states) solutions for the Schrödinger–Poisson system with $\varepsilon^0 = 1$ and $\varepsilon^1 = \varepsilon^4$ of

$$\begin{cases} -\frac{1}{2}\Delta u + (V + \phi(x))u = 0, & x \in \mathbb{R}^3, \\ -\text{div}[(1 + \varepsilon^4|\nabla\phi|^2)\nabla\phi] = |u|^2 - n^*, & x \in \mathbb{R}^3 \end{cases} \tag{4.126}$$

and with their asymptotic behavior when the nonlinearity coefficient in the Poisson equation ε goes to zero with suitable potential V.

From the mathematical view, if the Schrödinger equation with only one nonlinear nonlocal term $\phi(x)u$ in system (4.126) is replaced by the other different version of Schrödinger equations which have other nonlinear terms besides the nonlinear nonlocal term, we want to know that whether ground state solutions exist and if exists whether they converge to ones of the corresponding system for $\varepsilon = 0$.

When $\phi(x) \not\equiv 0$ and $\varepsilon \equiv 0$, Cerami and Vaira in [239] studied system (4.66) with $f(t) = |t|^{p-1}u(p \in (3,5))$ and obtained the existence of positive ground state solutions by minimizing the corresponding energy functional restricted to the Nehari manifold when K and a satisfy different assumptions, respectively. Sun, Chen and Nieto in [193] also studied system (4.66) with general f which is asymptotically linear at infinity and obtain the existence of a positive ground state solution under suitable assumptions about K and a by Mountain Pass Theorem.

When $\phi(x) \not\equiv 0$ and $\varepsilon \not\equiv 0$, there are some results with respect to Schrödinger–Poisson systems depending on a parameter ε, see [259–263] and the references therein. In [259–262], the perturbation parameter ε appears in the first Schrödinger equation of system, the domain is a flat domain or a Riemannian manifold in $\mathbb{R}^3$ and concentration of solutions were mainly studied. In [263], the perturbation parameter ε appears in the exponent of the nonlinearity ($f(t) = t^{6-\varepsilon}$) and multiplicity positive solutions are obtained. But there are few results for system (4.66) with $\phi(x) \not\equiv 0$ where the perturbation parameter $\varepsilon \not\equiv 0$ appears in the second equation. This section was obtained in [264].

4.3   Klein–Gordon–Maxwell Systems

This type of system arises in a very interesting physical context: as a model describing the nonlinear Klein–Gordon field interacting with the electromagnetic field. More specifically, it represents a solitary wave $\psi = u(x)e^{i\omega t}$ in equilibrium with a purely electrostatic field $\mathbf{E} = -\nabla\phi(x)$.

4.3.1   *Infinitely Many Solutions*

In this section we consider the following system

$$\begin{cases} -\Delta u + V(x)u - (2\omega + \phi)\phi u = f(x, u), & x \in \mathbb{R}^3, \\ \Delta\phi = (\omega + \phi)u^2, & x \in \mathbb{R}^3, \end{cases} \qquad (4.127)$$

where $\omega > 0$ is a constant, u, $\phi : \mathbb{R}^3 \to \mathbb{R}$, $V : \mathbb{R}^3 \to \mathbb{R}$ is a potential function.

In this section, we study problem (4.127) which is considered by X. He [265] at first. In that paper, he got infinitely many solutions for Klein-Gordon-Maxwell System with non-constant potential. In this section, we want to improve and complement his results and also remark that we can use similar arguments to get infinitely many solutions for Schrödinger–Poisson equation with non-constant potential. We study the superlinear case, firstly, then the sublinear one.

In order to prove our statements, we need the following assumptions: the potential $V : \mathbb{R}^3 \to \mathbb{R}$ is a continuous function satisfying

(C21) $V(x) \in C(\mathbb{R}^3, \mathbb{R})$, $\inf_{x \in \mathbb{R}^3} V(x) = V_0 > 0$ and there exists $v_0 > 0$ such that

$$\lim_{|y| \to \infty} \text{meas}\left\{x \in \mathbb{R}^3 : |x - y| \leq v_0, V(x) \leq M\right\} = 0. \quad \forall M > 0.$$

The condition (C21) was introduced by Bartsch *et al.* [189, Lemma 3.1] to guarantee the compactness of embeddings of the work spaces and generalizes a previous assumption for V given in [109].

Now, we make some assumptions on the nonlinear term.

(C22) $f \in C(\mathbb{R}^3 \times \mathbb{R}, \mathbb{R})$ and $|f(x,t)| \leq C(|t| + |t|^{p-1})$ for some $4 < p < 2^* = 6$, where C is a positive constant.

(C23) $\frac{F(x,t)}{t^4} \to +\infty$ as $|t| \to +\infty$ uniformly in $x \in \mathbb{R}^3$, where $F(x,t) = \int_0^t f(x,s)ds$.

(C24) Let $\mathcal{F}(x,t) := \frac{1}{4}f(x,t)t - F(x,t)$, there exist $C > 0$ and $r_0 > 0$ such that if $|t| \geq r_0$, then $\mathcal{F}(x,t) \geq -C|t|^2$ uniformly for $x \in \mathbb{R}^3$.

(C25) $f(x,-t) = -f(x,t)$, for all $x \in \mathbb{R}^3$, $t \in \mathbb{R}$.

The following result holds:

Theorem 4.6. *Under the assumptions* (C21), (C22)–(C25), *problem* (4.127) *has a sequence of solutions* $\{(u_n, \phi_n)\} \subset E \times D^{1,2}(\mathbb{R}^3)$ *satisfying*

$$\frac{1}{2} \int_{\mathbb{R}^3} \left(|\nabla u_n|^2 + V(x)u_n^2 - |\nabla \phi_n|^2 - (2\omega + \phi_n)\phi_n u_n^2\right) dx$$

$$- \int_{\mathbb{R}^3} F(x, u_n)dx \to +\infty.$$

Remark 4.7. In [265], problem (4.127) possesses infinitely many nontrivial solutions if $f(x,t)$ satisfies the following *global Ambrosetti–Rabinowitz condition* ((AR) for short):

$$\exists \theta > 4 \text{ such that } \theta F(x,t) \leq tf(x,t), \qquad \text{for } x \in \mathbb{R}^3 \text{ and } t \in \mathbb{R}.$$

The condition (AR) is widely assumed in the studies of elliptic problem by variational methods. As it is known, the condition (AR) is employed not only to show that the Euler–Lagrange functional associated has a mountain pass geometry, but also to guarantee that the Palais–Smale or Cerami sequences are bounded. Although (AR) is a quite natural condition, it is somewhat restrictive because it eliminates many nonlinearities such as $f(t) = t^3 \ln(1 + |t|)$. Observe that (AR) implies the weaker condition (C23), which is called f is superlinear at infinity.

We also obverse that problem (4.127) can be compared with recent interesting studies (see [218, 221, 222, 228, 236, 266–271] and references therein) on the Schrödinger–Poisson equation with non-constant potential

$$\begin{cases} -\Delta u + V(x)u + \phi u = f(x, u) & \text{in } \mathbb{R}^3, \\ -\Delta \phi = u^2, & \text{in } \mathbb{R}^3. \end{cases} \tag{4.128}$$

In [222], Yang and Han studied (4.128) and replaced the (AR) condition by (C24) and the following conditions:

(YH1) $\frac{f(x,t)}{t^3}$ is increasing in $t > 0$ and decreasing in $t < 0$, $\forall x \in \mathbb{R}^3$.
(YH2) There exists $\gamma > 0$ such that $F(x, t) \geq \gamma |t|^4$, for all $x \in \mathbb{R}^3$, $t \in \mathbb{R}$.

or

(YH3) There exists a constant $\frac{6}{5}(p-1) \leq \tau \leq 2(p-1)$ such that $\mathcal{F}(x, t) \geq c|t|^\tau$, for all $x \in \mathbb{R}^3$, $t \in \mathbb{R}$.

Notice that from (YH3), we have

$$\lim_{|t| \to +\infty} (f(x, t)t - 4F(x, t)) = +\infty, \text{ uniformly in } x \in \mathbb{R}^3,$$

used by He in [265]. Actually, due to the above conditions, we can get (C24) is satisfied.

In [236], Li and Chen use the following condition, first considered by Jeanjean [243] to replace (AR) condition, to get infinitely many solutions:

(LC) There exists a constant $\theta \geq 1$ such that

$$\theta H(x, t) \geq H(x, st)$$

for all $x \in \mathbb{R}^3$, $t \in \mathbb{R}$ and $s \in [0, 1]$, where $H(x, t) = tf(x, t) - 4F(x, t)$.

We can observe that when $s = 0$, then $H(x, t) = tf(x, t) - 4F(x, t) \geq 0$, but from our condition (C24), $H(x, t)$ may assume negative values.

In [221], Alves *et al.* used the following *global condition* to replace the (AR) condition:

(ASS) There exists $0 \leq \sigma < \alpha$ such that $tf(t) - 4F(t) \geq -\sigma t^2$, for all $t \in \mathbb{R}$, where $0 < \alpha \leq V(x)$, a constant.

But in our article, we only need the *local condition* (C24) to get the non-trivial solutions.

In [268], Huang and Tang use the following condition to replace the (AR) condition:

(HT) Let $\mathcal{F}(x,t) := \frac{1}{4}f(x,t)t - F(x,t)$, there exist $C > 0$ and $r_0 > 0$ such that if $|t| \geq r_0$, then $\mathcal{F}(x,t) \geq 0$ for all $x \in \mathbb{R}^3$.

But, in our theorem, $\mathcal{F}(x,t)$ may assume negative values.

Summing up, we can see that condition (C24) unifies all the conditions mentioned above which are crucial to get compactness of the functional I associated with the problem (4.127) or (4.128).

Remark 4.8. We also observe that we give a less order of functions f at zero. Indeed, if (C22) holds, certainly f has order less than $|t|^\alpha$, $0 < \alpha < 1$. Thus, we generalize condition (h_1) in [265] which is also a crucial condition to get infinitely many nontrivial solutions for the Schrödinger–Poisson with non-constant potential case, see references [218, 221, 222, 236].

Next, we study the sublinear case. Firstly, we assume that the nonlinear term satisfies the following hypothesis

(C26) $F(x,t) = b(x)|u|^p$, where $b : \mathbb{R}^3 \to \mathbb{R}^+$ is a positive continuous function such that $b \in L^{\frac{2}{2-p}}(\mathbb{R}^3)$ and $1 < p < 2$ is a constant.

We obtain the following results.

Theorem 4.7. *Under the assumptions* (C21) *and* (C26), *problem* (4.127) *has a sequence of solutions* $\{(u_n, \phi_n)\} \subset E \times D^{1,2}(\mathbb{R}^3)$ *satisfying*

$$\frac{1}{2}\int_{\mathbb{R}^3} \left(|\nabla u_n|^2 + V(x)u_n^2 - |\nabla \phi_n|^2 - (2\omega + \phi_n)\phi_n u_n^2 \right) dx$$

$$- \int_{\mathbb{R}^3} F(x, u_n)dx \to 0^-.$$

Actually, we shall prove a more general result than Theorem 4.7.

Theorem 4.8. *Assume that assumptions* (C21) *and* (C25) *are satisfied and:*

(C26′) *There exist* p, σ, $\gamma \in (1,2)$ *and* $\nu \in (2,6)$ *such that*

$$b(x)|t|^p \leq f(x,t)t \text{ and } |f(x,t)| \leq m(x)|t|^{\sigma-1} + h(x)|t|^{\gamma-1} + C|t|^{\nu-1}$$

for all $(x,t) \in \mathbb{R}^3 \times R$, *where* $b, m, h : \mathbb{R}^3 \to \mathbb{R}$ *are positive continuous functions satisfying* $b \in L^{\frac{2}{2-p}}(\mathbb{R}^3)$, $m \in L^{\frac{2}{2-\sigma}}(\mathbb{R}^3)$, $h \in L^{\frac{2}{2-\gamma}}(\mathbb{R}^3)$.

Then problem (4.127) *has a sequence of solutions* $\{(u_n, \phi_n)\} \subset E \times D^{1,2}(\mathbb{R}^3)$ *satisfying*

$$\frac{1}{2} \int_{\mathbb{R}^3} \left(|\nabla u_n|^2 + V(x)u_n^2 - |\nabla \phi_n|^2 - (2\omega + \phi_n)\phi_n u_n^2 \right) dx$$

$$- \int_{\mathbb{R}^3} F(x, u_n) dx \to 0^-.$$

Remark 4.9. We notice that our result improves and unites Theorems in [266, 269–271] because of the condition (C26′). Evidently, Theorem 4.8 generalizes Theorem 4.7. There are functions f satisfying Theorem 4.8 but not satisfying Theorem 4.7. For example, let

$$f(x, t) = b(x)|t|^{p-1} + h(x)|t|^{\gamma-1} + g(x)|t|^{\nu-1}, \quad \forall (x, t) \in \mathbb{R}^3 \times \mathbb{R},$$

where p, $\gamma \in (1, 2)$, $\nu \in (2, 6)$, b, $h : \mathbb{R}^3 \to \mathbb{R}$ are positive continuous functions with $b \in L^{\frac{2}{2-p}}(\mathbb{R}^3)$, $h \in L^{\frac{2}{2-\gamma}}(\mathbb{R}^3)$ and g is a bounded continuous function with $\inf_{x \in \mathbb{R}^3} g(x) > 0$. Furthermore, according to the example, we see that the nonlinear term $f(x, t)$ can be a combination of sublinear and superlinear terms as well as two sublinear terms.

Standard arguments prove that the weak solutions $(u, \phi) \in E \times D^{1,2}(\mathbb{R}^3)$ of system (4.127) are critical points of the functional given by

$$J(u, \phi) = \frac{1}{2} \int_{\mathbb{R}^3} \left(|\nabla u|^2 + V(x)u^2 - |\nabla \phi|^2 - (2\omega + \phi)\phi u^2 \right) dx - \int_{\mathbb{R}^3} F(x, u) dx.$$
$$(4.129)$$

The functional J is strongly indefinite, i.e. unbounded from below and from above on infinite dimensional spaces. To avoid this difficulty, we reduce the study of (4.129) to the study of a functional in the only variable u, as it has been done by the aforementioned authors.

Now we need the following technical results established in [272] (see also [228]).

Proposition 4.4. *For any fixed* $u \in H^1(\mathbb{R}^3)$, *there exists a unique* $\phi = \phi_u \in D^{1,2}(\mathbb{R}^3)$ *which solves equation*

$$- \Delta \phi + u^2 \phi = -\omega u^2. \quad (4.130)$$

Moreover, the map $\Phi : u \in H^1(\mathbb{R}^3) \mapsto \Phi[u] := \phi_u \in D^{1,2}(\mathbb{R}^3)$ *is continuously differentiable, and*

(i) $-\omega \leq \phi_u \leq 0$ *on the set* $\{x : u(x) \neq 0\}$;
(ii) $\|\phi_u\|_{D^{1,2}} \leq C\|u\|_E^2$ *and* $\int_{\mathbb{R}^3} |\phi_u| u^2 dx \leq C\|u\|_{12/5}^4 \leq C\|u\|_E^4$.

By (4.130), multiplying by ϕ_u and integrating by parts we obtain

$$\int_{\mathbb{R}^3} |\nabla \phi_u|^2 dx = -\int_{\mathbb{R}^3} \omega \phi_u u^2 dx - \int_{\mathbb{R}^3} \phi_u^2 u^2 dx. \qquad (4.131)$$

Using (4.131), we can rewrite J as a C^1 functional $I : E \to \mathbb{R}$ given by

$$I(u) = \frac{1}{2} \int_{\mathbb{R}^3} \left(|\nabla u|^2 + V(x)u^2 - \omega \phi_u u^2 \right) dx - \int_{\mathbb{R}^3} F(x, u)dx, \qquad (4.132)$$

while for I' we have

$$
\begin{aligned}
\langle I'(u), v \rangle &= \int_{\mathbb{R}^3} (\nabla u \cdot \nabla v + V(x)uv)dx - \int_{\mathbb{R}^3} \omega \phi_u uv dx \\
&\quad - \frac{1}{2} \int_{\mathbb{R}^3} \omega u^2 2(\Delta - u^2)^{-1}[(\omega + \phi_u)uv]dx - \int_{\mathbb{R}^3} f(x, u)v dx \\
&= \int_{\mathbb{R}^3} (\nabla u \cdot \nabla v + V(x)uv)dx - \int_{\mathbb{R}^3} \omega \phi_u uv dx \\
&\quad - \int_{\mathbb{R}^3} (\Delta - u^2)^{-1}[\omega u^2](\omega + \phi_u)uv dx - \int_{\mathbb{R}^3} f(x, u)v dx \\
&= \int_{\mathbb{R}^3} \left(\nabla u \cdot \nabla v + V(x)uv - (2\omega + \phi_u)\phi_u uv - f(x, u)v \right) dx,
\end{aligned}
$$

for all $v \in E$. Here we use the fact that $(\Delta - u^2)^{-1}[\omega u^2] = \phi_u$ (for details of the proof, see [272]). Now the functional I is not strongly indefinite anymore and we will look for its critical points since, as in [272], $(u, \phi) \in H^1(\mathbb{R}^3) \times D^{1,2}(\mathbb{R}^3)$ is a critical point for J, if and only if u is a critical point for I with $\phi = \phi_u$.

In order to obtain infinitely many solutions of (4.127), we shall use the following critical point theorem introduced by Bartsch in [27] and Kajikiya [29].

We need the following compactness result proved in [189].

Proposition 4.5. *Under the assumption* (C21), *the embedding* $E \hookrightarrow L^q(\mathbb{R}^3)$, $2 \le q < 2^* = 6$ *is compact.*

We also need the following important lemma to get our theorems.

Lemma 4.23. *For any* $2 \le q < 2^*$, *we have that*

$$\beta_q(k) := \sup_{u \in Z_k, \|u\|_E = 1} \|u\|_q \to 0, \text{ as } k \to \infty.$$

Proof. Obviously the sequence $\{\beta_q(k)\}$ is nonnegative and nonincreasing. Suppose that $\beta_q(k) \to \beta_q > 0$ as $k \to \infty$. Then for any k sufficiently large, there exists $u_k \in Z_k$ with $\|u_k\|_E = 1$ and $\|u_k\|_q \ge \beta_q/2$. For any $u \in E$, since $\{e_j\}$ is an orthogonal basis of E, there exists a sequence $\{\alpha_j\} \subset \mathbb{R}$ satisfying $u = \sum_{j=1}^{\infty} \alpha_j e_j$, thus by the Schwartz inequality and the Parseval equality we have

$$
|(u, u_k)_E| = \left| \left(\sum_{j=1}^{\infty} \alpha_j e_j, u_k \right)_E \right| = \left| \left(\sum_{j=k}^{\infty} \alpha_j e_j, u_k \right)_E \right|
$$

$$
\leq \left\| \sum_{j=k}^{\infty} \alpha_j e_j \right\|_E \|u_k\|_E = \sqrt{\sum_{j=k}^{\infty} \alpha_j^2} \to 0,
$$

as $k \to \infty$, where $(\, , \,)_E$ denotes the inner product in E. Using the Riesz–Fréchet representation theorem, we obtain that $u_k \rightharpoonup 0$ in E and thus $u_k \to 0$ in L^q because the embedding $E \hookrightarrow L^q$ is compact. This is a contradiction. $\qquad\square$

Consider $I : E \to \mathbb{R}$ defined by

$$
I(u) = \frac{1}{2} \int_{\mathbb{R}^3} \left(|\nabla u|^2 + V(x)u^2 - \omega \phi_u u^2 \right) dx - \int_{\mathbb{R}^3} F(x, u)dx,
$$

where F is even as a consequence of (C25), hence I is even. First we show that under the condition (C22), (C23) and (C21), the functional I satisfies the geometric condition of Theorem 1.10. We have the following lemma.

Lemma 4.24. *Assume that (C21) and (C22), (C23) hold. Then for every $k \in \mathbb{N}$, there exists $\rho_k > r_k > 0$, such that*

(i) $a_k := \max_{u \in Y_k, \|u\| = \rho_k} I(u) \leq 0$;

(ii) $b_k := \inf_{u \in Z_k, \|u\| = r_k} I(u) \to +\infty$ *as $k \to \infty$,*

where Y_k and Z_k are defined by (1.5).

Proof. It follows from (C23) that for any $M > 0$, there exists $\delta(M) > 0$, such that for all $x \in \mathbb{R}^3$, $|t| \geq \delta$, we have

$$
F(x, t) \geq \frac{1}{4}M|t|^4. \tag{4.133}
$$

By (C22), we have

$$
|F(x, t)| \leq \int_0^1 |f(x, st)t|ds \leq C \int_0^1 (|st| + |st|^{p-1})|t|ds \leq C(t^2 + |t|^p). \tag{4.134}
$$

From the above inequality, one has

$$
|F(x, t)| \leq C(1 + \delta^{p-2})t^2
$$

for all $x \in \mathbb{R}^3$ and $|t| \leq \delta$. Combined with (4.133), we obtain

$$F(x,t) \geq \frac{1}{4}M|t|^4 - \widetilde{M}|t|^2$$

for all $x \in \mathbb{R}^3$, and all $t \in \mathbb{R}$, where $\widetilde{M} = \frac{1}{4}M\delta^2 + C(1 + \delta^{p-2})$. Hence, due to Proposition 4.4, we have

$$I(u) \leq \frac{1}{2}\|u\|_E^2 + \frac{C}{4}\|u\|_E^4 - \frac{1}{4}M\|u\|_4^4 + \widetilde{M}\|u\|_2^2.$$

Since, on the finitely dimensional space Y_k all norms are equivalent, we have that

$$I(u) \leq \frac{1}{2}\|u\|_E^2 + \frac{C}{4}\|u\|_E^4 - \frac{1}{4}M\widetilde{C}\|u\|_E^4 + \widetilde{M}\widetilde{C}\|u\|_E^2.$$

Now since $\frac{C}{4} - \frac{1}{4}M\widetilde{C} < 0$, when M is large enough, it follows that

$$a_k := \max_{u \in Y_k, \|u\|_E = \rho_k} I(u) \leq 0$$

for some $\rho_k > 0$ large enough.

We prove that I satisfies (ii) of Theorem 1.10. By Lemma 4.23, we have

$$\beta_2(k) \to 0 \text{ and } \beta_p(k) \to 0$$

as $k \to \infty$. So, if C is a constant large enough such that Proposition 4.4 and (4.134) hold, there exists a positive integer k_1 such that

$$\beta_2(k) \leq \sqrt{\frac{1}{8C}}, \quad \forall k \geq k_1.$$

For each $k \geq k_1$, taking

$$r_k = \left(8C\beta_p^p(k)\right)^{1/(2-p)},$$

one has $r_k \to +\infty$ as $k \to \infty$ since $p > 4$. Now, due to Proposition 4.4 and (4.134), we have

$$I(u) \geq \frac{1}{2}\|u\|_E^2 - C\|u\|_2^2 - C\|u\|_p^p$$
$$\geq \frac{1}{2}\|u\|_E^2 - C\beta_2^2(k)\|u\|_E^2 - C\beta_p^p(k)\|u\|_E^p$$
$$\geq \frac{1}{4}\|u\|_E^2.$$

Thus, we obtain

$$b_k = \inf_{u \in Z_k, \|u\|_E = r_k} I(u) \geq \frac{1}{4}r_k^2 \to +\infty \text{ as } k \to \infty.$$

This proves (ii). $\qquad\qquad\qquad\qquad\qquad\qquad\qquad\qquad\qquad$ $\square$

Lemma 4.25. *Under the assumptions* (C21), (C22)–(C24), *the functional I satisfies the (PS) condition at any level* $c > 0$.

Proof. We suppose that $\{u_n\}$ is the (PS) sequence, that is for some $c \in \mathbb{R}^+$

$$I(u_n) = \frac{1}{2}\|u_n\|_E^2 - \frac{1}{2}\int_{\mathbb{R}^3}\omega\phi_{u_n}u_n^2 dx - \int_{\mathbb{R}^3}F(x,u_n)dx \to c \quad (n\to\infty)$$
(4.135)

and

$$I'(u_n) \to 0 \quad (n\to\infty).$$
(4.136)

We claim that $\{u_n\}$ is bounded. Otherwise, there exist a subsequence of $\{u_n\}$ still denoted by $\{u_n\}$ satisfying $\|u_n\|_E \to \infty$ as $n \to \infty$. Denote $w_n = \frac{u_n}{\|u_n\|_E}$, then $\{w_n\}$ is bounded. Up to a subsequence, for some $w \in E$, we get

$$w_n \rightharpoonup w \quad \text{in } E,$$
(4.137)

$$w_n \to w \quad \text{in } L^q(\mathbb{R}^3),\ 2 \le q < 2^*,$$
(4.138)

$$w_n(x) \to w(x) \text{ a.e. in } \mathbb{R}^3.$$
(4.139)

On one hand, $w \neq 0$ in E. Dividing both side of (4.135) by $\|u_n\|_E^4$ and using Proposition 4.4 we get for n sufficiently large that,

$$\int_{\mathbb{R}^3}\frac{F(x,u_n)}{\|u_n\|_E^4}dx = \frac{1}{2\|u_n\|_E^2} - \frac{\int_{\mathbb{R}^3}\omega\phi_{u_n}u_n^2 dx - c}{2\|u_n\|_E^4} + o(\|u_n\|_E^{-4}) \le C < \infty,$$
(4.140)

where C is a positive constant. Let $\Omega = \{x \in \mathbb{R}^3 : w(x) \neq 0\}$. Suppose that $|\Omega| > 0$,

$$\frac{F(x,u_n)}{\|u_n\|_E^4} = \frac{F(x,u_n)}{|u_n|^4}w_n^4(x) \to +\infty \quad (n\to\infty).$$

By (C23), there exists $L > 0$ such that

$$F(x,t) \ge 0, \quad \forall x \in \mathbb{R}^3, |t| \ge L.$$
(4.141)

Since $|u_n| = \|u_n\|_E|w_n| \to +\infty$ almost everywhere in Ω, by Fatou's Lemma, we have

$$\liminf_{n\to\infty}\int_\Omega \frac{F(x,u_n)}{\|u_n\|_E^4} \to +\infty.$$
(4.142)

From (4.134), it follows that

$$|F(x,t)| \le C't^2, \quad \forall x \in \mathbb{R}^3, |t| \le L,$$

where $C' = C(1 + L^{p-2})$. Combining with (4.141), we have

$$F(x,t) \ge -C't^2, \quad \forall(x,t) \in \mathbb{R}^3 \times \mathbb{R}.$$

Hence we obtain, using Sobolev embedding,

$$\int_{\mathbb{R}^3 \setminus \Omega} \frac{F(x, u_n)}{\|u_n\|_E^4} \geq -\frac{C' \int_{\mathbb{R}^3 \setminus \Omega} u_n^2}{\|u_n\|_E^4} \geq -\frac{C''}{\|u_n\|_E^2},$$

which implies that

$$\liminf_{n \to \infty} \int_{\mathbb{R}^3 \setminus \Omega} \frac{F(x, u_n)}{\|u_n\|_E^4} \geq 0. \tag{4.143}$$

Combining (4.142) and (4.143), we obtain $\liminf_{n \to \infty} \int_{\mathbb{R}^3} \frac{F(x, u_n)}{\|u_n\|_E^4} dx \to +\infty$. This contradicts (4.140). Hence, $|\Omega| = 0$, that is, $w(x) = 0$ almost everywhere $x \in \mathbb{R}^3$. By (C22) and (4.134), we have for all $x \in \mathbb{R}^3$ and $|t| \leq r_0$ with

$$\left| \frac{1}{4} f(x, t)t - F(x, t) \right| \leq 2C(t^2 + |t|^p) \leq C't^2,$$

where $C' = 2C(1 + r_0^{p-2})$. This, together with (C24), shows that

$$\frac{1}{4} tf(x, t) - F(x, t) \geq -C''|t|^2 \text{ for all } (x, t) \in \mathbb{R}^3 \times \mathbb{R}.$$

Thus, following Proposition 4.4,

$$I(u_n) - \frac{1}{4}\langle I'(u_n), u_n \rangle$$
$$= \frac{1}{4}\|u_n\|_E^2 + \frac{1}{4}\int_{\mathbb{R}^3} \phi_{u_n}^2 u_n^2 dx + \int_{\mathbb{R}^3} \left(\frac{1}{4} f(x, u_n)u_n - F(x, u_n) \right) dx$$
$$\geq \frac{1}{4}\|u_n\|_E^2 + \int \left(\frac{1}{4} f(x, u_n)u_n - F(x, u_n) \right)$$
$$\geq \frac{1}{4}\|u_n\|_E^2 - C'' \int_{\mathbb{R}^3} |u_n|^2 dx.$$

Thus,

$$\frac{I(u_n) - \frac{1}{4}\langle I'(u_n), u_n \rangle}{\|u_n\|_E^2} \geq \frac{1}{4} - C'' \int_{\mathbb{R}^3} \frac{|u_n|^2}{\|u_n\|_E^2} dx$$
$$= \frac{1}{4} - C'' \int_{\mathbb{R}^3} |w_n|^2 dx.$$

Using (4.137), we obtain a contradiction. Hence, $\{u_n\}$ is bounded. Now we shall prove $\{u_n\}$ contains a convergent subsequence. Without loss of generality, passing to a subsequence if necessary, there exists a $u \in E$ such that $u_n \rightharpoonup u$ in E, again by embedding, $u_n \to u$ in $L^q(\mathbb{R}^3)$ for $2 \leq q < 6$ and $u_n \to u$ almost everywhere $x \in \mathbb{R}^3$. So we can get

$$\int_{\mathbb{R}^3} (f(x, u_n) - f(x, u))(u_n - u)dx \to 0, \text{ as } n \to \infty.$$

We observe that

$$\langle I'(u_n) - I'(u), u_n - u \rangle \to 0, \text{ as } n \to \infty,$$

and we have

$$\int_{\mathbb{R}^3} [(2\omega + \phi_{u_n})\phi_{u_n} u_n - (2\omega + \phi_u)\phi_u u](u_n - u)dx$$

$$= 2\omega \int_{\mathbb{R}^3} (\phi_{u_n} u_n - \phi_u u)(u_n - u)dx + \int_{\mathbb{R}^3} (\phi_{u_n}^2 u_n - \phi_u^2 u)(u_n - u)dx \to 0,$$

as $n \to \infty$. Actually, by the Hölder inequality, Proposition 4.4 and the Sobolev inequality, we have

$$\left| \int_{\mathbb{R}^3} (\phi_{u_n} - \phi_u)u_n(u_n - u)dx \right| \leq \|(\phi_{u_n} - \phi_u)(u_n - u)\|_2 \|u_n\|_2$$

$$\leq \|\phi_{u_n} - \phi_u\|_6 \|u_n - u\|_3 \|u_n\|_2$$

$$\leq C\|\phi_{u_n} - \phi_u\|_{D^{1,2}} \|u_n - u\|_3 \|u_n\|_2,$$

where $C > 0$ is a constant. Because $u_n \to u$ in $L^q(\mathbb{R}^3)$ for any $2 \leq q < 6$, we have

$$\int_{\mathbb{R}^3} (\phi_{u_n} - \phi_u)u_n(u_n - u)dx \to 0, \text{ as } n \to \infty,$$

and

$$\int_{\mathbb{R}^3} \phi_u(u_n - u)(u_n - u)dx \leq \|\phi_u\|_6 \|u_n - u\|_3 \|u_n - u\|_2 \to 0, \text{ as } n \to \infty.$$

Thus, we get

$$\int_{\mathbb{R}^3} (\phi_{u_n} u_n - \phi_u u)(u_n - u)dx$$

$$= \int_{\mathbb{R}^3} (\phi_{u_n} - \phi_u)u_n(u_n - u)dx + \int_{\mathbb{R}^3} \phi_u(u_n - u)(u_n - u)dx \to 0,$$

as $n \to \infty$. Observe that the sequence $\{\phi_{u_n}^2 u_n\}$ is bounded in $L^{3/2}$, since

$$\|\phi_{u_n}^2 u_n\|_{3/2} \leq \|\phi_{u_n}\|_6^2 \|u_n\|_3,$$

so

$$\left| \int_{\mathbb{R}^3} (\phi_{u_n}^2 - \phi_u^2)(u_n - u)dx \right| \leq \|\phi_{u_n}^2 - \phi_u^2\|_{3/2} \|u_n - u\|_3$$

$$\leq (\|\phi_{u_n}^2\|_{3/2} + \|\phi_u^2\|_{3/2})\|u_n - u\|_3$$

$$\to 0 \text{ as } n \to \infty.$$

Now, we can get

$$\|u_n - u\|_E^2 = \langle I'(u_n) - I'(u), u_n - u \rangle$$

$$- \int_{\mathbb{R}^3} [(2\omega + \phi_{u_n})\phi_{u_n} u_n - (2\omega + \phi_u)\phi_u u](u_n - u)dx$$

$$+ \int_{\mathbb{R}^3} (f(x, u_n) - f(x, u))(u_n - u)dx \to 0,$$

as $n \to \infty$. That is $u_n \to u$ in E and the proof is complete. $\square$

Proof of Theorem 4.6. By Lemma 4.24, the functional I satisfies the geometric conditions of Theorem 1.10. Lemma 4.25 implies that I satisfies the (PS) condition. Hence, problem (4.127) has infinitely many nontrivial solutions $(u_n, \phi_n) \in E \times D^{1,2}(\mathbb{R}^3)$. This completes the proof. $\square$

Lemma 4.26. *Under* (C21) *and* (C26′), *the functional I defined by* (4.132) *is C^1 and the Fréchet derivative of functional $\psi(u) = \int_{\mathbb{R}^3} F(x, u)dx$ is compact.*

Proof. It is standard to prove that the functional I is continuous. Now, we prove the Fréchet derivative of functional $\psi(u) = \int_{\mathbb{R}^3} F(x, u)dx$ is compact. Simple calculations show that

$$\langle \psi'(u), v \rangle = \int_{\mathbb{R}^3} f(x, u)vdx.$$

Let $u_n \rightharpoonup u$. By Proposition 4.5, up to a subsequence, we can assume that

$$u_n \to u \quad \text{in } L^q(\mathbb{R}^3), 2 \le q < 2^*,$$

$$u_n(x) \to u(x) \text{ a.e. in } \mathbb{R}^3.$$

Note that

$$\|\psi'(u_n) - \psi'(u)\|_{E^*} = \sup_{\|v\|=1} |\langle \psi'(u_n) - \psi'(u), v \rangle|$$

$$\le \sup_{\|v\|=1} \int_{\mathbb{R}^3} |f(x, u_n) - f(x, u)||v|dx.$$

Setting $\lim_{n\to\infty} \sup_{\|v\|=1} \int_{\mathbb{R}^3} |f(x, u_n) - f(x, u)||v|dx = A$, we claim that $A = 0$. For otherwise, $A > 0$ and

$$\sup_{\|v\|=1} \int_{\mathbb{R}^3} |f(x, u_n) - f(x, u)||v|dx > \frac{A}{2}$$

for n sufficiently large. By the definition of supremum, there exists $v_0 \in E$ with $\|v_0\| = 1$ such that

$$\int_{\mathbb{R}^3} |f(x, u_n) - f(x, u)||v_0|dx \ge \frac{A}{2}.$$

For $q = 2$, ν, since $u_n \to u$ in $L^q(\mathbb{R}^N)$, by Lemma A.1 of [5], there exist a subsequence, still denoted by $(u_n$), and $g_1 \in L^2(\mathbb{R}^N)$, $g_2 \in L^\nu(\mathbb{R}^N)$, such that, almost everywhere on $\mathbb{R}^N$, $|u_n(x)|$, $|u(x)| \le g_1(x)$, $\forall n \in N$ and $|u_n(x)|$, $|u(x)| \le g_2(x)$, $\forall n \in N$. It follows from (C26') and the Hölder inequality that

$$
\begin{aligned}
&|f(x, u_n) - f(x, u)||v_0| \\
&\le \left[m(x)(|u_n|^{\sigma-1} + |u|^{\sigma-1}) + h(x)(|u_n|^{\gamma-1} + |u|^{\gamma-1}) \right. \\
&\quad \left. + C(|u_n|^{\nu-1} + |u|^{\nu-1}) \right] |v_0| \\
&\le \left[m(x)2|g_1|^{\sigma-1} + h(x)2|g_1|^{\gamma-1} + 2C|g_2|^{\nu-1} \right] |v_0|
\end{aligned}
$$

and

$$
\begin{aligned}
&\int_{\mathbb{R}^3} \left[m(x)2|g_1|^{\sigma-1} + h(x)2|g_1|^{\gamma-1} + 2C|g_2|^{\nu-1} \right] |v_0| dx \\
&\le 2\|m\|_{2/(2-\sigma)} \|g_1\|_2^{\sigma-1} \|v_0\|_2 + 2\|h\|_{2/(2-\gamma)} \|g_1\|_2^{\gamma-1} \|v_0\|_2 + 2C\|g_2\|_\nu^{\nu-1} \|v_0\|_\nu \\
&< +\infty.
\end{aligned}
$$

Therefore, using Lebesgue dominated convergence theorem, we have

$$
\frac{A}{2} \le \lim_{n \to \infty} \int_{\mathbb{R}^3} |f(x, u_n) - f(x, u)||v_0| dx = 0,
$$

a contradiction. Hence

$$
\|\psi'(u_n) - \psi'(u)\|_{E^*} \to 0 \text{ as } n \to \infty,
$$

and then ψ' is weakly continuous. Consequently, ψ' is continuous, $\psi \in C^1(E, \mathbb{R})$, and the compactness of ψ' follows from its weak continuity since E is a Hilbert space. $\qquad \square$

Now, the critical points of I are solutions of problem (4.127), so we can prove Theorem 4.8 in the following way.

Proof of Theorem 4.8. Choose $h \in C^\infty([0, \infty), \mathbb{R})$ such that $0 \le h(t) \le 1$ for $t \in [0, \infty)$, and $h(t) = 1$ for $0 \le t \le 1$, $h(t) = 0$ for $t \ge 2$. We consider the truncated functional

$$
\widetilde{I}(u) = \frac{1}{2} \int_{\mathbb{R}^3} \left(|\nabla u|^2 + V(x)u^2 - \omega\phi_u u^2 \right) dx - h(\|u\|) \int_{\mathbb{R}^3} F(x, u) dx.
$$

We know that $\widetilde{I} \in C^1(E, \mathbb{R})$. If we can prove that $\widetilde{I}(u)$ admits a sequence of nontrivial weak solutions $\{u_n\}$ with $u_n \to 0$ as $n \to \infty$ in E, Theorem 4.8 holds. Indeed, in this circumstance, it is not hard to see that $I(u) = \widetilde{I}(u)$ for $\|u\|_E < 1$. So the weak solutions of $\widetilde{I}$ are just weak solutions of problems

(4.127). By applying Theorem 1.12 we show $\widetilde{I}(u)$ admits a sequence of nontrivial weak solutions which converges to zero in E.

For $\|u\|_E \geq 2$, we have $\widetilde{I}(u) = \frac{1}{2}\int_{\mathbb{R}^3}\left(|\nabla u|^2 + V(x)u^2 - \omega\phi_u u^2\right)dx$, due to Proposition 4.4, which implies $\widetilde{I}(u_n) \to \infty$ as $\|u_n\|_E \to \infty$. Hence $\widetilde{I}$ is coercive on E. Thus $\widetilde{I}$ is bounded from below and satisfies the (PS) condition following Lemma 4.26 and Proposition 4.5. By (C25), it is easy to see that $\widetilde{I}$ is even and $\widetilde{I}(0) = 0$. This shows that (Φ_1) holds.

Now we claim that for any finite dimensional subspace $\widetilde{E} \subset E$, there exists $\varepsilon_1 > 0$ such that

$$\mathrm{meas}\{x \in \mathbb{R}^3 : b(x)|u(x)|^p \geq \varepsilon_1\|u\|_E^p\} \geq \varepsilon_1, \quad \forall u \in \widetilde{E} \setminus \{0\}. \qquad (4.144)$$

Otherwise, for any positive integer n, there exists $u_n \in \widetilde{E} \setminus \{0\}$ such that

$$\mathrm{meas}\left\{x \in \mathbb{R}^3 : b(x)|u(x)|^p \geq \frac{1}{n}\|u\|_E^p\right\} < \frac{1}{n}.$$

Set $v_n(x) := \frac{|u_n(x)|}{\|u_n\|_E} \in \widetilde{E} \setminus \{0\}$, then $\|v_n\|_E = 1$ for all $n \in \mathbb{N}$ and

$$\mathrm{meas}\left\{x \in \mathbb{R}^3 : b(x)|v(x)|^p \geq \frac{1}{n}\right\} < \frac{1}{n}. \qquad (4.145)$$

Since $\dim \widetilde{E} < \infty$, it follows from the compactness of the unit sphere of $\widetilde{E}$ that there exists a subsequence, still denoted by $\{v_n\}$, such that v_n converges to some v_0 in $\widetilde{E}$. Hence, we have $\|v_0\|_E = 1$. By the equivalence of the norms on the finite dimensional space $\widetilde{E}$, we have $v_n \to v_0$ in $L^2(\mathbb{R}^3)$, i.e.,

$$\int_{\mathbb{R}^3} |v_n - v_0|^2 dx \to 0 \text{ as } n \to \infty. \qquad (4.146)$$

By (4.146) and Hölder's inequality, we have

$$\int_{\mathbb{R}^3} b(x)|v_n - v_0|^p dx \leq \left(\int_{\mathbb{R}^3} b(x)^{\frac{2}{2-p}} dx\right)^{(2-p)/2}\left(\int_{\mathbb{R}^3}|v_n - v_0|^2 dx\right)^{p/2} \qquad (4.147)$$

$$= \|b\|_{\frac{2}{2-p}}\left(\int_{\mathbb{R}^3}|v_n - v_0|^2 dx\right)^{p/2} \to 0,$$

as $n \to \infty$. Thus, there exists $\xi_1, \xi_2 > 0$ such that

$$\mathrm{meas}\left\{x \in \mathbb{R}^3 : b(x)|v_0(x)|^p \geq \xi_1\right\} \geq \xi_2. \qquad (4.148)$$

In fact, if not, for all positive integer n, we have

$$\mathrm{meas}\left\{x \in \mathbb{R}^3 : b(x)|v_0(x)|^p \geq \frac{1}{n}\right\} = 0.$$

It implies that

$$0 \leq \int_{\mathbb{R}^3} b(x)|v_0(x)|^{p+2}dx < \frac{1}{n}\|v_0\|_2^2 \leq \frac{C^2}{n}\|v_0\|_E^2 = \frac{C^2}{n} \to 0,$$

as $n \to \infty$. Hence, $v_0 = 0$ which is contradicts $\|v_0\|_E = 1$. Therefore, (4.148) holds.

Now, let

$$\Omega_0 = \left\{ x \in \mathbb{R}^3 : b(x)|v_0|^p \geq \xi_1 \right\}, \quad \Omega_n = \left\{ x \in \mathbb{R}^3 : b(x)|v_n(x)|^p < \frac{1}{n} \right\}$$

and $\Omega_n^c = \mathbb{R}^3 \setminus \Omega_n = \{x \in \mathbb{R}^3 : b(x)|v_n(x)|^p \geq \frac{1}{n}\}$. By (4.145) and (4.148), we have

$$\begin{aligned} \operatorname{meas}(\Omega_n \cap \Omega_0) &= \operatorname{meas}(\Omega_0 \setminus (\Omega_n^c \cap \Omega_0)) \\ &\geq \operatorname{meas}(\Omega_0) - \operatorname{meas}(\Omega_n^c \cap \Omega_0) \\ &\geq \xi_2 - \frac{1}{n} \end{aligned}$$

for all positive integer n. Let n be large enough such that $\xi_2 - \frac{1}{n} \geq \frac{1}{2}\xi_2$ and $\frac{1}{2^p}\xi_1 - \frac{1}{n} \geq \frac{1}{2^{p+1}}\xi_1$. Then we have

$$\begin{aligned} \int_{\mathbb{R}^3} b(x)|v_n - v_0|^p dx &\geq \int_{\Omega_n \cap \Omega_0} b(x)|v_n - v_0|^p \\ &\geq \frac{1}{2^p} \int_{\Omega_n \cap \Omega_0} b(x)|v_0|^p - \int_{\Omega_n \cap \Omega_0} b(x)|v_n|^p \\ &\geq \left(\frac{1}{2^p}\xi_1 - \frac{1}{n} \right) \operatorname{meas}(\Omega_n \cap \Omega_0) \\ &\geq \frac{\xi_1}{2^{p+1}} \cdot \frac{\xi_2}{2} = \frac{\xi_1 \xi_2}{2^{p+2}} > 0 \end{aligned}$$

for n large enough, which is a contradiction to (4.147). Therefore, (4.144) holds.

Given any $k \in \mathbb{N}$, let $E_k := \operatorname{span}\{e_1, e_2, \ldots, e_k\}$ be a k-dimensional subspace of E, where $(e_i)_{j=1}^\infty$ is an orthogonal basis of E. Then there exists $\varepsilon_k > 0$ such that

$$\operatorname{meas}\{x \in \mathbb{R}^3 : b(x)|u(x)|^p \geq \varepsilon_k \|u\|_E^p\} \geq \varepsilon_k, \quad \forall u \in E_k \setminus \{0\}. \quad (4.149)$$

For $u \in E_k \setminus \{0\}$, take $\Omega_u = \{x \in \mathbb{R}^3 : b(x)|u(x)|^p \geq \varepsilon_k \|u\|_E^p\}$. By (4.149) and (C26′), we deduce, for $u \in E_k$ with $\|u\|_E \leq \min\{1, (\frac{\varepsilon_k^2}{p(1+C)})^{1/(2-p)}\}$,

$$
\widetilde{I}(u) = \frac{1}{2}\|u\|_E^2 - \int_{\mathbb{R}^3} \omega\phi_u u^2 dx - \int_{\mathbb{R}^3} F(x,u)dx
$$
$$
\leq \frac{1}{2}\|u\|_E^2 + \frac{1}{2}C\|u\|_E^4 - \frac{1}{p}\int_{\mathbb{R}^3} b(x)|u|^p dx
$$
$$
\leq \left(\frac{1+C}{2}\right)\|u\|_E^2 - \frac{1}{p}\int_{\Omega_u} b(x)|u|^p
$$
$$
\leq \left(\frac{1+C}{2}\right)\|u\|_E^2 - \frac{1}{p}\varepsilon_k\|u\|_E^p \operatorname{meas}\Omega_u
$$
$$
\leq -\left(\frac{1+C}{2}\right)\|u\|_E^2.
$$

Thus, choosing $0 < d_k \leq \min\{1, (\frac{\varepsilon_k^2}{p(1+C)})^{1/(2-p)}\}$, the above inequality implies that

$$
\{u \in E_k : \|u\|_E = d_k\} \subset \left\{u \in E : \widetilde{I}(u) \leq -\left(\frac{1+C}{2}\right)d_k^2\right\}.
$$

Therefore, letting $A_k = \{u \in E : \widetilde{I}(u) = -\left(\frac{1+C}{2}\right)d_k^2\}$, by genus proposition, we have $\gamma(A_k) \geq \gamma(\{u \in E_k : \|u\|_E = d_k\}) \geq k$ and $A_k \in \Gamma_k$, $\sup_{u \in A_k} \widetilde{I}(u) \leq -\left(\frac{1+C}{2}\right)d_k^2 < 0$. This shows ($\Phi_2$) holds. Hence, by Theorem 1.12, we complete the proof of Theorem 4.8. $\qquad\square$

Notes and Comments

Using variational method, as we know Benci and Fortunato [272,273] were the first one considered the following system

$$
\begin{cases}
-\Delta u + [m_0^2 - (\omega + \phi)^2]u = |u|^{p-2}u, & x \in \mathbb{R}^3, \\
-\Delta\phi + u^2\phi = -\omega u^2, & x \in \mathbb{R}^3.
\end{cases} \tag{4.150}
$$

where $4 < p < 6$, and established the existence of infinitely many solitary waves solutions. D'Aprile and Mugnai proved in [235] that there is no nontrivial finite energy solution of (4.127) if $p \geq 6$ and $m \geq \omega$ or $p \leq 2$ Furthermore, in [228] they proved that there are infinitely many finite energy radial solutions if one of the following conditions holds:

(i) $m_0 > \omega > 0$ and $p \in [4,6)$;
(ii) $m_0\sqrt{(p-2)/2} > \omega > 0$ and $p \in (2,4)$.

In [274], Azzollini *et al.* improved the existence range of (m_0, ω) for $p \in (2, 4)$ as follows:

$$0 < \omega < m_0 g(p), \qquad g(p) = \begin{cases} \sqrt{(p-2)(4-p)} & \text{if } 2 < p < 3, \\ 1 & \text{if } 3 \le p < 4. \end{cases}$$

They also dealt with a limit case $m_0 = \omega$ in [274]. Next, Mugnai [275] studied the existence of radially symmetric solitary waves for a system of a nonlinear Klein–Gordon equation coupled with Maxwell's equation in presence of a positive mass. The nonlinear potential appearing in the system is assumed to be positive and with more than quadratical growth at infinity. Later, ground state solutions [276, 277], semiclassical states [278] and nonradial solutions [279] are studied by different mathematicians. The critical exponent case also had been considered. Cassani [280] studied the critical exponent case and non-existence of solutions using the Brezis–Nirenberg method. Carrião *et al.* [281, 282] study the Klein–Gordon–Maxwell equations in higher dimensions with critical exponent and ground state solution, respectively. Wang [283] improved existence results in dimension three. In [267], Chen and Tang considered the nonhomogeneous Klein–Gordon–Maxwell System. Using the Ekeland's variational principle and the Mountain Pass Theorem in critical point theory they got two nontrivial solutions. Candela and Salvatore [251] used Bolle's method to get infinitely many solutions for the non-homogeneous Klein–Gordon–Maxwell or Schrödinger–Maxwell systems on a ball. Very recently, Jeong and Seok [284] established an abstract critical point theorem with a functional of the mountain-pass type with a small perturbation for the nonlocal term and studied Klein–Gordon–Maxwell System with a very general nonlinear term. This section was obtained in [285].

4.3.2 *Sign-Changing Potential*

In this section, we study the existence of solutions to the following Klein–Gordon–Maxwell system in $\mathbb{R}^3$:

$$\begin{cases} -\Delta u + V(x)u - (2\omega + \phi)\phi u = g(x, u), & x \in \mathbb{R}^3, \\ \Delta \phi = (\omega + \phi)u^2, & x \in \mathbb{R}^3, \end{cases} \qquad (4.151)$$

where $\omega > 0$ is a constant, $u, \phi : \mathbb{R}^3 \to \mathbb{R}$, $V : \mathbb{R}^3 \to \mathbb{R}$ is a potential function and $g(x, u) : \mathbb{R} \to \mathbb{R}$ is a continuous functions. The problem (4.151) has a variational structure. It is known that there is an energy functional I on

$H^1(\mathbb{R}^3)$,

$$I(u) = \frac{1}{2} \int_{\mathbb{R}^3} \left(|\nabla u|^2 + V(x)u^2 - \omega\phi_u u^2 \right) dx - \int_{\mathbb{R}^3} G(x, u)dx, \qquad (4.152)$$

where $G(x, u) = \int_0^\tau g(x, \tau)d\tau$, such that u solves (4.151) if and only if it is a critical point of I.

Motivated by [286], we want to show that problem (4.151) can get at least a nontrivial when nonlinear term g satisfies some assumptions and the potential function V is indefinite in sign. Due to the equations in $\mathbb{R}^3$ with potential indefinite in sign, the variational functional does not satisfy the mountain pass geometry. The nonlinearity considered here satisfies a condition which is much weaker than the classical (AR) condition. We obtain nontrivial solution via the local linking theorem. Our results obtained complement and improve the existing ones.

In order to reduce our statements, we need the following assumptions: the potential $V : \mathbb{R}^3 \to \mathbb{R}$ is a continuous function satisfying,

(C27) $V(x) \in C(\mathbb{R}^3, \mathbb{R})$ is bounded from below and there exists $v_0 > 0$ such that

$$\lim_{|y| \to \infty} \text{meas} \left\{ x \in \mathbb{R}^3 : |x - y| \leq v_0, V(x) \leq M \right\} = 0, \quad \forall M > 0.$$

Before stating our main results, we first make some assumptions on the nonlinear term.

(C28) $g \in C(\mathbb{R}^3 \times \mathbb{R}, \mathbb{R})$ and $|g(x, t)| \leq C(1 + |t|^{p-1})$ for some $4 \leq p < 2^* = 6$, where C is a positive constant.

(C29) $\frac{g(x,t)}{t} \to 0$ as $|t| \to 0$ uniformly for $x \in \mathbb{R}^3$.

(C30) $\frac{g(x,t)}{t^3} \to +\infty$ as $|t| \to +\infty$ uniformly for $x \in \mathbb{R}^3$.

(C31) Let $\mathcal{G}(x, t) := g(x, t)t - 4G(x, t)$, there exist $C > 0$ and $r_0 > 0$ such that if $|t| \geq r_0$, then $\mathcal{G}(t) \geq -C|t|^2$.

We next fix the following notations. Since V is bounded from below, we may chose $m > 0$ such that

$$\tilde{V}(x) := V(x) + m > 1, \qquad \text{for all } x \in \mathbb{R}^3.$$

A main difficulty to solve problem (4.151) is that the Sobolev embedding $H^1(\mathbb{R}^3) \hookrightarrow L^2(\mathbb{R}^3)$ is not compact. Thanks to the condition $(C27)$, this difficulty can be overcame by the compact embedding $X \hookrightarrow L^2(\mathbb{R}^3)$ of Bartsch *et al.* [189], where

$$E = \left\{ u \in H^1(\mathbb{R}^3) \,\middle|\, \int_{\mathbb{R}^3} V(x)u^2 dx < \infty \right\}$$

is a linear subspace of $H^1(\mathbb{R}^3)$ with the inner product $(u,v) := \int (\nabla u \cdot \nabla v + \tilde{V}(x)uv)$, and its norm is $\|u\| = (u,u)_E^{1/2}$. Since $V(x)$ is bounded form below, the embedding $E \hookrightarrow L^s(\mathbb{R}^3)$ is continuous for any $s \in [2,6]$.

According to the compact embedding $E \hookrightarrow L^2(\mathbb{R}^3)$ and the spectral theory of self-adjoint compact operators, it is easy to see that the eigenvalue problem

$$-\Delta u + V(x)u = \lambda u, \qquad u \in E \qquad (4.153)$$

possesses a complete sequence of eigenvalues

$$-\infty < \lambda_1 \leq \lambda_2 \leq \lambda_3 \leq \cdots, \qquad \lambda_k \to +\infty.$$

Each λ_k has been repeated in the sequence according to its finite multiplicity. We denote by φ_k the eigenfunction of λ_k, with $\|\varphi_k\|_2 = 1$. Note that the negative space E^- of the quadratic part of I is nontrivial, if and only if some λ_k is negative. Actually, E^- is spanned by the eigenfunctions corresponding to negative eigenvalues.

The main results of the present section can be stated as the following.

Theorem 4.9. *Suppose* (C27), (C28)–(C31) *are satisfied. If* 0 *is not an eigenvalue of* (4.153), *then the Klein–Gordon–Maxwell system* (4.151) *has at least one nontrivial solution* $u \in E$.

Remark 4.10. We emphasize that unlike all previous results about the equation (4.151), see e.g. [265, 282, 285, 287, 288], we have not assumed that the potential V is positive. When V is positive, the quadratic part of the functional I is positively definite, and I has a mountain pass geometry. Therefore, the Mountain Pass Theorem [172] can be applied. In our case, the quadratic part may possess a nontrivial negative space X^-, so I no longer possesses the mountain pass geometry. A natural idea is that we may try to apply the linking theorem (also called generalized Mountain Pass Theorem) [5, Theorem 2.12]. Unfortunately, due to the presence of the term involving ϕ_u, it turns out that I does not satisfy the required linking geometry either. To overcome this difficulty, we will employ the idea of local linking [23, 24].

Remark 4.11. In [265], problem (4.151) possesses infinitely many nontrivial solutions if $g(x,t)$ satisfies the following *global Ambrosetti–Rabinowitz condition* ((AR) for short):

$$\exists \theta > 4 \text{ such that } \theta G(x,t) \leq tg(x,t), \qquad \text{for } t \in \mathbb{R}, x \in \mathbb{R}^3.$$

The condition (AR) is widely assumed in the studies of elliptic equation by variational methods. As it is known, the condition (AR) is employed not only to show that the Euler–Lagrange functional associated has a mountain pass geometry, but also to guarantee that the Palais–Smale or Cerami sequences are bounded. Although (AR) is a quite natural condition, it is somewhat restrictive because it eliminates many nonlinearities such as $g(x, t) = t^3 \ln(1 + |t|)$. Observe that (AR) implies the weaker condition, there exist $\theta > 4$ and constant $C > 0$ such that $G(t) \geq C|t|^\theta$, for every $|t|$, sufficiently large, which, in its turn, implies another much weaker one (C30), which is called g is superlinear at infinity.

In case that g is odd, we can obtain an unbounded sequences of solutions.

Theorem 4.10. *Under the assumptions* (C27), (C28)–(C31) *and*

(C32) $g(x, u) = -g(x, -u)$ *uniformly for* $x \in \mathbb{R}^3$

the problem (4.151) *has a sequence of solutions* $u_n \in E$ *such that the energy*

$$I(u_n) \to +\infty.$$

Standard arguments prove that the weak solutions $(u, \phi) \in E \times D^{1,2}(\mathbb{R}^3)$ of system (4.151) are critical points of the functional given by

$$J(u, \phi) = \frac{1}{2} \int_{\mathbb{R}^3} \left(|\nabla u|^2 + V(x)u^2 - |\nabla \phi|^2 - (2\omega + \phi)\phi u^2 \right) dx - \int_{\mathbb{R}^3} G(x, u) dx.$$
$$(4.154)$$

As the same discussion in the previous section, we can rewrite I as a C^1 functional $I : E \to \mathbb{R}$ given by

$$I(u) = \frac{1}{2} \int_{\mathbb{R}^3} \left(|\nabla u|^2 + V(x)u^2 - \omega \phi_u u^2 \right) dx - \int_{\mathbb{R}^3} G(x, u) dx, \quad (4.155)$$

while for I' we have

$$\langle I'(u), v \rangle = \int_{\mathbb{R}^3} \left(\nabla u \cdot \nabla v + V(x)uv - (2\omega + \phi_u)\phi_u uv - g(x, u)v \right) dx,$$

for all $v \in E$. Now the functional I is not strongly indefinite anymore and we will look for its critical points since, as in [272], $(u, \phi) \in H^1(\mathbb{R}^3) \times D^{1,2}(\mathbb{R}^3)$ is a critical point for I, then u is a critical point for I with $\phi = \phi_u$.

For any $q \in [2, 6]$ we have a continuous embedding $E \hookrightarrow L^q(\mathbb{R}^3)$. Consequently there is a constant $\kappa_q > 0$ such that

$$\|u\|_q \leq \kappa_q \|u\|, \quad \text{for all } u \in E. \quad (4.156)$$

If $0 < \lambda_1$, it is easy to see that I has the mountain pass geometry. This case is simple and will be omitted here. For the proof of Theorem 4.9, since

0 is not an eigenvalue of (4.153), we may assume that $0 \in (\lambda_\ell, \lambda_{\ell+1})$ for some $\ell \geq 1$. Let

$$E^- = \text{span}\{\varphi_1, \ldots, \varphi_\ell\}, \qquad E^+ = (E^-)^\perp. \qquad (4.157)$$

Then E^- and E^+ are the negative space and positive space of the quadratic form

$$Q(u) = \frac{1}{2} \int_{\mathbb{R}^3} (|\nabla u|^2 + V(x)u^2)dx$$

respectively, note that $\dim E^- = \ell < \infty$. Moreover, there is a positive constants κ such that

$$\pm Q(u) \geq \kappa \|u\|^2, \qquad u \in E^\pm. \qquad (4.158)$$

As we have mentioned, because of the term involving ϕ_u in (4.152), our functional I does not satisfy the geometric assumption of the linking theorem. In fact, choose $\psi \in E^+$ with $\|\psi\| = 1$. For $R > r > 0$ set

$$N = \{u \in E^+ : \|u\| = r\}, \qquad M = \{u \in E^- \oplus \mathbb{R}^+\psi : \|u\| \leq R\},$$

then M is a submanifold of $E^- \oplus \mathbb{R}^+\psi$ with boundary ∂M. We do have

$$b = \inf_N I > 0, \qquad \sup_{u \in \partial M, \|u\| = R} I < 0$$

provided R is large enough. However, for $u \in E^-$ we have

$$I(u) = Q(u) - \frac{1}{2} \int_{\mathbb{R}^3} \omega \phi_u u^2 dx - \int_{\mathbb{R}^3} G(x, u)dx.$$

Because $-\frac{1}{2} \int_{\mathbb{R}^3} \omega \phi_u u^2 dx \geq 0$, this term involving may be very large and for some point $u \in \partial M \cap E^-$ we may have $I(u) > b$. Therefore the following geometric assumption of the linking theorem

$$b = \inf_N I > \sup_{\partial M} I$$

cannot be satisfied. Fortunately, we can apply the local linking theorem of Li and Willem [24] (see also Luan and Mao [23]) to overcome this difficulty and find critical points of I.

Next, we need the following compactness result in [189].

Proposition 4.6. *Under the assumption* (C27), *the embedding* $E \hookrightarrow L^q(\mathbb{R}^3)$, $2 \leq q < 2^* = 6$ *is compact.*

To study the functional I, it will be convenient to write it in a form in which the quadratic part is $\|u\|^2$. Obviously, due to (C28) and (C29), there exists a positive constant $C_1 > 0$, such that $|g(x,t)| \leq C_1|t|$ and $|G(x,t)| \leq \frac{C_1}{2}|t|^2|$, when $|t| < r_0$. So, there exists $C_2 > 0$, such that $tg(x,t) - 4G(x,t) \geq -C_2|t|^2$ for all $|t| < r_0$. Combining with (C31), we obtain

$$tg(x,t) - 4G(x,t) \geq -C_3|t|^2, \quad \forall t \in \mathbb{R}. \tag{4.159}$$

Let $f(x,t) = g(x,t) + mt$. Then, by a simple computation, we have

$$F(x,t) := \int_0^t f(x,\tau)\,d\tau \leq \frac{t}{4}f(x,t) + \frac{\tilde{b}}{4}t^2, \qquad \text{where } \tilde{b} = C_3 + m > 0. \tag{4.160}$$

Note that by (C30) we easily have

$$\lim_{|t|\to\infty} \frac{f(x,t)}{t^3} = +\infty, \tag{4.161}$$

uniformly in $x \in \mathbb{R}^3$. Moreover, using (C29) we get

$$\lim_{|t|\to 0} \frac{f(x,t)t}{t^4} = \lim_{|t|\to 0}\left(\frac{t^2}{t^4} \cdot \frac{g(x,t)t + mt^2}{t^2}\right) = +\infty.$$

Therefore, there is $\Lambda > 0$ such that

$$f(x,t)t \geq -\Lambda t^4, \tag{4.162}$$

for all $(x,t) \in \mathbb{R}^3 \times \mathbb{R}$.

With the modified nonlinearity f, our functional $I : E \to \mathbb{R}$ can be rewritten as follows:

$$I(u) = \frac{1}{2}\|u\|^2 - \frac{1}{2}\int_{\mathbb{R}^3} \omega\phi_u u^2\,dx - \int_{\mathbb{R}^3} F(x,u)\,dx. \tag{4.163}$$

Note that this does not imply that I has the mountain pass geometry, because unlike in (4.152), as $\|u\| \to 0$ the last term in (4.163) is not $o(\|u\|^2)$ anymore. The derivative of I is given below:

$$\langle I'(u), v\rangle = (u,v) - \int_{\mathbb{R}^3} ((2\omega + \phi_u)\phi_u uv + f(x,u)v)\,dx.$$

Lemma 4.27. *Suppose* (C27), (C28), (C30) *and* (C31) *are satisfied, then I satisfies the (PS) condition.*

Proof. Suppose $\{u_n\}$ is a sequence satisfying (1.3), where $\{\alpha_n\} \subset \mathbb{N}^2$ is admissible. We must prove that $\{u_n\}$ is bounded.

We may assume $\|u_n\| \to \infty$ for a contradiction. By (1.3) and noting that

$$\langle I'_{\alpha_n}(u_n), u_n \rangle = \langle I'(u_n), u_n \rangle \tag{4.164}$$

since $u_n \in E_{\alpha_n}$, for large n, using (4.160) we have

$$4 \cdot \sup_n I(u_n) + \|u_n\| \geq 4I(u_n) - \langle I'_{\alpha_n}(u_n), u_n \rangle$$

$$= \|u_n\|^2 + \int_{\mathbb{R}^3} (f(x, u_n)u_n - 4F(x, u_n))\, dx$$

$$\geq \|u_n\|^2 - \tilde{b} \int_{\mathbb{R}^3} u_n^2 dx. \tag{4.165}$$

Let $v_n = \|u_n\|^{-1} u_n$. Up to a subsequence, by the compact embedding $E \hookrightarrow L^2(\mathbb{R}^3)$ we deduce

$$v_n \rightharpoonup v \text{ in } E, \qquad v_n \to v \text{ in } L^2(\mathbb{R}^3) \qquad v_n(x) \to v(x) \text{ a.e. in } \mathbb{R}^3.$$

Multiplying by $\|u_n\|^{-2}$ to both sides of (4.165) and letting $n \to \infty$, we obtain

$$\tilde{b} \int_{\mathbb{R}^3} v^2 dx \geq 1.$$

Consequently, $v \neq 0$.

Using (4.162) and (4.156) with $q = 4$, we have

$$\int_{v=0} \frac{f(x, u_n)u_n}{\|u_n\|^4} dx = \int_{v=0} \frac{f(x, u_n)u_n}{u_n^4} v_n^4 dx$$

$$\geq -\Lambda \int_{v=0} v_n^4 dx$$

$$\geq -\Lambda \int_{\mathbb{R}^3} v_n^4 dx = -\Lambda \|v_n\|_4^4 \geq -\Lambda \kappa_4^4 > -\infty. \tag{4.166}$$

For $x \in \{x \in \mathbb{R}^3 : v \neq 0\}$, we have $|u_n(x)| \to +\infty$. By (4.161) we get

$$\frac{f(x, u_n(x))u_n(x)}{\|u_n\|^4} = \frac{f(x, u_n(x))u_n(x)}{u_n^4(x)} v_n^4(x) \to +\infty. \tag{4.167}$$

Consequently, using (4.166), (4.167) and the Fatou Lemma we obtain

$$\int_{\mathbb{R}^3} \frac{f(x, u_n)u_n}{\|u_n\|^4} dx \geq \int_{v \neq 0} \frac{f(x, u_n)u_n}{\|u_n\|^4} dx - \Lambda \kappa_4^4 \to +\infty. \tag{4.168}$$

Since $\{u_n\}$ is a sequence satisfying (1.3), using (4.168), for large n we obtain

$$C + 1 \geq \frac{1}{\|u_n\|^4} \left(\|u_n\|^2 + \int_{\mathbb{R}^3} \phi_{u_n}^2 u_n^2 dx - \langle I'(u_n), u_n \rangle \right)$$

$$= \int_{\mathbb{R}^3} \frac{f(x, u_n)u_n}{\|u_n\|^4} dx \to +\infty, \tag{4.169}$$

a contradiction.

Therefore, $\{u_n\}$ is bounded in E. Now, by the argument of [285] and using (4.164), the compact embedding $X \hookrightarrow L^2(\mathbb{R}^3)$ and

$$X = \overline{\bigcup_{n \in \mathbb{N}} X_{\alpha_n}},$$

we can easily prove that $\{u_n\}$ has a subsequence converging to a critical point of Φ. □

Lemma 4.28. *Under the assumptions* (V), *(C28) and (C29), the functional I has a local linking at 0 with respect to the decomposition* $E = E^- \oplus E^+$.

Proof. By (C28) and (C29), there exists $C > 0$ such that

$$|G(x, u)| \leq \frac{\kappa}{2\kappa_2^2} |u|^2 + C\kappa |u|^p . \tag{4.170}$$

If $u \in E^-$, then using (4.158) we deduce

$$I(u) = Q(u) - \int_{\mathbb{R}^3} \omega \phi_u u^2 dx - \int_{\mathbb{R}^3} G(x, u) dx$$

$$\leq -\kappa \|u\|^2 + \frac{1}{4} \|u\|^4 + \frac{\kappa}{2\kappa_2^2} |u|_2^2 + C\kappa |u|_p^p$$

$$\leq -\frac{\kappa}{2} \|u\|^2 + \frac{1}{4} \|u\|^4 + C_1 \|u\|^p , \tag{4.171}$$

where $C_1 = C\kappa\kappa_p^p$. Similarly, for $u \in E^+$ we have

$$I(u) \geq \frac{\kappa}{2} \|u\|^2 - C_1 \|u\|^p . \tag{4.172}$$

Since $p > 4$, the desired result (1.2) follows from (4.171) and (4.172). □

Lemma 4.29. *Let Y be a finite dimensional subspace of E, then I is anti-coercive on Y, that is*

$$J(u) \to -\infty, \qquad as \ \|u\| \to \infty, \ u \in E.$$

Proof. If the conclusion is not true, we can choose $\{u_n\} \subset Y$ and $\beta \in \mathbb{R}$ such that

$$\|u_n\| \to \infty, \qquad I(u_n) \geq \beta. \tag{4.173}$$

Let $v_n = \|u_n\|^{-1} u_n$. Since $\dim Y < \infty$, up to a subsequence we have

$$\|v_n - v\| \to 0, \qquad v_n(x) \to v(x) \text{ a.e. } \mathbb{R}^3$$

for some $v \in Y$, with $\|v\| = 1$. Since $v \neq 0$, similar to (4.168) we have

$$\int_{\mathbb{R}^3} \frac{F(x, u_n)}{\|u_n\|^4} dx \to +\infty. \tag{4.174}$$

Using (4.174) we deduce

$$I(u_n) = \|u_n\|^4 \left(\frac{1}{2 \|u_n\|^2} - \frac{1}{4 \|u_n\|^4} \int_{\mathbb{R}^3} \omega\phi_u u^2 dx - \int_{\mathbb{R}^3} \frac{F(u_n)}{\|u_n\|^4} dx \right) \to -\infty,$$

a contradiction with (4.173). ☐

Now, we are ready to prove our first results.

Proof of Theorem 4.9. We will find a nontrivial critical point of I via Theorem 1.5. It is easy to see that I maps bounded sets into bounded sets. In Lemmas 4.27 and 4.28 we see that I satisfies (PS) condition and I has a local linking at 0. It suffices to verify (1.4). Since $\dim(E^- \oplus E_m^+) < \infty$, this is a consequence of Lemma 4.29. ☐

Proof of Theorem 4.10. We define subspaces Y_k and Z_k of X as

$$Y_k = \text{span}\{\varphi_1, \varphi_2, \dots, \varphi_k\} \qquad Z_k = \overline{\text{span}\{\varphi_k, \varphi_{k+1}, \dots\}}.$$

Since g is odd, Φ is an even functional. By Lemma 4.27 we know that Φ satisfies (PS). It suffices to verify (i) and (ii) of Theorem 1.10.

Verification of (i). We assume that $0 \in [\lambda_\ell, \lambda_{\ell+1})$. Then if $k > \ell$ we have $Z_k \subset X^+$, where X^+ is defined in (4.157). Now, by (4.158), we have

$$Q(u) \geq \kappa \|u\|^2, \qquad u \in Z_k. \tag{4.175}$$

We recall that by [285, Lemma 2.8],

$$\beta_k = \sup_{u \in Z_k, \|u\|=1} |u|_p \to 0, \qquad \text{as } k \to \infty. \tag{4.176}$$

Let $r_k = (Cp\beta_k^p)^{1/(2-p)}$, where C is chosen in (4.170). For $u \in Z_k$ with $\|u\| = r_k$, using (4.170) and (4.175) we deduce

$$\Phi(u) = Q(u) - \frac{1}{2} \int_{\mathbb{R}^3} \omega\phi_u u^2 dx - \int_{\mathbb{R}^3} G(x, u) dx$$

$$\geq \kappa \|u\|^2 - \frac{\kappa}{2\kappa_2^2} |u|_2^2 - C\kappa |u|_p^p$$

$$\geq \kappa \left(\frac{1}{2} \|u\|^2 - C\beta_k^p \|u\|^p \right) = \kappa \left(\frac{1}{2} - \frac{1}{p} \right) (Cp\beta_k^p)^{2/(2-p)}.$$

Since $\beta_k \to 0$ and $p > 2$, it follows that

$$b_k = \inf_{u \in Z_k, \|u\|=r_k} \Phi(u) \to +\infty.$$

Verification of (ii). Since $\dim Y_k < \infty$, this is a consequence of Lemma 4.29. ☐

Notes and Comments

This section was obtained in [289].

4.3.3 Multiple Solutions without Odd Nonlinearities

In this section, we consider the following Klein–Gordon–Maxwell system

$$\begin{cases} -\Delta u + V(x)u - (2\omega + \phi)\phi u = f(x,u), & x \in \mathbb{R}^3, \\ \Delta\phi = (\omega + \phi)u^2, & x \in \mathbb{R}^3, \end{cases} \tag{4.177}$$

where ω is positive constant, the potential V and the nonlinearity f are allowed to be sign-changing. System (4.177) is a modified version of the classical Klein–Gordon–Maxwell system, which has a strong physical meaning since it appears in quantum mechanical models and in semiconductor theory. For more details about the physical background, we refer the reader to [228, 235, 272, 273, 280] and the references therein.

In this section, we consider another case of f being superlinear, that is, $f(x,u)/u \to +\infty$ as $u \to \infty$. Furthermore, the potential V and the primitive of f are also allowed to be sign-changing, which is quite different from the previous results. Before stating our main results, we give the following assumption on $V(x)$.

(C33) $V \in C(\mathbb{R}^3, \mathbb{R})$ and $\inf_{x \in \mathbb{R}^3} V(x) > -\infty$. Moreover, there exists a constant $d_0 > 0$ such that for any $M > 0$,

$$\lim_{|y| \to \infty} \text{meas}\left\{ x \in \mathbb{R}^3 : |x - y| \le d_0, V(x) \le M \right\} = 0,$$

where meas $(\cdot)$ denotes the Lebesgue measure in $\mathbb{R}^3$.

Inspired by Zhang and Xu [188], we can find a constant $V_0 > 0$ such that $\tilde{V}(x) := V(x) + V_0 \ge 1$ for all $x \in \mathbb{R}^3$ and let $\tilde{F}(x,u) := f(x,u) + V_0 u$, $\forall (x,u) \in \mathbb{R}^3 \times \mathbb{R}$. Then it is easy to verify the following lemma.

Lemma 4.30. *System* (4.177) *is equivalent to the following problem*

$$\begin{cases} -\Delta u + \tilde{V}(x)u - (2\omega + \phi)\phi u = \tilde{f}(x,u), & x \in \mathbb{R}^3, \\ \Delta\phi = (\omega + \phi)u^2, & x \in \mathbb{R}^3. \end{cases} \tag{4.178}$$

In what follows, we let $\mu > 4$ and impose some assumptions on $\tilde{f}$ and its primitive $\tilde{F}$ as follows:

(C34) $\tilde{f} \in C(\mathbb{R}^3 \times \mathbb{R}, \mathbb{R})$, and there exist constants c_1, $c_2 > 0$ and $q \in [4, 6)$ such that

$$|\tilde{f}(x,u)| \le c_1 |u|^3 + c_2 |u|^{q-1}.$$

(C35) $\lim_{|u| \to \infty} \frac{|\tilde{F}(x,u)|}{|u|^4} = \infty$ a.e. $x \in \mathbb{R}^3$ and there exists constants $c_3 \geq 0$, $r_0 \geq 0$ and $\tau \in (0,2)$ such that

$$\inf_{x \in \mathbb{R}^3} \tilde{F}(x,u) \geq c_3 |u|^\tau \geq 0, \quad \forall (x,u) \in \mathbb{R}^3 \times \mathbb{R}, \ |u| \geq r_0,$$

where and in the sequel, $\tilde{F}(x,u) = \int_0^u \tilde{f}(x,s)ds$.

(C36) $\tilde{\mathcal{F}}(x,u) := \frac{1}{4} u \tilde{f}(x,u) - \tilde{F}(x,u) \geq 0$, and there exists $c_4 > 0$ and $\kappa > 1$ such that

$$|\tilde{F}(x,u)|^\kappa \leq c_4 |u|^{2\kappa} \tilde{\mathcal{F}}(x,u), \quad \forall (x,u) \in \mathbb{R}^3 \times \mathbb{R}, \ |u| \geq r_0.$$

Now, we state our main results as follows:

Theorem 4.11. *Suppose that conditions* (C33), (C34), (C35) *and* (C36) *are satisfied. Then problem* (4.177) *possesses at least two different solutions.*

Remark 4.12. There are some functions not satisfying the condition (AR) for any $\mu > 4$. For example, the superlinear function $f(x,u) = \sin x \ln(1 + |u|)u^2$ does not satisfy condition (AR).

Remark 4.13. To the best of our knowledge, the condition (C33) is first given in [189], but $\inf_{x \in \mathbb{R}^3} V(x) > 0$ is required. From (C33), one can see that the potential $V(x)$ is allowed to be sign-changing. Therefore, the condition (C33) is weaken than (1.2) in $[265, 282, 285, 287, 288]$.

Remark 4.14. It is not difficult to find the functions V and f satisfying the above conditions. For example, let $V(x)$ be a zig-zag function with respect to $|x|$ defined by

$$V(x) = \begin{cases} 2n|x| - 2n(n-1) + a_0, & n-1 \leq |x| \leq (2n-1)/2, \\ -2n|x| + 2n^2 + a_0, & (2n-1)/2 \leq |x| \leq n, \end{cases}$$

where $n \in \mathbb{N}$ and $a_0 \in \mathbb{R}$.

Remark 4.15. Ding and Li [290] study (4.177) with sign-changing potential V. They got multiple solution with odd nonlinearity. Here we do not need the nonlinearity is odd and also got two solutions for problem (4.177).

Here, we give the sketch of how to look for two distinct critical points of the functional I (where I is defined by (4.181)). First, we consider a minimization of I constrained in a neighborhood of zero via the Ekeland's variational principle (see [5, 18]) and we can find a critical point of I which achieves the local minimum of I and the level of this local minimum is negative (see the Step 1 in the proof of Theorem 4.11); and then, around "zero" point, by using Mountain Pass Theorem (see [22]) we can also obtain other critical point of I with its positive level (see the Step 2 in the proof of Theorem 4.11). Obviously, these two critical points are different because they are in different levels.

Hereafter, we use the following notation. Denotes the space $E = \{u \in H^1(\mathbb{R}^3) : \int_{\mathbb{R}^3} (|\nabla u|^2 + \tilde{V}(x)|u|^2)dx < \infty\}$ with the norm

$$\|u\|_{H_{\tilde{V}}(\mathbb{R}^3)}^2 = \int_{\mathbb{R}^3} (|\nabla u|^2 + \tilde{V}(x)|u|^2)dx.$$

Next, we make the following assumption instead of (C33):

(C37) $\tilde{V} \in C(\mathbb{R}^3, \mathbb{R})$ and $\inf_{x \in \mathbb{R}^3} \tilde{V}(x) > 0$. Moreover, there exists a constant $d_0 > 0$ such that for any $M > 0$,

$$\lim_{|y| \to \infty} \text{meas} \{x \in \mathbb{R}^3 : |x - y| \le d_0, V(x) \le M\} = 0.$$

Remark 4.16. Under assumptions (C37), we know from Lemma 3.1 in [189] that the embedding $E \hookrightarrow L^s(\mathbb{R}^3)$ is compact for $s \in [2, 6)$.

Following technical results established in [272] (see also [228]), (4.178) can be reduced to a single equation with a nonlocal term.

Proposition 4.7. *For any fixed $u \in H^1(\mathbb{R}^3)$, there exists a unique $\phi = \phi_u \in D^{1,2}(\mathbb{R}^3)$ which solves equation*

$$- \Delta \phi + u^2 \phi = -\omega u^2. \tag{4.179}$$

Moreover, the map $\Phi : u \in H^1(\mathbb{R}^3) \mapsto \Phi[u] := \phi_u \in D^{1,2}(\mathbb{R}^3)$ is continuously differentiable, and

(i) $-\omega \le \phi_u \le 0$ on the set $\{x : u(x) \ne 0\}$;
(ii) $\|\phi_u\|_{D^{1,2}} \le C\|u\|_E^2$ and $\int_{\mathbb{R}^3} |\phi_u|u^2 dx \le C\|u\|_{12/5}^4 \le C\|u\|_E^4$.

By (4.179), multiplying by ϕ_u and integrating by parts we obtain

$$\int_{\mathbb{R}^3} |\nabla \phi_u|^2 dx = -\int_{\mathbb{R}^3} \omega \phi_u u^2 dx - \int_{\mathbb{R}^3} \phi_u^2 u^2 dx. \tag{4.180}$$

Using (4.180), we define a functional I on E by

$$I(u) = \frac{1}{2} \int_{\mathbb{R}^3} (|\nabla u|^2 + \tilde{V}(x)u^2 - \omega \phi_u u^2)dx - \int_{\mathbb{R}^3} \tilde{F}(x, u)dx, \tag{4.181}$$

for all $u \in E$. By condition (C34), we have

$$|\tilde{F}(x, u)| \le \frac{c_1}{4}|u|^4 + \frac{c_2}{q}|u|^q, \quad \forall (x, u) \in \mathbb{R}^3 \times \mathbb{R}. \tag{4.182}$$

Consequently, similar to the discussion in $[265, 285, 287]$, under assumptions (C37), Proposition 4.7 and (4.182), it is easy to prove that the functional I is of class $C^1(E, \mathbb{R})$. Moreover,

$$\langle I'(u), v \rangle = \int_{\mathbb{R}^3} (\nabla u \cdot \nabla v + \tilde{V}(x)uv - (2\omega + \phi_u)\phi_u uv - \tilde{f}(x, u)v)dx. \quad (4.183)$$

Hence, if $u \in E$ is a critical point of I, then the pair (u, ϕ_u) is a solution of system (4.178).

Lemma 4.31. *Assume that the conditions* (C37) *and* (C34) *hold. Then there exist* $\rho, \eta > 0$ *such that* $\inf\{I(u) : u \in E \text{ with } \|u\|_E = \rho\} > \eta$.

Proof. From (4.182) and the Sobolev inequality, we have

$$\left| \int_{\mathbb{R}^3} \tilde{F}(x, u)dx \right| \leq \int_{\mathbb{R}^3} \left| \frac{c_1}{4}|u|^4 + \frac{c_2}{q}|u|^q \right| dx$$

$$= \frac{c_1}{4}\|u\|_4^4 + \frac{c_2}{q}\|u\|_q^q$$

$$\leq S_4 \frac{c_1}{4}\|u\|_E^4 + S_q \frac{c_2}{q}\|u\|_E^q, \quad (4.184)$$

for any $u \in E$. Combining Proposition 4.7, (4.181) with (4.184), we have

$$I(u) = \frac{1}{2}\int_{\mathbb{R}^3}\left(|\nabla u|^2 + \tilde{V}(x)u^2 - \omega\phi_u u^2\right)dx - \int_{\mathbb{R}^3}\tilde{F}(x, u)dx$$

$$= \frac{1}{2}\|u\|_E^2 - \frac{1}{2}\int_{\mathbb{R}^3}\omega\phi_u u^2 dx - \int_{\mathbb{R}^3}\tilde{F}(x, u)dx$$

$$\geq \frac{1}{2}\|u\|_E^2 - \int_{\mathbb{R}^3}|\tilde{F}(x, u)|dx$$

$$\geq \frac{1}{2}\|u\|_E^2 - S_4 \frac{c_1}{4}\|u\|_E^4 - S_q \frac{c_2}{q}\|u\|_E^q$$

$$= \frac{1}{2}\|u\|_E^2 - C_1 4\|u\|_E^4 - C - 2\|u\|_E^q. \quad (4.185)$$

Since $q \in [4, 6)$, we can easily get that there exists $\eta > 0$ such that this lemma holds if we let $\|u\|_E = \rho > 0$ small enough. $\qquad \square$

Lemma 4.32. *Assume that the conditions* (C37) *and* (C35) *hold. Then there exists* $v \in E$ *with* $\|v\|_E = \rho$ *such that* $I(v) < 0$, *where* ρ *is given in Lemma 4.31.*

Proof. From (4.181), we have

$$\frac{I(tu)}{t^4} = \frac{t^2}{2}\left(\|u\|_E^2 - \int_{\mathbb{R}^3}\omega\phi_{tu}u^2 dx\right) - \frac{1}{t^4}\int_{\mathbb{R}^3}\tilde{F}(x, tu)dx.$$

Then, by Proposition 4.7, (C35) and Fatou's Lemma we deduce that

$$\lim_{t\to\infty} \frac{I(tu)}{t^4} = \lim_{t\to\infty}\left[\frac{1}{2t^2}\left(\|u\|_E^2 - \int_{\mathbb{R}^3}\omega\phi_t uu^2 dx\right) - \frac{1}{t^4}\int_{\mathbb{R}^3}\tilde{F}(x,tu)dx\right]$$

$$\leq \lim_{t\to\infty}\left[\frac{1}{2t^2}\left(\|u\|_E^2 - \int_{\mathbb{R}^3}\omega^2 u^2 dx\right) - \frac{1}{t^4}\int_{\mathbb{R}^3}\tilde{F}(x,tu)dx\right]$$

$$= -\lim_{t\to\infty}\int_{\mathbb{R}^3}\frac{\tilde{F}(x,tu)}{t^4u^4}u^4 dx$$

$$= -\int_{\mathbb{R}^3}\lim_{t\to\infty}\frac{\tilde{F}(x,tu)}{t^4u^4}u^4 dx$$

$$= -\infty \text{ as } n\to\infty.$$

Thus, the lemma is proved by taking $v = t_0 u$ with $t_0 > 0$ large enough. $\square$

Based on Lemmas 4.31 and 4.32, Theorem 1.4 implies that there is a sequence $\{u_n\} \subset E$ such that

$$I(u_n)\to c > 0 \text{ and } (1 + \|u_n\|_E)\|I'(u_n)\|_{E^*}\to 0, \text{ as } n\to\infty. \tag{4.186}$$

Lemma 4.33. *Assume that the conditions* (C37), (C34), (C35) *and* (C36) *hold. Then the sequence* $\{u_n\}$ *defined in* (4.186) *is bounded in* E.

Proof. Arguing by contradiction, we can assume $\|u_n\|_E \to \infty$. Define $v_n := \frac{u_n}{\|u_n\|_E}$. Clearly, $\|v_n\|_E = 1$ and $\|v_n\|_s \leq S_s\|v_n\|_E = S_s$, for $2 \leq s < 6$. Observe that for n large enough, from (4.186) and (C36) we have

$$c + 1 \geq I(u_n) - \frac{1}{4}\langle I'(u_n), u_n\rangle \tag{4.187}$$

$$= \frac{1}{4}\|u_n\|_E^2 + \frac{1}{4}\int_{\mathbb{R}^3}\phi_{u_n}^2 u_n^2 dx + \int_{\mathbb{R}^3}\left(\frac{1}{4}\tilde{f}(x,u_n)u_n - \tilde{F}(x,u_n)\right)dx$$

$$\geq \int_{\mathbb{R}^3}\tilde{\mathcal{F}}(x,u_n)dx.$$

In view of Proposition 4.7, (4.181) and (4.186), we have

$$\frac{1}{2} = \frac{I(u_n)}{\|u_n\|_E^2} + \frac{1}{\|u_n\|_E^2}\int_{\mathbb{R}^3}\tilde{F}(x,u_n)dx + \frac{1}{2\|u_n\|_E^2}\int_{\mathbb{R}^3}\omega\phi_{u_n}u_n^2 dx \tag{4.188}$$

$$\leq \frac{I(u_n)}{\|u_n\|_E^2} + \frac{1}{\|u_n\|_E^2}\int_{\mathbb{R}^3}|\tilde{F}(x,u_n)|dx$$

$$\leq \limsup_{n\to\infty}\left[\frac{I(u_n)}{\|u_n\|_E^2} + \frac{1}{\|u_n\|_E^2}\int_{\mathbb{R}^3}|\tilde{F}(x,u_n)|dx\right]$$

$$\leq \limsup_{n\to\infty}\int_{\mathbb{R}^3}\frac{|\tilde{F}(x,u_n)|}{\|u_n\|_E^2}dx.$$

For $0 \le a < b$, let $\Omega_n(a,b) := \{x \in \mathbb{R}^3 : a \le |u_n(x)| < b\}$. Going if necessary to a subsequence, we may assume that $v_n \rightharpoonup v$ in E. Then by Remark 4.16, we have $v_n \to v$ in $L^s(\mathbb{R}^3)$ for $2 \le s < 6$, and $v_n \to v$ a.e. on $\mathbb{R}^3$.

We now consider the following two possible cases about v.

Case 1: If $v = 0$, then $v_n \to 0$ in $L^s(\mathbb{R}^3)$ for $2 \le s < 6$, and $v_n \to 0$ a.e. on $\mathbb{R}^3$. Hence, it follows from (4.182) and $v_n := \frac{u_n}{\|u_n\|_E^2}$ that

$$
\int_{\Omega_n(0,r_0)} \frac{|\tilde{F}(x,u_n)|}{\|u_n\|_E^2} dx = \int_{\Omega_n(0,r_0)} \frac{|\tilde{F}(x,u_n)|}{|u_n|^2} |v_n|^2 dx
$$
$$
\le \left(\frac{c_1}{4} r_0^2 + \frac{c_2}{q} r_0^{q-2} \right) \int_{\Omega_n(0,r_0)} |v_n|^2 dx
$$
$$
\le C_4 \int_{\mathbb{R}^3} |v_n|^2 dx \to 0, \ \text{as} \ n \to \infty. \tag{4.189}
$$

From (C36), we know that $\kappa > 1$. Thus, if we set $\kappa' = \kappa/(\kappa1)$, then $2\kappa' \in (2,6)$. Hence, it follows from (C36), Proposition 4.7 and (4.187) that

$$
\int_{\Omega_n(r_0,\infty)} \frac{|\tilde{F}(x,u_n)|}{\|u_n\|_E^2} dx = \int_{\Omega_n(r_0,\infty)} \frac{|\tilde{F}(x,u_n)|}{|u_n|^2} |v_n|^2 dx
$$
$$
\le \left[\int_{\Omega_n(r_0,\infty)} \left(\frac{|\tilde{F}(x,u_n)|}{|u_n|^2} \right)^\kappa dx \right]^{1/\kappa} \left[\int_{\Omega_n(r_0,\infty)} |v_n|^{2\kappa'} dx \right]^{1/\kappa'}
$$
$$
\le c_4^{1/\kappa} \left[\int_{\Omega_n(r_0,\infty)} \tilde{\mathcal{F}}(x,u_n) dx \right]^{1/\kappa} \left[\int_{\Omega_n(r_0,\infty)} |v_n|^{2\kappa'} dx \right]^{1/\kappa'}
$$
$$
\le c_4^{1/\kappa} (c+1)^{1/\kappa} \left[\int_{\Omega_n(r_0,\infty)} |v_n|^{2\kappa'} dx \right]^{1/\kappa'}
$$
$$
\le C_5 \left[\int_{\Omega_n(r_0,\infty)} |v_n|^{2\kappa'} dx \right]^{1/\kappa'} \to 0, \ \text{as} \ n \to \infty. \tag{4.190}
$$

Combining (4.189) with (4.190), we have

$$
\int_{\mathbb{R}^3} \frac{|\tilde{F}(x,u_n)|}{\|u_n\|_E^2} dx = \int_{\Omega_n(0,r_0)} \frac{|\tilde{F}(x,u_n)|}{\|u_n\|_E^2} dx + \int_{\Omega_n(r_0,\infty)} \frac{|\tilde{F}(x,u_n)|}{\|u_n\|_E^2} dx \to 0
$$

as $n \to \infty$, which contradicts (4.188).

Case 2: If $v \ne 0$, we set $A := \{x \in \mathbb{R}^3 : v(x) \ne 0\}$. Then meas$(A) > 0$. For a.e. $x \in A$, we have $\lim_{n\to\infty} |u_n(x)| = \infty$. Hence $A \subset \Omega_n(r_0,\infty)$

for $n \in \mathbb{N}$ large enough. It follows from Proposition 4.7, (4.181), (4.182), (4.186) and Fatou's Lemma that

$$0 = \lim_{n \to \infty} \frac{c + o(1)}{\|u_n\|_E^4} = \lim_{n \to \infty} \frac{I(u_n)}{\|u_n\|_E^4} \tag{4.191}$$

$$= \lim_{n \to \infty} \left[\frac{1}{2\|u_n\|_E^2} - \frac{1}{2\|u_n\|_E^4} \int_{\mathbb{R}^3} \omega \phi_{u_n} u_n^2 dx - \int_{\mathbb{R}^3} \frac{\tilde{F}(x, u_n)}{\|u_n\|_E^4} dx \right]$$

$$\leq \left[\frac{1}{2\|u_n\|_E^2} - \frac{1}{2\|u_n\|_E^4} \int_{\mathbb{R}^3} \omega^2 u_n^2 dx - \int_{\Omega_n(0, r_0)} \frac{|\tilde{F}(x, u_n)|}{|u_n|^4} |v_n|^4 dx \right.$$

$$\left. - \int_{\Omega_n(r_0, \infty)} \frac{|\tilde{F}(x, u_n)|}{|u_n|^4} |v_n|^4 dx \right] \tag{4.192}$$

$$\leq \limsup_{n \to \infty} \int_{\Omega_n(0, r_0)} \left(\frac{c_1}{4} + \frac{c_2}{q} |u_n|^{q-4} \right) |v_n|^4 dx$$

$$- \liminf_{n \to \infty} \left[\int_{\Omega_n(r_0, \infty)} \frac{|\tilde{F}(x, u_n)|}{|u_n|^4} |v_n|^4 dx \right] \tag{4.193}$$

$$\leq \left(\frac{c_1}{4} + \frac{c_2}{q} |r_0|^{q-4} \right) \limsup_{n \to \infty} \int_{\Omega_n(0, r_0)} |v_n|^4 dx$$

$$- \liminf_{n \to \infty} \left[\int_{\Omega_n(r_0, \infty)} \frac{|\tilde{F}(x, u_n)|}{|u_n|^4} |v_n|^4 dx \right] \tag{4.194}$$

$$\leq C_7 - \liminf_{n \to \infty} \int_{\Omega_n(r_0, \infty)} \frac{|\tilde{F}(x, u_n)|}{|u_n|^4} |v_n|^4 dx$$

$$= C_7 - \liminf_{n \to \infty} \int_{\mathbb{R}^3} \frac{|\tilde{F}(x, u_n)|}{|u_n|^4} [\chi_{\Omega_n(r_0, \infty)}(x)] |v_n|^4 dx$$

$$= C_7 - \int_{RR^3} \liminf_{n \to \infty} \frac{|\tilde{F}(x, u_n)|}{|u_n|^4} [\chi_{\Omega_n(r_0, \infty)}(x)] |v_n|^4 dx \to -\infty \text{ as } n \to \infty,$$

which is a contradiction. Thus $\{u_n\}$ is bounded in E. The proof is completed. □

To complete our proof, we have to cite a result in [190].

Lemma 4.34. *Assume that p_1, $p_2 > 1$, r, $q \geq 1$ and $\Omega \subseteq \mathbb{R}^N$. Let $g(x, t)$ be a Carathéodory function on $\Omega \times \mathbb{R}$ and satisfy*

$$|g(x, t)| \leq a_1 |t|^{(p_1 - 1)/r} + a_2 |t|^{(p_2 - 1)/r}, \quad \forall (x, t) \in \Omega \times \mathbb{R},$$

where a_1, $a_2 \geq 0$. If $u_n \to u$ in $L^{p_1}(\Omega) \cap L^{p_2}(\Omega)$, and $u_n \to u$ a.e. $x \in \Omega$, then for any $v \in L^{p_1 q}(\Omega) \cap L^{p_2 q}(\Omega)$,

$$\lim_{n \to \infty} \int_{\Omega} |g(x, u_n) - g(x, u)|^r |v|^q dx \to 0. \tag{4.195}$$

Lemma 4.35. *If the conditions* (C37) *and* (C34) *hold. Then any bounded sequence* $\{u_n\}$ *satisfying* (4.186) *has a convergent subsequence in* E.

Proof. Going if necessary to a subsequence, we may assume that $u_n \rightharpoonup u$ in E. Then by Remark 4.16, we have $v_n \to v$ in $L^s(\mathbb{R}^3)$, for $2 \le s < 6$. Let us take $r \equiv 1$ in Lemma 4.34 and combine with $u_n \to u$ in $L^s(\mathbb{R}^3)$ for $2 \le s < 6$, one can get

$$\lim_{n\to\infty} |\tilde{f}(x, u_n) - \tilde{f}(x, u)||u_n - u|dx \to 0, \text{ as } n \to \infty. \tag{4.196}$$

We observe that

$$\langle I'(u_n) - I'(u), u_n - u \rangle \to 0, \text{ as } n \to \infty,$$

and we have

$$\int_{\mathbb{R}^3} [(2\omega + \phi_{u_n})\phi_{u_n} u_n - (2\omega + \phi_u)\phi_u u](u_n - u)dx$$

$$= 2\omega \int (\phi_{u_n} u_n - \phi_u u)(u_n - u) + \int_{\mathbb{R}^3} (\phi_{u_n}^2 u_n - \phi_u^2 u)(u_n - u)dx \to 0,$$

as $n \to \infty$. Actually, by Hölder's inequality, Proposition 4.7 and the Sobolev inequality, we have

$$\left| \int_{\mathbb{R}^3} (\phi_{u_n} - \phi_u)u_n(u_n - u)dx \right| \le \|(\phi_{u_n} - \phi_u)(u_n - u)\|_2 \|u_n\|_2$$

$$\le \|\phi_{u_n} - \phi_u\|_6 \|u_n - u\|_3 \|u_n\|_2$$

$$\le C\|\phi_{u_n} - \phi_u\|_{D^{1,2}} \|u_n - u\|_3 \|u_n\|_2,$$

where $C > 0$ is a constant. Because $u_n \to u$ in $L^s(\mathbb{R}^3)$ for any $2 \le s < 6$, we have

$$\int_{\mathbb{R}^3} (\phi_{u_n} - \phi_u)u_n(u_n - u)dx \to 0, \text{ as } n \to \infty,$$

and

$$\int_{\mathbb{R}^3} \phi_u(u_n - u)(u_n - u)dx \le \|\phi_u\|_6 \|u_n - u\|_3 \|u_n - u\|_2 \to 0, \text{ as } n \to \infty.$$

Thus, we get

$$\int_{\mathbb{R}^3} (\phi_{u_n} u_n - \phi_u u)(u_n - u)dx = \int_{\mathbb{R}^3} (\phi_{u_n} - \phi_u)u_n(u_n - u)dx$$

$$+ \int_{\mathbb{R}^3} \phi_u(u_n - u)(u_n - u)dx \to 0,$$

as $n \to \infty$. Observe that the sequence $\{\phi_{u_n}^2 u_n\}$ is bounded in $L^{3/2}(\mathbb{R}^3)$, since

$$\|\phi_{u_n}^2 u_n\|_{3/2} \le \|\phi_{u_n}\|_6^2 \|u_n\|_3,$$

so

$$\left| \int_{\mathbb{R}^3} (\phi_{u_n}^2 - \phi_u^2)(u_n - u)dx \right| \leq \|\phi_{u_n}^2 - \phi_u^2\|_{3/2}\|u_n - u\|_3$$

$$\leq (\|\phi_{u_n}^2\|_{3/2} + \|\phi_u^2\|_{3/2})\|u_n - u\|_3$$

$$\to 0 \text{ as } n \to \infty.$$

Now, using (4.196), we can get

$$\|u_n - u\|_E^2 = \langle I'(u_n) - I'(u), u_n - u \rangle$$

$$- \int_{\mathbb{R}^3} [(2\omega + \phi_{u_n})\phi_{u_n}u_n - (2\omega + \phi_u)\phi_u u](u_n - u)dx$$

$$+ \int_{\mathbb{R}^3} (\tilde{f}(x, u_n) - \tilde{f}(x, u))(u_n - u)dx \to 0,$$

as $n \to \infty$. That is $u_n \to u$ in E and the proof is complete. $\square$

Proof of Theorem 4.11. To complete the proof of Theorem 4.11, we need to consider the following two steps.

Step 1. We first show that there exists a function $u_0 \in E$ such that $I'(u_0) = 0$ and $I(u_0) < 0$. Let $r_0 = 1$, for any $|u| \geq 1$, from (C35), we have

$$\tilde{F}(x, u_n) \geq c_3|u|^\tau > 0. \tag{4.197}$$

By (C34), for a.e. $x \in \mathbb{R}^3$ and $0 \leq |u| \leq 1$, there exists $M > 0$ such that

$$\left| \frac{\tilde{f}(x, u)u}{u^2} \right| \leq \left| \frac{(c_1|u|^3 + c_2|u|^{q-1})|u|}{|u|^2} \right| \leq M,$$

which implies that

$$\tilde{f}(x, u)u \geq -M|u|^2.$$

Using the equality $\tilde{F}(x, u) = \int_0^1 \tilde{f}(x, tu)dt$, for a.e. $x \in \mathbb{R}^3$ and $0 \leq |u| \leq 1$, we obtain

$$\tilde{F}(x, u) > -\frac{1}{2}M|u|^2. \tag{4.198}$$

In view of (4.197) and (4.198), we have for a.e. $x \in \mathbb{R}^3$ and all $u \in \mathbb{R}$ that

$$\tilde{F}(x, u) \geq -\frac{1}{2}M|u|^2 + c_3|u|^\tau.$$

Then, we have

$$\tilde{F}(x, t\psi) \geq -\frac{1}{2}Mt^2|\psi|^2 + t^\tau c_3|\psi|^\tau. \tag{4.199}$$

Combining Proposition 4.7, (4.181) with (4.199), we have

$$I(tu) = \frac{t^2}{2}\|u\|_E^2 - \frac{t^2}{2}\int_{\mathbb{R}^3} \omega\phi_t uu^2 dx - \int_{\mathbb{R}^3} \tilde{F}(x, tu)dx$$

$$\leq \frac{t^2}{2}\|u\|_E^2 - \frac{t^2}{2}\int_{\mathbb{R}^3} \omega^2 u^2 dx + \frac{t^2 M}{2}\int_{\mathbb{R}^3} |u|^2 dx - t^\tau c_3 \int_{\mathbb{R}^3} |u|^\tau dx.$$

Since $\mu > 4$, $\tau \in (0, 2)$, for t small enough, we can get that $I(tu) < 0$. Thus, we obtain

$$c_0 = \inf\{I(u) : u \in \bar{B}_\rho\} < 0,$$

where $\rho > 0$ is given by Lemma 4.31, $B_\rho = \{u \in E : \|u\|_E < rho\}$. By the Ekeland's variational principle, there exists a sequence $\{u_n\} \subset B_\rho$ such that

$$c_0 \leq I(u_n) \leq c_0 + \frac{1}{n},$$

and

$$I(w) \geq I(u_n) - \frac{1}{n}\|w - u_n\|_E,$$

for all $w \in B_\rho$. Then, following the idea of [5], we can show that $\{u_n\}$ is a bounded Cerami sequence of I. Therefore, Lemma 4.35 implies that there exists a function $u_0 \in E$ such that $I'(u_0) = 0$ and $I(u_0) = c_0 < 0$.

   Step 2. We now show that there exists a function $\tilde{u}_0 \in E$ such that $I'(\tilde{u}_0) = 0$ and $I(\tilde{u}_0) = \tilde{c}_0 > 0$. By Lemmas 4.31, 4.32 and Theorem 1.4, there is a sequence $\{u_n\} \in E$ satisfies (4.186). Moreover, Lemma 4.33 and 4.35 shows that this sequence has a convergent subsequence and is bounded in E. So, we complete the Step 2.

   Therefore, combining the above two Steps and Lemma 4.30, we complete the proof of Theorem 4.11.   □

Notes and Comments

This type of system is settled on the whole space $\mathbb{R}^3$, the Sobolev embedding is not compact for the whole space. A natural idea is study this system on the radial space. Interesting reader can see the references [228, 235, 267, 272–277, 280, 281, 283, 284]. Recently, Carrião, Cunha and Miyagaki [282] studied this type of system with positive periodic potential V. They proved the existence of positive ground state solutions for this system when a periodic potential V is introduced. The method combined the minimization of the corresponding Euler–Lagrange functional on the Nehari manifold with the Brézis and Nirenberg technique. Later, Cunha [288] presented some results on the existence of positive and ground state solutions for the nonlinear (4.177). She introduced a general nonlinearity with subcritical and supercritical growth which does not require the usual Ambrosetti–Rabinowitz condition. Another situation for potential V is considered by He [265] (see also [285, 287]). He used coercive potential V which is introduced by Rabinowitz [178]. By means of a variant fountain theorem and the Symmetric Mountain Pass Theorem, he obtained the existence of infinitely many large energy solutions. Recently, Li and Tang [285] generalized He's result.

In some of the aforementioned references, the potential V is always assumed to be positive or vanish at infinity and the following famous Ambrosetti–Rabinowitz condition ((AR) for short) is usually required.

(AR) There exists $\mu > 4$ such that

$$0 < \mu F(x, u) \leq u f(x, u), \quad u \neq 0.$$

It is well known that the role of (AR) is to ensure the boundedness of the Palais–Smale (PS) sequences of the energy functional, which is very crucial in applying the critical point theory.

This section was obtained in [291].

4.3.4 *Partially Sublinear Nonlinearities*

In this section we study the existence and multiplicity of solutions for the following Klein–Gordon–Maxwell systems:

$$\begin{cases} -\Delta u + V(x)u - (2\omega + \phi)\phi u = f(x, u), & x \in \mathbb{R}^3, \\ \Delta \phi = (\omega + \phi)u^2, & x \in \mathbb{R}^3, \end{cases} \tag{4.200}$$

where $\omega > 0$ is a constant, $u, \phi : \mathbb{R}^3 \to \mathbb{R}$, $V : \mathbb{R}^3 \to \mathbb{R}$ is a potential function.

Theorem 4.12 ([285]). *Assume that V and f satisfy the following conditions:*

(A_1) $V \in C(\mathbb{R}^3, \mathbb{R})$ *and* $\inf_{x \in \mathbb{R}^3} V(x) \geq V_0 > 0$;

(A_2) *For every $M > 0$,* $\text{meas}\{x \in \mathbb{R}^3 : V(x) \leq M\} < \infty$, *where* $\text{meas}(\cdot)$ *denotes the Lebesgue measure in $\mathbb{R}^3$;*

(B_1) *There exist $p, \sigma, \gamma \in (1, 2)$ and $\nu \in (2, 6)$ such that*

$$b(x)|t|^p \leq f(x, t)t \text{ and } |f(x, t)| \leq m(x)|t|^{\sigma-1} + h(x)|t|^{\gamma-1} + C|t|^{\nu-1}$$

for all $(x, t) \in \mathbb{R}^3 \times R$, where $b, m, h : \mathbb{R}^3 \to \mathbb{R}$ are positive continuous functions satisfying $b \in L^{\frac{2}{2-p}}(\mathbb{R}^3)$, $m \in L^{\frac{2}{2-\sigma}}(\mathbb{R}^3)$, $h \in L^{\frac{2}{2-\gamma}}(\mathbb{R}^3)$;

(B₂) $f(x,-z) = -f(x,z)$, $(x,z) \in \mathbb{R}^3 \times \mathbb{R}$.

Then (4.200) *possesses infinitely many solutions.*

In this section, we will complement Theorem 4.12. Our main result is as follows.

Theorem 4.13. *Assume that V satisfies (A_1) and (A_2) and f satisfies (B₂) and the following conditions:*

(C38) *There exist $\delta > 0$, $1 \leq \gamma < 2$, $C > 0$ such that $f \in C(\mathbb{R}^3 \times [-\delta, \delta], \mathbb{R})$ and $|f(x,z)| \leq C|z|^{\gamma-1}$.*

(C39) $\lim_{z \to 0} F(x,z)/|z|^2 = +\infty$ *uniformly in some ball $B_r(x_0) \subset \mathbb{R}^3$, where $F(x,z) = \int_0^z f(x,s)ds$.*

(C40) $K : \mathbb{R}^3 \to \mathbb{R}^+$ *is a positive continuous function such that $K \in L^{2/(1-\gamma)}(\mathbb{R}^3) \cap L^\infty(\mathbb{R}^3)$.*

Then (4.200) *possesses infinitely many solutions $\{u_k\}$ such that $\|u_k\|_{L^\infty} \to 0$ as $k \to \infty$.*

Remark 4.17. In [285], we obtain infinitely many solutions when nonlinearity is sublinear at zero. Using the truncation techniques, nonlinearity is subcritical at infinity. In this section, we can see that when we give a control at infinity, we can generalize the nonlinearity to partially sublinear and get more information about the solutions (such as the solutions of L^∞ norm is convergent to zero).

We need the following compactness result proved in [189].

Proposition 4.8. *Under the assumption (A_1) and (A_2), the embedding $E \hookrightarrow L^q(\mathbb{R}^3)$, $2 \leq q < 2^* = 6$ is compact.*

Now we need the following technical results established in [272] (see also [228]).

Proposition 4.9. *For any fixed $u \in H^1(\mathbb{R}^3)$, there exists a unique $\phi = \phi_u \in D^{1,2}(\mathbb{R}^3)$ which solves equation*

$$- \Delta\phi + u^2\phi = -\omega u^2. \tag{4.201}$$

Moreover, the map $\Phi : u \in H^1(\mathbb{R}^3) \mapsto \Phi[u] := \phi_u \in D^{1,2}(\mathbb{R}^3)$ is continuously differentiable, and

(i) $-\omega \leq \phi_u \leq 0$ *on the set $\{x : u(x) \neq 0\}$;*
(ii) $\|\phi_u\|_{D^{1,2}} \leq C\|u\|_E^2$ *and $\int |\phi_u|u^2 \leq C\|u\|_{12/5}^4 \leq C\|u\|_E^4$.*

Proof of Theorem 4.13. Choose $\hat{f} \in C(\mathbb{R}^3 \times \mathbb{R}, \mathbb{R})$ such that $\hat{f}$ is odd in $u \in \mathbb{R}$, $\hat{f}(x, u) = f(x, u)$ for $x \in \mathbb{R}^3$ and $|u| < \delta/2$, and $\hat{f}(x, u) = 0$ for $x \in \mathbb{R}^3$ and $|u| > \delta$. In order to obtain solutions of (4.200), we consider the following problem

$$\begin{cases} -\Delta u + V(x)u - (2\omega + \phi)\phi u = K(x)\hat{f}(x, u), & x \in \mathbb{R}^3, \\ \Delta\phi = (\omega + \phi)u^2, & x \in \mathbb{R}^3. \end{cases} \quad (4.202)$$

Moreover, (4.202) is variational and its solutions are the critical points of the functional defined in E by

$$I(u) = \frac{1}{2} \int_{\mathbb{R}^3} \left(|\nabla u|^2 + V(x)u^2 - \omega\phi_u u^2 \right) dx - \int_{\mathbb{R}^3} F(x, u)dx.$$

From (C38), it is easy to check that I is well defined on E and $I \in C^1(E, \mathbb{R})$ (see [285] for more detail), and

$$I'(u)v = \int_{\mathbb{R}^3} (\nabla u \cdot \nabla v + V(x)uv - (2\omega + \phi_u)\phi_u uv - \hat{f}(x, u)v)dx, \quad v \in E.$$

Note that I is even, and $I(0) = 0$. For $u \in E$,

$$\int_{\mathbb{R}^3} K(x)|\hat{F}(x, u)|dx \leq C \int_{\mathbb{R}^3} K(x)|u|^\gamma dx \leq C\|K\|_{L^{\frac{2}{2-\gamma}}(\mathbb{R}^3)}\|u\|_{L^2(\mathbb{R}^3)}^\gamma \leq C\|u\|^\gamma.$$

Hence, it follows from Proposition 4.9 that

$$I(u) \geq \frac{1}{2}\|u\|^2 - C\|u\|^\gamma, \quad u \in E. \quad (4.203)$$

Now, we prove the (PS) condition. Let $\{u_n\}$ be a sequence in E, so that $I(u_n)$ is bounded and $I'(u_n) \to 0$. We shall prove that $\{u_n\}$ converges strongly. By (4.203), we claim that $\{u_n\}$ is bounded. Assume without loss of generality that $\{u_n\}$ converges to u weakly in E. Observe that

$$\|u_n - u\|^2 = \langle I'(u_n) - I'(u), u_n - u\rangle - \int_{\mathbb{R}^3} [(2\omega + \phi_{u_n})\phi_{u_n}u_n$$

$$+ (2\omega + \phi_u)\phi_u u](u_n - u)dx$$

$$+ \int_{\mathbb{R}^3} K(x)(\hat{f}(x, u_n) - \hat{f}(x, u))(u_n - u)dx$$

$$\equiv I_1 + I_2 + I_3,$$

It is clear that $I_1 \to 0$ as $n \to \infty$. In the following, we will estimate I_2. Actually, by the Hölder inequality, Proposition 4.9 and the Sobolev inequality, we have

$$\left| \int_{\mathbb{R}^3} (\phi_{u_n} - \phi_u)u_n(u_n - u)dx \right| \leq \|(\phi_{u_n} - \phi_u)(u_n - u)\|_2\|u_n\|_2$$

$$\leq \|\phi_{u_n} - \phi_u\|_6\|u_n - u\|_3\|u_n\|_2$$

$$\leq C\|\phi_{u_n} - \phi_u\|_{D^{1,2}}\|u_n - u\|_3\|u_n\|_2,$$

where $C > 0$ is a constant. Because $u_n \to u$ in $L^q(\mathbb{R}^3)$ for any $2 \le q < 6$, we have

$$\int_{\mathbb{R}^3} (\phi_{u_n} - \phi_u)u_n(u_n - u)dx \to 0, \text{ as } n \to \infty,$$

and

$$\int_{\mathbb{R}^3} \phi_u(u_n - u)(u_n - u)dx \le \|\phi_u\|_6\|u_n - u\|_3\|u_n - u\|_2 \to 0, \text{ as } n \to \infty.$$

Thus, we get

$$\int_{\mathbb{R}^3} (\phi_{u_n}u_n - \phi_u u)(u_n - u)dx = \int_{\mathbb{R}^3} (\phi_{u_n} - \phi_u)u_n(u_n - u)dx$$

$$+ \int_{\mathbb{R}^3} \phi_u(u_n - u)(u_n - u)dx \to 0,$$

as $n \to \infty$. Observe that the sequence $\{\phi_{u_n}^2 u_n\}$ is bounded in $L^{3/2}$, since

$$\|\phi_{u_n}^2 u_n\|_{3/2} \le \|\phi_{u_n}\|_6^2\|u_n\|_3,$$

so

$$\left| \int_{\mathbb{R}^3} (\phi_{u_n}^2 - \phi_u^2)(u_n - u)dx \right| \le \|\phi_{u_n}^2 - \phi_u^2\|_{3/2}\|u_n - u\|_3$$

$$\le (\|\phi_{u_n}^2\|_{3/2} + \|\phi_u^2\|_{3/2})\|u_n - u\|_3$$

$$\to 0 \text{ as } n \to \infty.$$

We estimate I_3, by using (f$_3$), for any $R > 0$,

$$\int_{\mathbb{R}^3} K(x)|\hat{f}(x, u_n) - \hat{f}(x, u)|\,|u_n - u|dx$$

$$\le C \int_{\mathbb{R}^3 \backslash B_R(0)} K(x)(|u_n|^\gamma + |u|^\gamma)dx + C \int_{B_R(0)} (|u_n|^{\gamma-1} + |u|^{\gamma-1})|u_n - u|dx$$

$$\le C \left(\|u_n\|_{L^2(\mathbb{R}^3 \backslash B_R(0))}^\gamma + \|u\|_{L^2(\mathbb{R}^3 \backslash B_R(0))}^\gamma \right) \|K\|_{L^{\frac{2}{2-\gamma}}(\mathbb{R}^3 \backslash B_R(0))}$$

$$+ C \left(\|u_n\|_{L^\gamma(B_R(0))}^{\gamma-1} + \|u\|_{L^\gamma(B_R(0))}^{\gamma-1} \right) \|u_n - u\|_{L^\gamma(B_R(0))}$$

$$\le C\|K\|_{L^{\frac{2}{2-\gamma}}(\mathbb{R}^3 \backslash B_R(0))} + C\|u_n - u\|_{L^\gamma(B_R(0))},$$

which implies

$$\lim_{n \to +\infty} \int_{\mathbb{R}^3} K(x)|\hat{f}(x, u_n) - \hat{f}(x, u))|\,|u_n - u|dx = 0.$$

Therefore, $\{u_n\}$ converges strongly in E and the (PS) condition holds for I. By (f$_2$) and (f$_3$), for any $L > 0$, there exists $\delta = \delta(L) > 0$ such that if $u \in C_0^\infty(B_r(x_0))$ and $|u|_\infty < \delta$ then $K(x)\hat{F}(x, u(x)) \geq L|u(x)|^2$, and it follows from Proposition 4.9 that

$$I(u) \leq \frac{1}{2}\|u\|^2 + \frac{1}{4}\|u\|^4 - L\|u\|_{L^2(\mathbb{R}^3)}^2.$$

This implies, for any $k \in \mathbb{N}$, if X^k is a k–dimensional subspace of $C_0^\infty(B_r(x_0))$ and ρ_k is sufficiently small then $\sup_{X^k \cap S_{\rho_k}} I(u) < 0$, where $S_\rho = \{u \in E : \|u\| = \rho\}$. Now we apply Theorem 1.13 to obtain infinitely many solutions $\{u_k\}$ for (4.202) such that

$$\|u_k\| \to 0, \ k \to \infty. \tag{4.204}$$

Finally we show that $\|u_k\|_{L^\infty} \to 0$ as $k \to \infty$. Let u be a solution of (4.202) and $\alpha > 0$. Let $M > 0$ and set $u^M(x) = \max\{-M, \min\{u(x), M\}\}$. Multiplying both sides of (4.202) with $|u^M|^\alpha u^M$ implies

$$\frac{4}{(\alpha+2)^2}\int_{\mathbb{R}^3} |\nabla|u^M|^{\frac{\alpha}{2}+1}|^2 dx \leq C \int_{\mathbb{R}^3} |u^M|^{\alpha+1} dx.$$

By using the iterating method in [30], we can get the following estimate

$$\|u\|_{L^\infty(\mathbb{R}^3)} \leq C_1 \|u\|_{L^6(\mathbb{R}^3)}^\nu,$$

where ν is a number in $(0, 1)$ and $C_1 > 0$ is independent of u and α. By (4.204) and Sobolev embedding Theorem [30], we derive that $\|u_k\|_{L^\infty(\mathbb{R}^3)} \to 0$ as $k \to \infty$. Therefore, u_k are the solutions of (4.200) as k sufficiently large. The proof is completed. $\square$

Notes and Comments

This section was obtained in [292].

4.3.5 Klein–Gordon Equation Coupled with Born–Infeld Theory

This section was motivated by the search of the existence of solitary wave solutions for the following Klein–Gordon equation coupled with Born-Infeld theory

$$\begin{cases} -\Delta u + [m^2 - (\omega + \phi)^2]u = f(x, u), & \text{in } \mathbb{R}^3, \\ \Delta\phi + \beta\Delta_4\phi = 4\pi(\omega + \phi)u^2, & \text{in } \mathbb{R}^3, \end{cases} \tag{4.205}$$

where $\Delta_4 u = \text{div}(|\nabla u|^2 \nabla u)$. This system arises in an interesting physical context (we refer the reader to [293] and the references therein for more details on the physical aspects). It can be deduced by coupling the Klein–Gordon equation

$$\psi_{tt} - \Delta\psi + m^2\psi - f(x, \psi) = 0,$$

with Born–Infeld theory

$$\mathcal{L}_{\text{BI}} = \frac{b^2}{4\pi}(1 - \sqrt{1 - \frac{1}{b^2}(|\mathbb{E}|^2 - |\mathbb{B}|^2)}),$$

where $\psi = \psi(x, t) \in \mathbb{C}$, $x \in \mathbb{R}^3$, $t \in \mathbb{R}$, m is a real constant, $p \in (2, 6]$, $\mathbb{E}$ is the electric field and $\mathbf{B}$ is the magnetic induction field. As done in [293,294], they set

$$\beta = \frac{1}{2b^2}$$

and consider the second-order expansion of $\mathcal{L}_{\text{BI}}$ for $\beta \to 0$, in this way the Lagrangian density takes the form

$$\mathcal{L}_1 = \frac{1}{4\pi}\left[\frac{1}{2}(|\mathbb{E}|^2 - |\mathbb{B}|^2) + \frac{\beta}{4}(|\mathbb{E}|^2 - |\mathbb{B}|^2)^2\right].$$

In this case, under the same electrostatic solitary wave ansatz as in [293, 294], they prove that the critical point of $\mathcal{F}$ defined as

$$\mathcal{F}(u, \phi) = \int_{\mathbb{R}^3}\left[\frac{1}{2}\left(|\nabla u|^2 + [m^2 - (\omega + \phi)^2]u^2\right) - \frac{1}{8\pi}|\nabla\phi|^2\right.$$

$$\left. - \frac{\beta}{16\pi}|\nabla\phi|^4 - f(x, u)\right]dx.$$

In recent years, lots of existence results for problem (4.205) had got via modern variational methods under various hypotheses on the nonlinear term f, see [294–296] and the references therein. We recall some of them as follows.

Case 1: $f(x, u) = |u|^{q-2}u$, in [294] the authors found the existence of infinitely many radially symmetric solutions for system (4.205) when $4 < q < 6$ and $|\omega| < |m|$; in [295] the range $q \in (2, 4]$ was also covered provided $\sqrt{(\frac{q}{2} - 1)}|m| > \omega > 0$.

Case 2: $f(x, u) = |u|^{q-2}u + |u|^{2^*-2}u$, only Teng and Zhang in [296] get that problem (4.205) has at least a nontrivial solution when $4 < q < 6$ and $\omega < m$.

The object of this section is to consider the following nonhomogeneous Klein-Gordon equation coupled with Born–Infeld theory

$$\begin{cases} -\Delta u + [m^2 - (\omega + \phi)^2]u = |u|^{q-2}u + h(x), & \text{in } \mathbb{R}^3, \\ \Delta\phi + \beta\Delta_4\phi = 4\pi(\omega + \phi)u^2, & \text{in } \mathbb{R}^3, \end{cases} \tag{4.206}$$

Now, we state the main results:

Theorem 4.14. *Suppose that $h \in C^1(\mathbb{R}^3) \cap L^2(\mathbb{R}^3)$ is a radial function and $h \not\equiv 0$. If $m > \omega > 0$ and $4 < q < 6$, problem (4.206) admits at least two different solutions (u, ϕ) such that*

$$\int_{\mathbb{R}^3} |u|^2 dx + \int_{\mathbb{R}^3} |\nabla u|^2 dx < +\infty, \tag{4.207}$$

$$\int_{\mathbb{R}^3} |\nabla \phi|^2 dx + \int_{\mathbb{R}^3} |\nabla \phi|^4 dx < +\infty. \tag{4.208}$$

Moreover the field u and ϕ are radially symmetric.

Remark 4.18. To the best of our knowledge, it seems that Theorem 4.14 is the first result about the existence of multiple solutions for the non-homogeneous Klein–Gordon equation coupled with Born–Infeld theory on $\mathbb{R}^3$.

Theorem 4.15. *Suppose that $h \in C^1(\mathbb{R}^3) \cap L^2(\mathbb{R}^3)$ is a radial function and $h \not\equiv 0$. If $\sqrt{\left(\frac{q}{2} - 1\right)} m > \omega > 0$ and $2 < q \le 4$, problem (4.206) admits at least two different solutions (u, ϕ) satisfying (4.207) and (4.208), and the field u and ϕ are also radially symmetric.*

Remark 4.19. The assumption $\sqrt{\left(\frac{q}{2} - 1\right)} m > \omega > 0$ is merely technical, since it is needed only to prove the boundedness of Palais–Smale sequence (see the proof of Theorem 4.15). Of course it implies the assumption $m > \omega > 0$ if $2 < q \le 4$.

Before beginning the proofs of our main results, we give several notations. Define the function space $D(\mathbb{R}^3)$ the completion of $C_0^\infty(\mathbb{R}^3, \mathbb{R})$ with respect to the norm

$$\|\phi\|_D = \|\nabla \phi\|_{L^2} + \|\nabla \phi\|_{L^4}.$$

Denoting by $D^{1.2}(\mathbb{R}^3)$ the completion of $C_0^\infty(\mathbb{R}^3, \mathbb{R})$ with respect to the norm

$$\|\phi\|_{D^{1.2}} = \|\nabla \phi\|_{L^2}.$$

It is clear that $D(\mathbb{R}^3)$ is continuously embedded in $D^{1.2}(\mathbb{R}^3)$. Moreover $D^{1.2}(\mathbb{R}^3)$ is continuously embedded in $L^6(\mathbb{R}^3)$ by Sobolev inequality and $D(\mathbb{R}^3)$ is continuously embedded in $L^\infty(\mathbb{R}^3)$ by Proposition 8 in [293].

System (4.206) has a variational structure. Indeed we consider the functional

$$J : H^1(\mathbb{R}^3) \times D(\mathbb{R}^3) \to \mathbb{R}$$

defined by

$$J(u, \phi) = \frac{1}{2} \int_{\mathbb{R}^3} \left(|\nabla u|^2 + [m^2 - (\omega + \phi)^2] u^2 \right) dx - \frac{1}{8\pi} \int_{\mathbb{R}^3} |\nabla \phi|^2 dx$$

$$- \frac{\beta}{16\pi} \int_{\mathbb{R}^3} |\nabla \phi|^4 dx - \frac{1}{q} \int_{\mathbb{R}^3} |u|^q dx - \int_{\mathbb{R}^3} h(x) u dx.$$

Evidently, the action functional J belongs to $C^1(H^1(\mathbb{R}^3) \times D(\mathbb{R}^3), \mathbb{R})$ and the partial derivatives in (u, ϕ) are given, for $\zeta \in H^1(\mathbb{R}^3), \eta \in D(\mathbb{R}^3)$, by

$$\left\langle \frac{\partial J}{\partial u}(u, \phi), \zeta \right\rangle = \int_{\mathbb{R}^3} \left(\nabla u \cdot \nabla \zeta + [m^2 - (\omega + \phi)^2] u \zeta - |u|^{q-2} u \zeta - h(x) \zeta \right) dx,$$

$$\left\langle \frac{\partial J}{\partial \phi}(u, \phi), \eta \right\rangle = - \int_{\mathbb{R}^3} \left(\frac{1}{4\pi} \left((1 + \beta |\nabla \phi|^2) \nabla \phi \cdot \nabla \eta \right) + (\phi + \omega) u^2 \eta \right) dx.$$

Thus, we have the following result:

Proposition 4.10. *The pair (u, ϕ) is a weak solution of system (4.206) if and only if it is a critical point of J in $H^1(\mathbb{R}^3) \times D(\mathbb{R}^3)$.*

Moreover, we have the following lemma.

Lemma 4.36. *For every $u \in H^1(\mathbb{R}^3)$,*

(1) there exists a unique function $\phi_u \in D(\mathbb{R}^3)$ which solves the second equation of system (4.206);
(2) if u is radially symmetric, ϕ_u is radial too;
(3) $\phi_u \leq 0$. Moreover, $-\omega \leq \phi_u$, if $u(x) \neq 0$.

The first result is proved in Lemma 3 of [294], while the second one, though not explicitly stated, is proved in Lemma 5 of [294]. The third result can be found in Lemma 2.3 of [295].

So, we can consider the functional $I : H^1(\mathbb{R}^3) \to \mathbb{R}$ defined by $I(u) = J(u, \phi_u)$, i.e.

$$I(u) = \frac{1}{2} \int_{\mathbb{R}^3} \left(|\nabla u|^2 + [m^2 - (\omega + \phi_u)^2] u^2 \right) dx - \frac{1}{8\pi} \int_{\mathbb{R}^3} |\nabla \phi_u|^2 dx$$

$$- \frac{\beta}{16\pi} \int_{\mathbb{R}^3} |\nabla \phi_u|^4 dx - \frac{1}{q} \int_{\mathbb{R}^3} |u|^q dx - \int_{\mathbb{R}^3} h(x) u dx. \qquad (4.209)$$

Furthermore I is C^1 and we have, for any $u, v \in H^1(\mathbb{R}^3)$ (see the proof of Lemma 6 in [294]),

$$\langle I'(u), v \rangle = \int_{\mathbb{R}^3} \left(\nabla u \cdot \nabla v + [m^2 - (\omega + \phi_u)^2]uv - |u|^{q-2}uv - h(x)v \right) dx$$

(4.210)

$$= \int_{\mathbb{R}^3} \left(\nabla u \cdot \nabla v + [m^2 - \omega^2]uv - 2\omega\phi_u uv - \phi_u^2 uv - |u|^{q-2}uv - h(x)v \right) dx.$$

Now, we have:

Proposition 4.11 ([295] Lemma 2.4, or [272] Proposition 3.5).
The following statements are equivalent:
(1) $(u, \phi) \in H^1(\mathbb{R}^3) \times D(\mathbb{R}^3)$ is a critical point of J (i.e. (u, ϕ) is a solution of system (4.205));
(2) u is a critical point of I and $\phi = \phi_u$.

After multiplying $\Delta\phi_u + \beta\Delta_4\phi_u = 4\pi(\omega + \phi_u)u^2$ by ϕ_u and integration by parts, we obtain

$$-\int_{\mathbb{R}^3} \left(|\nabla\phi_u|^2 + \beta|\nabla\phi_u|^4 \right) dx = 4\pi \int_{\mathbb{R}^3} (\omega + \phi_u)\phi_u u^2 dx. \qquad (4.211)$$

Therefore, the reduced functional takes also the form

$$I(u) = \frac{1}{2} \int_{\mathbb{R}^3} \left(|\nabla u|^2 + [m^2 - \omega^2]u^2 + \phi_u^2 u^2 \right) dx + \frac{1}{8\pi} \int_{\mathbb{R}^3} |\nabla\phi_u|^2 dx$$

(4.212)

$$+ \frac{3\beta}{16\pi} \int_{\mathbb{R}^3} |\nabla\phi_u|^4 dx - \frac{1}{q} \int_{\mathbb{R}^3} |u|^q dx - \int_{\mathbb{R}^3} h(x)u dx.$$

Since system (4.205) is set on $\mathbb{R}^3$, it is well known that the Sobolev embedding $H^1(\mathbb{R}^3) \hookrightarrow L^s(\mathbb{R}^3)$ ($2 \le s \le 6$) is not compact, and then it is usually difficult to prove that a minimizing sequence or a Palais-Smale sequence is strongly convergent when we seek solutions of system (4.205) by variational methods. To overcome this difficulty we restrict ourselves to radial functions $u = u(r)$, $r = |x|$. More precisely we shall consider the functional I on the space of the radial functions

$$H_r^1(\mathbb{R}^3) := \{u \in H^1(\mathbb{R}^3) : u = u(r), \ r = |x|\}.$$

$H_r^1(\mathbb{R}^3)$ is a natural constraint for I, namely any critical point $u \in H_r^1(\mathbb{R}^3)$ of $I_{|H_r^1(\mathbb{R}^3)}$ is also a critical point of I (see [294], Lemma 5). Then we are reduced to look for critical points of $I_{|H_r^1(\mathbb{R}^3)}$. We recall (see [297] or [169]) that, for $2 < s < 6$, $H^1_r(\mathbb{R}^3)$ is compactly embedded into $L^s(\mathbb{R}^3)$. In the following, we still denote $I_{|H^1_r(\mathbb{R}^3)}$ by I.

Now we give a sketch of how to look for two distinct critical points of the functional I. First, we consider a minimization of I constrained in a neighborhood of zero, by using the Ekeland's variational principle (see [18], or [171]), we can find a critical point of I which achieves the local minimum of I. Moreover, this local minimum is negative, see the Step 1 in the proof of Theorem 4.14. Next, around "zero", by the Mountain Pass Theorem (see [172], or [21]), we can also get a critical point of I and its level is positive (see Step 2 in the proof of Theorem 4.14). Because these two critical points are in different level, they must be distinct.

Before going to the proof of Theorem 4.14, we give some useful preliminary results.

Lemma 4.37. *Let $m > \omega > 0$, $2 < q < 6$, and $h \in L^2(\mathbb{R}^3) \cap C^1(\mathbb{R}^3)$ is a radial function. Then there exist some constants $\rho_1, \alpha_1, m_1 > 0$ such that $I(u)\big|_{\|u\|_{H^1}=\rho_1} \geq \alpha_1$ for all h satisfying $\|h\|_{L^2} < m_1$.*

Proof. Using (4.209), $\phi_u \leq 0$ (see Lemma 4.36), the Hölder inequality and $H^1_r(\mathbb{R}^3) \hookrightarrow L^s(\mathbb{R}^3)$ for any $s \in [2, 2^*]$, we have

$$I(u) \geq \frac{1}{2} \int_{\mathbb{R}^3} \left(|\nabla u|^2 + [m^2 - \omega^2]u^2 \right) dx - \frac{1}{q}\|u\|^q_{L^q} - \|h\|_{L^2}\|u\|_{L^2}$$

$$\geq \frac{1}{2} \int_{\mathbb{R}^3} \left(|\nabla u|^2 + [m^2 - \omega^2]u^2 \right) dx - a_1\|u\|^q_{H^1} - \|h\|_{L^2}\|u\|_{H^1}$$

$$= \|u\|_{H^1}[a_2\|u\|_{H^1} - a_1\|u\|^{q-1}_{H^1} - \|h\|_{L^2}],$$

where $a_2 = \frac{\min\{1, m^2 - \omega^2\}}{2}$ and a_1 are positive constants. Setting

$$g_1(t) = a_2 t - a_1 t^{q-1}$$

for $t \geq 0$, we see that there exists a constant $\rho_1 > 0$ such that $\max_{t \geq 0} g_1(t) = g_1(\rho_1) > 0$. Taking $m_1 := \frac{1}{2}g_1(\rho_1)$, then it follows that there exists a constant $\alpha_1 > 0$ such that $I(u)\big|_{\|u\|_{H^1}=\rho_1} \geq \alpha_1$ for all h satisfying $\|h\|_{L^2} < m_1$. $\qquad\square$

Lemma 4.38. *Let $m > \omega > 0$ and $2 < q < 6$. Then there exists a function $v_1 \in E$ with $\|v_1\|_{H^1} > \rho_1$ such that $I(v_1) < 0$, where ρ_1 is given by Lemma 4.37.*

Proof. By (4.207), (i) of Lemma 4.36 and $q > 2$, we have

$$I(tu) = \frac{t^2}{2} \int_{\mathbb{R}^3} \left(|\nabla u|^2 + [m^2 - (\omega + \phi_{tu})^2]u^2 \right) dx - \frac{1}{8\pi} \int_{\mathbb{R}^3} |\nabla \phi_{tu}|^2 dx$$

$$- \frac{\beta}{16\pi} \int_{\mathbb{R}^3} |\nabla \phi_{tu}|^4 dx - \frac{t^q}{q} \int_{\mathbb{R}^3} |u|^q dx - t \int_{\mathbb{R}^3} h(x) u \, dx$$

$$\leq \frac{t^2}{2} \int_{\mathbb{R}^3} \left(|\nabla u|^2 + m^2 u^2 \right) dx - \frac{t^q}{q} \int_{\mathbb{R}^3} |u|^q dx - t \int_{\mathbb{R}^3} h(x) u \, dx$$

$$\to -\infty,$$

as $t \to +\infty$ for $u \in H_r^1(\mathbb{R}^3), u \neq 0$. The lemma is proved by taking $v = t_1 u$ with $t_1 > 0$ large enough and $u \neq 0$. $\square$

Lemma 4.39. *Assume that $m > \omega > 0$, $2 < q < 6$, and $\{u_n\} \subset H_r^1(\mathbb{R}^3)$ is a bounded Palais-Smale sequence of I, then $\{u_n\}$ has a strongly convergent subsequence in $H_r^1(\mathbb{R}^3)$.*

Proof. Consider a sequence $\{u_n\}$ in $H_r^1(\mathbb{R}^3)$ which satisfies

$$I(u_n) \to c, \ I'(u_n) \to 0, \ \text{and} \ \sup_n \|u_n\|_{H^1} < +\infty.$$

Going if necessary to a subsequence, we can assume that $u_n \rightharpoonup u$ in $H_r^1(\mathbb{R}^3)$. Since the embedding $H_r^1(\mathbb{R}^3) \hookrightarrow L^s(\mathbb{R}^3)$ is compact for any $s \in (2, 2^*)$, we have $u_n \to u$ in $L^s(\mathbb{R}^3)$ for any $s \in (2, 2^*)$. By (4.210), we easily get

$$\int_{\mathbb{R}^3} \left(|\nabla(u_n - u)|^2 + [m^2 - \omega^2](u_n - u)^2 \right) dx$$

$$= \langle I'(u_n) - I'(u), u_n - u \rangle + \int_{\mathbb{R}^3} \left(|u_n|^{q-2} u_n - |u|^{q-2} u \right) (u_n - u) dx$$

$$+ 2\omega \int_{\mathbb{R}^3} (\phi_{u_n} u_n - \phi_u u)(u_n - u) dx + \int_{\mathbb{R}^3} (\phi_{u_n}^2 u_n - \phi_u^2 u)(u_n - u) dx.$$

It is clear that

$$\langle I'(u_n) - I'(u), u_n - u \rangle \to 0, \ \text{as} \ n \to \infty.$$

Since $u_n \to u$ in $L^s(\mathbb{R}^3)$ for any $s \in (2, 2^*)$, we have

$$\int_{\mathbb{R}^3} \left(|u_n|^{q-2} u_n - |u|^{q-2} u \right) (u_n - u) dx \to 0, \ \text{as} \ n \to \infty.$$

From (4.211) and the boundedness of $\{u_n\} \subset H_r^1(\mathbb{R}^3)$, we get

$$\int_{\mathbb{R}^3} \left(|\nabla \phi_{u_n}|^2 + \beta |\nabla \phi_{u_n}|^4 \right) dx = -4\pi \int_{\mathbb{R}^3} (\omega + \phi_{u_n}) \phi_{u_n} u_n^2 dx$$

$$\leq -4\pi\omega \int_{\mathbb{R}^3} \phi_{u_n} u_n^2 dx$$

$$\leq 4\pi\omega \|\phi_{u_n}\|_{L^6} \|u_n\|_{L^{\frac{12}{5}}}^2$$

$$\leq a_3 \|\phi_{u_n}\|_D \|u_n\|_{H^1}^2$$

$$\leq a_4 \|\phi_{u_n}\|_D$$

and so $\{\phi_{u_n}\}$ is bounded. By the Hölder inequality and the Sobolev inequality, we have

$$\left| \int_{\mathbb{R}^3} (\phi_{u_n} - \phi_u) u_n (u_n - u) dx \right| \leq \|(\phi_{u_n} - \phi_u)(u_n - u)\|_{L^2} \|u_n\|_{L^2}$$

$$\leq \|\phi_{u_n} - \phi_u\|_{L^6} \|u_n - u\|_{L^3} \|u_n\|_{L^2}$$

$$\leq a_5 \|\phi_{u_n} - \phi_u\|_D \|u_n - u\|_{L^3} \|u_n\|_{L^2},$$

where $a_5 > 0$ is a constant. Because $u_n \to u$ in $L^s(\mathbb{R}^3)$ for any $s \in (2, 2^*)$, we have

$$\int_{\mathbb{R}^3} (\phi_{u_n} - \phi_u) u_n (u_n - u) dx \to 0, \text{ as } n \to \infty,$$

and

$$\int_{\mathbb{R}^3} \phi_u (u_n - u)(u_n - u) dx \leq \|\phi_u\|_{L^6} \|u_n - u\|_{L^3} \|u_n - u\|_{L^2} \to 0, \text{ as } n \to \infty.$$

Thus, we get

$$\int_{\mathbb{R}^3} (\phi_{u_n} u_n - \phi_u u)(u_n - u) dx$$

$$= \int_{\mathbb{R}^3} (\phi_{u_n} - \phi_u) u_n (u_n - u) dx + \int_{\mathbb{R}^3} \phi_u (u_n - u)(u_n - u) dx \to 0, \text{ as } n \to \infty.$$

Again by the Hölder inequality and the Sobolev inequality, we have

$$\left| \int_{\mathbb{R}^3} (\phi_{u_n}^2 - \phi_u^2) u_n (u_n - u) dx \right| \leq \|(\phi_{u_n}^2 - \phi_u^2)(u_n - u)\|_{L^2} \|u_n\|_{L^2}$$

$$\leq \|\phi_{u_n}^2 - \phi_u^2\|_{L^6} \|u_n - u\|_{L^3} \|u_n\|_{L^2}$$

$$\leq a_6 \|\phi_{u_n}^2 - \phi_u^2\|_D \|u_n - u\|_{L^3} \|u_n\|_{L^2},$$

where $a_6 > 0$ is a constant. Because $u_n \to u$ in $L^s(\mathbb{R}^3)$ for any $s \in (2, 2^*)$, we have

$$\int_{\mathbb{R}^3} (\phi_{u_n}^2 - \phi_u^2) u_n (u_n - u) dx \to 0, \text{ as } n \to \infty,$$

and

$$\int_{\mathbb{R}^3} \phi_u^2(u_n - u)(u_n - u)dx \leq \|\phi_u^2\|_{L^6} \|u_n - u\|_{L^3} \|u_n - u\|_{L^2} \to 0, \quad \text{as } n \to \infty.$$

Thus, we get

$$\int_{\mathbb{R}^3} (\phi_{u_n}^2 u_n - \phi_u^2 u)(u_n - u)dx$$

$$= \int_{\mathbb{R}^3} (\phi_{u_n}^2 - \phi_u^2)u_n(u_n - u)dx + \int_{\mathbb{R}^3} \phi_u^2(u_n - u)(u_n - u)dx \to 0, \quad \text{as } n \to \infty.$$

So that

$$a_7\|u_n - u\|_{H^1} \leq \int_{\mathbb{R}^3} \left(|\nabla(u_n - u)|^2 + [m^2 - \omega^2](u_n - u)^2 \right) dx \to 0, \quad \text{as } n \to \infty,$$

where $a_7 = \min\{1, m^2 - \omega^2\}$. $\qquad\square$

Proof of Theorem 4.14. The proof of this theorem is divided in two steps.

Step 1 There exists $u_1 \in E$ such that $I'(u_1) = 0$ and $I(u_1) < 0$.

Since $h \in L^2(\mathbb{R}^3) \cap C^1(\mathbb{R}^3)$ is a radial function and $h \not\equiv 0$, we can choose a function $\psi \in H_r^1(\mathbb{R}^3)$ such that

$$\int_{\mathbb{R}^3} h(x)\psi(x)dx > 0.$$

Hence, by (4.207) we have

$$I(tu) \leq \frac{t^2}{2} \int_{\mathbb{R}^3} \left(|\nabla\psi|^2 + m^2\psi^2 \right) dx - \frac{t^q}{q} \int_{\mathbb{R}^3} |\psi|^q dx - t \int_{\mathbb{R}^3} h(x)\psi dx < 0$$

for $t > 0$ small enough, thus, we obtain

$$c_1 = \inf\{I(u) : u \in \overline{B}_{\rho_1}\} < 0,$$

where $\rho_1 > 0$ is given by Lemma 4.37, $B_{\rho_1} = \{u \in H_r^1(\mathbb{R}^3) : \|u\|_{H^1} < \rho_1\}$. By the Ekeland's variational principle, there exists a sequence $\{u_n\} \subset \overline{B}_{\rho_1}$ such that

$$c_1 \leq I(u_n) < c_1 + \frac{1}{n},$$

and

$$I(w) \geq I(u_n) - \frac{1}{n}\|w - u_n\|_{H^1}$$

for all $w \in \overline{B}_{\rho_1}$. Then by a standard procedure, we can show that $\{u_n\}$ is bounded Palais-Smale sequence of I. Thus Lemma 4.39 implies that there exists a function $u_1 \in H_r^1(\mathbb{R}^3)$ such that $I'(u_1) = 0$ and $I(u_1) = c_1 < 0$.

Step 2 There exists $\widetilde{u}_1 \in E$ such that $I'(\widetilde{u}_1) = 0$ and $I(\widetilde{u}_1) > 0$.

From Lemma 4.37, Lemma 4.38 and the Mountain Pass Theorem, there is a sequence $\{u_n\} \subset H_r^1(\mathbb{R}^3)$ such that

$$I(u_n) \to \widetilde{c}_1 > 0, \quad \text{and} \quad I'(u_n) \to 0.$$

In view of Lemma 4.39, we only need to check that $\{u_n\}$ is bounded in $H_r^1(\mathbb{R}^3)$.

By (4.209), (4.210), (4.211) and Sobolev inequality, we have

$$q(\widetilde{c}_1 + 1) + \|u_n\|_{H^1}$$
$$\geq qI(u_n) - \langle I'(u_n), u_n \rangle$$
$$= \left(\frac{q}{2} - 1\right) \int_{\mathbb{R}^3} \left(|\nabla u_n|^2 + [m^2 - \omega^2]u_n^2\right) dx + \left(\frac{q}{2} + 1\right) \int_{\mathbb{R}^3} \phi_{u_n}^2 u_n^2 dx$$
$$+ 2 \int_{\mathbb{R}^3} \omega \phi_{u_n} u_n^2 dx + \frac{q}{8\pi} \int_{\mathbb{R}^3} |\nabla \phi_{u_n}|^2 dx + \frac{3\beta q}{16\pi} \int_{\mathbb{R}^3} |\nabla \phi_{u_n}|^4 dx$$
$$- (q - 1) \int_{\mathbb{R}^3} h(x)u_n dx$$
$$= \left(\frac{q}{2} - 1\right) \int_{\mathbb{R}^3} \left(|\nabla u_n|^2 + [m^2 - \omega^2]u_n^2\right) dx + 2 \int_{\mathbb{R}^3} (\phi_{u_n}^2 u_n^2 + \omega \phi_{u_n} u_n^2) dx$$
$$+ \frac{q}{8\pi} \int_{\mathbb{R}^3} |\nabla \phi_{u_n}|^2 dx + \frac{3\beta q}{16\pi} \int_{\mathbb{R}^3} |\nabla \phi_{u_n}|^4 dx - (q - 1) \int_{\mathbb{R}^3} h(x)u_n dx$$
$$\geq \left(\frac{q}{2} - 1\right) \int_{\mathbb{R}^3} \left(|\nabla u_n|^2 + [m^2 - \omega^2]u_n^2\right) dx - (q - 1) \|h\|_{L^2} \|u_n\|_{L^2}$$

for n large enough. Since $q > 4$ and $\|h\|_{L^2} < m_1$, it follows that $\{u_n\}$ is bounded in $H_r^1(\mathbb{R}^3)$. $\qquad\square$

Next, we will give the proof of Theorem 4.15.

Proof of Theorem 4.15. By Remark 4.19, we know that Lemma 4.37, Lemma 4.38, Lemma 4.39 and the proof of step 1 of Theorem 4.14 are still hold when $\sqrt{\left(\frac{q}{2} - 1\right)}m > \omega > 0$ and $2 < q \leq 4$. So we also only need to check that $\{u_n\}$ is bounded in $H_r^1(\mathbb{R}^3)$.

By (4.209) and (4.211), we obtain

$$I(u) = \frac{1}{2} \int_{\mathbb{R}^3} \left(|\nabla u|^2 + [m^2 - (\omega + \phi_u)^2] u^2 \right) dx - \frac{1}{8\pi} \int_{\mathbb{R}^3} |\nabla \phi_u|^2 dx$$

$$- \frac{\beta}{16\pi} \int_{\mathbb{R}^3} |\nabla \phi_u|^4 dx - \frac{1}{q} \int_{\mathbb{R}^3} |u|^q dx - \int_{\mathbb{R}^3} h(x) u dx$$

$$= \frac{1}{2} \int_{\mathbb{R}^3} \left(|\nabla u|^2 + [m^2 - \omega^2] u^2 - \omega \phi_u u^2 \right) dx - \frac{1}{2} \int_{\mathbb{R}^3} (\omega + \phi_u) \phi_u u^2 dx$$

$$- \frac{1}{8\pi} \int_{\mathbb{R}^3} |\nabla \phi_u|^2 dx - \frac{\beta}{16\pi} \int_{\mathbb{R}^3} |\nabla \phi_u|^4 dx - \frac{1}{q} \int_{\mathbb{R}^3} |u|^q dx - \int_{\mathbb{R}^3} h(x) u dx$$

$$= \frac{1}{2} \int_{\mathbb{R}^3} \left(|\nabla u|^2 + [m^2 - \omega^2] u^2 - \omega \phi_u u^2 \right) dx + \frac{\beta}{16\pi} \int_{\mathbb{R}^3} |\nabla \phi_u|^4 dx$$

$$- \frac{1}{q} \int_{\mathbb{R}^3} |u|^q dx - \int_{\mathbb{R}^3} h(x) u dx.$$

Thus, by $2 < q \leq 4$ and $\left(\frac{q}{2} - 1 \right) m^2 > \omega^2$,

$$q(\widetilde{c}_1 + 1) + \|u_n\|_{H^1}$$

$$\geq qI(u_n) - \langle I'(u_n), u_n \rangle$$

$$= \left(\frac{q}{2} - 1 \right) \int_{\mathbb{R}^3} \left(|\nabla u_n|^2 + [m^2 - \omega^2] u_n^2 \right) dx + \int_{\mathbb{R}^3} \phi_{u_n}^2 u_n^2 dx$$

$$- \left(\frac{q}{2} - 2 \right) \int_{\mathbb{R}^3} \omega \phi_{u_n} u_n^2 dx + \frac{\beta q}{16\pi} \int_{\mathbb{R}^3} |\nabla \phi_{u_n}|^4 dx - (q - 1) \int_{\mathbb{R}^3} h(x) u_n dx$$

$$\geq \left(\frac{q}{2} - 1 \right) \int_{\mathbb{R}^3} \left(|\nabla u_n|^2 + m^2 u_n^2 \right) dx - \omega^2 \left(\frac{q}{2} - 1 \right) \int_{\mathbb{R}^3} u_n^2 dx$$

$$- \left(\frac{q}{2} - 2 \right) \int_{\mathbb{R}^3} \omega \phi_{u_n} u_n^2 dx - (q - 1) \int_{\mathbb{R}^3} h(x) u_n dx$$

$$\geq \left(\frac{q}{2} - 1 \right) \int_{\mathbb{R}^3} \left(|\nabla u_n|^2 + m^2 u_n^2 \right) dx - \omega^2 \int_{\mathbb{R}^3} u_n^2 dx - (q - 1) \int_{\mathbb{R}^3} h(x) u_n dx$$

$$\geq \left(\frac{q}{2} - 1 \right) \int_{\mathbb{R}^3} \left(|\nabla u_n|^2 + m^2 u_n^2 \right) dx - \omega^2 \|u_n\|_{H^1} - (q - 1) \|h\|_{L^2} \|u_n\|_{H^1}$$

for n large enough. Since $q > 2$ and $\|h\|_{L^2} < m_1$, it follows that $\{u_n\}$ is bounded in $H_r^1(\mathbb{R}^3)$. $\qquad\square$

Notes and Comments

This section was obtained in [298].

Chapter 5

Gradient Systems

5.1 Introduction

In this chapter, we will examine the existence of solutions for several types of gradient systems both in one dimension and N dimension.

5.2 One Dimension Systems

5.2.1 Two-Point Boundary Value Systems

In this section, we study the Neumann boundary value problems:

$$
\begin{cases}
-(|u_1'(x)|^{p_1-2}u_1'(x))' + |u_1(x)|^{p_1-2}u_1(x) = \lambda F_{u_1}(x, u_1, \ldots, u_m) & x \in]a, b[, \\
-(|u_2'(x)|^{p_2-2}u_2'(x))' + |u_2(x)|^{p_2-2}u_2(x) = \lambda F_{u_2}(x, u_1, \ldots, u_m) & x \in]a, b[, \\
\vdots \\
-(|u_m'(x)|^{p_m-2}u_m'(x))' + |u_m(x)|^{p_m-2}u_m(x) = \lambda F_{u_m}(x, u_1, \ldots, u_m) \\
x \in]a, b[, \\
u_i'(a) = u_i'(b) = 0
\end{cases}
$$

$$(5.1)$$

where $p_i > 1$ are constants, for $1 \leq i \leq m$, λ is a positive parameter, $F : [a, b] \times \mathbb{R}^m \to \mathbb{R}$ is a function such that $F(\cdot, t_1, \ldots, t_m)$ is measurable in $[a, b]$ for all $(t_1, \ldots, t_m) \in \mathbb{R}^m$, $F(x, \cdot, \ldots, \cdot)$ is C^1 in $\mathbb{R}^m$ for every $x \in [a, b]$ and for every $\varrho > 0$,

$$
\sup_{|(t_1, \ldots, t_m)| \leq \varrho} \sum_{i=1}^{m} |F_{t_i}(x, t_1, \ldots, t_m)| \in L^1([a, b]),
$$

and F_{u_i} denotes the partial derivative of F with respect to u_i for $1 \leq i \leq m$.

Let us introduce notation that will be used later. Let Y_i be the Sobolev space $W^{1,p_i}([a,b])$ endowed with the norm

$$\|u\|_{p_i} := \left(\int_a^b |u'(x)|^{p_i} dx + \int_a^b |u(x)|^{p_i} dx \right)^{1/p_i},$$

and let

$$k_i = 2^{(p_i-1)/p_i} \max\{(b-a)^{-1/p_i}; (b-a)^{(p_i-1)/p_i}\}.$$

We recall the following inequality which we use in the sequel

$$|u(x)| \le k_i \|u\|_{p_i} \tag{5.2}$$

for all $u \in Y_i$, and for all $x \in [a,b]$. Let $K = \max\{k_i^{p_i}\}$, for $1 \le i \le m$. Here and in the sequel, $X := Y_1 \times \cdots \times Y_m$.

We say that $u = (u_1, \ldots, u_m)$ is a weak solution to the (5.1) if $u = (u_1, \ldots, u_m) \in X$ and

$$\sum_{i=1}^m \int_a^b \left(|u_i'(x)|^{p_i-2} u_i'(x) v_i'(x) + |u_i(x)|^{p_i-2} u_i(x) v_i(x) \right) dx$$

$$- \lambda \sum_{i=1}^m \int_a^b F_{u_i}(x, u_1, \ldots, u_m) v_i(x) dx = 0$$

for every $v = (v_1, \ldots, v_m) \in X$. For $\gamma > 0$ we denote the set

$$\Theta(\gamma) = \left\{ (t_1, \ldots, t_m) \in \mathbb{R}^m : \sum_{i=1}^m \frac{|t_i|^{p_i}}{p_i} \le \frac{\gamma}{\prod_{i=1}^m p_i} \right\}. \tag{5.3}$$

Let

$$\Phi(u) = \sum_{i=1}^m \frac{\|u_i\|_{p_i}^{p_i}}{p_i}, \tag{5.4}$$

$$\Psi(u) = \int_a^b F(x, u_1(x), \ldots, u_m(x)) dx. \tag{5.5}$$

It is well known that Φ and Ψ are well defined and continuously differentiable functionals whose derivatives at the point $u = (u_1, \ldots, u_m) \in X$ are the functionals $\Phi'(u), \Psi'(u) \in X^*$, given by

$$\Phi'(u)(v) = \sum_{i=1}^m \int_a^b \left(|u_i'(x)|^{p_i-2} u_i'(x) v_i'(x) + |u_i(x)|^{p_i-2} u_i(x) v_i(x) \right) dx,$$

$$\Psi'(u)(v) = \int_a^b \sum_{i=1}^m F_{u_i}(x, u_1(x), \ldots, u_m(x)) v_i(x) dx$$

for every $v = (v_1, \ldots, v_m) \in X$, respectively. Moreover, Φ is sequentially weakly lower semicontinuous, Φ' admits a continuous inverse on X^* as well as Ψ is sequentially weakly upper semicontinuous. Furthermore, $\Psi' : X \to X^*$ is a compact operator. Indeed, it is enough to show that Ψ' is strongly continuous on X. For this, for fixed $(u_1, \ldots, u_m) \in X$ let $(u_{1n}, \ldots, u_{mn}) \to (u_1, \ldots, u_m)$ weakly in X as $n \to +\infty$, then we have $(u_{1n}, \ldots, u_{mn})$ converges uniformly to $(u_1, \ldots, u_m)$ on $[a, b]$ as $n \to +\infty$(see [299]). Since $F(x, \cdot, \ldots, \cdot)$ is C^1 in $\mathbb{R}^m$ for every $x \in [a, b]$, the derivatives of F are continuous in $\mathbb{R}^m$ for every $x \in [a, b]$, so for $1 \le i \le m$, $F_{u_i}(x, u_{1n}, \ldots, u_{mn}) \to F_{u_i}(x, u_1, \ldots, u_m)$ strongly as $n \to +\infty$ which follows $\Psi'(u_{1n}, \ldots, u_{mn}) \to \Psi'(u_1, \ldots, u_m)$ strongly as $n \to +\infty$. Thus we proved that Ψ' is strongly continuous on X, which implies that Ψ' is a compact operator by Proposition 26.2 of [299].

Before our proof, we first list nonlinear term F satisfies the following hypotheses, where μ_1, μ_2 and ν are some constants.

(D1) $F(x, 0, \ldots, 0) = 0$ for a.e. $x \in [a, b]$,

(D2) $a_\nu(\mu_2) < a_\nu(\mu_1)$, where

$$a_\nu(\mu) := $$

$$K \prod_{i=1}^m p_i \frac{\int_a^b \sup_{(t_1, \ldots, t_m) \in \Theta(\mu)} F(x, t_1, \ldots, t_m) dx - \int_a^b F(x, \nu, \ldots, \nu) dx}{\mu - K \sum_{i=1}^m (\prod_{j=1, j\neq i}^m p_j) \nu^{p_i}},$$

(D3) $\dfrac{\int_a^b \sup_{(t_1, \ldots, t_m) \in \Theta(\mu)} F(x, t_1, \ldots, t_m) dx}{\mu} < \dfrac{\int_a^b F(x, \nu, \ldots, \nu) dx}{K \sum_{i=1}^m (\prod_{j=1, j\neq i}^m p_j) \nu^{p_i}}$,

(D4) $\liminf\limits_{\mu \to +\infty} \dfrac{\int_a^b \sup_{(t_1, \ldots, t_m) \in \Theta(\mu)} F(x, t_1, \ldots, t_m) dx}{\mu}$

$$< \frac{1}{K \prod_{i=1}^m p_i (b-a)} \limsup\limits_{|t_1| \to +\infty, \ldots, |t_m| \to +\infty} \frac{\int_a^b F(x, t_1, \ldots, t_m) dx}{\sum_{i=1}^m \frac{|t_i|^{p_i}}{p_i}}.$$

We formulate our main result as follows:

Theorem 5.1. *Assume that there exist a non-negative constant c_1 and two positive constants c_2 and d with $c_1 < K(b-a) \sum_{i=1}^m (\prod_{j=1, j\neq i}^m p_j) d^{p_i} < c_2$ such that (D1) and (D2) are satisfied. Then, for each $\lambda \in \left] \frac{1}{a_d(c_1)}, \frac{1}{a_d(c_2)} \right[$, system (5.1) admits at least one non-trivial weak solution $u_0 = (u_{01}, \ldots, u_{0m}) \in X$ such that*

$$\frac{c_1}{K \prod_{i=1}^m p_i} < \sum_{i=1}^m \frac{\|u_{0i}\|_{p_i}^{p_i}}{p_i} < \frac{c_2}{K \prod_{i=1}^m p_i}.$$

Proof. To apply Theorem 1.19 to our problem, we introduce the functionals $\Phi, \Psi : X \to \mathbb{R}$ for each $u = (u_1, \ldots, u_n) \in X$, as (5.4) and (5.5). Moreover, Φ is sequentially weakly lower semicontinuous, Φ' admits a continuous inverse on X^* as well as $\Psi' : X \to X^*$ is a compact operator. Set $w(x) = (w_1(x), \ldots, w_m(x))$ such that for $1 \leq i \leq m$,

$$w_i(x) = d$$

$r_1 = \frac{c_1}{K \prod_{i=1}^m p_i}$ and $r_2 = \frac{c_2}{K \prod_{i=1}^m p_i}$. It is easy to verify that $w = (w_1, \ldots, w_m) \in X$, and in particular, one has

$$\|w_i\|_{p_i}^{p_i} = (b-a)d^{p_i}$$

for $1 \leq i \leq m$. So, from the definition of Φ, we have

$$\Phi(w) = (b-a) \sum_{i=1}^m \frac{d^{p_i}}{p_i}.$$

From the conditions $c_1 < K \sum_{i=1}^m (\prod_{j=1, j \neq i}^m p_j)(b-a)d^{p_i} < c_2$, we obtain

$$r_1 < \Phi(w) < r_2.$$

Moreover, from (5.2) one has

$$\sup_{x \in [a,b]} |u_i(x)|^{p_i} \leq k_i^{p_i} \|u_i\|_{p_i}^{p_i}$$

and

$$\sup_{x \in [a,b]} |u_i(x)|^{p_i} \leq K \|u_i\|_{p_i}^{p_i}$$

for each $u = (u_1, \ldots, u_m) \in X$, so from the definition of Φ, we observe that

$$\Phi^{-1}(]-\infty, r_2[) = \{(u_1, \ldots, u_n) \in X : \Phi(u_1, \ldots, u_n) < r_2\}$$

$$= \left\{ (u_1, \ldots, u_n) \in X : \sum_{i=1}^m \frac{\|u_i\|_{p_i}^{p_i}}{p_i} < r_2 \right\}$$

$$\subseteq \left\{ (u_1, \ldots, u_n) \in X : \sum_{i=1}^m \frac{|u_i(x)|^{p_i}}{p_i} \leq \frac{c_2}{\prod_{i=1}^m p_i} \text{ for all } x \in [a, b] \right\}$$

from which it follows

$$\sup_{(u_1, \ldots, u_m) \in \Phi^{-1}(]-\infty, r_2[)} \Psi(u) = \sup_{(u_1, \ldots, u_m) \in \Phi^{-1}(]-\infty, r_2[)} \int_a^b F(x, u_1(x), \ldots, u_m(x))dx$$

$$\leq \int_a^b \sup_{(t_1, \ldots, t_m) \in \Theta(c_2)} F(x, t_1, \ldots, t_m)dx.$$

Since for $1 \leq i \leq m$, for each $x \in [a, b]$, the condition (A1) ensures that

$$\beta(r_1, r_2) \leq \frac{\sup_{u \in \Phi^{-1}(]-\infty, r_2[)} \Psi(u) - \Psi(w)}{r_2 - \Phi(w)}$$

$$\leq \frac{\int_a^b \sup_{(t_1, \ldots, t_m) \in \Theta(c_2)} F(x, t_1, \ldots, t_m) dx - \Psi(w)}{r_2 - \Phi(w)}$$

$$\leq a_d(c_2).$$

On the other hand, by similar reasoning as before, one has

$$\rho(r_1, r_2) \geq \frac{\Psi(w) - \sup_{u \in \Phi^{-1}(]-\infty, r_1[)} \Psi(u)}{\Phi(w) - r_1}$$

$$\geq \frac{\Psi(w) - \int_a^b \sup_{(t_1, \ldots, t_m) \in \Theta(c_1)} F(x, t_1, \ldots, t_m) dx}{\Phi(w) - r_1}$$

$$\geq a_d(c_1).$$

Hence, from assumption (A2), one has $\beta(r_1, r_2) < \rho(r_1, r_2)$. Therefore, from Theorem 1.19, taking into account that the weak solutions of the system (5.1) are exactly the solutions of the equation $\Phi'(u) - \lambda \Psi'(u) = 0$, we have the conclusion. □

Now we point out the following consequence of Theorem 5.1.

Theorem 5.2. *Suppose that there exist two positive constants c and d with $c > K(b-a) \sum_{i=1}^m (\prod_{j=1, j \neq i}^m p_j) d^{p_i}$ such that (D1) and (D3) are satisfied. Then, for each*

$$\lambda \in \left] \frac{K(b-a) \sum_{i=1}^m (\prod_{j=1, j \neq i}^m p_j) d^{p_i}}{\int_a^b F(x, d, \ldots, d) dx}, \frac{c}{\int_a^b \sup_{(t_1, \ldots, t_m) \in \Theta(c)} F(x, t_1, \ldots, t_m) dx} \right[,$$

system (5.1) admits at least one non-trivial weak solution $u_0 = (u_{01}, \ldots, u_{0n}) \in X$ such that $\sum_{i=1}^m \frac{\|u_{0i}\|_\infty^{p_i}}{p_i} < \frac{c}{K \prod_{i=1}^m p_i}$.

Proof. The conclusion follows from Theorem 5.1, by taking $c_1 = 0$ and $c_2 = c$. Indeed, owing to our assumptions, one has

$$a_d(c_2) = K \prod_{i=1}^m p_i \frac{\int_a^b \sup_{(t_1, \ldots, t_m) \in \Theta(c)} F(x, t_1, \ldots, t_m) dx - \int_a^b F(x, d, \ldots, d) dx}{c - K(b-a) \sum_{i=1}^m (\prod_{j=1, j \neq i}^m p_j) d^{p_i}}$$

$$\leq K \prod_{i=1}^m p_i \frac{\int_a^b \sup_{(t_1, \ldots, t_m) \in \Theta(c)} F(x, t_1, \ldots, t_m) dx - \frac{\int_a^b \sup_{(t_1, \ldots, t_m) \in \Theta(c)} F(x, t_1, \ldots, t_m) dx}{K \sum_{i=1}^m (\prod_{j=1, j \neq i}^m p_j) d^{p_i}}}{c - K(b-a) \sum_{i=1}^m (\prod_{j=1, j \neq i}^m p_j) d^{p_i}}$$

$$= \frac{\int_a^b \sup_{(t_1, \ldots, t_m) \in \Theta(c)} F(x, t_1, \ldots, t_m) dx}{c}.$$

On the other hand, taking assumption (A1) into account, one has

$$\frac{\int_a^b F(x, d, \ldots, d)dx}{K(b-a)\sum_{i=1}^m (\prod_{j=1, j\neq i}^m p_j)d^{p_i}} = a_d(c_1).$$

Moreover, since

$$\sup_{x\in[a,b]} |u_i(x)|^{p_i} \leq K\|u_i\|_{p_i}^{p_i}$$

for each $u = (u_1, \ldots, u_m) \in X$, an easy computation ensures that $\sum_{i=1}^m \frac{\|u_{0i}\|_\infty^{p_i}}{p_i} < \frac{c}{K\prod_{i=1}^m p_i}$ whenever $\Phi(u) < r_2$. Now, owing to Assumption (A3), it is sufficient to invoke Theorem 5.1 for concluding the proof. □

Theorem 5.3. *Assume that* (D1) *and* (D4) *are satisfied.*
Then, for every $\lambda \in \Lambda :=]\lambda_1, \lambda_2[$, *where*

$$\lambda_1 = \frac{(b-a)}{\limsup_{|t_1|\to+\infty, \ldots, |t_m|\to+\infty} \frac{\int_a^b F(x,t_1,\ldots,t_m)dx}{\sum_{i=1}^m \frac{|t_i|^{p_i}}{p_i}}}$$

and

$$\lambda_2 = \frac{1}{K\prod_{i=1}^m \liminf_{\mu\to+\infty} \frac{\int_a^b \sup_{(t_1,\ldots,t_m)\in\Theta(\mu)} F(x,t_1,\ldots,t_m)dx}{\mu}}$$

the problem (5.1) *admits an unbounded sequence of weak solutions which is unbounded in* X.

Proof. Our goal is to apply Theorem 1.18. Now, as has been pointed out before, the functionals Φ and Ψ satisfy the regularity assumptions required in Theorem 1.18. Let $\{c_n\}$ be a real sequence such that $\lim_{n\to+\infty} c_n = +\infty$ and

$$\liminf_{n\to+\infty} \frac{\int_a^b \sup_{(t_1,\ldots,t_m)\in\Theta(c_n)} F(x,t_1,\ldots,t_m)dx}{c_n} = \mathcal{A}. \tag{5.6}$$

Taking into account (5.2) for every $u \in X$ one has

$$|u(x)| \leq K\|u\|_{p_i}.$$

Also note

$$\sum_{i=1}^m \frac{|u_i(x)|^{p_i}}{p_i} \leq K\left(\sum_{i=1}^m \frac{\|u_i(x)\|_{p_i}^{p_i}}{p_i}\right).$$

Hence, an easy computation ensures that $\sum_{i=1}^m u \leq c_n$ whenever $u \in \Phi^{-1}(]-\infty, r_n[)$, where

$$r_n = \frac{1}{K}\frac{c_n}{\prod_{i=1}^m p_i}.$$

Taking into account that $\|u_i^0\|_{p_i} = 0$ (where $u_i^0(x) = 0$ for every $x \in [a,b]$) and that $\int_a^b F(t,0,\ldots,0)dx = 0$ for all $x \in [a,b]$, for every n large enough, one has

$$\varphi(r_n) = \inf_{u \in \Phi^{-1}(]-\infty,r_n[)} \frac{(\sup_{v \in \Phi^{-1}(]-\infty,r_n[)} \Psi(v)) - \Psi(u)}{r_n - \Phi(u)} = \inf_{\sum_{i=1}^m \frac{\|u_i\|_{p_i}^{p_i}}{p_i} < r_n}$$

$$\times \frac{\sup_{\sum_{i=1}^m \frac{\|v_i\|_{p_i}^{p_i}}{p_i} < r_n} \int_a^b F(t,v_1(x),\ldots,v_m(x))dx - \int_a^b F(t,u_1(x),\ldots,u_m(x))dx}{r_n - \sum_{i=1}^m \frac{\|u_i\|_{p_i}^{p_i}}{p_i}}$$

$$\leq \frac{\sup_{\sum_{i=1}^m \frac{\|v_i\|_{p_i}^{p_i}}{p_i} < r_n} \int_a^b F(t,v_1(x),\ldots,v_m(x))dx}{r_n}$$

$$\leq K \prod_{i=1}^m p_i \liminf_{n \to +\infty} \frac{\int_a^b \sup_{(t_1,\ldots,t_m) \in \Theta(c_n)} F(x,t_1,\ldots,t_m)dx}{c_n}.$$

Therefore, since from assumption (D4) one has $\mathcal{A} < +\infty$, we obtain

$$\gamma = \liminf_{n \to +\infty} \varphi(r_n) \leq K \prod_{i=1}^m p_i \mathcal{A} < +\infty. \tag{5.7}$$

Now, fix $\lambda \in]\lambda_1, \lambda_2[$ and let us verify that the functional I_λ is unbounded from below. Let $\{\xi_{i,n}\}$ be m positive real sequences such that $\lim_{n \to +\infty} \sqrt{\sum_{i=1}^m \xi_{i,n}^2} = +\infty$, and

$$\limsup_{n \to +\infty} \frac{\int_a^b F(x,\xi_{1,n},\ldots,\xi_{m,n})dx}{\sum_{i=1}^m \frac{|\xi_{i,n}|^{p_i}}{p_i}} = \mathcal{B} \tag{5.8}$$

For each $n \in \mathbb{N}$ define

$$w_{i,n}(x) := \xi_{i,n}$$

and put $w_n := (w_{1,n},\ldots,w_{m,n})$.

We easy get that

$$\|w_{i,n}\|_{p_i}^{p_i} = (b-a)|\xi_{i,n}|^{p_i}.$$

At this point, bearing in mind (i), we infer

$$\Phi(w_n) - \lambda\Psi(w_n) = \sum_{i=1}^m \frac{|\xi_{i,n}|^{p_i}}{p_i} - \lambda \int_a^b F(x,\xi_{1,n},\ldots,\xi_{m,n})dx, \qquad \forall n \in \mathbb{N}.$$

If $\mathcal{B} < +\infty$, let $\epsilon \in]\frac{1}{\lambda\mathcal{B}}, 1[$. By (5.8) there exists v_ϵ such that

$$\int_a^b F(x,\xi_{1,n},\ldots,\xi_{m,n})dx > \epsilon\mathcal{B}\sum_{i=1}^m \frac{|\xi_{i,n}|^{p_i}}{p_i}, \qquad \forall n > v_\epsilon.$$

Moreover

$$\Phi(w_n) - \lambda\Psi(w_n) \leq \sum_{i=1}^{m} \frac{|\xi_{i,n}|^{p_i}}{p_i} - \lambda\epsilon\mathcal{B}\sum_{i=1}^{m} \frac{|\xi_{i,n}|^{p_i}}{p_i}, \qquad \forall n > v_\epsilon.$$

Taking into account the choice of ϵ, one has

$$\lim_{n\to+\infty}[\Phi(w_n) - \Psi(w_n)] = -\infty.$$

If $\mathcal{B} = +\infty$, let us consider $M > \frac{1}{\lambda}$. By (5.8) there exist v_m such that

$$\int_a^b F(x, \xi_{1,n}, \ldots, \xi_{m,n}) dx > M \sum_{i=1}^{m} \frac{|\xi_{i,n}|^{p_i}}{p_i}, \qquad \forall n > v_m.$$

Moreover

$$\Phi(w_n) - \lambda\Psi(w_n) \leq \sum_{i=1}^{m} \frac{|\xi_{i,n}|^{p_i}}{p_i} - \lambda M \sum_{i=1}^{m} \frac{|\xi_{i,n}|^{p_i}}{p_i}, \qquad \forall n > v_\epsilon.$$

Taking into account the choice of M, also in this case, one has

$$\lim_{n\to+\infty}[\Phi(w_n) - \Psi(w_n)] = -\infty.$$

Applying Theorem 1.18, we deduce that the functional $\Phi - \lambda\Psi$ admits a sequence of critical points which is unbounded in X. Hence, our claim is proved and the conclusion is achieved. $\square$

Remark 5.1. If

$$\liminf_{\mu\to+\infty} \frac{\int_a^b \sup_{(t_1,\ldots,t_m)\in\Theta(\mu)} F(x, t_1, \ldots, t_m) dx}{\mu} = 0$$

and

$$\limsup_{|t_1|\to+\infty,\ldots,|t_m|\to+\infty} \frac{\int_a^b F(x, t_1, \ldots, t_m) dx}{\sum_{i=1}^{m} \frac{|t_i|^{p_i}}{p_i}} = +\infty,$$

clearly, hypothesis (D4) is verified and Theorem 5.3 guarantees the existence of infinitely many weak solutions for problem (5.1), for every $\lambda \in]0, +\infty[$, the main result ensures the existence of infinitely many weak solutions for the problem (5.1).

Notes and Comments

In the last decade or so, many authors applied the variational method to study the existence or multiplicity solutions of Neumann problem of its variations; see, for example, [300–306] and the references therein. We noted that the main tools in these cited papers are several critical points theorems due to Bonanno [42], Bonanno and Bisci [41], Bonanno and Marano [39]. A Neumann boundary value problem for a class of gradient systems had already studied by Afrouzi, Hadjian and Heidarkhani [307] and Hedarkhani and Tian [84] in the ODE case and Afrouzi, Heidarkhani and O'Regan [83] in the PDE case. In that papers at least three solutions were established. The aim of this article is to prove the existence of at least one non-trivial solution and infinitely many solutions for (5.1) for appropriate values of the parameter λ belonging to a precise real interval. Our motivation comes from the recent paper [41, 300]. We want to systematically study a class of gradient systems under Neumann boundary by using Bonanno's critical point theorems. For basic notation and definitions, and also for a thorough account on the subject, we refer the reader to [299, 308]. This section was obtained in [309].

5.2.2 Hamiltonian Systems

We consider the following problem. One wishes to solve

$$- \ddot{x}(t) = B(t)x(t) + \nabla_x V(t, x(t)), \qquad (5.9)$$

where

$$x(t) = (x_1(t), \ldots, x_n(t))$$

is a map from $I = [0, T]$ to $\mathbb{R}^n$ such that each component $x_j(t)$ is a periodic function in H^1 with period T, and the function $V(t, x) = V(t, x_1, \cdots, x_n)$ is continuous from $\mathbb{R}^{n+1}$ to $\mathbb{R}$ with

$$\nabla_x V(t, x) = (\partial V/\partial x_1, \cdots, \partial V/\partial x_n) \in C(\mathbb{R}^{n+1}, \mathbb{R}^n). \qquad (5.10)$$

For each $x \in \mathbb{R}^n$, the function $V(t, x)$ is periodic in t with period T. The elements of the symmetric matrix $B(t)$ are to be real-valued functions $b_{jk}(t) = b_{kj}(t)$, and each function is to be periodic with period T. We will consider each function to be defined on the interval I.

In this section, we shall study this problem under the following assumptions. Our assumption on $B(t)$ is:

(D5) Each component of $B(t)$ is an integrable function on I, i.e., for each j and k, $b_{jk}(t) \in L^1(I)$.

Although this assumption is very weak, it is sufficient to allow us to find an extension $\mathcal{D}$ of the operator

$$\mathcal{D}_0 x := -\ddot{x}(t) - B(t)x(t)$$

which have a discrete, countable spectrum consisting of isolated eigenvalues of finite multiplicity with a finite lower bound $-L$

$$-\infty < -L < \lambda_1 < \lambda_2 < \cdots < \lambda_l < \cdots .$$

($\mathcal{D}$ is defined in the next section.) Let λ_l be the first positive eigenvalue of $\mathcal{D}$. We allow $\lambda_{l-1} = 0$. For the superquadratic case, we assume:

(D6) Assume

$$2V(t,x) \geq \lambda_{l-1}|x|^2, \qquad t \in I, x \in \mathbb{R}^n$$

and there are positive constants $\mu < \lambda_l$ and δ such that

$$2V(t,x) \leq \mu|x|^2, \qquad |x| \leq \delta, x \in \mathbb{R}^n.$$

(D7) There exists $q > 2$ such that

$$\lim_{|x| \to \infty} \frac{V(t,x)}{|x|^q} < +\infty$$

uniformly for all $t \in I$.

(D8) $\lim_{|x| \to \infty} \frac{V(t,x)}{|x|^2} = +\infty$ uniformly for all $t \in I$.

(D9) There exists $\theta \geq 1$ such that

$$\theta F(t,x) \geq F(t,sx) \quad \forall (t,x) \in I \times \mathbb{R}^n, \forall s \in [0,1],$$

where $F(t,x) = (x, \nabla_x V(t,x)) - 2V(t,x)$.

Then we have

Theorem 5.4. *If the functions $B(t)$ and $V(t,x)$ satisfy assumptions* (D5), (D6)–(D9), *then there exists a T-periodic weak solution of* (5.9) *whose weak second derivative is an element of $L^1(I)$.*

If the potential V is even, we can get infinitely many solutions.

Theorem 5.5. *If the functions $B(t)$ and $V(t,x)$ satisfy assumptions* (D5), (D7)–(D9) *and $V(t,x)$ is even in x for all $t \in I$, then there exists infinitely many nontrivial T-periodic weak solution to* (5.9) *whose weak second derivative is an element of $L^1(I)$.*

If we change hypothesis (D9) to (V4′) below, we also have

Theorem 5.6. *The conclusion of Theorem 5.4 holds if we replace hypothesis* (D9) *with*

(V4′) There exists $p > q - 2$ such that

$$\liminf_{|x| \to \infty} \frac{(\nabla_x V(t,x), x) - 2V(t,x)}{|x|^p} > 0 \text{ uniformly for } t \in I.$$

Theorem 5.7. *The conclusion of Theorem 5.5 holds if we replace hypothesis* (D9) *with* (V4′).

For the subquadratic case, we assume:

(D10) There exists $r > 0$ such that

$$V(t, -x) = V(t, x), \quad \forall |x| \leq r \text{ and } t \in I.$$

(D11) $V(t, 0) = 0$ for $t \in I$ and

$$\lim_{|x| \to 0} \frac{V(t,x)}{|x|^2} = +\infty \text{ uniformly for } t \in I.$$

Now, we have

Theorem 5.8. *Suppose that* $V(t,x)$ *satisfies* (D10), (D11), *then problem* (5.9) *possesses infinitely many solutions.*

In proving the previous theorems we shall make use of the following considerations. For convenience, we define the bilinear forms

$$a(u, v) = (\dot{u}, \dot{v}) + (u, v),$$

$$b(u, v) = -\sum_{j=1}^{n} \sum_{k=1}^{n} \int_0^T (b_{jk}(t) + \delta_{jk}) u_k(t) v_j(t) dt,$$

and

$$d(u, v) = a(u, v) + b(u, v).$$

Thus, $\|u\|_H^2 = a(u, u)$.

Lemma 5.1 ([310]). *The operator* $\mathcal{D}$ *associated with the bilinear form* $d(\cdot, \cdot)$ *under assumption* (D5) *is self-adjoint. Its essential spectrum is the null set and there exists a finite real value* L *such that* $\sigma(\mathcal{D}) \subset [-L, \infty)$. $\mathcal{D}$ *has a discrete, countable spectrum consisting of isolated eigenvalues of finite multiplicity with a finite lower bound* $-L$

$$-\infty < -L < \lambda_1 < \lambda_2 < \cdots < \lambda_l < \cdots .$$

Moreover, the corresponding system of eigenfunctions $\{e_i : i \in \mathbb{N}\}(-\ddot{e}_i - B(t)e_i = \lambda_i e_i)$, *forms an orthogonal basis of* H.

Using assumption (D5), since each diagonal element $b_{jj}(t)$ of the matrix is in $L^1(I, \mathbb{R}^n)$ and I has finite measure, the diagonal elements of the matrix $B(t) + I$, $b_{jj}(t) + 1$, are also in $L^1(I, \mathbb{R}^n)$ and we can find a constant which bounds the magnitude of $b(u, u)$ on the set H,

$$|b(u, u)| \leq K_B \|u\|_\infty^2.$$

Since,

$$\|u\|_\infty \leq C\|u\|_H, \quad u \in H,$$

we have the following estimate

$$|b(u, u)| \leq K_B \|u\|_\infty^2 \leq K_B (C\|u\|_H)^2 \leq K\|u\|_H^2. \tag{5.11}$$

Finding a weak solution to (5.9) is equivalent to finding a solution to the following operator equation in $L^2(I, \mathbb{R}^n)$,

$$\mathcal{D}x = -\partial_x V(t, x(t)).$$

Define the real-valued functional G on the function space H by

$$G(x) = d(x, x) - 2 \int_0^T V(t, x(t))dt, \tag{5.12}$$

where

$$d(x, x) = (\dot{x}, \dot{x}) - \int_0^T (B(t)x(t), x(t))_{\mathbb{R}^n}\, dt. \tag{5.13}$$

The first term on the right side of (5.13) is bounded by the square of the H norm of x, and because the norm on the space H is an upper bound for the L^∞ norm, the matrix term is also bounded by the square of the H norm. In the potential term in (5.12), the potential V is continuously differentiable in x, so this term is a continuous map from H to $\mathbb{R}$. Consequently, the functional G is a continuously differentiable map from H to $\mathbb{R}$.

For any $y \in H$,

$$\frac{1}{2}(G'(x), y) = d(x, y) - (\nabla_x V(t, x), y).$$

Now, we prove an important lemma to show that we only need to show that the Cerami sequence is bounded when we want to get a solution. It is well known (cf., e.g. [25, Theorem 3.4.1, p. 64]).

Lemma 5.2. *If the Cerami sequence of G is bounded, then a subsequence converges weakly to a solution of problem (5.9).*

Proof. Suppose this sequence $\{x_k\}$ is bounded in H. Let $C > 0$ be such that for any k we have $\|x_k\|_H \leq C$. Then there must exist a subsequence (x_k) (still denote by $\{x_k\}$) which converges weakly in H and strongly in $L^\infty(I, \mathbb{R}^n)$ and $L^2(I, \mathbb{R}^n)$ to a function $x \in H$. Since $\{x_k\}$ is Cerami sequence of G, we see that we have

$$(1 + \|x_k\|_H)\|G'(x_k)\|_H \leq C'$$

when k is big enough. So, we also have

$$\frac{1}{2}(G'(x_k), y) = d(x_k, y) - (\nabla_x V(t, x_k), y)$$

$$= (x_k, y)_H - (x_k, y) - \int_0^T (B(t)x_k(t), y(t))_{\mathbb{R}^n} dt$$

$$- \int_0^T (\nabla_x V(t, x_k), y)_{\mathbb{R}^n} dt, \quad y \in H.$$

By the weak convergence of $\{x_k\}$ in H, the first term on the right side of the above equation converges in $\mathbb{R}$. By the convergence of $\{x_k\}$ in $L^2(I, \mathbb{R}^n)$ and the Schwartz inequality, the second term converges. By the convergence of $\{x_k\}$ in $L^\infty(I, \mathbb{R}^n)$, the last two terms converge. Thus,

$$\frac{1}{2}(G'(x), y) = d(x, y) - (\nabla_x V(t, x), y) = 0, \quad y \in H.$$

Hence, x is a weak solution of (5.9). □

Now, we will show the functional satisfies the geometric structure of linking theorem. By this we mean that there are two sets, A, B which link and separate the functional.

Lemma 5.3. *If $V(t, x)$ satisfies (D6), the functional $-G$ satisfies the geometric structure of linking theorem.*

Proof. Define the subspaces M and N of H as

$$N = \bigoplus_{k<l} E(\lambda_k), \qquad M = N^\perp, \qquad H = M \bigoplus N.$$

We note that (D6) implies

$$G(v) \leq 0, \qquad v \in N.$$

In fact, we have

$$G(x) = d(x, x) - 2 \int_I V(t, x)dt \leq \int_I [\lambda_{l-1}|x|^2 - 2V(t, x)]dt \leq 0, \quad x \in N.$$

Note that there is a positive $\rho > 0$ such that

$$|x(t)| < \delta$$

when $\|x\|_H = \rho$. In fact, we have $|x(t)| \le c_0 \|x\|_H$. If $x \in M$, then

$$G(x) = d(x) - 2\int_I V(t,x)dt \ge d(x)\left[1 - \frac{2\mu\|x\|^2}{d(x)}\right] > \varepsilon > 0.$$

Take

$$A = \partial B_\rho \cap M$$
$$B = N,$$

where

$$B_\rho = \{x \in H : \|x\|_H < \rho\}.$$

By Example 8, p. 22 of [311], A links B. Moreover,

$$\sup_A[-G] \le 0 \le \inf_B[-G].$$

$\square$

Next, we will prove the Cerami sequence of functional G is bounded.

Lemma 5.4. *Suppose that* (D5), (D7)–(D9) *hold. Then the Cerami sequence of functional G is bounded.*

Proof. Let $\{x_k\}$ be a Cerami sequence in H. If $\{x_k\}$ is unbounded, up to a subsequence we may assume, for some $C \in \mathbb{R}$, that

$$G(x_k) \to c, \quad (1 + \|x_k\|_H)\|G(x_k)\| \to 0 \text{ and } \|x_k\|_H \to \infty \quad (5.14)$$

as $n \to \infty$. In particular,

$$\lim_{n\to\infty} \int_0^T ((\nabla_x V(t,x_k), x_k) - 2V(t,x_k))dt$$
$$= \lim_{n\to\infty} \left(G(x_k) - \frac{1}{2}(G'(x_k), x_k)\right) = c.$$

We consider $w_k = x_k/\|x_k\|_H$. Passing, if necessary, to a subsequence, we get that w_k converges to w weakly in H, strongly in $L^\infty(I, \mathbb{R}^n)$ and $L^2(I, \mathbb{R}^n)$.

If $w = 0$, we choose a sequence $\{s_k\} \subset \mathbb{R}$ such that

$$G(s_k x_k) = \max_{s\in[0,1]} G(sx_k).$$

For any $m > 0$, letting $v_k = \sqrt{m}w_k$, one has that v_k converges to 0 strongly in $L^\infty(I, \mathbb{R}^n)$ and $L^2(I, \mathbb{R}^n)$.

Now, for k large enough, $\sqrt{m}\|x_k\|_H^{-1} \in (0, 1)$, we deduce that

$$G(s_k x_k) \geq G(v_k) = m - \int_0^T |v_k|^2 dt - \int_0^T (B(t)v_k, v_k)_{\mathbb{R}^n} dt - 2 \int_0^T V(t, v_k) dt,$$

which implies that

$$\liminf_{k\to\infty} G(s_k x_k) \geq m - 2 \int_0^T V(t, 0) dt \geq m - C_8,$$

where C_8 is a positive constant. Since m is arbitrary, we have that $\lim_{k\to\infty} G(s_k x_k) = +\infty$. Since $G(0) < +\infty$ and $G(x_k) \to c(n \to \infty)$, we see that $s_k \in (0, 1)$, and

$$\int_0^T |s_k \dot{x}_k|^2 dt - \int_0^T (B(t)s_k x_k, s_k x_k)_{\mathbb{R}^n} dt$$

$$- \int_0^T (\nabla V(t, s_k x_k), s_k x_k) dt = \frac{1}{2}(G'(s_k x_k), s_k x_k)$$

$$= s_k \frac{d}{ds}\Big|_{s=s_k} G(s x_k)$$

$$= 0$$

for large k. Therefore, using (D9),

$$\int_0^T ((\nabla V(t, x_k), x_k) - 2V(t, x_k)) \geq \frac{1}{\theta} \int_0^T F(t, s_k x_k) dt$$

$$= \frac{1}{\theta} \int_0^T ((\nabla V(t, s_k x_k), x_k) - 2V(t, s_k x_k))$$

$$= \frac{1}{\theta} \int_0^T (|s_k \dot{x}_k|^2 dt - (B(t)s_k x_k, s_k x_k)_{\mathbb{R}^n} dt - 2V(t, s_k x_k)) dt$$

$$= \frac{1}{\theta} G(s_k x_k)$$

$$\to +\infty,$$

which contradicts (5.14).

If $w \neq 0$, since

$$G(x_k) = \|x_k\|_H^2 - \int_0^T |x_k|^2 dt - \int_0^T (B(t)x_k, x_k)_{\mathbb{R}^n} dt - 2 \int_0^T V(t, x_k) dt,$$

dividing both sides by $\|x_k\|_H^2$ and letting $k \to \infty$, we obtain

$$1 = \int_0^T |w|^2 dt + \int_0^T (B(t)w, w)_{\mathbb{R}^n} dt + \lim_{k\to\infty} \int_0^T \frac{2V(t, x_k)}{\|x_k\|_H} dt \tag{5.15}$$

$$= \int_0^T |w|^2 dt + \int_0^T (B(t)w, w)_{\mathbb{R}^n} dt + \lim_{k\to\infty} \left(\int_{w=0} + \int_{w\neq 0} \right) \frac{2V(t, x_k)}{\|x_k\|_H} dt,$$

since (w_k) converges to w in the $L^\infty(I, \mathbb{R}^n)$ norm. Each matrix element $b_{jk}(t)$ is integrable, so the sequence of functions in $\int_0^T (B(t)w_k, w_k)_{\mathbb{R}^n} dt$ converges in $L^1(I, \mathbb{R}^n)$.

From (D8), we see that there exists a constant $C_9 > 0$ such that $V(t, x) \geq 0$ for all $|x| > C_9$ and all $t \in I$. Since V is continuous, one sees that $|V(t, x)| \leq C_{10}$ for all $|x| \leq C_9$ and all $t \in I$. Hence, we have that $V(t, x) \geq -C_{10}$ for all $x \in \mathbb{R}^n$ and all $t \in I$, and thus

$$\int_{w=0} \frac{2V(t, x_k)}{\|x_k\|_H} dt \geq -\frac{1}{\|x_k\|_H} \int_{w=0} 2C_{10} dt \to 0$$

for all k. This implies that

$$\liminf_{k \to \infty} \int_{w=0} \frac{2V(t, x_k)}{\|x_k\|_H} dt \geq 0.$$

On the other hand, for $t \in \Omega := \{t \in I : w(t) \neq 0\}$, one has that $|x_k(t)| \to +\infty$ as $k \to \infty$, so that, using (D8),

$$\frac{V(t, x_k(t))}{|x_k(t)|^2} |w_k(t)|^2 \to +\infty \text{ as } k \to \infty.$$

Noting that meas $\Omega > 0$ and using the Fatou Lemma, we deduce that

$$\int_{w \neq 0} \frac{V(t, x_k)}{\|x_k\|_H^2} = \int_{w \neq 0} \frac{V(t, x_k(t))}{|x_k(t)|^2} |w_k(t)|^2 \to +\infty \text{ as } k \to \infty.$$

This contradicts (5.15). Hence, (x_k) is a bounded sequence in H. $\qquad\square$

Proof of Theorem 5.4. Due to Lemmas 5.2, 5.3, 5.4, we know that G satisfies all the assumptions of Theorem 1.7. This problem (5.9) has at least one solution. $\qquad\square$

In order to obtain infinitely many solutions of (5.9), we shall use the critical point theorems [27]. For simplicity, we use the eigenfunctions $\{e_i\}$ of $\mathcal{D}$ as an orthogonal basis of H. Put

$$H_k = \text{span}\{e_k\}, \quad Y_k = \bigoplus_{i=1}^k H_i, \quad Z_k = \overline{\bigoplus_{i=k}^\infty H_i}. \qquad (5.16)$$

For the space H, let $\{e_s\}_{s \in \mathbb{N}}$ be a basis for H and define Y_k and Z_k as in (5.16).

For this space split, we have this lemma

Lemma 5.5. *If* $Z_k = \overline{\bigoplus_{j \geq k} H_j}$, *then*

$$\beta_k := \sup_{u \in Z_k, \|u\|_H = 1} \|u\|_\infty \to 0 \text{ as } k \to \infty.$$

Proof. It is clear that $0 < \beta_{k+1} < \beta_k$, so $\beta_k \to \beta \geq 0$, $k \to \infty$. For every $k \in \mathbb{N}$, there exists $u_k \in Z_k$ such that $\|u_k\|_H = 1$ and

$$\|u_k\|_\infty > \frac{1}{2}\beta_k. \tag{5.17}$$

Going to the components of u_k, we can assume that

$$\int_0^T u_k^i(t)dt = 0, i = 1, 2, \ldots, n.$$

It follows from the mean value theorem that

$$u_k^i(\xi_i) = 0, i = 1, 2, \ldots, n,$$

for some $\xi_i \in I$. Hence, for $t \in I$, by the Hölder inequality,

$$|u_k^i(t)|^2 = 2\int_{\xi_i}^t (\dot{u}_k^i(s), u_k^i(s))ds \leq 2\|\dot{u}_k^i\|_2\|u_k^i\|_2, i = 1, 2, \ldots, n,$$

and then, using the Cauchy–Schwartz inequality, for $t \in I$,

$$|u_k(t)|^2 = \sum_{i=1}^n |u_k^i(t)|^2$$

$$\leq 2\sum_{i=1}^n \|\dot{u}_k^i\|_2\|u_k^i\|_2$$

$$\leq 2\left(\sum_{i=1}^n \|\dot{u}_k^i\|_2^2\right)^{1/2}\left(\sum_{i=1}^n \|u_k^i\|_2^2\right)^{1/2}$$

$$= 2\|\dot{u}_k\|_2\|u_k\|_2,$$

i.e.

$$\|u_k\|_\infty^2 \leq 2\|\dot{u}_k\|_2\|u_k\|_2. \tag{5.18}$$

Since each diagonal element $b_{jj}(t)$ of the matrix is in $L^1(I, \mathbb{R}^n)$ and I has finite measure, the diagonal elements of the matrix $B(t)$ are also in $L^1(I, \mathbb{R}^n)$ and we can find a constant which bounds the magnitude of $B(t)$, By the definition of Z_k, $\|u\|_H = 1$ and (5.11), we obtain

$$\lambda_k \int_0^T |u_k|^2 dt \leq \int_0^T |\dot{u}_k|^2 dt - \int_0^T (B(t), u_k, u_k)_{\mathbb{R}^n} dt$$

$$\leq \|u_k\|_H^2 + K\|u_k\|_H^2$$

$$\leq K + 1,$$

which implies that

$$\|u_k\|_2^2 \leq \frac{K+1}{\lambda_k} \to 0$$

since $\lambda_k \to +\infty$ as $k \to \infty$. Combining this with (5.18) we have that

$$\|u_k\|_\infty^2 \leq 2\|u_k\|_H \|u_k\|_2 = 2\|u_k\|_2 \to 0 \text{ as } k \to \infty.$$

Thus, letting $k \to \infty$ in (5.17), we get that $\beta = 0$, which completes the proof. $\qquad\square$

Since $V(t, \cdot)$ is even, we know that $G(-x) = G(x)$. It remains to verify conditions ($\Phi 1$) and ($\Phi 2$) of Theorem 1.11. We have the following lemma.

Lemma 5.6. *Under assumptions* (D5) *and* (D7), (D8), *the functional G satisfies the geometric condition of Theorem 1.11. That is to say, the following assertion is satisfied. For every $k \in \mathbb{N}$, there exists $\rho_k > r_k > 0$, such that*

($\Phi 1$) $a_k := \max_{x \in Y_k, \|x\|_H = \rho_k} G(x) \leq 0$;
($\Phi 2$) $b_k := \inf_{x \in Z_k, \|x\|_H = r_k} G(x) \to +\infty$ *as* $k \to \infty$.

Proof. Taking $r_k = \beta_k^{-1}$ and using Lemma 5.5, one has that

$$r_k \to +\infty \text{ as } k \to \infty.$$

Choose k large enough such that $k > l$ and $r_k \geq \left(\frac{C_0}{C'''}\right)^{q-2}$. Now, for $x \in Z_k$ with $\|x\|_H = r_k$, we have that

$$
\begin{aligned}
G(x) &\geq \lambda_l \|x\|^2 - \int_0^T 2V(t, x)dt \\
&\geq C_0 \|x\|_H^2 - C' \int_0^T |x|^q dt \\
&\geq C_0 \|x\|_H^2 - C'' \int_0^T \|x\|_H^q dt \\
&\geq C_0 \|x\|_H^2 - C''' \|x\|_H^q \\
&= (C_0 - C''' \|x\|_H^{q-2}) \|x\|_H^2 \\
&\geq \frac{C_0 r_k^2}{2},
\end{aligned}
$$

by (D7), which implies that

$$\inf_{x \in Z_k, \|x\|_H = r_k} G(x) \geq \frac{C_0 r_k^2}{2} \to +\infty \text{ as } k \to \infty.$$

Now, we verify (ii) of Theorem 1.11. Since Y_k is finite dimensional, there exists a constant $C_k > 0$ such that

$$C_k \|x\|_2 \geq \|x\|_H, \quad x \in Y_k. \tag{5.19}$$

From (D8) we deduce that there exists $M > 0$ such that

$$V(t, x) \geq (1 + K)C_k^2 |x|^2$$

for all $|x| \geq M$ and $t \in I$. Due to the continuity of V, one also has

$$|V(t, x)| \leq C$$

for all $|x| \leq M$ and $t \in I$, which yields

$$V(t, x) \geq (1 + K)C_k^2 |x|^2 - C' \tag{5.20}$$

for all $x \in \mathbb{R}^n$ and $t \in I$. Hence, using (5.19) and (5.20), we obtain

$$
\begin{aligned}
G(x) &= a(x, x) + b(x, x) - \int_0^T 2V(t, x)dt \\
&\leq \|x\|_H^2 + K\|x\|_H^2 - 2(1 + K)C_k^2 \|x\|_2^2 + 2C'T \\
&\leq (1 + K)\|x\|_H^2 - 2(1 + K)\|x\|_H^2 + 2C'T \\
&= -(1 + K)\|x\|_H^2 + 2C'T
\end{aligned}
\tag{5.21}
$$

for $x \in Y_k$. So, choosing $\rho_k > \max\{(2C'T/(1 + K))^{1/2}, r_k\}$, (5.21) implies that

$$\max_{x \in Y_k, \|x\| = \rho_k} G(x) \leq -\frac{\rho_k}{2} < 0.$$

The proof is complete. $\qquad\square$

Proof of Theorem 5.5. Due to Lemma 5.2, 5.4 and 5.6, we know that G satisfies all the assumptions of Theorem 1.11, so problem (5.9) has infinitely many nontrivial solutions. $\qquad\square$

Lemma 5.7. *Suppose that* (D5), (D7), (D8) *and* (V4$'$) *hold. Then the Cerami sequence of G is bounded.*

Proof. Let $\{x_k\}$ be a sequence in H such that $\{G(x_k)\}$ is bounded and $(1 + \|x_k\|_H)\|G'(x_k)\|_H \to 0$ as $k \to \infty$. There then exists a constant $M > 0$ such that

$$|G(x_k)| \leq M \text{ and } (1 + \|x_k\|_H)\|G'(x_k)\|_H \leq M \tag{5.22}$$

for all k. It follows from (V4$'$) that there exist $c_1 > 0$ and $M_1 > 0$ such that

$$(\nabla_x V(t, x), x) - 2V(t, x) \geq c_1 |x|^p \tag{5.23}$$

for all $|x| \geq M_1$ and $t \in I$. By the continuity of V and $\nabla_x V$, one has that

$$|(\nabla_x V(t, x), x) - 2V(t, x)| \leq C$$

for all $|x| \leq M$ and $t \in I$. Combining this and (5.23) we get that

$$(\nabla_x V(t,x) - 2V(t,x)) \geq c_1 |x|^p - C$$

for all $x \in \mathbb{R}^n$ and $t \in I$. Hence, we have that

$$3M \geq G(x_k) - \frac{1}{2}(G'(x_k), x_k)$$

$$= \int_0^T [(\nabla_x V(t,x_k), x_k) - 2V(t,x)]dt$$

$$\geq c_1 \int_0^T |x_k|^p dt - CT$$

for all k, which implies that

$$\int_0^T |x_k|^p dt < +\infty \quad \forall k \in \mathbb{N}. \tag{5.24}$$

It follows from (D7) that there exist $c_3 > 0$ and $M_2 > 0$ such that

$$V(t,x) \leq c_3 |x|^q$$

for all $|x| \geq M_2$ and $t \in I$. By the continuity of V, one has that

$$|V(t,x)| \leq C$$

for all $|x| \leq M_2$ and $t \in I$. Hence, we obtain

$$|V(t,x)| \leq c_3 |x|^q + C \tag{5.25}$$

for all $x \in \mathbb{R}^n$ and $t \in I$.

By the assumption (D5) and the finite measure of I, we deduce that for any there exists a constant L such that $|\int_0^T (B(t)x(t), x(t))_{\mathbb{R}^n} dt| \leq K_B \|x\|_\infty^2 \leq L \int_0^T |x|^2 dt + \epsilon \|x\|_H^2$ for all $x \in H$. Combining (5.25), (5.22) and Hölder's inequality, we obtain

$$\|x_k\|_H^2 = G(x_k) - b(x_k, x_k) + \int_0^T 2V(t, x_k)dt \tag{5.26}$$

$$= G(x_k) - \int_0^T |x_k|^2 dt - \int_0^T (B(t)x_k(t), x_k(t))_{\mathbb{R}^n} dt + \int_0^T 2V(t, x_k)dt$$

$$\leq M + (1+L) \int_0^T |x_k|^2 dt + \int_0^T 2V(t, x_k)dt + \epsilon \|x_k\|_H^2$$

$$\leq M + (1+L)T^{(q-2)/q} \left(\int_0^T |x_k|^q dt \right)^{2/q} + 2c_3 \int_0^T |x_k|^q dt + 2CT + \epsilon \|x_k\|_H^2$$

for all k. If $p > q$, the Hölder inequality and (5.24) imply that

$$\int_0^T |x_k|^q dt \leq T^{(p-q)/p} \left(\int_0^T |x_k|^p dt \right)^{q/p} < +\infty \quad \forall k \in \mathbb{N},$$

and hence the boundedness of $\{x_k\}$ follows from (5.26). If $p \leq q$, we also have

$$\int_0^T |x_k|^q dt = \int_0^T |x_k|^p |x_k|^{q-p} dt$$

$$\leq \|x_k\|_\infty^{q-p} \int_0^T |x_k|^p dt$$

$$\leq C^{q-p} \|x_k\|_H^{q-p} \int_0^T |x_k|^p dt \quad \forall k \in \mathbb{N}.$$

Noting the fact that $q - p < 2$, we see that (5.26) still yields that (x_k) is bounded. □

Proof of Theorem 5.7. Due to Lemma 5.2, 5.6, Lemma 5.7, we know G satisfies all the assumptions of Theorem 1.11, so problem (5.9) has infinitely many nontrivial solutions. □

Proof of Theorem 5.8. We consider the truncated functional

$$I(x) = \|x\|_H^2 - \left(b(x,x) + 2 \int_0^T V(t,x) dt \right) h(\|x\|_H)$$

for all $x \in H$, where $h : \mathbb{R}^+ \to [0,1]$ is a non-increasing C^1 function such that $h(s) = 1$ for $0 \leq s \leq \delta/(2C)$ and $h(s) = 0$ for $s \geq \delta/C$. Obviously, $I \in C^1(H, \mathbb{R})$ and $I(0) = 0$. If we can show that I satisfies the hypotheses of Kajikiya's critical point theorem (Theorem 1.12), it will follow that I admits a sequence of critical points $\{x_k\}$ such that $I(x_k) \leq 0$, $x_k \neq 0$ and $x_k \to 0$ as $k \to \infty$ to get the desire results.

When $\|x\|_H \leq \delta/C$, by the Sobolev embedding, one has

$$|x(t)| \leq \|x\|_\infty \leq C\|x\|_H \leq \delta \quad \forall t \in I.$$

Using (D10),

$$V(t, -x) = V(t, x), \quad t \in I,$$

and then $I(x) = I(-x)$.

For $\|x\|_H \geq \delta/C$, we have that

$$I(x) = \|x\|_H^2,$$

which shows that

$$I(x) \to +\infty, \text{ as } \|x\|_H \to \infty.$$

This implies that I is bounded from below and the (PS) sequence is bounded. Due to Lemma 5.2, we know that this enough to get a solution of problem (5.9).

Given any $k \in \mathbb{N}$, let $E_k = \bigoplus_{j=1}^{k} X_j$ where $X_j = \text{span}(e_j)$. There exists a constant $c_k > 0$ such that

$$c_k \|x\|_2 \geq \|x\|_H, \quad \forall x \in E_k,$$

by the equivalence of the norms on the finite dimensional spaces E_k. Using (D11), we see that there exists $0 < r_1$ such that

$$V(t, x) \geq (1 + K)c_k^2 |x|^2$$

for all $|x| \leq r_1$ and $t \in I$. Therefore, for $x \in E_k$ with $\|x\|_H = l_k := \frac{1}{2}\min\{1, \frac{r_1}{c_k}\}$, we obtain

$$
\begin{aligned}
I(x) &= \|x\|_H^2 + b(x, x) - \int_0^T 2V(t, x)dt \\
&\leq (1 + K)\|x\|_H^2 - \int_0^T 2V(t, x)dt \\
&\leq (1 + K)\|x\|_H^2 - 2(1 + K)c_k^2 \int_0^T |x|^2 dt \\
&\leq (1 + K)\|x\|_H^2 - 2(1 + K)\|x\|_H^2 \\
&= -(1 + K)l_k^2,
\end{aligned}
$$

which implies that

$$\left\{x \in E_k : \|x\|_H = l_k\right\} \subset \left\{x \in H : I(x) \leq -(1 + K)l_k^2\right\}.$$

Now taking $A_k = \left\{x \in H : I(x) \leq -(1 + K)l_k^2\right\}$, by Theorem 1.12, we get that

$$\gamma(A_k) \geq \gamma\left(\{x \in E_k : \|x\|_H = l_k\}\right) \geq k.$$

That is $A_k \in \Gamma_k$ and

$$\sup_{x \in A_k} I(x) \leq -(1 + K)l_k^2 < 0.$$

The proof is complete. $\qquad\square$

Notes and Comments

The periodic non-autonomous problem

$$- \ddot{x}(t) = \nabla_x V(t, x(t)) \qquad (5.27)$$

has an extensive history in the case of singular systems (cf., e.g., Ambrosetti–Coti Zelati [312]). The first to consider it for potentials satisfying (5.10) were Berger and Schechter [313] in 1977. They proved the existence of solutions to (5.27) under the condition that

$$V(t, x) \to \infty \text{ as } |x| \to \infty$$

uniformly for a.e. $t \in I$. Subsequently, Mawhin–Willem [314], Tang [315–319] Tang–Wu [320–322] and others proved existence under various conditions (cf. the references given in these publications). Most previous work considered the case when $B(t) = 0$. Ding and Girardi [323] considered the case of (5.9) when the potential oscillates in magnitude and sign, and found conditions for solutions when the matrix $B(t)$ is symmetric and negative definite and the function $V(x)$ grows superquadratically and satisfies a homogeneity condition. Antonacci [324, 325] gave conditions for existence of solutions with stronger constraints on the potential but without the homogeneity condition, and without the negative definite condition on the matrix. Generalizations of the above results are given by [324, 326–337]. This section was obtained in [338].

5.2.3 Quasilinear Hamiltonian Systems

Consider the existence of periodic solutions for the p-Hamiltonian systems

$$\begin{cases} -(|u'|^{p-2}u')' + A(t)|u|^{p-2}u = \lambda \nabla F(t, u), \\ u(T) - u(0) = u'(T) - u'(0) = 0, \end{cases} \qquad (5.28)$$

where $\lambda \in]0, +\infty[$, $p > 1$, $T > 0$, $F : [0, T] \times \mathbb{R}^N \to \mathbb{R}$ is a function satisfies the following assumption:
$F(t, u)$ is measurable with respect to t, for all $u \in \mathbb{R}^N$, continuously differentiable in u, for a.e. $t \in [0, T]$, and there exist $a \in C(\mathbb{R}^+; \mathbb{R}^+)$ and $b \in L^1([0, T]; \mathbb{R}^+)$ such that

$$|F(t, u)| \leq a(|u|)b(t), \qquad |\nabla F(t, u)| \leq a(|u|)b(t)$$

for all $u \in \mathbb{R}^N$ and a.e. $t \in [0, T]$. $A = (a_{ij}(t))_{N \times N}$ is a symmetric matrix valued function with $a_{ij} \in L^\infty[0, T]$ and there exists a positive constant $\underline{\lambda}$ such that $(A(t)|x|^{p-2}x, x) \geq \underline{\lambda}|x|^p$ for all $x \in \mathbb{R}^N$ and $t \in [0, T]$, that is, $A(t)$ is positive definite for all $t \in [0, T]$.

Here and in the sequel, the Sobolev space $W_T^{1,p}$ is defined by

$$W_T^{1,p} = \left\{ u : [0,T] \to \mathbb{R}^N \left| \begin{array}{l} u \text{ is absolutely continuous,} \\ u(0) = u(T) \text{ and } u' \in L^p(0,T;\mathbb{R}^N) \end{array} \right. \right\},$$

and endowed with the norm

$$\|u\|_A = \left(\int_0^T |u'(t)|^p dt + \int_0^T (A(t)|u(t)|^{p-2} u(t), u(t)) dt \right)^{1/p}.$$

Observe that

$$(A(t)|x|^{p-2}x, x) = |x|^{p-2} \sum_{i,j=1}^N a_{ij}(t) x_i x_j$$

$$\leq |x|^{p-2} \sum_{i,j=1}^N |a_{ij}(t)||x_i||x_j|$$

$$\leq \left(\sum_{i,j=1}^N \|a_{ij}\|_\infty \right) |x|^p,$$

then there exists a constant $\overline{\lambda} \leq \sum_{i,j=1}^N \|a_{ij}\|_\infty$ such that $(A(t)|x|^{p-2}x, x) \leq \overline{\lambda}|x|^p$ for all $x \in \mathbb{R}^N$. So, we have

$$\min\{1,\underline{\lambda}\}\|u\|^p \leq \|u\|_A^p \leq \max\{1,\overline{\lambda}\}\|u\|^p, \qquad (5.29)$$

where

$$\|u\| = \left(\int_0^T |u(t)|^p dt + \int_0^T |u'(t)|^p dt \right)^{1/p}.$$

Let

$$k_0 = \sup_{u \in W_T^{1,p} \setminus \{0\}} \frac{\|u\|_\infty}{\|u\|_A}, \qquad \|u\|_\infty = \sup_{t \in [0,T]} |u(t)|, \qquad (5.30)$$

where $|\cdot|$ is the usual norm in $\mathbb{R}^N$. Since $W_T^{1,p} \hookrightarrow C^0$ is compact, one has $k_0 < +\infty$ and for each $u \in W_T^{1,p}$, there exists $\xi \in [0,T]$ such that $|u(\xi)| = \min_{t \in [0,T]} |u(t)|$. Hence, by Hölder's inequality, one has

$$|u(t)| \leq \sqrt[q]{2} \max\{T^{\frac{1}{q}}, T^{-\frac{1}{p}}\}\|u\| \qquad (5.31)$$

for each $t \in [0,T]$ and $q = \frac{p}{p-1}$(see [339] for more details). So, by (5.30) and the above expression, we have

$$\|u\|_\infty \leq \sqrt[q]{2} \max\{T^{\frac{1}{q}}, T^{-\frac{1}{p}}\}\|u\| \leq \sqrt[q]{2} \max\{T^{\frac{1}{q}}, T^{-\frac{1}{p}}\} (\min\{1,\underline{\lambda}\})^{-\frac{1}{p}} \|u\|_A,$$

then from this and (5.29) it follows that

$$k_0 \leq k := \sqrt[q]{2} \max\{T^{\frac{1}{q}}, T^{-\frac{1}{p}}\} (\min\{1, \underline{\lambda}\})^{-\frac{1}{p}}. \tag{5.32}$$

As usual, a weak solution of problem (5.28) is any $u \in W_T^{1,p}$ such that

$$\int_0^T ((|u'(t)|^{p-2}u'(t), v'(t)) + (A(t)|u(t)|^{p-2}u(t), v(t)))dt \tag{5.33}$$

$$= \lambda \int_0^T (\nabla F(t, u(t)), v(t))dt$$

for all $v \in W_T^{1,p}$.

In this section, we will give out more piecewise results than that in [339]. Now, we state our main results.

Theorem 5.9. *Assume that there exist a constant vector* $d = (d_1, \cdots, d_N) \in \mathbb{R}^N$, *a positive constant* c *with* $c < k|d|(\underline{\lambda}T)^{1/p}$, *such that*

(D12) $\dfrac{\int_0^T \max_{x \in B} F(t,x)dt}{c^p} < \dfrac{\int_0^T F(t,d)dt}{\overline{\lambda}T(k|d|)^p}$, *where* $B = \{x \in \mathbb{R}^N : 0 \leq |x| \leq c\}$;

(D13) $\limsup_{|x| \to +\infty} \dfrac{F(t,x)}{|x|^p} < \dfrac{\int_0^T \max_{x \in B} F(t,x)dt}{c^p}$ *for a.e.* $t \in [0, T]$ *and all* $x \in \mathbb{R}^N$;

(D14) $F(t, 0) = 0$ *for all* $t \in [0, T]$.

Then, there exist a non-empty open interval $\Lambda :=\,]\dfrac{\overline{\lambda}T(k|d|)^p}{pk^p \int_0^T F(t,d)dt}$, $\dfrac{c^p}{pk^p \int_0^T \max_{x \in B} F(t,x)dt}[$ *with the following property: for each* $\lambda \in \Lambda$ *problem* (5.28) *has at least three solutions.*

Remark 5.2. Under the more weak assumptions as for Theorem 2 of [339], Theorem 5.9 ensures a more precise conclusion. In fact, our condition (D13) is more weak than condition (i_3) in Theorem 2 of [339]. For example, let $F(s) = \frac{s^p}{\ln(2+s^2)}$. Clearly, function F satisfies our condition (D13) but doesn't satisfy (i_3) in Theorem 2 of [339]. Furthermore, Theorem 5.9 give out a larger interval Λ than Theorem 2 of [339].

Now, assume F is a C^1 function on $\mathbb{R}^N$, with $F(0) = 0$, we can get the following result.

Corollary 5.1. *Assume that there exist a positive constant* c *and a vector* $d \in \mathbb{R}^N$ *with* $c < |d|$, *such that*

(D15) $\dfrac{\max_{|x| \leq c} F(x)}{c^p} < \dfrac{1}{\overline{\lambda}k^p} \dfrac{F(d)}{|d|^p}$;

(D16) $\limsup_{|x| \to \infty} \dfrac{F(x)}{|x|^p} < \dfrac{\max_{|x| \leq c} F(x)}{c^p}$.

Then, for every function $b \in L^1([0,T]) \setminus \{0\}$ *that is a.e. nonnegative and for every* $\lambda \in]\frac{\overline{\lambda} k^p}{p\|b\|_1} \frac{|d|^p}{F(d)}, \frac{1}{p\|b\|_1 k^p} \frac{c^p}{\max_{|x| \le c} F(x)}[$, *problem*

$$\begin{cases} -(|u'|^{p-2}u')' + A(t)|u|^{p-2}u = \lambda b(t)\nabla F(u), \\ u(T) - u(0) = u'(T) - u'(0) = 0, \end{cases} \tag{5.34}$$

admits at least three solutions.

Remark 5.3. Corollary 5.1 improves Theorem 3 of [339] since condition (D15) in our results is more general than conditions (j_1). We also obtain a larger interval of parameter than the interval insured by Theorem 3 in [339].

Now, we want to get infinitely many periodic solutions for the perturbed Hamiltonian system. Put

$$\mathcal{A} := \liminf_{|x| \to +\infty} \frac{\int_0^T \max_{|\xi| \le x} F(t,\xi)dt}{|x|^p},$$

$$\mathcal{B} := \limsup_{|x| \to +\infty} \frac{\int_0^T F(t,x)dt}{|x|^p},$$

$$\lambda_1 := \frac{\overline{\lambda}T}{pk^p\mathcal{B}},$$

$$\lambda_2 := \frac{1}{pk^p\mathcal{A}}.$$

Theorem 5.10. *Assume that,*

(D17) F *is non-negative in* $[0,T] \times \mathbb{R}^N$;
(D18) $\mathcal{A} < \overline{\lambda}T\mathcal{B}$.

Then, for every $\lambda \in \Lambda :=]\lambda_1, \lambda_2[$, *the problem* (5.28) *admits an unbounded sequence of periodic solutions which is unbounded in* $W_T^{1,p}$.

Remark 5.4. If

$$\liminf_{|x| \to +\infty} \frac{\int_0^T \max_{|\xi| \le x} F(t,\xi)dt}{|x|^p} = 0 \text{ and } \limsup_{|x| \to +\infty} \frac{\int_0^T F(t,x)dt}{|x|^p} = +\infty,$$

clearly, hypothesis (D18) is verified and Theorem 5.10 guarantees the existence of infinitely many weak solutions for problem (5.28), for every $\lambda \in]0, +\infty[$.

Replacing the conditions at infinity of the potential by a similar at zero, the same result holds and, in addition, the sequence of pairwise distinct solutions uniformly converges to zero. Precisely, set

$$\lambda_1^* := \frac{\overline{\lambda}T}{pk^p \limsup_{|x| \to 0} \frac{\int_0^T F(t,x)dt}{|x|^p}}, \qquad \lambda_2^* := \frac{1}{pk^p \liminf_{|x| \to 0} \frac{\int_0^T \max_{|\xi| \le x} F(t,\xi)dt}{|x|^p}}.$$

Arguing as in the proof of Theorem 5.10, we obtain the following result.

Theorem 5.11. *Assume that F is non-negative in $[0,T] \times \mathbb{R}^N$ and satisfies the following conditions,*

$$\liminf_{|x| \to 0} \frac{\int_0^T \max_{|\xi| \le x} F(t,\xi)dt}{|x|^p} < \overline{\lambda}T \limsup_{|x| \to 0} \frac{\int_0^T F(t,x)dt}{|x|^p}. \tag{5.35}$$

Then, for every $\lambda \in]\lambda_1^*, \lambda_2^*[$, the problem (5.28) *admits an unbounded sequence of non-zero weak solutions which strongly converges to 0 in $W_T^{1,p}$.*

Proof of Theorem 5.9. For each $u \in X$, let

$$\Phi(u) = \frac{\|u\|_A^p}{p}, \qquad \Psi(u) = \int_0^T F(t,u)dt.$$

Under the conditions of Theorem 5.9, Φ is a continuously Gâteaux differentiable and sequentially weakly lower semicontinuous functional. Moreover, the Gâteaux derivative of Φ admits a continuous inverse on X^* (we can get this result due to Proposition 2.4 in [340]). Ψ is continuously Gâteaux differential functional whose Gâteaux derivative is compact. Obviously, Φ is bounded on each bounded subset of X.

In particular, for each $u, \xi \in X$,

$$\langle \Phi'(u), \xi \rangle = \int_0^T ((|u'(t)|^{p-2}u'(t), \xi'(t)) + (A(t)|u(t)|^{p-2}u(t), \xi(t)))dt,$$

$$\langle \Psi'(u), \xi \rangle = \int_0^T (\nabla F(t,u), \xi)dt.$$

Hence, it follows from (5.33) that the weak solutions of equation (5.28) are exactly the solutions of the equation

$$\Phi'(u) - \lambda \Psi'(u) = 0.$$

Furthermore from (D13) there exist two constants $\gamma, \tau \in \mathbb{R}$ with

$$\limsup_{|x| \to +\infty} \frac{F(t,x)}{|x|^p} < \gamma < \frac{\int_0^T \max_{x \in B} F(t,x)dt}{c^p}$$

such that

$$F(t, x) \leq \gamma |x|^p + \tau$$

for a.e. $t \in [0, T]$ and all $x \in \mathbb{R}$. Fix $u \in X$. Then

$$F(t, u(t)) \leq \gamma |u(t)|^p + \tau$$

for all $t \in [0, T]$. Then, for any fixed $\lambda \in]\frac{\overline{\lambda}(k|d|)^p T}{pk^p \int_0^T F(t,d)dt}$, $\frac{c^p}{pk^p \int_0^T \max_{x \in B} F(t,x)dt}[$, since $\|u\|_\infty \leq k\|u\|_A$, we have

$$
\begin{aligned}
\Phi(u) - \lambda \Psi(u) &= \frac{\|u\|_A^p}{p} - \lambda \int_0^T F(t, u(t))dt \\
&\geq \frac{\|u\|_A^p}{p} - \lambda \int_0^T (\gamma |u(t)|^p + \tau)dt \\
&\geq \frac{\|u\|_A^p}{p} - \lambda \gamma \|u\|_\infty^p T - \lambda T \tau \\
&\geq \left(\frac{1}{p} - \lambda \gamma k^p T \right) \|u\|_A^p - \lambda T \tau \\
&> \frac{1}{p} \left(1 - \gamma \frac{c^p}{\int_0^T \max_{x \in B} F(t, x)dt} \right) \|u\|_A^p - \lambda T \tau,
\end{aligned}
$$

and so,

$$\lim_{\|u\| \to +\infty} (\Phi(u) - \lambda \Psi(u)) = +\infty.$$

So, hypothesis (a_2) of Theorem 1.17 is satisfied.

Let $u^*(t) = d$. Then, we have $(u^*(t))' = 0$. It is easy to verify that $u^* \in W_T^{1,p}$, and in particular, one has

$$\|u^*\|^p = T|d|^p. \tag{5.36}$$

Hence, we obtain from $\underline{\lambda}|d|^p \leq (A(t)|d|^{p-2}d, d) \leq \overline{\lambda}|d|^p$ and (5.36) that

$$(\underline{\lambda}T)^{1/p}|d| \leq \|u^*\|_A \leq (\overline{\lambda}T)^{1/p}|d|. \tag{5.37}$$

By $c < k|d|(\underline{\lambda}T)^{1/p}$, it follows from (5.37) that

$$\|u^*\|_A > \frac{c}{k}. \tag{5.38}$$

Hence, thanks to the condition (D12) and (5.37), we have

$$
\begin{aligned}
\int_0^T \max_{x \in B} F(t, x)dt &< \frac{1}{\overline{\lambda}T} \left(\frac{c}{k|d|} \right)^p \int_0^T F(t, d)dt \\
&\leq \left(\frac{c}{k\|u^*\|_A} \right)^p \int_0^T F(t, u^*)dt. \tag{5.39}
\end{aligned}
$$

Now, put

$$r = \frac{1}{p} \left(\frac{c}{k} \right)^p .$$

Thanks to (5.38), there exists $u^* \in X$ such that

$$\Phi(u^*) = \frac{\|u^*\|_A^p}{p} > r > 0 = \Phi(0). \tag{5.40}$$

We obtain from (5.30) and $k_0 \le k$ that

$$\sup_{t \in [0,T]} |u(t)| \le k \|u\|_A \tag{5.41}$$

for each $u \in X$. Hence, for each $u \in X$ such that

$$\Phi(u) = \frac{\|u\|_A^p}{p} \le r,$$

by (5.41) one has

$$\sup_{t \in [0,T]} |u(t)| \le c. \tag{5.42}$$

It follows from (5.39) and (5.42) that

$$\sup_{u \in \Phi^{-1}(]-\infty,r])} \Psi(u) = \sup_{\{u \mid \Phi(u) \le r\}} \int_0^T F(t, u) dt$$

$$= \sup_{\left\{ u \mid \|u\|_A^p \le pr \right\}} \int_0^T F(t, u) dt$$

$$\le \int_0^T \sup_{\{x \mid 0 \le |x| \le c\}} F(t, x) dt$$

$$= \int_0^T \max_{x \in B} F(t, x) dt$$

$$< \frac{1}{p} \left(\frac{c}{k} \right)^p \frac{p}{\|u^*\|_A^p} \int_0^T F(t, u^*(t)) dt$$

$$= r \frac{p}{\|u^*\|_A^p} \int_0^T F(t, u^*(t)) dt$$

$$= r \frac{\Psi(u^*)}{\Phi(u^*)}.$$

So, one has

$$\sup_{u \in \Phi^{-1}(]-\infty,r])} \Psi(u) < r \frac{\Psi(u^*)}{\Phi(u^*)}. \tag{5.43}$$

and (a_1) of Theorem 1.17 is satisfied.

From Theorem 1.17, for each $\lambda \in \Lambda_r: =]\frac{\Phi(u^*)}{\Psi(u^*)}, \frac{r}{\sup_{\Phi(x) \le r} \Psi(x)}[$, the functional $\Phi - \lambda \Psi$ has at least three distinct critical points which are the periodic solutions of (5.28) and the conclusion is achieved. $\qquad \square$

Proof of Theorem 5.10. Our goal is to apply Theorem 1.18. Let $\{c_n\}$ be a real sequence such that $\lim_{n \to +\infty} c_n = +\infty$ and

$$\liminf_{n \to +\infty} \frac{\int_0^T \max_{|\xi| \le c_n} F(t,\xi)dt}{c_n^p} = \mathcal{A}. \tag{5.44}$$

Taking into account (4) for every $u \in X$ one has

$$\sup_{t \in [0,T]} |u(t)| \le k\|u\|_A.$$

Hence, an easy computation ensures that $|u| \le c_n$ whenever $u \in \Phi^{-1}(]-\infty, r_n[)$, where

$$r_n = \frac{1}{p}\left(\frac{c_n}{k}\right)^p, \qquad \forall n \in \mathbb{N}.$$

Taking into account that $\|u^0\|_A = 0$ (where $u^0(t) = 0$ for every $t \in [0,T]$) and that $\int_0^T F(t,0)dt \ge 0$ for all $t \in [0,T]$, for every n large enough, one has

$$
\begin{aligned}
\varphi(r_n) &= \inf_{u \in \Phi^{-1}(]-\infty, r_n[)} \frac{(\sup_{v \in \Phi^{-1}(]-\infty, r_n[)} \Psi(v)) - \Psi(u)}{r_n - \Phi(u)} \\
&= \inf_{\|u\|_A^p/p < r_n} \frac{\sup_{\|v\|_A^p/p < r_n} \int_0^T F(t,v)dt - \int_0^T F(t,u)dt}{r_n - \|u\|_A^p/p} \\
&\le \frac{\sup_{\|v\|_A^p/p < r_n} \int_0^T F(t,v)dt}{r_n} \\
&\le pk^p \frac{\int_0^T \max_{|x| \le c_n} F(t,x)dt}{c_n^p}.
\end{aligned}
$$

Therefore, since from assumption (D18) one has $\mathcal{A} < +\infty$, we obtain

$$\gamma = \liminf_{n \to +\infty} \varphi(r_n) \le pk^p\mathcal{A} < +\infty. \tag{5.45}$$

Moreover, we can also observe that, owing to (5.44) and (5.45),

$$\Lambda \subseteq\ ]0, 1/\gamma[.$$

Now, fix $\lambda \in]\lambda_1, \lambda_2[$ and let us verify that the functional $\Phi(u) - \lambda\Psi(u)$ is unbounded from below. Indeed, we can consider a N positive real sequence $\{d_{i,n}\}$ such that $|d_{i,n}| \to +\infty$ as $n \to \infty$. For each $n \in \mathbb{N}$ define

$$w_{i,n}(t) = d_{i,n}$$

and put $w_n := (w_{1,n}, \dots, w_{N,n})$. Since $1/\lambda < \frac{pk^p}{\overline{\lambda}T}\mathcal{B}$, we can get that there exists a positive constant $\eta > 0$ and

$$\frac{1}{\lambda} < \eta < \frac{pk^p}{\overline{\lambda}T} \frac{\int_0^T F(t,d_n)dt}{|d_n|^p}. \tag{5.46}$$

Fix $n \in \mathbb{N}$, a simple computation shows that

$$(\underline{\lambda}T)^{1/p}|d_n| \leq \|w_n\|_A \leq (\overline{\lambda}T)^{1/p}|d_n|. \tag{5.47}$$

On the other hand, thanks to (D17), (5.46) and (5.47), we achieve

$$\Phi(w_n) - \lambda\Psi(w_n) \leq \frac{\overline{\lambda}T|d_n|^p}{p} - \lambda \int_0^T F(t, d_n)dt < \frac{|d_n|^p\overline{\lambda}T}{pk^p}(1 - \lambda\eta)$$

for every $n \in \mathbb{N}$ large enough. Hence, $\Phi(u) - \lambda\Psi(u)$ is unbounded from below.

Applying Theorem 1.18 we deduce that the functional $\Phi(u) - \lambda\Psi(u)$ admits a sequence of critical points which is unbounded in X. Hence, our claim is proved and the conclusion is achieved.   □

Notes and Comments

Contrast to Hamiltonian systems, for the general case $p > 1$, the study on the existence and multiplicity of periodic solutions are new, see [341–350]. Lü *et al.* [342] dealt with the existence of infinitely many periodic solutions for the ordinary p-Laplacian system

$$\begin{cases} -(|u'|^{p-2}u')' = \nabla F(t, u), \\ u(T) - u(0) = u'(T) - u'(0) = 0, \end{cases} \tag{5.48}$$

where $F: [0, T] \times \mathbb{R}^N \to \mathbb{R}$ is a function continuously differentiable with respect to the second variable and satisfying some usual growth conditions. They stated that if

$$\limsup_{R \to +\infty} \inf_{|a|=R} \int_0^T F(s, a)ds = +\infty$$

and

$$\liminf_{r \to +\infty} \sup_{|b|=r} \int_0^T F(s, b)ds = -\infty,$$

then

(i) there exists a sequence (u_n) of periodic solutions of (5.48) such that u_n is a critical point of φ and $\lim_{n \to \infty} \varphi(u_n) = +\infty$,

(ii) there exists a sequence (u_n^*) of periodic solutions of (5.48) such that u_n^* is a local minimum point of φ and $\lim_{n \to \infty} \varphi(u_n^*) = -\infty$, where $\varphi(u) = \frac{1}{p}\int_0^T |u'(t)|^p dt - \int_0^T F(t,u(t))dt$.

The same variational methods as were used there, which are based on the seminal paper of Bonanno and Marano [39], have been applied to obtain multiple solutions (see, for instance, [302, 304, 331, 351–356]). In particular, Li *et al.* [339] had proved the existence of at least three solutions for problem (5.28). The technical approach is based on the three critical points theorem of Averna and Bonanno [357]. Their theorem under novel assumptions ensures the existence of an open interval $\Lambda \subseteq [0, +\infty)$ such that, for each $\lambda \in \Lambda$, problem (5.28) admits at least three weak solutions. Very recently in [41], presenting a version of the infinitely many critical points theorem of Ricceri (see [36, Theorem 2.5]), the existence of an unbounded sequence of weak solutions for a Sturm–Liouville problem, having discontinuous non-linearities, had been established. In a such approach, an appropriate oscillating behavior of the nonlinear term either at infinity or at zero was required. This type of methodology had been used then in several works in order to obtain existence results for different kinds of problems (see, for instance, [306, 331, 351, 358–361]). This section was obtained in [362].

5.3 *N* Dimension Systems

5.3.1 *Resonance Elliptic Systems*

We consider the following elliptic system

$$\begin{cases} -\Delta u = \lambda u + \delta v + f(x,u,v), & \text{in } \Omega, \\ -\Delta v = \delta u + \gamma v + g(x,u,v), & \text{in } \Omega, \\ u = v = 0, & \text{on } \partial\Omega, \end{cases} \qquad (5.49)$$

where Ω is a bounded smooth domain in $\mathbb{R}^N$ ($N \geq 3$) and λ, δ, $\gamma \in \mathbb{R}$. The nonlinearities (f,g) are the gradient of some function, that is, there exists a function $F \in C^1(\overline{\Omega} \times \mathbb{R}, \mathbb{R})$ such that $\nabla F = (f,g)$. The system is resonance if the following conditions holds:

$$(D19) \quad \sigma(A^*) \cap \sigma(-\Delta) \neq \emptyset,$$

where

$$A^* = \begin{pmatrix} \lambda & \delta \\ \delta & \gamma \end{pmatrix}; \qquad \sigma(A^*) = \{\xi, \zeta\} = \left\{ \frac{\lambda+\gamma}{2} \pm \sqrt{\left(\frac{\lambda-\gamma}{2}\right)^2 + \delta^2} \right\}$$

denotes the spectrum of the matrix A^* and $\sigma(-\Delta) = \{\lambda_k : k = 1, 2, \ldots$ and $0 < \lambda_1 < \lambda_2 < \cdots\}$ denotes the eigenvalues of the Laplacian on Ω with zero boundary conditions.

By using the minimax methods in critical point theory, we obtain the multiplicity results for subquadratic cases, which generalize and sharply improve the results in [363–365]. Furthermore, compared with their proofs, ours are much simpler. For more general operator, we refer the reader to see the paper [361, 366, 367].

Let $|\cdot|$ and $(\cdot, \cdot)$ denote respectively the usual norm and inner product in $\mathbb{R}^2$. We consider the subquadratic case and make the following assumptions:

(D20) There exists constants $c > 0$ and $1 < p < 2^* := \frac{2N}{N-2}$ such that

$$|\nabla F(x, U)| \leq c(1 + |U|^{p-1}), \quad \forall (x, U) \in \Omega \times \mathbb{R}^2.$$

(D21) $F(x, 0) = 0$, for all $x \in \Omega$, and

$$\lim_{|U| \to 0} \frac{F(x, U)}{|U|^2} = +\infty \quad \text{uniformly for a.e. } x \in \Omega.$$

(D22) $F(x, U) = F(x, -U), \forall (x, U) \in \Omega \times \mathbb{R}^2$.

Our main results are as follows:

Theorem 5.12. *Suppose that* (V), (D20)–(D22) *hold, then problem* (5.49) *possesses infinitely many nontrivial solutions.*

Remark 5.5. Zou [368] studied systems (5.49) which are asymptotically linear with resonance. Applying the minimax technique, they obtained the following theorem.

Theorem 5.13. *Assume the following conditions are satisfied:*

(Z1) $|\nabla F(x, U)| \leq c(1 + |U|^\sigma)$ *for almost all* $x \in \Omega$ *and* $U \in \mathbb{R}^2$, *where* $\sigma \in (0, 1)$ *is a constant.*

(Z2) $\liminf_{|U| \to +\infty} \frac{\pm F(x, U)}{|U|^{1+\sigma}} := a^\pm(x) \succeq 0$ *uniformly for almost all* $x \in \Omega$, *where* $a^\pm(x) \succeq 0$ *indicates that* $a^\pm(x) \geq 0$ *with strict inequality holding on a set of positive measure.*

(Z3) *There exist* $\delta_1, \delta_2 \in (1, 2)$, $c_1 > 0$, $c_2 > 0$, $t_0 > 0$ *such that*

$$c_1 |U|^{\delta_1} \leq F(x, U) \leq c_2 |U|^{\delta_2}$$

for almost all $x \in \Omega$ *and* $|U| \leq t_0$.

(Z4) $F(x, -U) = F(x, U)$ *for a.e.* $x \in \Omega$ *and* $U \in \mathbb{R}^2$.

Then, (5.49) *has infinitely many small-energy solutions.*

Subsequently, Chen and Ma [363] considered the subquadratic case and proved the following theorem.

Theorem 5.14. *Suppose that* (V) *and the following conditions are satisfied:*

(CM1) $F(x, U) \geq 0$, $\forall (x, U) \in \Omega \times \mathbb{R}^2$, *and there exist constants* $\mu \in [1, 2)$ *and* $R_1 > 0$ *such that*

$$(\nabla F(x, U), U) \leq \mu F(x, U), \quad \forall x \in \Omega \text{ and } |U| \geq R_1.$$

(CM2) *There exist constants* $a \in [1, 2)$ *and* c_2, c_2', $R_2 > 0$ *such that*

$$F(x, U) \geq c_2' |U|^a, \quad \forall x \in \Omega \text{ and } U \in \mathbb{R}^2$$

and

$$F(x, U) \leq c_2 |U|, \quad \forall x \in \Omega \text{ and } |U| \leq R_2.$$

(CM3) $\liminf_{|U| \to \infty} \frac{F(x, U)}{|U|} \geq d > 0$ *uniformly for* $x \in \Omega$.

and $F(x, U)$ *is even in* U. *Then* (5.49) *possesses infinitely many nontrivial solutions.*

Theorem 5.12 unifies and greatly extends theorems 5.13 and 5.14. (Z2) and (Z3) in Theorem 5.13 and (CM2) and (CM3) in Theorem 5.14 are completely removed. Hence, Theorem 5.12 generalizes and significantly improves upon Theorems 5.13 and 5.14. There exist functions F satisfying our theorem 5.12 and not satisfying Theorems 5.13 and 5.14. For example, let

$$F(x, U) = h(x)|U|^{3/2} \left(-\ln \left(\frac{1 + |U|^2}{4} \right) \right)$$

for all $x \in \Omega$ and $U \in \mathbb{R}^2$, where $h \in L^1(\Omega; \mathbb{R}^+)$ with $\inf_{x \in \Omega} h(x) > 0$. A straightforward computation shows that F satisfies the conditions of Theorem 5.12, but it does not satisfy the corresponding conditions of Theorems 5.13 and 5.14, since $F(x, U) < 0$ for all $|U| > \sqrt{3}$, and $\lim_{|U| \to \infty} |\nabla F(x, U)| = +\infty$ uniformly for $x \in \Omega$.

Now, we give the variational framework of our problem and some related preliminary lemmas.

In the following, we use $\|\cdot\|_2$ and $\|\cdot\|_{L^2}$ to denote the norms of $L^2(\Omega)$ and $L^2(\Omega) \times L^2(\Omega)$, respectively. Let $E := H_0^1(\Omega)$ and $W := H_0^1(\Omega) \times H_0^1(\Omega)$, where $H_0^1(\Omega)$ is the usual Sobolev space with the norm $\|\cdot\|_E$ generated by the inner product

$$\langle u, v \rangle_E = \int_\Omega \nabla u \cdot \nabla v \, dx, \quad \forall u, v \in H_0^1(\Omega).$$

Then for $U = (u_1, u_2)$ and $V = (v_1, v_2)$ in W, the induced inner product and norm on W are given respectively by

$$\langle U, V \rangle_W = \langle u_1, v_1 \rangle_E + \langle u_2, v_2 \rangle_E \text{ and } \|U\|_W^2 = \|u_1\|_E^2 + \|u_2\|_E^2.$$

Let $\overrightarrow{e_1} := (e_{11}, e_{12})$, $\overrightarrow{e_2} := (e_{21}, e_{22}) \in \mathbb{R}^2$ be the normalized eigenvectors of A^* such that

$$A^* \overrightarrow{e_1} = \xi \overrightarrow{e_1}, \quad A^* \overrightarrow{e_2} = \zeta \overrightarrow{e_2}, \quad \overrightarrow{e_1} \cdot \overrightarrow{e_2} = 0, \quad |\overrightarrow{e_1}| = |\overrightarrow{e_2}| = 1.$$

For any $\alpha \in \mathbb{R}$. Let H_α^+, H_α^-, H_α^0 be the subspaces of $H_0^1(\Omega)$, where the quadratic form

$$u \to \|u\|_E^2 - \alpha \|u\|_2^2$$

is positive definite, negative definite and zero, respectively. Let

$$W^0 := H_\xi^0 \times H_\zeta^0, \quad W^+ := H_\xi^+ \times H_\zeta^+ \text{ and } W^- := H_\xi^- \times H_\zeta^-.$$

Set

$$A_1 := \mathrm{id} - \xi(-\Delta)^{-1} \text{ and } A_2 := \mathrm{id} - \zeta(-\Delta)^{-1},$$

where id denotes the identity on $H_0^1(\Omega)$. We introduce an operator

$$A : W \to W, A = (A_1, A_2),$$

which is defined by

$$AU = (A_1 u_1, A_2 u_2), \quad \forall U = (u_1, u_2) \in W.$$

Then A is a bounded self-adjoint operator from W to W and $\ker A = W^0$ with $\dim W^0 < \infty$. The space W is spitted as

$$W = W^- \oplus W^0 \oplus W^+,$$

where W^- and W^+ are invariant under A, $A|_{W^-}$ is negative, and $A|_{W^+}$ is positive definite. More precisely, there exists a positive constant c_0 such that

$$\pm \langle AU^\pm, U^\pm \rangle_W \geq c_0 \|U^\pm\|_W^2, \quad \forall U^\pm \in W^\pm.$$

Here and in what follows, for any $U \in W$, we always denote by U^0, U^+ and U^- the vectors in W with $U = U^0 + U^- + U^+$, $U^0 \in W^0$ and $U^\pm \in W^\pm$. Note that $\dim W^0$ and $\dim W^-$ are finite. Furthermore, $\sigma(A^*) \cap \sigma(-\Delta) \neq \emptyset$ implies $\dim W^0 \neq 0$. For problem (5.49), we consider the following functional:

$$\Phi(U) = \frac{1}{2} \langle AU, U \rangle_W - \int_\Omega \tilde{F}(x, U) dx, \quad U = (u_1, u_2) \in W,$$

where $\tilde{F}(x, s, t) = F(x, s\vec{e_1} + t\vec{e_2})$. In view of the assumptions of F and the definition of $\tilde{F}$, we know that the (weak) solutions of system (5.49) are the critical points of the functional Φ by the discussion of [369].

Next, we define an equivalent inner product $\langle \cdot, \cdot \rangle$ and the corresponding norm $\| \cdot \|$ on W given respectively by

$$\langle U, V \rangle = \langle AU^+, V^+ \rangle_W - \langle AU^-, V^- \rangle_W \text{ and } \|U\| = \langle U, U \rangle^{1/2},$$

where $U, V \in W^0 \oplus W^- \oplus W^+$ with $U = U^0 + U^- + U^+$ and $V = V^0 + V^- + V^+$. Therefore, Φ can be rewritten as

$$\Phi(U) = \frac{1}{2}\|U^+\|^2 - \frac{1}{2}\|U^-\|^2 - \int_\Omega \tilde{F}(x, U)dx.$$

Furthermore, $\Phi \in C^1(W, \mathbb{R})$ and the derivatives are given by

$$\Phi'(U)V = \langle U^+, V^+ \rangle - \langle U^-, V^- \rangle - \int_\Omega \left(\nabla \tilde{F}(x, U), V \right) dx,$$

for any $U, V \in W^0 \oplus W^- \oplus W^+$ with $U = U^0 + U^- + U^+$ and $V = V^0 + V^- + V^+$.

Note that dim W^0 and dim W^- are finite. We choose an orthonormal basis $\{e_m\}_{m=1}^k$ for W^0, an orthonormal basis $\{e_m\}_{m=k+1}^{l_0}$ for W^- and an orthonormal basis $\{e_m\}_{m=l_0+1}^\infty$ for W^+, where $1 \leq k < \infty$ and $k+1 \leq l_0 < \infty$. Then $\{e_m\}_{m=1}^\infty$ is an orthonormal basis of W.

Proof of Theorem 5.12. We consider the truncated functional

$$I(U) = \frac{1}{2}\|U^+\|^2 - \left(\frac{1}{2}\|U^-\|^2 + \int_\Omega \tilde{F}(x, U)dx \right) h(\|U\|)$$

for all $U \in W$, where $h : \mathbb{R}^+ \to [0, 1]$ is a non-increasing C^1 function such that $h(t) = 1$ for $0 \leq t \leq 1$ and $h(t) = 0$ for $t \geq 2$. Obviously, $I \in C^1(W, \mathbb{R})$ and $I(0) = 0$. If we can prove that I admits a sequence of critical points $\{U_k\}$ such that $I(U_k) \leq 0$, $U_k \neq 0$ and $U_k \to 0$ as $k \to \infty$, then we can apply Kajikiya's critical point theorem (Theorem 1.12) to get the desire results. Due to (D22), $F(x, -U) = F(x, U)$ for all $(x, U) \in \Omega \times \mathbb{R}^2$, so $I(U) = I(-U)$, that is I is even.

For $\|U\| \geq 2$, we have that

$$I(U) = \frac{1}{2}\|U^+\|^2,$$

which shows that

$$I(U) \to +\infty, \text{ as } \|U\| \to \infty.$$

This implies I is bounded from below and satisfies the (PS) condition. Actually, due to the coercivity of the functional I, we can get a (PS) sequence $\{U_j\}$ bounded. By the fact of $\dim(W^0 \oplus W^-) < \infty$, without loss of generality, we may assume

$$U_j^- \to U^-, \quad U_j^0 \to U^0, \quad U_j^+ \rightharpoonup U^+ \text{ and } U_j \rightharpoonup U, \text{ as } j \to \infty, \quad (5.50)$$

for some $U = U^0 + U^- + U^+ \in W = W^0 \oplus W^- \oplus W^+$. By virtue of the Riesz Representation Theorem, $I' : W \to W^*$ and $G' : W \to W^*$ can be viewed as $I' : W \to W$ and $G' : W \to W$ respectively, where W^* is the dual space of W and $G(U) := \int_\Omega \tilde{F}(x, U) dx$. Note that

$$o(1) = I'(U_j) = U_j^+ - \left(U_j^- + G'(U_j)\right), \quad \forall j \in \mathbb{N},$$

that is,

$$U_j^+ = U_j^- + G'(U_j) + o(1), \quad \forall j \in \mathbb{N}. \tag{5.51}$$

Note that the assumptions of F and the definition of $\tilde{F}$, the Sobolev embedding, by the standard argument (see [370]), imply $G' : W \to W^*$ is compact. Therefore, $G' : W \to W$ is also compact. Due to the compactness of G' and (5.50), the right hand side of (5.51) is converge strongly in W, and hence $U_j^+ \to U^+$ in W. Combining this with (5.50), we have $U_j \to U$ in W. Therefore, I satisfies the (PS) condition.

Given any $k \geq k_1 := l_0 + 1$, where l_0 is defined in Section 2. Let $E_k = \bigoplus_{j=1}^k X_j$ where $X_j = \text{span}(e_j)$, where $\{e_j\}$ is an orthogonal basis of W. There exists a constant $c_k > 0$ such that

$$\|U\|_{L^2} \geq c_k \|U\|, \quad \forall U \in E_k, \quad \forall k \in \mathbb{N},$$

by the equivalence of the norms on the finite dimensional spaces E_k. Using (D21), there exists $0 < r_1 < 1$ such that

$$F(x, U) \geq \frac{1}{c_k^2} |U|^2,$$

for all $|U| \leq r_1$ and a.e. $x \in \Omega$. Therefore, for $U \in E_k$ with $\|U\| = l_k :=$

$\frac{1}{2}\min\{1, \frac{r_1}{c_k}\}$, we obtain

$$
\begin{aligned}
I(U) &= \frac{1}{2}\|U^+\|^2 - \frac{1}{2}\|U^-\|^2 - \int_\Omega \tilde{F}(x, U)dx \\
&\leq \frac{1}{2}\|U^+\|^2 - \int_\Omega \tilde{F}(x, U)dx \\
&\leq \frac{1}{2}\|U\|^2 - \frac{1}{c_k^2}\|U\|_{L^2}^2 \\
&\leq \frac{1}{2}\|U\|^2 - \frac{1}{c_k^2}c_k^2\|U\|^2 \\
&= -\frac{1}{2}\|U\|^2 \\
&= -\frac{1}{2}l_k^2,
\end{aligned}
$$

which implies that

$$
\{U \in E_k : \|U\| = l_k\} \subset \left\{U \in W : I(U) \leq -\frac{1}{2}l_k^2\right\}.
$$

Now taking $A_k = \left\{U \in W : I(U) \leq -\frac{1}{2}l_k^2\right\}$, by Theorem 1.12, we get that

$$
\gamma(A_k) \geq \gamma\left(\{U \in E_k : \|U\| = l_k\}\right) \geq k,
$$

so $A_k \in \Gamma_k$ and

$$
\sup_{U \in A_k} I(U) \leq -\frac{1}{2}l_k^2 < 0.
$$

The proof is complete. □

Notes and Comments

A vast literature on the study of the existence and multiplicity of solutions for resonance elliptic systems via the critical point theory had grown since Costa and Magalhães published their paper [369]; see [363–365,368,371–373] and the references therein. In [369], Costa and Magalhães considered subquadratic perturbations of semilinear elliptic systems which are in variational form. Later, Zou [365] presented two different theorems. If ∇F is not odd, is sublinear and satisfies certain assumptions at infinity (and near the origin), classical linking theorems with the Cerami condition can be used to prove the existence of at least one (nontrivial) solution of (5.49). Furthermore, if the nonlinear term ∇F is odd and (5.49) is a strongly resonant problem, one can obtain solutions under suitable hypotheses on F

and by using a multiplicity theorem due to Fei [374]. In particular, this result holds for a single elliptic equation at resonance. Zou *et al.* [373] got existence of one and of two nonzero solutions in the case where the problem is resonant and F is sublinear at zero and infinity. Zou [368] considered cooperative and noncooperative elliptic systems that are asymptotically linear at infinity. He obtained infinitely many solutions with small energy if the potential is even. Pomponio [372] considered an asymptotically linear cooperative elliptic system at resonance. Recently, Ma had generalized Zou's results [365, 368, 373] in [364] and [371], respectively. Very recently, Chen and Ma [363] studied a class of resonant cooperative elliptic systems with sublinear or superlinear terms and obtain infinitely many nontrivial solutions by two variant fountain theorems developed by Zou [220]. In the present section, we studied the existence of infinitely many non-trivial solutions of (5.49) under the symmetric condition. This section was obtained in [375].

5.3.2 *Quasilinear Fourth-Order Elliptic Systems*

Consider the Navier boundary value problem involving the (p,q)-biharmonic systems

$$
\begin{cases}
\Delta\left(|\Delta u|^{p-2}\Delta u\right) = \lambda F_u(x, u, v) + \mu G_u(x, u, v), & \text{in } \Omega, \\
\Delta\left(|\Delta v|^{q-2}\Delta v\right) = \lambda F_v(x, u, v) + \mu G_v(x, u, v), & \text{in } \Omega, \\
u = \Delta u = v = \Delta v = 0, & \text{on } \partial\Omega,
\end{cases}
\tag{5.52}
$$

where λ, $\mu \in [0, +\infty)$, $\Omega \subset \mathbb{R}^N (N \geq 1)$ is a non-empty bounded open set with a sufficient smooth boundary $\partial\Omega$, $p > \max\left\{1, \frac{N}{2}\right\}$, $q > \max\left\{1, \frac{N}{2}\right\}$, $F, G \colon \Omega \times \mathbb{R} \times \mathbb{R} \mapsto \mathbb{R}$ are functions that $F(\cdot, s, t)$, $G(\cdot, s, t)$ are measurable in Ω for all $(s, t) \in \mathbb{R} \times \mathbb{R}$ and $F(x, \cdot, \cdot)$, $G(x, \cdot, \cdot)$ are continuously differentiable in $\mathbb{R} \times \mathbb{R}$ for a.e. $x \in \Omega$, F_i denotes the partial derivative of F with respect to i, $i = u, v$, so does G.

Here in the sequel, X will be denoted the Cartesian product of two Sobolev spaces $W^{2,p}(\Omega) \cap W_0^{1,p}(\Omega)$ and $W^{2,q}(\Omega) \cap W_0^{1,q}(\Omega)$, i.e., $X = (W^{2,p}(\Omega) \cap W_0^{1,p}(\Omega)) \times (W^{2,q}(\Omega) \cap W_0^{1,q}(\Omega))$. The space X will be endowed with the norm

$$
\|(u,v)\| = \|u\|_p + \|v\|_q, \quad \|u\|_p = \left(\int_\Omega |\Delta u|^p \, dx\right)^{\frac{1}{p}}, \quad \|v\|_q = \left(\int_\Omega |\Delta v|^q \, dx\right)^{\frac{1}{q}}.
$$

Let

$$K = \max \left\{ \sup_{u \in W^{2,p}(\Omega) \cap W_0^{1,p}(\Omega) \setminus \{0\}} \frac{\sup_{x \in \Omega} |u(x)|^p}{\|u\|_p^p}, \right.$$
$$\left. \sup_{v \in W^{2,q}(\Omega) \cap W_0^{1,q}(\Omega) \setminus \{0\}} \frac{\sup_{x \in \Omega} |v(x)|^q}{\|v\|_q^q} \right\}. \qquad (5.53)$$

Since $p > \max\left\{1, \frac{N}{2}\right\}$, $q > \max\left\{1, \frac{N}{2}\right\}$, $W^{2,p}(\Omega) \cap W_0^{1,p}(\Omega) \hookrightarrow C^0(\overline{\Omega})$ and $W^{2,q}(\Omega) \cap W_0^{1,q}(\Omega) \hookrightarrow C^0(\overline{\Omega})$ are compact, and one has $K < +\infty$. As usual, a weak solution of problem (5.52) is any $(u, v) \in X$ such that

$$\int_\Omega |\Delta u|^{p-2} \Delta u \Delta \xi dx + \int_\Omega |\Delta v|^{q-2} \Delta v \Delta \eta dx$$
$$= \lambda \int_\Omega F_u(x, u, v)\xi dx + \lambda \int_\Omega F_v(x, u, v)\eta dx \qquad (5.54)$$
$$+ \mu \int_\Omega G_u(x, u, v)\xi dx + \mu \int_\Omega G_v(x, u, v)\eta dx$$

for every $(\xi, \eta) \in X$.

To our best of knowledge, there are few results about multiple solutions to (p, q)-biharmonic systems. In this paper, we prove the existence of at least three solutions of problem (5.52). The technical approach is based on the three critical points theorem of Ricceri [31]. Our Theorem under novel assumptions ensures the existence of an open interval $\Lambda \subseteq [0, +\infty)$ and a positive real number ρ such that, for each $\lambda \in \Lambda$, problem (5.52) admits at least three weak solutions whose norms in X are less than ρ.

Now, for every $x^0 \in \Omega$ and pick r_1, r_2 with $r_2 > r_1 > 0$, such that $B(x^0, r_1) \subset B(x^0, r_2) \subseteq \Omega$, where $B(x^0, r_1)$ denote the ball with center at x^0 and radius of r_1. Put

$$\sigma_1 = \sigma_1(N, p, r_1, r_2) = \frac{12(N+2)^2(r_1 + r_2)}{(r_2 - r_1)^3} \left(\frac{K\pi^{\frac{N}{2}}(r_2^N - r_1^N)}{\Gamma\left(1 + \frac{N}{2}\right)} \right)^{\frac{1}{p}}, \qquad (5.55)$$

$$\sigma_2 = \sigma_2(N, q, r_1, r_2) = \frac{12(N+2)^2(r_1 + r_2)}{(r_2 - r_1)^3} \left(\frac{K\pi^{\frac{N}{2}}(r_2^N - r_1^N)}{\Gamma\left(1 + \frac{N}{2}\right)} \right)^{\frac{1}{q}}, \qquad (5.56)$$

$$\theta_1 = \begin{cases} \dfrac{3N}{(r_2 - r_1)(r_2 + r_1)} \left(\dfrac{K\pi^{\frac{N}{2}}((r_2 + r_1)^N - (2r_1)^N)}{2^N \Gamma\left(1 + \frac{N}{2}\right)} \right)^{\frac{1}{p}}, & N < \dfrac{4r_1}{r_2 - r_1}, \\[4mm] \dfrac{12r_1}{(r_2 - r_1)^2(r_2 + r_1)} \left(\dfrac{K\pi^{\frac{N}{2}}((r_2 + r_1)^N - (2r_1)^N)}{2^N \Gamma\left(1 + \frac{N}{2}\right)} \right)^{\frac{1}{p}}, & N \geq \dfrac{4r_1}{r_2 - r_1}, \end{cases}$$
$$(5.57)$$

$$
\theta_2 = \begin{cases} \dfrac{3N}{(r_2 - r_1)(r_2 + r_1)} \left(\dfrac{K\pi^{\frac{N}{2}}((r_2+r_1)^N - (2r_1)^N)}{2^N \Gamma\left(1+\frac{N}{2}\right)} \right)^{\frac{1}{q}}, & N < \frac{4r_1}{r_2 - r_1}, \\[3mm] \dfrac{12r_1}{(r_2 - r_1)^2(r_2 + r_1)} \left(\dfrac{K\pi^{\frac{N}{2}}((r_2+r_1)^N - (2r_1)^N)}{2^N \Gamma\left(1+\frac{N}{2}\right)} \right)^{\frac{1}{q}}, & N \geq \frac{4r_1}{r_2 - r_1}, \end{cases}
$$

$$(5.58)$$

where $\Gamma(\cdot)$ is the Gamma function. Our main results are the following theorems.

Theorem 5.15. *Suppose that $r_2 > r_1 > 0$, such that $B(x^0, r_2) \subset \Omega$ and assume that there exist four positive constants c, d, γ and β with $\gamma < p$, $\beta < q$, $\frac{d^p \theta_1^p}{p} + \frac{d^q \theta_2^q}{q} > \frac{c}{pq}$, and a function $\alpha \in L^1(\Omega)$ such that*

(D23) $F(x, s, t) \geq 0$ *for a.e.* $x \in \Omega \backslash B(x^0, r_1)$ *and all* $(s, t) \in [0, d] \times [0, d]$;

(D24) $(\frac{d^p \sigma_1^p}{p} + \frac{d^q \sigma_2^q}{q}) m(\Omega) \sup_{(x, s, t) \in \Omega \times A} F(x, s, t) < \frac{c}{pq} \int_{B(x^0, r_1)} F(x, d, d)$ dx, *where* $A = \{(s, t) | \frac{|s|^p}{p} + \frac{|t|^q}{q} \leq \frac{c}{pq}\}$;

(D25) $F(x, s, t) \leq \alpha(x)(1 + |s|^\gamma + |t|^\beta)$ *for a.e.* $x \in \Omega$ *and all* $(s, t) \in \mathbb{R} \times \mathbb{R}$.

Then there exist an open interval $\Lambda \subseteq [0, +\infty)$ and a positive real number ρ with the following property: for each $\lambda \in \Lambda$ and for two Carathéodory functions $G_u, G_v \colon \Omega \times \mathbb{R} \times \mathbb{R} \mapsto \mathbb{R}$, satisfying

(D26) $\sup_{\{|s| \leq \zeta, |t| \leq \zeta\}}(|G_u(\cdot, s, t)| + |G_v(\cdot, s, t)|) \in L^1(\Omega)$, *for all $\zeta > 0$,*

there exists $\delta > 0$ such that, for each $\mu \in [0, \delta]$, problem (5.52) has at least three solutions whose norms in X are less than ρ.

Remark 5.6. If F and G are independent of s, we reduce problem (5.52) to a biharmonic equation. Hence our Theorem 5.15 generalizes Theorem 1 in [67].

Now we give an example to illustrate the result of Theorem 5.15.

Example 5.1. Consider the problem

$$
\begin{cases} \Delta\left(|\Delta u|\Delta u\right) = \lambda(e^{-u}u^{17}(18 - u)) + \mu G_u(x, u, v), & \text{in } \Omega, \\ \Delta\left(|\Delta v|\Delta v\right) = \lambda(e^{-v}v^{11}(12 - u)) + \mu G_v(x, u, v), & \text{in } \Omega, \qquad (P') \\ u = \Delta u = v = \Delta v = 0, & \text{on } \partial\Omega, \end{cases}
$$

where $\Omega = \{(x, y) \in \mathbb{R} \times \mathbb{R} \colon x^2 + y^2 < 9\}$. Taking into account $K = \frac{4}{\pi}$, choosing $x^0 = (0, 0), r_1 = 1, r_2 = 2$ and $F_u(x, u, v) = e^{-u}u^{17}(18 - u)$, $F_v(x, u, v) = e^{-v}v^{11}(12 - u)$ for each $(u, v) \in \mathbb{R} \times \mathbb{R}$. So that $\sigma_1 = \sigma_2 = 576\sqrt[3]{3}$, $\theta_1 = \theta_2 = 2\sqrt[3]{5}$ and for every Carathéodory functions $G_u(x, u, v) \colon \Omega \times \mathbb{R} \times \mathbb{R} \mapsto \mathbb{R}$, $G_v(x, u, v) \colon \Omega \times \mathbb{R} \times \mathbb{R} \mapsto \mathbb{R}$ satisfying (j_4), all the assumptions of Theorem 5.15, with $p = q = 3$, are satisfied by choosing, for instance $c = \frac{1}{4}, d = 2, \gamma = 2$ and $\alpha(x)$ sufficient large. So there exist an open interval $\Lambda \in [0, +\infty]$ and a positive real number ρ such that, for each $\lambda \in \Lambda$, Problem (P') admits at least three solutions in $W^{2,3}(\Omega) \cap W_0^{1,3}(\Omega) \times W^{2,3}(\Omega) \cap W_0^{1,3}(\Omega)$ whose norms are less than ρ.

If $N = 1$, we can get a better result than Theorem 5.15. For simplicity, fixing $\Omega =]0, 1[$, $p > 1$, $q > 1$, we have the following result.

Theorem 5.16. *Let $F \colon \mathbb{R}^2 \mapsto \mathbb{R}$ be a C^1 function and assume that there exist five positive constants c, d, a, γ, β with $\frac{(32d)^p}{2Kp} + \frac{(32d)^q}{2Kq} > \frac{c}{pq}$, where K is given by (5.53), such that*

(D27) $F(s,t) \geq 0$ for all $(s,t) \in [0,d] \times [0,d]$;

(D28) $\{\frac{(32d)^p}{2Kp} + \frac{(32d)^q}{2Kq}\} \sup_{(s,t)\in A} F(s,t) < \frac{c}{2pq} F(d,d),$ where $A = \{(s,t) | \frac{|s|^p}{p} + \frac{|t|^q}{q} \leq \frac{c}{pq}\}$;

(D29) $F(s,t) \leq a(1 + |s|^\gamma + |t|^\beta)$ for all $(s,t) \in \mathbb{R} \times \mathbb{R}$.

Then there exist an open interval $\Lambda \subseteq [0, +\infty)$ and a positive real number ρ with the following property: for each $\lambda \in \Lambda$ and for two Carathéodory functions $G_u, G_v \colon \Omega \times \mathbb{R} \times \mathbb{R} \mapsto \mathbb{R}$, satisfying

(D30) $\sup_{\{|s|\leq\zeta,|t|\leq\zeta\}}(|G_u(\cdot,s,t)| + |G_v(\cdot,s,t)|) \in L^1(\Omega),$ for all $\zeta > 0$,

there exists $\delta > 0$ such that, for each $\mu \in [0, \delta]$, problem

$$\begin{cases} \left(|u''|^{p-2}u''\right)'' = \lambda F_u(u,v) + \mu G_u(x,u,v), & in\,]0,1[, \\ \left(|v''|^{q-2}v''\right)'' = \lambda F_v(u,v) + \mu G_v(x,u,v), & in\,]0,1[, \\ u(0) - u(1) = u''(0) - u''(1) = 0, \\ v(0) - v(1) = v''(0) - v''(1) = 0, \end{cases}$$

has at least three solutions whose norms in $W^{2,p}(0,1) \cap W_0^{1,p}(0,1) \times W^{2,q}(0,1) \cap W_0^{1,q}(0,1)$ are less than ρ.

Remark 5.7. Our Theorem 5.16 generalizes Theorem 2 in [67] as the same reason in Remark 1.

Before giving the proof of Theorem 5.15, let us see the following two lemmas.

Lemma 5.8. *Assume that there exist two positive constants c, d with $\frac{d^p \theta_1^p}{p} + \frac{d^q \theta_2^q}{q} > \frac{c}{pq}$, such that*

(D23) $F(x,s,t) \geq 0$ *for a.e.* $x \in \Omega \backslash B(x^0, r_1)$ *and all* $(s,t) \in [0,d] \times [0,d]$;

(D24) $(\frac{d^p \sigma_1^p}{p} + \frac{d^q \sigma_2^q}{q}) m(\Omega) \sup_{(x,s,t) \in \Omega \times A} F(x,s,t) < \frac{c}{pq} \int_{B(x^0, r_1)} F(x,d,d)$ dx, *where* $A = \{(s,t) | \frac{|s|^p}{p} + \frac{|t|^q}{q} \leq \frac{c}{pq}\}$.

Then there $u^* \in W^{2,p}(\Omega) \cap W_0^{1,p}(\Omega)$ *and* $v^* \in W^{2,q}(\Omega) \cap W_0^{1,q}(\Omega)$ *such that*

$$\frac{\|u^*\|_p^p}{p} + \frac{\|v^*\|_q^q}{q} > \frac{1}{pq} \frac{c}{K},$$

and

$$m(\Omega) \sup_{(x,s,t) \in \Omega \times A} F(x,s,t) < \frac{c}{K} \frac{\int_\Omega F(x, u^*(x), v^*(x)) dx}{q \|u^*\|_p^p + p \|v^*\|_q^q},$$

where $A = \{(s,t) | \frac{|s|^p}{p} + \frac{|t|^q}{q} \leq \frac{c}{pq}\}$.

Proof. Let

$$w(x) = \begin{cases} 0, & x \in \Omega \backslash B(x^0, r_2), \\ \frac{d(3(l^4 - r_2^4) - 4(r_1 + r_2)(l^3 - r_2^3) + 6r_1 r_2(l^2 - r_2^2))}{(r_2 - r_1)^3 (r_1 + r_2)}, & x \in B(x^0, r_2) \backslash B(x^0, r_1), \\ d, & x \in B(x^0, r_1), \end{cases}$$

(5.59)

where $u^*(x) = v^*(x) = w(x)$ *and* $l = \text{dist}(x, x^0) = \sqrt{\sum_{i=1}^N (x_i - x_i^0)^2}$.
We have

$$\frac{\partial u^*(x)}{\partial x_i} = \begin{cases} 0, & x \in \Omega \backslash B(x^0, r_2) \cup B(x^0, r_1), \\ \frac{12d(l^2(x_i - x_i^0) - (r_1 + r_2)l(x_i - x_i^0) + r_1 r_2(x_i - x_i^0))}{(r_2 - r_1)^3 (r_1 + r_2)}, & x \in B(x^0, r_2) \backslash B(x^0, r_1), \end{cases}$$

$$\frac{\partial^2 u^*(x)}{\partial^2 x_i} = \begin{cases} 0, & x \in \Omega \backslash B(x^0, r_2) \cup B(x^0, r_1), \\ \frac{12d(r_1 r_2 + (2l - r_1 - r_2)(x_i - x_i^0)^2 / l - (r_2 + r_1 - l)l)}{(r_2 - r_1)^3 (r_1 + r_2)}, & x \in B(x^0, r_2) \backslash B(x^0, r_1), \end{cases}$$

$$\sum_{i=1}^N \frac{\partial^2 u^*(x)}{\partial^2 x_i} = \begin{cases} 0, & x \in \Omega \backslash B(x^0, r_2) \cup B(x^0, r_1), \\ \frac{12d((N+2)l^2 - (N+1)(r_1 + r_2)l + Nr_1 r_2)}{(r_2 - r_1)^3 (r_1 + r_2)}, & x \in B(x^0, r_2) \backslash B(x^0, r_1). \end{cases}$$

It is easy to verify that $u^* \in W^{2,p}(\Omega) \cap W_0^{1,p}(\Omega)$, and in particular, one has

$$\|u^*\|_p^p = \frac{(12d)^p 2 \prod^{\frac{N}{2}}}{(r_2 - r_1)^{3p}(r_1 + r_2)^p \Gamma(\frac{N}{2})}$$

$$\times \int_{r_1}^{r_2} |(N+2)r^2 - (N+1)(r_1 + r_2)r + Nr_1 r_2|^p r^{N-1} dr. \quad (5.60)$$

Similarly we also have

$$\|v^*\|_q^q = \frac{(12d)^q 2 \prod^{\frac{N}{2}}}{(r_2 - r_1)^{3q}(r_1 + r_2)^q \Gamma(\frac{N}{2})}$$

$$\times \int_{r_1}^{r_2} |(N+2)r^2 - (N+1)(r_1 + r_2)r + Nr_1 r_2|^q r^{N-1} dr. \quad (5.61)$$

Here, we obtain from (5.55), (5.56), (5.57), (5.58), (5.60) and (5.61) that

$$\frac{\theta_1^p d^p}{K} < \|u^*\|_p^p < \frac{\sigma_1^p d^p}{K}, \qquad \frac{\theta_2^q d^q}{K} < \|v^*\|_q^q < \frac{\sigma_2^q d^q}{K}. \quad (5.62)$$

By the assumption

$$\frac{d^p \theta_1^p}{p} + \frac{d^q \theta_2^q}{q} > \frac{c}{pq},$$

it follows from (5.62) that

$$\frac{\|u^*\|_p^p}{p} + \frac{\|v^*\|_q^q}{q} > \frac{1}{K}\left(\frac{d^p \theta_1^p}{p} + \frac{d^q \theta_2^q}{q}\right) > \frac{1}{pq}\frac{c}{K}.$$

Since, $0 \le u^* \le d, 0 \le v^* \le d$ for each $x \in \Omega$, the condition (D23) ensures that

$$\int_{\Omega \setminus B(x^0, r_2)} F(x, u^*(x), v^*(x))dx + \int_{B(x^0, r_2)\setminus B(x^0, r_1)} F(x, u^*(x), v^*(x))dx \ge 0.$$

Hence, by condition (D24) and (5.62), we have

$$m(\Omega) \sup_{(x,s,t)\in\Omega\times A} F(x,s,t) < \frac{c}{pq\left(\frac{d^p \sigma_1^p}{p} + \frac{d^q \sigma_2^q}{q}\right)} \int_{B(x^0, r_1)} F(x, d, d)dx$$

$$< \frac{c}{Kpq} \frac{1}{\frac{\|u^*\|_p^p}{p} + \frac{\|v^*\|_q^q}{q}} \int_{B(x^0, r_1)} F(x, d, d)dx$$

$$\le \frac{c}{K} \frac{\int_\Omega F(x, u^*(x), v^*(x))dx}{q\|u^*\|_p^p + p\|v^*\|_q^q}.$$

$$\square$$

Lemma 5.9. *Let* $T: X \mapsto X^*$ *be the operator defined by*

$$\langle T(u,v),(\xi,\eta)\rangle = \int_\Omega |\Delta u(x)|^{p-2}\Delta u(x)\Delta\xi(x)dx + \int_\Omega |\Delta v(x)|^{q-2}\Delta v(x)\Delta\eta(x)dx$$

for all $(u,v),(\eta,\xi) \in X$, *where* X^* *denote the dual of* X. *Then* T *admits a continuous inverse on* X^*.

Proof. Let us denote by T_p the operator defined from $W^{2,p}(\Omega) \cap W_0^{1,p}(\Omega)$ into $(W^{2,p}(\Omega) \cap W_0^{1,p}(\Omega))^*$ by

$$\langle T_p(u), \xi \rangle = \int_\Omega |\Delta u(x)|^{p-2} \Delta u(x) \Delta \xi(x)\, dx \quad \forall u, \xi \in W^{2,p}(\Omega) \cap W_0^{1,p}(\Omega)$$

and T_q the corresponding one with p replaced by q.

Observe that

$$\langle T(u,v), (\xi, \eta) \rangle = \langle T_p(u), \xi \rangle + \langle T_q(v), \eta \rangle, \quad \forall (u,v), (\xi, \eta) \in X.$$

T_p, T_q are duality mappings on $W^{2,p}(\Omega) \cap W_0^{1,p}(\Omega)$ and $W^{2,q}(\Omega) \cap W_0^{1,q}(\Omega)$ corresponding to the gauge functions $\phi_p(t) = t^{p-1}$ and $\phi_q(t) = t^{q-1}$, respectively. Hence T_p and T_q are monotone and demicontinuous.

For $p \geq 2$ there exists a positive constant C_p such that the following inequality (see [97])

$$(|x|^{p-2}x - |y|^{p-2}y, x - y) \geq C_p |x - y|^p$$

holds for all $x, y \in \mathbb{R}^N$. Consequently, for every $(u_1, v_1), (u_2, v_2) \in X$, we have

$$\langle (T(u_1, v_1) - T(u_2, v_2)), (u_1 - u_2, v_1 - v_2) \rangle$$
$$= \langle (T_p(u_1) - T_p(u_2)), (u_1 - u_2) \rangle + \langle (T_q(v_1) - T_q(v_2)), (v_1 - v_2) \rangle$$
$$\geq C_p \|u_1 - u_2\|_p^p + C_q \|v_1 - v_2\|_q^q > 0.$$

So T is a strict monotone operator.

Moreover, for $(u_n, v_n) \to (u, v)$ in X, we have $T_p(u_n) \rightharpoonup T_p(u)$ in $(W^{2,p}(\Omega) \cap W_0^{1,p}(\Omega))^*$ and $T_q(v_n) \rightharpoonup T_q(v)$ in $(W^{2,q}(\Omega) \cap W_0^{1,q}(\Omega))^*$.

Since $W^{2,p}(\Omega) \cap W_0^{1,p}(\Omega)$ and $W^{2,q}(\Omega) \cap W_0^{1,q}(\Omega)$ are reflexive, we get $T(u_n, v_n) \rightharpoonup T(u, v)$ in X^*. Hence T is demicontinuous. On the other hand, T is coercive since $\langle T(u, v), (u, v) \rangle = \|u\|_p^p + \|v\|_q^q$.

Now, we show that T satisfies property:

(S₂) if $(u_n, v_n) \rightharpoonup (u, v)$ and $T(u_n, v_n) \to T(u, v)$, $(u_n, v_n) \to (u, v)$.

Let us take a sequence $(u_n, v_n) \in X$ such that $(u_n, v_n) \rightharpoonup (u, v)$ in X and $T(u_n, v_n) \to T(u, v)$ in X^*. Then

$$\langle T(u_n, v_n), (u_n, v_v) \rangle \to \langle T(u, v), (u, v) \rangle.$$

So

$$\|u_n\|_p^p + \|v_n\|_q^q \to \|u\|_p^p + \|v\|_q^q.$$

According to the uniform convexity of X, $(u_n, v_n) \to (u, v)$ in X.

Note that the strict monotonicity of T implies its injectivity. Moreover, T is a coercive operator and demicontinuous, so it's semicontinuous. Consequently, thanks to Minty–Browder theorem (see [299]), the operator T is an surjection and admits an inverse mapping.

It suffices then to show the continuity of T^{-1}. Let $((f_n)_n, (g_n)_n)$ be a sequence of X^* such that $f_n \to f$ in $(W^{2,p}(\Omega) \cap W_0^{1,p}(\Omega))^*$ and $g_n \to g$ in $(W^{2,q}(\Omega) \cap W_0^{1,q}(\Omega))^*$. Let (u_n, v_n) and (u, v) in X such that

$$T_p^{-1}(f_n) = u_n, T_q^{-1}(g_n) = v_n \quad \text{and} \quad T_p^{-1}(f) = u, T_q^{-1}(g) = v.$$

By the coercivity of T, one deducts that the sequence (u_n, v_n) is bounded in the reflexive space X. For a subsequence, we have $(u_n, v_n) \rightharpoonup (\widehat{u}, \widehat{v})$ in X, which implies

$$\lim_{n \to +\infty} \langle T(u_n, v_n) - T(u, v), (u_n, v_n) - (\widehat{u}, \widehat{v}) \rangle$$
$$= \lim_{n \to +\infty} \langle (f_n, g_n) - (f, g), (u_n, v_n) - (\widehat{u}, \widehat{v}) \rangle = 0.$$

It follows by the property of (S_2) and the continuity of T that

$$(u_n, v_n) \to (\widehat{u}, \widehat{v}) \quad \text{in } X \quad \text{and} \quad T(u_n, v_n) \to T(\widehat{u}, \widehat{v}) = T(u, v) \quad \text{in } X^*.$$

Moreover, since T is an injection, we conclude that $(u, v) = (\widehat{u}, \widehat{v})$. □

Now we can give the proof of our main results.

Proof of Theorem 5.15. For each $(u, v) \in X$, let

$$\Phi(u, v) = \frac{\|u\|_p^p}{p} + \frac{\|v\|_q^q}{q}, \qquad \Psi(u, v) = -\int_\Omega F(x, u, v) dx,$$

$$J(u, v) = -\int_\Omega G(x, u, v) dx.$$

Under the condition of Theorem 5.15, Φ is a continuously Gâteaux differentiable and sequentially weakly lower semicontinuous functional. Moreover, from Lemma 2 the Gâteaux derivative of Φ admits a continuous inverse on X^*. Ψ and J are continuously Gâteaux differential functional whose Gâteaux derivative is compact. Obviously, Φ is bounded on each bounded subset of X. In particular, for each $(u, v), (\xi, \eta) \in X$,

$$\langle \Phi'(u, v), (\xi, \eta) \rangle = \int_\Omega |\Delta u(x)|^{p-2} \Delta u(x) \Delta \xi(x) dx + \int_\Omega |\Delta v(x)|^{q-2} \Delta v(x) \Delta \eta(x) dx,$$

$$\langle \Psi'(u, v), (\xi, \eta) \rangle = -\int_\Omega F_u(x, u, v) \xi(x) dx - \int_\Omega F_v(x, u, v) \eta(x) dx,$$

$$\langle J'(u,v),(\xi,\eta)\rangle = -\int_\Omega G_u(x,u,v)\xi(x)dx - \int_\Omega G_v(x,u,v)\eta(x)dx.$$

Hence, it follows from (5.54) that the weak solutions of systems (5.52) are exactly the solutions of the equation

$$\Phi'(u,v) + \lambda\Psi'(u,v) + \mu J'(u,v) = 0.$$

Thanks to (D25), for each $\lambda > 0$, one has that

$$\lim_{\|(u,v)\|\to+\infty} (\Phi(u,v) + \lambda\Psi(u,v)) = +\infty, \tag{5.63}$$

and so the first assumption of Theorem 1.14 holds.

Thanks to Lemma 5.8, there exist $(u^*,v^*) \in X$ such that

$$\Phi(u^*,v^*) = \frac{\|u^*\|_p^p}{p} + \frac{\|v^*\|_q^q}{q} > \frac{1}{pq}\frac{c}{K} > 0 = \Phi(0,0) \tag{5.64}$$

and

$$m(\Omega)\sup_{(x,s,t)\in\Omega\times A} F(x,s,t) < \frac{c}{K}\frac{\int_\Omega F(x,u^*(x),v^*(x))dx}{q\|u^*\|_p^p + p\|v^*\|_q^q}, \tag{5.65}$$

where $A = \{(s,t)| \frac{|s|^p}{p} + \frac{|t|^q}{q} \le \frac{c}{pq}\}$.

Now, we obtain from (5.53) that

$$\sup_{x\in\Omega}|u(x)|^p \le K\|u\|_p^p, \qquad \sup_{x\in\Omega}|v(x)|^q \le K\|v\|_q^q$$

for each $(u,v) \in X$, then we have

$$\sup_{x\in\Omega}\left\{\frac{|u(x)|^p}{p} + \frac{|v(x)|^q}{q}\right\} \le K\left\{\frac{\|u\|_p^p}{p} + \frac{\|v\|_q^q}{q}\right\} \tag{5.66}$$

for each $(u,v) \in X$. Let $r = \frac{1}{pq}\frac{c}{K}$, for each $(u,v) \in X$ such that

$$\Phi(u,v) = \frac{\|u\|_p^p}{p} + \frac{\|v\|_q^q}{q} \le r,$$

by (5.66) one has

$$\sup_{x\in\Omega}\left\{\frac{|u(x)|^p}{p} + \frac{|v(x)|^q}{q}\right\} \le \frac{c}{pq}. \tag{5.67}$$

So, it follows from (5.67) and (5.65) that

$$\sup_{\{(u,v)|\Phi(u,v)\le r\}} (-\Psi(u,v)) = \sup_{\{(u,v)|\frac{\|u\|_p^p}{p}+\frac{\|v\|_q^q}{q}\le r\}} \int_\Omega F(x,u,v)dx$$

$$\le \int_\Omega \sup_{(s,t)\in A} F(x,s,t)dx$$

$$\le m(\Omega) \sup_{(x,s,t)\in\Omega\times A} F(x,s,t)$$

$$< \frac{c}{Kpq} \frac{\int_\Omega F(x,u^*(x),v^*(x))dx}{\frac{\|u^*\|_p^p}{p}+\frac{\|v^*\|_q^q}{q}}$$

$$= r\frac{\int_\Omega F(x,u^*(x),v^*(x))dx}{\frac{\|u^*\|_p^p}{p}+\frac{\|v^*\|_q^q}{q}}$$

$$= r\frac{-\Psi(u^*,v^*)}{\Phi(u^*,v^*)}.$$

So, one has

$$\sup_{\{(u,v)|\Phi(u,v)\le r\}} (-\Psi(u,v)) < r\frac{-\Psi(u^*,v^*)}{\Phi(u^*,v^*)}. \tag{5.68}$$

Fix h such that

$$\sup_{\{(u,v)|\Phi(u,v)\le r\}} (-\Psi(u,v)) < h < r\frac{-\Psi(u^*,v^*)}{\Phi(u^*,v^*)},$$

from (5.64), (5.68) and Proposition 1.14, with $(u_0,v_0) = (0,0)$ and $(u_1,v_1) = (u^*,v^*)$, we obtain

$$\sup_{\lambda\ge 0}\inf_{x\in X}(\Phi(x)+\lambda(h+\Psi(x))) < \inf_{x\in X}\sup_{\lambda\ge 0}(\Phi(x)+\lambda(h+\Psi(x))), \tag{5.69}$$

and so the assumption of Theorem 1.14 holds.

Now, set $I = [0,+\infty)$, by (5.63) and (5.69), all the assumptions of Theorem 1.14 are satisfied. Hence, our conclusion follows from Theorem 1.14. □

Proof of Theorem 5.16. For each $(u,v) \in W^{2,p}(0,1) \cap W_0^{1,p}(0,1) \times W^{2,q}(0,1) \cap W_0^{1,q}(0,1)$, let

$$\Phi(u,v) = \frac{\|u\|_p^p}{p} + \frac{\|v\|_q^q}{q}, \qquad \Psi(u,v) = -\int_0^1 F(u,v)dx,$$

$$J(u,v) = -\int_\Omega G(x,u,v)dx.$$

Under the condition, Φ is a continuously Gâteaux differentiable and sequentially weakly lower semicontinuous functional. Moreover, from Lemma 2 the Gâteaux derivative of Φ admits a continuous inverse on X^*. Ψ and J are continuously Gâteaux differential functional whose Gâteaux derivative is compact. Obviously, Φ is bounded on each bounded subset of $W^{2,p}(0,1) \cap W_0^{1,p}(0,1) \times W^{2,q}(0,1) \cap W_0^{1,q}(0,1)$. Hence, it is well known that the weak solutions of systems are exactly the solutions of the equation

$$\Phi'(u,v) + \lambda\Psi'(u,v) + \mu J'(u,v) = 0.$$

Thanks to (D29), for each $\lambda > 0$, one has that

$$\lim_{\|(u,v)\| \to +\infty} (\Phi(u,v) + \lambda\Psi(u,v)) = +\infty, \tag{5.70}$$

and so the first assumption of Theorem holds.

Now, let

$$r = \frac{1}{pq}\frac{c}{K}.$$

We obtain from (5.53) that

$$\sup_{x \in (0,1)} |u(x)|^p \leq K\|u\|_p^p, \qquad \sup_{x \in (0,1)} |v(x)|^q \leq K\|v\|_q^q$$

for each $(u,v) \in W^{2,p}(0,1) \cap W_0^{1,p}(0,1) \times W^{2,q}(0,1) \cap W_0^{1,q}(0,1)$, then we have

$$\sup_{x \in (0,1)} \left\{ \frac{|u(x)|^p}{p} + \frac{|v(x)|^q}{q} \right\} \leq K \left\{ \frac{\|u\|_p^p}{p} + \frac{\|v\|_q^q}{q} \right\} \tag{5.71}$$

for each $(u,v) \in W^{2,p}(0,1) \cap W_0^{1,p}(0,1) \times W^{2,q}(0,1) \cap W_0^{1,q}(0,1)$. Hence, for each $(u,v) \in W^{2,p}(0,1) \cap W_0^{1,p}(0,1) \times W^{2,q}(0,1) \cap W_0^{1,q}(0,1)$ such that

$$\Phi(u,v) = \frac{\|u\|_p^p}{p} + \frac{\|v\|_q^q}{q} \leq r,$$

by (5.71) one has

$$\sup_{x \in (0,1)} \left\{ \frac{|u(x)|^p}{p} + \frac{|v(x)|^q}{q} \right\} \leq \frac{c}{pq}. \tag{5.72}$$

Now if we put $u^*(x) = v^*(x) = w(x)$, where

$$w(x) = \begin{cases} d - 16d(\frac{1}{4} - |x - \frac{1}{2}|)^2, & x \in [0, \frac{1}{4}] \cup]\frac{3}{4}, 1], \\ d, & x \in]\frac{1}{4}, \frac{3}{4}], \end{cases} \tag{5.73}$$

it is easy to verify that $(u^*, v^*) \in W^{2,p}(0,1) \cap W_0^{1,p}(0,1) \times W^{2,q}(0,1) \cap W_0^{1,q}(0,1)$ and get

$$\|u^*\|_p^p = \frac{(32d)^p}{2}, \qquad\qquad \|v^*\|_q^q = \frac{(32d)^q}{2} \qquad (5.74)$$

Now, under the assumption of $\frac{(32d)^p}{2Kp} + \frac{(32d)^q}{2Kq} > \frac{c}{pq}$, we have

$$\Phi(u^*, v^*) = \frac{\|u^*\|_p^p}{p} + \frac{\|v^*\|_q^q}{q} > r > 0 = \Phi(0,0). \qquad (5.75)$$

Moreover, $0 \le u^* \le d, 0 \le v^* \le d$, it follows from (D27), (D28) and (5.74) that

$$\sup_{(s,t)\in A} F(s,t) < \frac{c}{2pq(\frac{(32d)^p}{2Kp} + \frac{(32d)^q}{2Kq})} F(d,d)$$

$$< \frac{c}{Kpq} \frac{1}{\frac{\|u^*\|_p^p}{p} + \frac{\|v^*\|_q^q}{q}} \int_0^1 F(u^*(x), v^*(x))dx \qquad (5.76)$$

$$\le \frac{c}{K} \frac{\int_0^1 F(u^*(x), v^*(x))dx}{q\|u^*\|_p^p + p\|v^*\|_q^q},$$

where $A = \{(s,t) | \frac{|s|^p}{p} + \frac{|t|^q}{q} \le \frac{c}{pq}\}$.

So, it follows from (5.72) and (5.76) that

$$\sup_{\{(u,v)|\Phi(u,v)\le r\}} (-\Psi(u,v)) = \sup_{\{(u,v)|\frac{\|u\|_p^p}{p} + \frac{\|v\|_q^q}{q} \le r\}} \int_0^1 F(u,v)dx$$

$$\le \int_0^1 \sup_{(s,t)\in A} F(s,t)dx$$

$$\le \sup_{(s,t)\in A} F(s,t)$$

$$< \frac{c}{Kpq} \frac{\int_0^1 F(u^*(x), v^*(x))dx}{\frac{\|u^*\|_p^p}{p} + \frac{\|v^*\|_q^q}{q}} \qquad (5.77)$$

$$= r \frac{\int_0^1 F(u^*(x), v^*(x))dx}{\frac{\|u^*\|_p^p}{p} + \frac{\|v^*\|_q^q}{q}}$$

$$= r \frac{-\Psi(u^*, v^*)}{\Phi(u^*, v^*)}.$$

Fix h such that

$$\sup_{\{(u,v)|\Phi(u,v)\le r\}} (-\Psi(u,v)) < h < r \frac{-\Psi(u^*, v^*)}{\Phi(u^*, v^*)},$$

from (5.75), (5.77) and Proposition 1.14, with $(u_0, v_0) = (0,0)$ and $(u_1, v_1) = (u^*, v^*)$, we obtain

$$\sup_{\lambda \geq 0} \inf_{x \in X} (\Phi(x) + \lambda(h + \Psi(x))) < \inf_{x \in X} \sup_{\lambda \geq 0} (\Phi(x) + \lambda(h + \Psi(x))), \quad (5.78)$$

and so the assumption of Theorem 1.14 holds.

Now, set $I = [0, +\infty)$, by (5.70) and (5.78), all the assumptions of Theorem 1.14 are satisfied. Hence, our conclusion follows from Theorem 1.14. $\square$

Notes and Comments

Recently, some authors consider the elliptic systems. In particular, in [376], Boccardo and de Figueiredo study critical points of functionals corresponding to solutions of quasilinear elliptic systems involving the p-Laplacian. They consider subcritical or mixed subcritical growth conditions. Under suitable conditions on the nonlinearities, they show the existence of non-trivial solutions of the systems according to various cases (sublinear, superlinear and resonant case). The p-Laplacian with weighted is study by Miyagaki and Rodrigues in [377]. Using the three critical points theorem, Kristály [378] get two non-trivial solutions for a class of quasilinear elliptic systems on strip-like domains. In [379], Cammaroto, Chinnì and Di Bella also consider this problem. In [380], Li and Tang get three solutions for a class of quasilinear systems involving the (p, q)-Laplacian with Dirichlet boundary condition. Afrouzi and Heidarkhani [381] unify and generalize Li and Tang's problem. In [79], El Manouni and Kbiri Alaoui consider the (p, q)-Laplacian systems with Neumann conditions using Ricceri's three three critical points theorem. This section was obtained in [85].

Chapter 6

Variable Exponent Problems

6.1 Introduction

Differential equations with non-standard growth and corresponding function spaces with variable exponents have been a very active field of investigation in recent years. The theory of variable exponent Lebesgue and Sobolev spaces has been surveyed in [382]; see also the monograph [383]. Fan [384] summarized some results of his research group on the existence and multiplicity of solutions of eigenvalue problems. Variable exponent Lebesgue and Sobolev spaces are natural extension of classical constant exponent L^p-spaces. Such kind of theory finds many applications for example in nonlinear elastic mechanics [385], electrorheological fluids [386] or image restoration [387]. During the last decade Lebesgue and Sobolev spaces with variable exponent have been intensively studied. In particular, the Sobolev inequalities have been shown for variable exponent spaces on Euclidean spaces (see [8, 9]).

6.2 $p(x)$-Laplacian Problems

In this section, we consider the Neumann problems involving the $p(x)$-Laplacian operator

$$\begin{cases} -\text{div}\left(|\nabla u|^{p(x)-2}\nabla u\right) + a(x)|u|^{p(x)-2}u = \lambda f(x, u) + \mu g(x, u), & \text{in } \Omega, \\ \dfrac{\partial u}{\partial \nu} = 0, & \text{on } \partial\Omega, \end{cases}$$
$$\tag{6.1}$$

where $\Omega \subset \mathbb{R}^N$ ($N \geq 3$) is a bounded domain with smooth boundary, $\lambda, \mu > 0$ are real numbers, $p(x)$ is a continuous function on $\overline{\Omega}$ with $\inf_{x \in \overline{\Omega}} p(x) > N$ and $a \in L^\infty(\Omega)$ with $\text{essinf}_{x \in \Omega} a(x) = a_0 > 0$. We denote by ν the outward unit normal to $\partial\Omega$. The main interest in studying such problems arises from the presence of the $p(x)$-Laplacian operator div $\left(|\nabla u|^{p(x)-2}\nabla u\right)$, which is a generalization of the classical p-Laplacian operator div $\left(|\nabla u|^{p-2}\nabla u\right)$ obtained in the case when p is a positive constant.

When $\mu = 0$, in [388], Mashiyev studied the particular case

$$f(t) = b|t|^{q-2}t - d|t|^{s-2}t$$

where b and d are positive constants, $2 < s < q < \inf_{x\in\overline{\Omega}}p(x)$ and $N < \inf_{x\in\overline{\Omega}}p(x)$; and

$$f(x,t) = |t|^{q(x)-2}t - |t|^{s(x)-2}t$$

where $2 < \inf_{x\in\overline{\Omega}}s(x) \leq \sup_{x\in\overline{\Omega}}s(x) < \inf_{x\in\overline{\Omega}}q(x) \leq \sup_{x\in\overline{\Omega}}q(x) < \inf_{x\in\overline{\Omega}}p(x)$ and $N < \inf_{x\in\overline{\Omega}}p(x)$ for all $x \in \overline{\Omega}$. He established the existence of at least three weak solutions by using the Ricceri's variational principle.

In this section, we assume $f(x,u)$ and $g(x,u)$ satisfy the following general conditions:

(E1) $f, g : \Omega \times \mathbb{R} \to \mathbb{R}$ are Carathéodory function and satisfies

$$|f(x,t)| \leq c_1 + c_2|t|^{\alpha(x)-1}, \qquad \forall (x,t) \in \Omega \times \mathbb{R},$$

$$|g(x,t)| \leq c_1' + c_2'|t|^{\beta(x)-1}, \qquad \forall (x,t) \in \Omega \times \mathbb{R},$$

where $\alpha(x), \beta(x) \in C(\overline{\Omega})$, $\alpha(x), \beta(x) > 1$ and $1 < \alpha^+ = \max_{x\in\overline{\Omega}}\alpha(x) < p^- = \min_{x\in\overline{\Omega}}p(x)$, $1 < \beta^+ = \max_{x\in\overline{\Omega}}\beta(x) < p^- = \min_{x\in\overline{\Omega}}p(x)$ and c_1, c_2, c_1', c_2' are positive constants.

(E2) There exist a constant t_0 and the following conditions

$$f(x,t) < 0 \qquad \text{when} |t| \in (0,t_0)$$
$$f(x,t) > M > 0 \qquad \text{when} |t| \in (t_0, +\infty),$$

where M is a positive constant, is satisfied.

Following along the same lines as in [388], we will prove that there also exist three weak solutions for such a general problem for λ sufficiently large and requiring μ small enough.

Next, the weighted-variable-exponent Sobolev space $W_a^{1,p(x)}(\Omega)$ is defined by

$$W_a^{1,p(x)}(\Omega) = \{u \in L_a^{p(x)}(\Omega) : |\nabla u| \in L_a^{p(x)}(\Omega)\},$$

with the norm

$$\|u\|_a = \inf\left\{\beta > 0 : \int_\Omega \left(\left|\frac{\nabla u(x)}{\beta}\right|^{p(x)} + a(x)\left|\frac{u(x)}{\beta}\right|^{p(x)}\right) dx \leq 1\right\},$$

for all $u \in W_a^{1,p(x)}(\Omega)$. Then the norms $\| \cdot \|_a$ and $\| \cdot \|$ in $W_a^{1,p(x)}(\Omega)$ are equivalent. If $1 < p^- \leq p^+ < \infty$, then the space $W_a^{1,p(x)}(\Omega)$ is a separable and reflexive Banach space.

Now, we will prove that for problem (6.1) there also exist three weak solutions for the general case.

Definition 6.1. We say $u \in W_a^{1,p(x)}$ is a weak solution of problem (6.1) if

$$\int_\Omega (|\nabla u|^{p(x)-2}\nabla u \nabla v + a(x)|u|^{p(x)-2}u)dx - \lambda \int_\Omega f(x,u)vdx$$

$$- \mu \int_\Omega g(x,u)vdx = 0$$

for any $v \in W_a^{1,p(x)}$

Theorem 6.1. *Assume that $p^- > N$ and $f(x,u)$ satisfies* (E1), (E2). *Then there exist an open interval $\Lambda \in (0,\infty)$ and a positive real number $q > 0$ such that each $\lambda \in \Lambda$ and every function $g : \Omega \times \mathbb{R} \to \mathbb{R}$ which is satisfied* (E1), *there exists $\delta > 0$ such that for each $\mu \in [0,\delta]$ problem (6.1) has at least three solutions whose norms are less than q.*

Proof. Let X denote the weighted variable exponent Lebesgue space $W_a^{1,p(x)}(\Omega)$. Define

$$F(x,t) = \int_0^t f(x,s)ds \text{ and } G(x,t) = \int_0^t g(x,s)ds.$$

In order to use Theorem 1.15, we define the functions $\Phi, \Psi, J : X \to \mathbb{R}$ by

$$\Phi(u) = \int_\Omega \frac{1}{p(x)}(|\nabla u|^{p(x)} + a(x)u^{p(x)})dx$$

$$\Psi(u) = -\int_\Omega F(x,u)dx$$

$$J(u) = -\int_\Omega G(x,u)dx.$$

Arguments similar to those used in the proof of Proposition 3.1 in [389], we know $\Phi, \Psi, J \in C^1(X,\mathbb{R})$ with the derivatives given by

$$\langle \Phi'(u), v \rangle = \int_\Omega (|\nabla u|^{p(x)-2}\nabla u \nabla v + a(x)u^{p(x)-2}uv)dx$$

$$\langle \Psi'(u), v \rangle = -\int_\Omega f(x,u)vdx$$

$$\langle J'(u), v \rangle = -\int_\Omega g(x,u)vdx$$

for any $u, v \in X$. Thus, there exists $\lambda, \mu > 0$ such that u is a critical point of the operator $\Phi(u) + \lambda\Psi(u) + \mu J(u)$, that is $\Phi'(u) + \lambda\Psi'(u) + \mu J'(u) = 0$. For proving our result, it is enough to verify that Φ, Ψ and J satisfy the hypotheses of Theorem 1.15.

It is obvious that $(\Phi')^{-1} : X^* \to X$ exists and continuous, because $\Phi' : X \to X^*$ is a homeomorphism by Lemma 2.2 in [388]. Moreover, $\Psi', J' : X \to X^*$ are completely continuous because of the assumption (E1) and [6], which imply Ψ' and J' are compact.

Next, we will verify that condition(i) of Theorem 1.15 is fulfilled. In fact, by Proposition 1.7, we have

$$\Phi(u) \geq \frac{1}{p^+} \int_\Omega (|\nabla u|^{p(x)} + a(x)|u|^{p(x)})dx = \frac{1}{p^+}\rho(u) \geq \frac{1}{p^+}\|u\|_a^{p^-},$$

$u \in X, \|u\|_a > 1$. On the other hand, due to the assumption (E1), we have

$$\Psi(u) = - \int_\Omega F(x, u)dx = \int_\Omega -F(x, u)dx$$

and

$$|F(x, t)| \leq c_1|t| + c_2\frac{1}{\alpha(x)}|t|^{\alpha(x)}.$$

Therefore,

$$\Psi(u) \geq -c_1 \int_\Omega |u|dx - c_2 \int_\Omega \frac{1}{\alpha(x)}|u|^{\alpha(x)}dx$$

$$\geq -c_3\|u\|_a - \frac{c_2}{\alpha^+} \int_\Omega (|u|^{\alpha^+} + |u|^{\alpha^-})dx$$

$$= -c_3\|u\|_a - c_4(|u|_{\alpha^+}^{\alpha^+} + |u|_{\alpha^-}^{\alpha^-}).$$

Using Remark 1.1, we know that X is continuously embedded in L^{α^+} and L^{α^-}. Furthermore, we can find two positive constants $d_1, d_2 > 0$ such that

$$|u|_{\alpha^+} \leq d_1\|u\|_a \text{ and } |u|_{\alpha^-} \leq d_2\|u\|_a \quad \forall u \in X.$$

Moreover

$$\Psi(u) \geq -c_3\|u\|_a - c_4d_1\|u\|_a^{\alpha^+} - c_4d_2\|u\|_a^{\alpha^-}.$$

It follows that

$$\Phi(u) + \lambda\Psi(u) \geq \left(\frac{1}{p^+} - \lambda c_3\right)\|u\|_a^{p^-} - \lambda c_4(d_1\|u\|_a^{\alpha^+} + d_2\|u\|_a^{\alpha^-}), \forall u \in X.$$

Since $1 < \alpha^+ < p^-$, then $\lim_{\|u\|_a \to \infty} \Phi(u) + \lambda\Psi(u) = \infty$ and (i) is verified.

In the following, we will verify the conditions (ii) and (iii) in Theorem 1.15. By $F_t'(x, t) = f(x, t)$ and assumption (E2), it follows that $F(x, t)$

is increasing for $t \in (t_0, \infty)$ and decreasing for $t \in (0, t_0)$, uniformly with respect to x. Obviously, $F(x, 0) = 0$. $F(x, t) \to \infty$ when $t \to \infty$, because of assumption (E2). Then there exists a real number $\delta > t_0$ such that

$$F(x, t) \geq 0 = F(x, 0) \geq F(x, \tau), \quad \forall x \in X, t > \delta, \tau \in (0, t_0).$$

Let a, b be two real numbers such that $0 < a < \min\{t_0, k\}$ with k given in Remark 1 and $b > \delta$ satisfies

$$b^{p^-} \|a\|_{L^1(\Omega)} > 1$$

and

$$b^{p^+} \|a\|_{L^1(\Omega)} > 1.$$

Let $b > 1$. When $t \in [0, a]$, we have $F(x, t) \leq F(x, 0)$, it follows that

$$\int_\Omega \sup_{0 \leq t \leq a} F(x, t) dx \leq \int_\Omega F(x, 0) dx = 0.$$

Furthermore, we can get $\int_\Omega F(x, b) dx > 0$ because of $b > \delta$. Moreover,

$$\frac{1}{k^{p^+}} \frac{a^{p^+}}{b^{p^-}} \int_\Omega F(x, b) dx > 0.$$

The above two inequalities imply

$$\int_\Omega \sup_{0 \leq t \leq a} F(x, t) dx \leq 0 < \frac{1}{k^{p^+}} \frac{a^{p^+}}{b^{p^-}} \int_\Omega F(x, b) dx.$$

Consider u_0, $u_1 \in X$ with $u_0(x) = 0$ and $u_1(x) = b$ for any $x \in \Omega$. We define $r = \frac{1}{p^+} \left(\frac{a}{k}\right)^{p^+}$. Clearly, $r \in (0, 1)$. A simple computation implies

$$\Phi(u_0) = \Psi(u_0) = 0$$

and

$$\Phi(u_1) = \int_\Omega \frac{1}{p(x)} a(x) b^{p(x)} dx \geq \frac{1}{p^+} b^{p^-} \|a\|_{L^1(\Omega)} > \frac{1}{p^+} > \frac{1}{p^+} \left(\frac{a}{k}\right)^{p^+}$$

$$\Psi(u_1) = -\int_\Omega F(x, u_1(x)) dx = -\int_\Omega F(x, b) dx.$$

Similarly for $b < 1$, by help of Proposition 1.7, we get the desired result.

Thus, we obtain

$$\Phi(u_0) < r < \Phi(u_1)$$

and (ii) in Theorem 1.15 is verified.

On the other hand, we have

$$-\frac{(\Phi(u_1) - r)\Psi(u_0) + (r - \Phi(u_0))\Psi(u_1)}{\Phi(u_1) - \Phi(u_0)} = -r\frac{\Psi(u_1)}{\Phi(u_1)}$$

$$= r\frac{\int_\Omega F(x,b)dx}{\int_\Omega \frac{1}{p(x)}a(x)b^{p(x)}dx} > 0.$$

Next, we consider the case $u \in X$ with $\Phi(u) \leq r < 1$. Since $\frac{1}{p(x)}\rho(u) \leq \Phi(u) \leq r$, we obtain $\rho(u) \leq p^+ r = \left(\frac{a}{k}\right)^{p^+} < 1$, it follows that $\|u\|_a < 1$. Furthermore, it is clear that

$$\frac{1}{p^+}\|u\|_a^{p^+} \leq \frac{1}{p^+}\rho(u) \leq \Phi(u) \leq r.$$

Thus, using Remark 1.1, we have

$$|u(x)| \leq k\|u\|_a \leq k(p^+r)^{\frac{1}{p^+}} = a, \quad \forall x \in \Omega, u \in X, \Phi(u) \leq r.$$

The above inequality shows that

$$-\inf_{u\in\Phi^{-1}([-\infty,r])}\Psi(u) = \sup_{u\in\Phi^{-1}([-\infty,r])}-\Psi(u) \leq \int_\Omega \sup_{0\leq t\leq a} F(x,t)dx \leq 0.$$

It follows that

$$-\inf_{u\in\Phi^{-1}([-\infty,r])}\Psi(u) < r\frac{\int_\Omega F(x,b)dx}{\int_\Omega \frac{1}{p(x)}a(x)b^{p(x)}dx}.$$

That is

$$\inf_{u\in\Phi^{-1}([-\infty,r])}\Psi(u) > \frac{(\Phi(u_1) - r)\Psi(u_0) + (r - \Phi(u_0))\Psi(u_1)}{\Phi(u_1) - \Phi(u_0)}$$

which means that condition (iii) in Theorem 1.15 is verified. Then the proof of Theorem 6.1 is achieved. □

Remark 6.1. Applying ([32], Theorem 2.1) in the proof of Theorem 6.1, an upper bound of the interval of parameters λ for which (6.1) has at least three weak solutions is obtained when $\mu = 0$. To be precise, in the conclusion of Theorem 6.1 one has

$$\Lambda \subseteq \left]0, h\frac{\int_\Omega \frac{1}{p(x)}a(x)b^{p(x)}dx}{\int_\Omega F(x,b)dx}\right[$$

for each $h > 1$ and b as in the proof of Theorem 6.1.

Notes and Comments

Let us point out that when $p(x) = p =$ constant, there is a large literature which deal with problems involving the p-Laplacian with Dirichlet boundary conditions both in bounded or unbounded domains, which we do not need to cite here since the reader may easily find such papers.

Note that many papers deal with problems related to the p-Laplacian with Neumann conditions in the scalar case. We can cite, among others, the articles [390, 391] and refer to the references therein for details. The case of $p(x)$-Laplacian with Neumann conditions had been studied by Dai [392], Mihailescu [393] and Liu [394]. This section was obtained in [395].

6.3 $p(x)$-Laplacian-Like Problems

The aim of this section is to discuss the existence and multiplicity of solutions of the following $p(x)$-Laplacian-like equation in $\mathbb{R}^N$:

$$-\mathbf{div}\left(\left(1 + \frac{|\nabla u|^{p(x)}}{\sqrt{1 + |\nabla u|^{2p(x)}}}\right)|\nabla u|^{p(x)-2}\nabla u\right) + |u|^{p(x)-2}u = K(x)f(u),$$

$$u \in W^{1,p(x)}(\mathbb{R}^N),$$

(6.2)

where $p(x) = p(|x|) \in C((\mathbb{R}^N))$ with $2 \leq N < p^- := \inf_{\mathbb{R}^N} p(x) \leq p^+ := \sup_{\mathbb{R}^N} p(x) < +\infty$, $K : \mathbb{R}^N \to \mathbb{R}$ is a measurable function and $f \in C(\mathbb{R}, \mathbb{R})$.

Recently, the following equation also has been studied very well

$$-\Delta_{p(x)}u + |u|^{p(x)-2}u = f(x, u), \quad \text{in } \mathbb{R}^N,$$

$$u \in W^{1,p(x)}(\mathbb{R}^N).$$

(6.3)

When $p(x) = p(|x|) \in C(\mathbb{R}^N)$ with $2 \leq N < p^- \leq p^+ < +\infty$, the authors in [396] proved the existence of infinitely many distinct homoclinic radially symmetric solutions for (6.3), under adequate hypotheses about the nonlinearity at zero (and at infinity). For $p(x)$-Laplacian-like operator, Rodrigues [397] established the existence of nontrivial solutions for problem (6.2) on bounded area under the case of superlinear, by assuming the following key condition:

(AR) there exist $\theta > p^+$ and $M > 0$ such that

$$0 < \theta F(t) := \theta \int_0^t f(s)ds \leq f(t)t, \quad \forall |t| \geq M.$$

This condition is originally due to Ambrosetti and Rabinowitz [172] in the case $p(x) \equiv 2$. Actually, condition (AR) is quite natural and important not only to ensure that the Euler–Lagrange functional associated to problem (6.3) has a mountain pass geometry, but also to guarantee that Palais–Smale sequence of the Euler–Lagrange functional is bounded. But this condition is very restrictive eliminating many nonlinearities. In this section, we introduce a new condition (E3) (motivated by [398]), below, which is different from the Ambrosetti–Rabinowitz condition (AR).

(E3) there exist a constant $M \geq 0$ and a decreasing function τ in the space $C(\mathbb{R} \setminus (-M, M), \mathbb{R})$, such that

$$0 < (p^+ + \tau(t))F(t) := (p^+ + \tau(t)) \int_0^t f(s)ds \leq f(t)t, \quad |t| \geq M,$$

where $\tau(t) > 0$, $\lim_{|t| \to +\infty} |t| \tau(t) = +\infty$ and $\lim_{|t| \to +\infty} \int_M^{|t|} \frac{\tau(s)}{s} ds = +\infty$.

Remark 6.2. Obviously, when $\inf_{|t| \geq M} \tau(t) > 0$, condition (E3) and (AR) are equivalent. However, condition (E3) is weaker than (AR) when $\inf_{|t| \geq M} \tau(t) = 0$. For example, let $|t| \geq M = 2$, and assume that $F(t) = |t|^{p^+} \ln|t|$. Then $f(t) = (p^+ + \tau(t))\mathrm{sgn}(t)|t|^{p^+ - 1}\ln|t|$ satisfies condition (E3) not (AR), where $\tau(t) = \frac{1}{\ln t} \in C(\mathbb{R} \setminus (-M, M), \mathbb{R})$.

Remark 6.3. Condition (E3) was introduced in [398] to study p-Laplacian equation in $\mathbb{R}^N$. We can see that this new condition (E3) can also study $p(x)$-Laplacian-like equation and another situation with $p^- > N$ when compared with the reference [397].

The aim of this section is twofold. First, we want to handle the case $p^- > N$ and the area is the whole space. Although important problems can be treated within this framework, only a few works are available in this direction, see [396]. The main difficulty in studying problem (6.2) lies in the fact that no compact embedding is available for $W^{1,p(x)}(\mathbb{R}^N) \hookrightarrow L^\infty(\mathbb{R}^N)$. However, the subspace of radially symmetric functions of $W^{1,p(x)}(\mathbb{R}^N)$, denoted further by $W_r^{1,p(x)}(\mathbb{R}^N)$, can be embedded compactly into $L^\infty(\mathbb{R}^N)$ whenever $N < p^- \leq p^+ < +\infty$ (cf. [396, Theorem 2.1]). Second, instead of some usual assumption on the nonlinear term f, we assume that it satisfies a modified Ambrosetti–Rabinowitz condition (E3).

To state our results, we first introduce the following assumptions:

(E4) $K \in L^1(\mathbb{R}^N) \cap L^\infty(\mathbb{R}^N)$ is radial, $K(x) \geq 0$ for any $x \in \mathbb{R}^N$ and $\sup_{d>0} \operatorname{ess\,inf}_{|x| \leq d} K(x) > 0$.

(E5) $f(t) = o(t^{p^+ - 1})$ for t near 0.

Now, we are ready to state the main result of this section.

Theorem 6.2. *Suppose that* (E4), (E5), (E3) *hold. Then problem* (6.2) *has a nontrivial radially symmetric solution. Furthermore, if $f(t) = f(-t)$, then problem* (6.2) *has infinitely many pairs of radially symmetric solutions.*

In this section we prove Theorem 6.2 when $\inf_{|t| \geq M} \tau(t) = 0$. If $\inf_{|t| \geq M} \tau(t) > 0$, then conditions (AR) and (E3) are equivalent, and the proof is rather standard. We may assume that $M \geq 1$, and that there is constant $N_0 > 0$ such that

$$|\tau(t)| \leq N_0 \qquad (6.4)$$

for all $t \in \mathbb{R} \setminus (-M, M)$.

We introduce the energy functional φ associated to problem (6.2) defined by

$$\varphi(u) = \int_{\mathbb{R}^N} \frac{1}{p(x)} (|\nabla u(x)|^{p(x)} + \sqrt{1 + |\nabla u(x)|^{2p(x)}} + |u(x)|^{p(x)}) dx$$
$$- \int_{\mathbb{R}^N} K(x) F(u) dx \quad u \in W_r^{1,p(x)}(\mathbb{R}^N).$$

Due to the principle of symmetric criticality of Palais (see [308]), the critical points of $\varphi|_{W_r^{1,p(x)}(\mathbb{R}^N)}$ are critical points of φ as well, so radially symmetric weak solutions of problem (6.2).

Claim 6.1. *Let $W = \{w \in W_r^{1,p(x)}(\mathbb{R}^N) : \|w\| = 1\}$. Then, for any $w \in W$, there exist $\delta_w > 0$ and $\lambda_w > 0$, such that*

$$\varphi(\lambda v) < 0, \quad \forall v \in W \cap B(w, \delta_w), \forall |\lambda| \geq \lambda_w,$$

where $B(w, \delta_w) = \{v \in W_r^{1,p(x)}(\mathbb{R}^N) : \|v - w\| < \delta_w\}$.

Proof. Since the embedding $W_r^{1,p(x)}(\mathbb{R}^N) \hookrightarrow L^\infty(\mathbb{R}^N)$ is compact, there is constant $C > 0$ such that $|u|_\infty \leq C\|u\|$. Thus, for all $w \in W$ and a.e. $x \in \mathbb{R}^N$, we have $|w(x)| \leq C$. By the definition of $\tau(t)$ and decreasing property of $\tau(t)$, we deduce that there exists $t_\lambda \in \{t \in \mathbb{R} : M \leq |t| \leq |\lambda|C\}$ such that $\tau(t_\lambda) = \min_{M \leq |t| \leq |\lambda|C} \tau(t)$. Then $|\lambda| \geq \frac{t_\lambda}{C}$ and $\lim_{|\lambda| \to +\infty} |t_\lambda| \to +\infty$. From condition (E3), we conclude that $F(t) \geq C_1 |t|^{p^+} H(|t|)$ for all $|t| \geq M$, where $H(t) = \exp(\int_M^{|t|} \frac{\tau(s)}{s} ds)$. Hence, using $\lim_{|t| \to +\infty} \int_M^{|t|} \frac{\tau(s)}{s} ds = +\infty$, it follows that $H(|t|)$ increases when $|t|$ increases, and $\lim_{|t|\to+\infty} H(|t|) = +\infty$.

Fix $w \in W$. By $\|w\| = 1$, we deduce that $\mu(\{x \in \mathbb{R}^N : w(x) \neq 0\}) > 0$, and that there exists a $\bar{t}_w > M$ such that $\mu(\{x \in \mathbb{R}^N : |\bar{t}_w w(x)| \geq M\}) > 0$, where μ is the Lebesgue measure.

Set $\Omega_1 := \{x \in \mathbb{R}^N : |\bar{t}_w w(x)| \geq M\}$ and $\Omega_2 := \mathbb{R}^N \backslash \Omega_1$. Then $\mu(\Omega_1) > 0$. Therefore, for any $x \in \Omega_1$, we have that $|w(x)| \geq \frac{M}{\bar{t}_w}$. Now take $\delta_w = \frac{M}{2C\bar{t}_w}$. Then, for any $v \in W \cap B(w, \delta_w)$, $|v - w|_\infty \leq C\|v - w\| < \frac{M}{2\bar{t}_w}$. Hence, for all $x \in \Omega_1$, we deduce that $|v(x)| \geq \frac{M}{2\bar{t}_w}$ and $|\lambda v(x)| \geq M$ for any $x \in \Omega_1$ and $\lambda \in \mathbb{R}$ with $|\lambda| \geq 2\bar{t}_w$. Thus, for $|\lambda| \geq 2\bar{t}_w$, by the above estimates and $H(|t|)$ increases when $|t|$ increases, we have

$$\int_{\Omega_1} K(x)F(\lambda v(x))dx \geq C_1|\lambda|^{p^+} \int_{\Omega_1} K(x)|v(x)|^{p^+} H(|\lambda v(x)|)dx$$

$$\geq C_1|\lambda|^{p^+} \left(\frac{M}{2\bar{t}_w}\right)^{p^+} H\left(|\lambda|\frac{M}{2\bar{t}_w}\right) \int_{\Omega_1} K(x)dx.$$

$$(6.5)$$

On the other hand, by continuity, we deduce that there exists a $C_2 > 0$ such that $F(t) \geq -C_2$ when $|t| \leq M$. Note that $F(t) > 0$ if $|t| \geq M$. Hence,

$$\int_{\Omega_2} K(x)F(\lambda v(x))dx = \int_{\Omega_2 \cup \{x \in \mathbb{R}^N : |\lambda v(x)| \geq M\}} K(x)F(\lambda v(x))dx$$

$$+ \int_{\Omega_2 \cup \{x \in \mathbb{R}^N : |\lambda v(x)| \leq M\}} K(x)F(\lambda v(x))dx$$

$$\geq \int_{\Omega_2 \cup \{x \in \mathbb{R}^N : |\lambda v(x)| \leq M\}} K(x)F(\lambda v(x))dx$$

$$\geq -C_2|K|_1.$$

$$(6.6)$$

Hence, for $v \in W \cap B(w, \delta_w)$ and $|\lambda| > 1$, from (6.5) and (6.6), we have

$$\varphi(\lambda v) = \int_{\mathbb{R}^N} \frac{|\lambda|^{p(x)}}{p(x)} (|\nabla v|^{p(x)} + \sqrt{1 + |\nabla v|^{2p(x)}} + |v|^{p(x)})dx$$

$$- \int_{\mathbb{R}^N} K(x)F(\lambda v(x))dx$$

$$\leq 2|\lambda|^{p^+} - C_1|\lambda|^{p^+} \left(\frac{M}{2\bar{t}_w}\right)^{p^+} H\left(|\lambda|\frac{M}{2\bar{t}_w}\right) \int_{\Omega_1} K(x)dx + C_2|K|_1$$

$$= |\lambda|^{p^+} \left[2 - C_1\left(\frac{M}{2\bar{t}_w}\right)^{p^+} H\left(|\lambda|\frac{M}{2\bar{t}_w}\right) \int_{\Omega_1} K(x)dx\right] + C_2|K|_1$$

$$\to -\infty,$$

as $|\lambda| \to +\infty$, because $\lim_{|t|\to+\infty} H(|t|) = +\infty$. $\qquad\square$

Claim 6.2. *There exist $\nu > 0$ and $\rho > 0$ such that $\inf_{\|u\|=\nu} \varphi(u) \geq \rho > 0$.*

Proof. Note that $|u|_\infty \to 0$ if $\|u\| \to 0$. Then, by hypothesis (E5), we have

$$\int_{\mathbb{R}^N} K(x)F(u)dx = |K|_1 o(|u|_\infty^{p^+}) = |K|_1 o(\|u\|^{p^+}),$$

which implies

$$\varphi(u) = \int_{\mathbb{R}^N} \frac{1}{p(x)}(|\nabla u(x)|^{p(x)} + \sqrt{1 + |\nabla u|^{2p(x)}} + |u(x)|^{p(x)})dx$$

$$- \int_{\mathbb{R}^N} K(x)F(u)dx$$

$$\geq \frac{2}{p^+}\|u\|^{p^+} - |K|_1 o(\|u\|^{p^+}).$$

Therefore, there exist $1 > \nu > 0$ and $\rho > 0$ such that $\inf_{\|u\|=\nu} \varphi(u) \geq \rho > 0$. $\qquad\square$

Claim 6.3. *The functional φ satisfies the (PS) condition.*

Proof. Let $\{u_n\} \subset W_r^{1,p(x)}(\mathbb{R}^N)$ be a (PS) sequence of the functional φ; that is, $|\varphi(u_n)| \leq c$ and $|\langle \varphi'(u_n), h \rangle| \leq \varepsilon_n \|h\|$ with $\varepsilon_n \to 0$, for all $h \in W_r^{1,p(x)}(\mathbb{R}^N)$. We will prove that the sequence $\{u_n\}$ is bounded in $W_r^{1,p(x)}(\mathbb{R}^N)$. Indeed, if $\{u_n\}$ is unbounded in $W_r^{1,p(x)}(\mathbb{R}^N)$, we may assume that $\|u_n\| \to \infty$ as $n \to \infty$. Let $u_n = \lambda_n w_n$, where $\lambda_n \in \mathbb{R}$, $w_n \in W$. It follows that $|\lambda_n| \to \infty$.

Let $\Omega_1^n := \{x \in \mathbb{R}^N : |\lambda_n w_n(x)| \geq M\}$ and $\Omega_2^n := \mathbb{R}^N \backslash \Omega_1^n$. Then

$$-\varepsilon_n|\lambda_n| = -\varepsilon_n\|u_n\|$$

$$\leq \langle \varphi'(u_n), u_n \rangle$$

$$= \int_{\mathbb{R}^N} \left(|\nabla u_n|^{p(x)} + \frac{|\nabla u_n|^{2p(x)}}{\sqrt{1 + |\nabla u_n|^{2p(x)}}} + |u_n|^{p(x)} \right) dx - \int_{\mathbb{R}^N} K(x)f(u_n)u_n dx$$

$$\leq \int_{\mathbb{R}^N} |\lambda_n|^{p(x)} \left(|\nabla w_n|^{p(x)} + \frac{|\nabla w_n|^{2p(x)}}{\sqrt{1 + |\nabla w_n|^{2p(x)}}} + |w_n|^{p(x)} \right)$$

$$- \int_{\Omega_1^n} K(x)f(\lambda_n w_n)\lambda_n w_n dx - \int_{\Omega_2^n} K(x)f(\lambda_n w_n)\lambda_n w_n dx,$$

which implies that

$$\int_{\Omega_1^n} K(x)f(\lambda_n w_n)\lambda_n w_n \, dx$$

$$\leq \int_{\mathbb{R}^N} |\lambda_n|^{p(x)}\left(|\nabla w_n|^{p(x)} + \frac{|\nabla w_n|^{2p(x)}}{\sqrt{1+|\nabla w_n|^{2p(x)}}} + |w_n|^{p(x)}\right) dx$$

$$+ \varepsilon_n|\lambda_n| - \int_{\Omega_2^n} K(x)f(\lambda_n w_n)\lambda_n w_n dx.$$

Note that $0 < (p^+ + \tau(t_{\lambda_n}))F(\lambda_n w_n) \leq f(\lambda_n w_n)\lambda_n w_n$ in Ω_1^n. So,

$$\int_{\Omega_1^n} K(x)F(\lambda_n w_n)dx \leq \frac{1}{p^+ + \tau(t_{\lambda_n})} \int_{\Omega_1^n} K(x)f(\lambda_n w_n)\lambda_n w_n dx.$$

Then, by (6.4), it follows that

$$\varphi(u_n) = \varphi(\lambda_n w_n)$$

$$= \int_{\mathbb{R}^N} \frac{|\lambda_n|^{p(x)}}{p(x)}(|\nabla w|^{p(x)} + \sqrt{1+|\nabla w|^{2p(x)}} + |w|^{p(x)})dx$$

$$- \int_{\mathbb{R}^N} K(x)F(\lambda_n w_n)dx$$

$$= \int_{\mathbb{R}^N} \frac{|\lambda_n|^{p(x)}}{p(x)}(|\nabla w|^{p(x)} + \sqrt{1+|\nabla w|^{2p(x)}} + |w|^{p(x)})dx$$

$$- \int_{\Omega_1^n} K(x)F(\lambda_n w_n)dx - \int_{\Omega_2^n} K(x)F(\lambda_n w_n)dx$$

$$\geq \frac{1}{p^+}\int_{\mathbb{R}^N} |\lambda_n|^{p(x)}(|\nabla w|^{p(x)} + \sqrt{1+|\nabla w|^{2p(x)}} + |w|^{p(x)})dx$$

$$- \frac{1}{p^+ + \tau(t_{\lambda_n})}\int_{\Omega_1^n} K(x)f(\lambda_n w_n)\lambda_n w_n dx - \int_{\Omega_2^n} K(x)F(\lambda_n w_n)dx$$

$$\geq \frac{1}{p^+} \int_{\mathbb{R}^N} |\lambda_n|^{p(x)} (2|\nabla w_n|^{p(x)} + |w_n|^{p(x)}) dx$$

$$- \frac{1}{p^+ + \tau(t_{\lambda_n})} \left[\int_{\mathbb{R}^N} |\lambda_n|^{p(x)} (2|\nabla w_n|^{p(x)} + |w_n|^{p(x)}) dx + \varepsilon_n |\lambda_n| \right]$$

$$+ \frac{1}{p^+ + \tau(t_{\lambda_n})} \int_{\Omega_2^n} K(x) f(\lambda_n w_n) \lambda_n w_n dx - \int_{\Omega_2^n} K(x) F(\lambda_n w_n) dx$$

$$= \frac{\tau(t_{\lambda_n})}{p^+(p^+ + \tau(t_{\lambda_n}))} \int_{\mathbb{R}^N} |\lambda_n|^{p(x)} (2|\nabla w_n|^{p(x)} + |w_n|^{p(x)}) dx$$

$$- \frac{1}{p^+ + \tau(t_{\lambda_n})} \varepsilon_n |\lambda_n| + T(\lambda_n w_n)$$

$$\geq \frac{\tau(t_{\lambda_n})}{p^+(p^+ + N_0)} |\lambda_n|^{p^-} - \frac{1}{p^+} \varepsilon_n |\lambda_n| + T(\lambda_n w_n)$$

$$= |\lambda_n| \left[\frac{|\lambda_n|^{p^- - 1} \tau(t_{\lambda_n})}{p^+(p^+ + N_0)} - \frac{\varepsilon_n}{p^+} \right] + T(\lambda_n w_n)$$

$$\geq |\lambda_n| \left[\frac{|\lambda_n|^{p^- - 1} \tau(t_{\lambda_n})}{p^+(p^+ + N_0)} - \frac{\varepsilon_n}{p^+} \right] - C_2,$$

where

$$T(\lambda_n w_n) = \frac{1}{p^+ + \tau(t_{\lambda_n})} \int_{\Omega_2^n} K(x) f(\lambda_n w_n) \lambda_n w_n \, dx - \int_{\Omega_2^n} K(x) F(\lambda_n w_n) \, dx$$

is bounded from below. We know that $|\lambda_n| \to +\infty$, and so $|t_{\lambda_n}| \to +\infty$, as $n \to +\infty$. It follows from (E3) and $p^- > N \geq 2$ that

$$\lim_{n \to +\infty} |\lambda_n|^{p^- - 1} \tau(t_{\lambda_n}) \geq \lim_{n \to +\infty} \frac{|t_{\lambda_n}| \tau(t_{\lambda_n})}{M} = +\infty.$$

This means that $\lim_{n \to +\infty} \varphi(u_n) \to +\infty$. This is a contradiction. So, the sequence $\{u_n\}$ is bounded in $W_r^{1,p(x)}(\mathbb{R}^N)$. Note that the embedding

$W_r^{1,p(x)}(\mathbb{R}^N) \hookrightarrow L^\infty(\mathbb{R}^N)$ is compact, there exists a $u \in W_r^{1,p(x)}(\mathbb{R}^N)$ such that passing to subsequence, still denoted by $\{u_n\}$, it converges strongly to u in $L^\infty(\mathbb{R}^N)$, and in the same way as the proof of [399, Proposition 3.1] we can conclude that u_n converges strongly also in $W_r^{1,p(x)}(\mathbb{R}^N)$. Thus, φ satisfies the (PS) condition. □

Proof of Theorem 6.2. Due to Claims 6.1, 6.2 and 6.3, we know that φ satisfies the conditions of the classical Mountain Pass Theorem due to Ambrosetti and Rabinowitz [172]. Hence, we obtain a nontrivial critical point, which gives rise to a nontrivial radially symmetric solution to problem (6.2).

From Claims 6.1 and 6.2, φ satisfies (a) and the (PS) condition. For any finite-dimensional subspace $\hat{E} \subset E$, $S \cap \hat{E} = \{w \in \hat{E} : \|w\| = 1\}$ is compact. By Claim 6.1 and the finite covering theorem, it is easy to verify that φ satisfies condition (b). Hence, by the $\mathbb{Z}_2$ version of the Mountain Pass Theorem, φ has a sequence of critical points $\{u_n\}_{n=1}^\infty$. That is, problem (6.2) has infinitely many pairs of radially symmetric solutions. □

Notes and Comments

Capillarity can be briefly explained by considering the effects of two opposing forces: adhesion, i.e. the attractive (or repulsive) force between the molecules of the liquid and those of the container; and cohesion, i.e. the attractive force between the molecules of the liquid. The study of capillary phenomena has gained some attention recently. This increasing interest is motivated not only by fascination in naturally-occurring phenomena such as motion of drops, bubbles, and waves but also its importance in applied fields ranging from industrial and biomedical and pharmaceutical to microfluidic systems. Ni and Serrin [400] initiated the study of ground states for equations of the form

$$-\mathbf{div}\left(\frac{|\nabla u|}{\sqrt{1 + |\nabla u|^2}}\right) = f(u), \quad \text{in } \mathbb{R}^N$$

with very general right hand side f.

Recently, the study of various mathematical problems with variable exponent growth condition had received considerable attention in recent years;

see e.g. [395, 401–404]. For background information, we refer the reader to [385, 405]. This section was obtained in [406].

6.4 $p(x)$-Biharmonic Problems

Consider the fourth-order quasilinear elliptic equation

$$\begin{cases} \Delta(|\Delta u|^{p(x)-2}\Delta u) = \lambda |u|^{p(x)-2}u + f(x,u) & \text{in } \Omega, \\ u = \Delta u = 0 & \text{on } \partial\Omega, \end{cases} \quad (6.7)$$

where Ω is a bounded domain in $\mathbb{R}^N$ with smooth boundary $\partial\Omega$, $N \geq 1$. $\Delta(|\Delta u|^{p(x)-2}\Delta u)$ is the $p(x)$-biharmonic operator, $\lambda \leq 0$, p is a continuous function on $\overline{\Omega}$ with $\inf_{x\in\overline{\Omega}} p(x) > 1$ and $f: \Omega \times \mathbb{R} \to \mathbb{R}$ is a Carathéodory function. Denote $p \in C(\overline{\Omega})$, $1 < p^- = \min_{x\in\overline{\Omega}} p(x) \leq p^+ = \max_{x\in\overline{\Omega}} p(x) < +\infty$.

In the sequel, X will denote the Sobolev space $W^{2,p(x)}(\Omega) \cap W_0^{1,p(x)}(\Omega)$, $|E|$ be the Lebesgue measure of E, C_i, $i \in \mathbb{N}$, will denote some positive constants. Let $F(x,t) = \int_0^t f(x,s)\,ds$ and function $\mathcal{F}(x,t) = f(x,t)t - p^+ F(x,t)$.

The energy functional corresponding to problem (6.7) is defined on X as

$$I(u) = \int_\Omega \frac{1}{p(x)}\left(|\Delta u|^{p(x)} - \lambda|u|^{p(x)}\right)dx - \int_\Omega F(x,u)\,dx.$$

Let us recall that weak solution of problem (6.7) is any $u \in X$ such that

$$\int_\Omega \left(|\Delta u|^{p(x)-2}\Delta u \Delta v - \lambda|u|^{p(x)-2}uv\right)dx = \int_\Omega f(x,u)v\,dx \quad \text{for any } v \in X.$$

The aim of this section is to show the existence of nontrivial solutions of problem (6.7). First we give the following assumptions on the function f.

(E6) $|f(x,t)| \leq a + b|t|^{\alpha(x)-1}$ for all $(x,t) \in \Omega \times \mathbb{R}$, with $a, b \geq 0$ and $1 < \alpha(x) < p_2^*(x)$, where

$$p_2^*(x) = \begin{cases} \dfrac{Np(x)}{N - 2p(x)} & p(x) < N/2, \\ \infty & p(x) \geq N/2, \end{cases}$$

is the critical exponent just as in many papers..

(E7) $\lim_{|t|\to\infty} \frac{f(x,t)t}{|t|^{p^+}} = +\infty$ uniformly for a.e. $x \in \Omega$.

(E8) There exists a constant $\theta \geq 1$, such that for any $s \in [0,1]$ and $t \in \mathbb{R}$, the inequalities $\theta\mathcal{F}(x,t) \geq \mathcal{F}(x,st)$ hold for a.e. $x \in \Omega$.

(E9) $f(x,t) = o(|t|^{p^+ - 1})$ as $t \to 0$ and uniformly for $x \in \Omega$, with $\alpha^- > p^+$, where $1 < \alpha^- = \min_{x \in \overline{\Omega}} \alpha(x) \leq \alpha^+ = \max_{x \in \overline{\Omega}} \alpha(x) < +\infty$.

We can state the following results.

Theorem 6.3. *If f satisfies* (E6)–(E9), *then problem* (6.7) *has at least a nontrivial solution.*

Remark 6.4. Beginning with [172], many authors have obtained nontrivial solutions of superlinear problems,

$$\begin{cases} -\Delta_p u = f(x, u) & \text{in } \Omega, \\ u = 0 & \text{on } \partial\Omega, \end{cases}$$

under various assumptions of the behavior of f near zero and infinity, in the semilinear case $p = 2$ and quasilinear case $p \neq 2$. Recently, Navier problems involving the biharmonic have been studied by some authors, see [407–409] and references therein. El Amrouss *et al.* [16] study a class of $p(x)$-biharmonic, they obtained the existence and multiplicity of solutions for (6.7) under (E6), (E9) and

(AR) There exist $M > 0$, $\theta > p^+$ such that for all $|t| \geq M$ and $x \in \Omega$,

$$0 < F(x,t) \leq \frac{t}{\theta} f(x,t).$$

Obviously, (E7) can be derived from (AR). Under (AR), any (PS) sequence of the corresponding energy functional is bounded, which plays an important role of the application of variational methods. Indeed, there are many superlinear functions which do not satisfy the (AR) condition. For instance when $p(x) = 2$, $\theta = 1$, the function below does not satisfy the (AR) condition

$$f(x,t) = 2t \ln(1 + |t|). \tag{6.8}$$

But it is easy to see the function above (6.8) satisfies (E6)-(E9).

Remark 6.5. When $p(x) \equiv 2$ and for the Laplacian equation, condition (E8) was first introduce by Jeanjean [243]. Recently, many authors solve superlinear problem without (AR) condition using this hypothesis. In [410], Miyagaki and Souto discuss many different superlinear conditions introduced by Costa and Magalhães [411], Willem and Zou [412], Schechter and Zou [413]. There is an interesting paper [414], in which Li and Yang considered the relations among many different superlinear conditions. For other superlinear problem, we refer the reader to see [415–417] and reference therein.

Next, if the nonlinear term is odd, we can obtain infinitely many pairs of solutions.

Theorem 6.4. *Suppose that f satisfies the conditions (E6)–(E8) and the following condition*

(E10) $f(x, -t) = -f(x, t)$, $x \in \Omega$, $t \in \mathbb{R}$.

Then problem (6.7) has infinitely many weak solutions.

For problem (6.7), if $p(x) = p$, a constant and $\lambda = 0$, we can get nontrivial solutions for a p-biharmonic equation under weaker conditions, that is, we will study the following equation

$$\begin{cases} \Delta(|\Delta u|^{p-2}\Delta u) = f(x, u) & \text{in } \Omega, \\ u = \Delta u = 0 & \text{on } \partial\Omega. \end{cases} \tag{6.9}$$

At first, we give a remark about the eigenvalue of p-biharmonic operator.

Remark 6.6. Notice that for the eigenvalue problem

$$\begin{cases} \Delta(|\Delta u|^{p-2}\Delta u) = \lambda |u|^{p-2}u & \text{in } \Omega, \\ \Delta u = u = 0 & \text{on } \partial\Omega, \end{cases} \tag{6.10}$$

as for the p-Laplacian eigenvalue problem with Dirichlet boundary data,

$$\lambda_n = \inf_{A \in \Gamma_n} \sup_{u \in A} \int_\Omega |u|^p dx, \qquad n = 1, 2, \ldots$$

is the sequence of eigenvalues, where

$$\Gamma_n = \{A \subseteq W^{2,p}(\Omega) \cap W_0^{1,p}(\Omega) \backslash \{0\} : A = -A, \gamma(A) \geq n\},$$

$\gamma(A)$ being the Krasnoselski's genus of set A. It has been recently proved by Drábek and Ôtani [418] that (6.10) has the least eigenvalue

$$\lambda_1 = \left\{ \int_\Omega |\Delta u|^p dx : u \in W^{2,p}(\Omega) \cap W_0^{1,p}(\Omega), \|u\|^p = 1 \right\} \tag{6.11}$$

which is simple, positive, and isolated in the sense that the solutions of (6.10) with $\lambda = \lambda_1$ form a one-dimensional linear space spanned by a positive eigenfunction ϕ_1 associated with λ_1 and there exists $\varrho > 0$ so that $(\lambda_1, \lambda_1 + \varrho)$ does not contain other eigenvalues. Let $V = \text{span}\{\phi_1\}$ be the eigenspace associated with λ_1, where $0 < \lambda_1$ denotes the first eigenvalues of $(\Delta(|\Delta u|^{p-2}\Delta u), W^{2,p} \cap W_0^{1,p}(\Omega))$ and $\|\phi_1\| = 1$. Taking a subspace $W \subset W^{2,p}(\Omega) \cap W_0^{1,p}(\Omega)$ complementing V, that is, $W^{2,p}(\Omega) \cap W_0^{1,p}(\Omega) = V \oplus W$, there exists $\widehat{\lambda} > \lambda_1$ with

$$\int_\Omega |\Delta u|^p dx \geq \widehat{\lambda} \int_\Omega |u|^p dx \tag{6.12}$$

for each $u \in W$ (in case $p = 2$, one may take $\widehat{\lambda} = \lambda_2$).

Now, we give another result about p-biharmonic equation.

Theorem 6.5. *Suppose $f(x,t)$ satisfies (E6) (for p-biharmonic equation, $\alpha(x)$ independent of x), (E8) (where $\mathcal{F}(x,t) = f(x,t)t - pF(x,t)$) and the following conditions hold:*

(E11) *There exists $\delta > 0$ and $\overline{\lambda} \in]\lambda_1, \widehat{\lambda}[$, such that $\lambda_1 |t|^p \leq pF(x,t) \leq \overline{\lambda}|t|^p$, for $|t| \leq \delta$, for a.e. $x \in \Omega$ and $t \in \mathbb{R}$;*

(E12) $\lim_{|t| \to +\infty} \frac{f(x,t)}{|t|^{p-2}t} = \infty$ *uniformly for a.e. $x \in \Omega$.*

Then problem (6.9) has at least one nontrivial solution.

Remark 6.7. In [419], Liu and Squassina had already studied problem (6.9) using local linking theorem, but they assume f satisfied the (AR) condition. We can give out a function f satisfies our theorem but does not satisfies their theorem. The function $f(x,t) = p|t|^{p-2}t\ln(1+|t|^p)$ is desired. So, our Theorem 6.5 generalizes Theorem 1.3 of Liu and Squassina [419].

In problem (6.9), if we let $p = 2$, that is we consider the following equation.

$$\begin{cases} \Delta^2 u = f(x,u) & \text{in } \Omega, \\ u = \Delta u = 0 & \text{on } \partial\Omega. \end{cases} \tag{6.13}$$

Considering $f(x,t)$ satisfies (E6), (E10) and drop the condition (E7) and (E8), we can also obtain the existence of infinitely many solutions for problem (6.13) under the following assumption via the Symmetric Mountain Pass Theorem of Kajikiya [29].

(E13) $\lim_{t \to 0} \frac{f(x,t)}{t} = +\infty$.

Theorem 6.6. *Suppose $f(x,t)$ satisfies (E6), (E10) and (E13), problem (6.13) admits a sequence of nontrivial weak solutions $\{u_n\}$ with $u_n \to 0$ in X as $n \to \infty$.*

Remark 6.8. In our Theorem 6.6, we can observe that contrary to most of the known results no condition on the behavior of the involved nonlinearities near infinity is assumed.

Let us define the functional

$$J(u) = \int_\Omega \frac{1}{p(x)}(|\Delta u|^{p(x)} - \lambda|u|^{p(x)})\,dx.$$

It is well known that J is well defined, even and C^1 in X. Moreover, the operator $L := J' : X \to X^*$ defined as

$$\langle L(u), v \rangle = \int_\Omega (|\Delta u|^{p(x)-2} \Delta u \Delta v - \lambda |u|^{p(x)-2} uv) \, dx \quad \text{for all } u, v \in X$$

satisfies the assertions of the following theorem.

Proposition 6.1 (see El Amrouss *et al.* [16]). (1) L *is continuous, bounded and strictly monotone.*
(2) L *is of* (S_+) *type i.e. if* $u_n \rightharpoonup u$ *in* X *and* $\limsup_{n \to \infty} \langle L(u_n) - L(u), u_n - u \rangle \le 0$ *then* $u_n \to u$ *in* X.
(3) L *is a homeomorphism.*

At first, we show that the functional I satisfies the Cerami condition. Before that, we deduce from (E8) that

$$f(x, t)t - p^+ F(x, t) \ge 0, \text{ for all } (x, t) \in \Omega \times \mathbb{R}. \tag{6.14}$$

Let $t > 0$. For $x \in \Omega$, we have

$$\frac{\partial}{\partial t} \left(\frac{F(x, t)}{t^{p^+}} \right) = \frac{t^{p^+} f(x, t) - p t^{p^+ - 1} F(x, t)}{t^{2p^+}} \ge 0. \tag{6.15}$$

By (E9),

$$\lim_{t \to 0^+} \frac{F(x, t)}{t^{p^+}} = 0. \tag{6.16}$$

From (6.15) and (6.16) we conclude that $F(x, t) \ge 0$ for all $x \in \Omega$ and $t \ge 0$. Arguing similarly for the case $t \le 0$, eventually we obtain

$$F(x, t) \ge 0 \quad \text{for all } (x, t) \in \Omega \times \mathbb{R}.$$

Lemma 6.1. *Under the assumptions* (E6)–(E8), I *satisfies the Cerami condition.*

Proof. For all $c \in \mathbb{R}$, we show that I satisfies (i) of Cerami condition. Let $\{u_n\} \subset X$ be bounded, $I(u_n) \to c$ and $I'(u_n) \to 0$. Hence $\{u_n\}$ has a weakly convergent subsequence in X. Without loss of generality, we assume that $u_n \rightharpoonup u$, $\Psi(u) = \int_\Omega F(x, u) \, dx$, then $\Psi' : X \to X^*$ is completely continuous because of assumption (E6) and f is a Carathéodory function. Hence $\Psi'(u_n) \to \Psi'(u)$. Since $I'(u_n) = J'(u_n) - \Psi'(u_n) \to 0$ and J' is a homeomorphism, in view of Proposition 6.1, we know $u_n \to u$ in X.

Now check that I satisfies (ii) of Cerami condition too. We argue by contradiction. There exist $c \in \mathbb{R}$ and $\{u_n\} \subset X$ satisfying:

$$I(u_n) \to c, \qquad \|u_n\| \to +\infty, \qquad \|I'(u_n)\|\|u_n\| \to 0. \qquad (6.17)$$

We can choose $\|u_n\| > 1$, for $n \in \mathbb{N}$, thus

$$\int_\Omega \left(\frac{1}{p(x)} - \frac{1}{p^+} \right) (|\Delta u|^{p(x)} - \lambda |u|^{p(x)}) dx + \frac{1}{p^+} \int_\Omega \mathcal{F}(x, u_n) dx$$

$$= I(u_n) - \frac{1}{p^+} \langle I'(u_n), u_n \rangle. \qquad (6.18)$$

From (6.17), we known that $I(u_n) - \frac{1}{p^+} \langle I'(u_n), u_n \rangle \to c$, when $n \to \infty$. Denote $w_n = \frac{u_n}{\|u_n\|}$, then $\|w_n\| = 1$, so $\{w_n\}$ is bounded. Up to a subsequence, for some $w \in X$, we get

$$w_n \rightharpoonup w \text{ in } X,$$
$$w_n \to w \text{ in } L^{r(x)}, r(x) \le p_k^*(x)$$
$$w_n(x) \to w(x) \text{ a.e. in } \Omega.$$

If $w(x) \equiv 0$, we can define a sequence $\{t_n\} \subset \mathbb{R}$:

$$I(t_n u_n) = \max_{t \in [0,1]} I(t u_n).$$

Fix any $m > 0$, let $\overline{w}_n = (2p^+ m)^{1/p^-} w_n$, since $w_n \rightharpoonup w \equiv 0$ and $\Psi(u)$ is weakly continuous,

$$\lim_{n \to \infty} \int_\Omega F(x, \overline{w}_n) \, dx = \lim_{n \to \infty} \int_\Omega F(x, (2p^+ m)^{1/p^-} w_n) \, dx = 0.$$

Then for n large enough,

$$\begin{aligned}
I(t_n u_n) \ge & I(\overline{w}_n) \\
= & \int_\Omega \frac{1}{p(x)} (|\Delta (2p^+ m)^{1/p^-} w_n|^{p(x)} - \lambda |(2p^+ m)^{1/p^-} w_n|^{p(x)}) \, dx \\
& - \int_\Omega F(x, \overline{w}_n) \, dx \\
\ge & \int_\Omega \frac{1}{p^+} (2p^+ m)(|\Delta w_n|^{p(x)} - \lambda |w_n|^{p(x)}) \, dx - \int_\Omega F(x, \overline{w}_n) \, dx \\
= & 2m - \int_\Omega F(x, \overline{w}_n) \, dx \\
\ge & m.
\end{aligned}$$

$$(6.19)$$

Due to (6.19), $\lim_{n \to \infty} I(t_n u_n) = +\infty$. Since $I(0) = 0$, and $I(u_n) \to c$, then $0 < t_n < 1$, If n large enough, we have

$$\int_{\Omega} (|\Delta t_n u_n|^{p(x)} - \lambda|t_n u_n|^{p(x)} - f(x, t_n u_n) t_n u_n) \, dx = \langle I'(t_n u_n), t_n u_n \rangle$$

$$= t_n \frac{d}{dt}\bigg|_{t=t_n} I(t u_n) = 0.$$

Now using (6.18) and (E8), we obtain

$$\frac{1}{\theta} I(t_n u_n) = \frac{1}{\theta} \left(I(u_n) - \frac{1}{p^+} \langle I'(u_n), u_n \rangle \right)$$

$$\leq \int_{\Omega} \left(\frac{1}{p(x)} - \frac{1}{p^+} \right) t_n^{p(x)} (|\Delta u|^{p(x)} - \lambda|u|^{p(x)}) dx$$

$$+ \frac{1}{p^+} \int_{\Omega} \frac{\mathcal{F}(x, u_n)}{\theta} dx$$

$$\leq \int_{\Omega} \left(\frac{1}{p(x)} - \frac{1}{p^+} \right) (|\Delta u|^{p(x)} - \lambda|u|^{p(x)}) dx + \frac{1}{p^+} \int_{\Omega} \mathcal{F}(x, u_n) dx$$

$$\to c,$$

which contradicts (6.19).

If $w(x) \not\equiv 0$ we have $|u_n(x)| \to +\infty$. Using (E7) we obtain

$$\frac{F(x, u_n)}{|u_n|^{p^+}} |w_n|^{p^+} \to +\infty. \tag{6.20}$$

Since the set $\Theta = \{x \in \Omega : w(x) \neq 0\}$ has positive Lebesgue measure and $\|u_n\| > 1$ for n large, using (6.14), (6.20), Proposition 1.11 and the Fatou's Lemma we have

$$\frac{1}{p^-} \geq \frac{1}{p^- \|u_n\|^{p^+}} \int_{\Omega} (|\Delta u_n|^{p(x)} - \lambda|u_n|^{p(x)}) \, dx$$

$$\geq \frac{1}{\|u_n\|^{p^+}} \int_{\Omega} \frac{1}{p(x)} (|\Delta u_n|^{p(x)} - \lambda|u_n|^{p(x)}) \, dx$$

$$= \frac{1}{\|u_n\|^{p^+}} \left(I(u_n) + \int_{\Omega} F(x, u_n) \, dx \right)$$

$$\geq \frac{1}{\|u_n\|^{p^+}} \int_{\Omega} F(x, u_n) dx - 1 = \int_{\Omega} \frac{F(x, u_n)}{\|u_n\|^{p^+}} dx - 1$$

$$\geq \int_{w \neq 0} \frac{F(x, u_n)}{|u_n|^{p^+}} |w_n|^{p^+} \, dx - 1$$

$$\to +\infty,$$

which is impossible.

Now, we had proved that every $(C)_c$ sequence is bounded. Since the operator L is (S_+) type (see Proposition 6.1 for more details), a standard argument shows that I satisfies Cerami condition. □

Next lemma has already proved by El Amrouss *et al.* [16]. For completeness, we give out the proof.

Lemma 6.2 (see El Amrouss *et al.* [16]). *There exist $r, \beta > 0$ such that $I(u) \geq \beta$ for all $u \in X$ such that $\|u\| = r$.*

Proof. Conditions (E6) and (E9) assure that
$$|F(x,t)| \leq \varepsilon |t|^{p^+} + C_3(\varepsilon) |t|^{\alpha(x)} \quad \text{for all } (x,t) \in \Omega \times \mathbb{R}.$$
For $\|u\|$ small enough, we have
$$
\begin{aligned}
I(u) &\geq \frac{1}{p^+} \int_\Omega (|\Delta u|^{p(x)} - \lambda |u|^{p(x)}) \, dx. - \int_\Omega F(x, u) \, dx, \\
&\geq \frac{1}{p^+} \|u\|^{p^+} - \varepsilon \int_\Omega |u|^{p^+} \, dx - C_3(\varepsilon) \int_\Omega |u|^{\alpha(x)} \, dx.
\end{aligned}
\tag{6.21}
$$
By condition (E9), it follows that
$$p^- \leq p \leq p^+ < \alpha^- \leq \alpha < p_2^*$$
then $X \subset L^{p^+}(\Omega)$, $X \subset L^{\alpha(x)}(\Omega)$, with a continuous and compact embedding, what implies the existence of $C_4, C_5 > 0$ such that
$$|u|_{p^+} \leq C_4 \|u\| \quad \text{and} \quad |u|_{\alpha(x)} \leq C_5 \|u\|$$
for all $u \in X$. Since $\|u\|$ is small enough, we deduce
$$\int_\Omega |u|^{\alpha(x)} \, dx \leq \max(|u|_{\alpha(x)}^{\alpha^-}, |u|_{\alpha(x)}^{\alpha^+}) \leq C_6 \|u\|^{\alpha^-}.$$
Replacing in (6.21), it results that
$$I(u) \geq \frac{1}{p^+} \|u\|^{p^+} - \varepsilon C_4^{p^+} \|u\|^{p^+} - C_7 \|u\|^{\alpha^-}.$$
Let us choose $\varepsilon > 0$ such that $\varepsilon C_4^{p^+} \leq \frac{1}{2p^+}$, we obtain
$$
\begin{aligned}
I(u) &\geq \frac{1}{2p^+} \|u\|^{p^+} - C_7 \|u\|^{\alpha^-} \\
&\geq \|u\|^{p^+} \left(\frac{1}{2p^+} - C_7 \|u\|^{\alpha^- - p^+} \right).
\end{aligned}
$$
Since $p^+ < \alpha^-$, the function $t \mapsto (\frac{1}{2p^+} - C_7 t^{\alpha^- - p^+})$ is strictly positive in a neighborhood of zero. It follows that there exist $r > 0$ and $\beta > 0$ such that
$$I(u) \geq \beta \quad \forall u \in X, \|u\| = r.$$
The proof is complete. □

Proof of Theorem 6.3. To apply the Mountain Pass Theorem, we must prove that

$$I(tu) \to -\infty \quad \text{as } t \to +\infty,$$

for a certain $u \in X$. From condition (E7), for any $\varepsilon > 0$, there exists $M > 0$ such that $\frac{f(x,t)}{t^{p^+-1}} \geq \frac{1}{\varepsilon}$ for all $t > M$ and $x \in \Omega$. Set $C_8(\varepsilon) = \frac{1}{\varepsilon}M^{p^+-1}$, consequently,

$$f(x,t) \geq \frac{1}{\varepsilon}t^{p^+-1} - C_8(\varepsilon)$$

for all $t \geq 0$ and $x \in \Omega$, which implies that

$$F(x,t) \geq \frac{1}{p^+\varepsilon}t^{p^+} - C_8(\varepsilon)t \tag{6.22}$$

for all $t \geq 0$. It follows from (6.22) that

$$F(x,t\varphi_1) \geq \frac{1}{p^+\varepsilon}t^{p^+}\varphi_1^{p^+} - C_8(\varepsilon)t\varphi_1,$$

where $\varphi_1 \in X$, $t > 0$ and $\|\varphi_1\| \geq 1$. Dividing by t^{p^+}, one has

$$\frac{F(x,t\varphi_1)}{t^{p^+}} \geq \frac{1}{p^+\varepsilon}\varphi_1^{p^+} - \frac{C_8(\varepsilon)\varphi_1}{t^{p^+-1}},$$

thus we have

$$\int_\Omega \frac{F(x,t\varphi_1)}{t^{p^+}}dx \geq \int_\Omega \left(\frac{1}{p^+\varepsilon}\varphi_1^{p^+} - \frac{C_8(\varepsilon)\varphi_1}{t^{p^+-1}}\right)dx. \tag{6.23}$$

Letting $t \to \infty$ in (6.23), it follows that

$$\liminf_{t \to +\infty}\int_\Omega \frac{F(x,t\varphi_1)}{t^{p^+}}dx \geq \frac{1}{p^+\varepsilon}\int_\Omega \varphi_1^{p^+}dx$$

for all $\varepsilon > 0$. For $\varepsilon > 0$ is arbitrary, let $\varepsilon \to 0$, then one obtains

$$\lim_{t \to \infty}\int_\Omega \frac{F(x,t\varphi_1)}{t^{p^+}}dx = +\infty.$$

Consequently, let $\|t\varphi_1\| > 1$,

$$\frac{I(t\varphi_1)}{t^{p^+}} \leq \frac{1}{p^+}\|\varphi_1\|^{p^+} - \int_\Omega \frac{F(x,t\varphi_1)}{t^{p^+}}dx \to -\infty \quad (t \to +\infty).$$

It follows that there exists t_0 big enough and $e = t_0\varphi_1$, such that $I(e) < 0$. According to the Mountain Pass Theorem, I has at least one nontrivial critical point, so problem (6.7) has a nontrivial solution. $\square$

We use Bartsch's fountain theorem [27] under Cerami condition to prove Theorem 6.4. Before proving the theorem, let us show the following lemma. The space $W^{2,p(x)}(\Omega) \cap W_0^{1,p(x)}(\Omega)$ is reflexive and separable, then there exists $\{e_i\} \subset W^{2,p(x)}(\Omega) \cap W_0^{1,p(x)}(\Omega)$ and $\{f_i\} \subset (W^{2,p(x)}(\Omega) \cap W_0^{1,p(x)}(\Omega))^*$ such that

$$W^{2,p(x)}(\Omega) \cap W_0^{1,p(x)}(\Omega) = \overline{\langle e_i, i \in \mathbb{N}\rangle},$$

$$(W^{2,p(x)}(\Omega) \cap W_0^{1,p(x)}(\Omega))^* = \overline{\langle f_i, i \in \mathbb{N}\rangle}, \quad \langle e_i, f_j\rangle = \delta_{i,j},$$

where $\delta_{i,j}$ denotes the Kronecker symbol. Put

$$X_k = \text{span}\{e_k\}, \quad Y_k = \bigoplus_{i=1}^{k} X_i, \quad Z_k = \overline{\bigoplus_{i=k}^{\infty} X_i}. \tag{6.24}$$

Lemma 6.3 (see El Amrouss *et al.* [16]). *If* $\alpha(x) < p_2^*(x)$ *for all* $x \in \overline{\Omega}$, *then* $\lim_{k\to+\infty} \beta_k = 0$, *where* $\beta_k = \sup\{|u|_{\alpha(x)}/\|u\| = 1 : u \in Z_k\}$ *and* Z_k *defined by* (6.24).

Proof of Theorem 6.4. From conditions (E6)–(E8) and (E10), I is an even functional satisfies the Cerami condition. We will prove that for k large enough, there exists $\rho_k > r_k > 0$, such that assertion (Φ1) and (Φ2) of Theorem 1.11 satisfied.

(i) For $u \in Z_k$ such that $\|u\| = r_k > 1$, we have by condition (E6)

$$I(u) = \int_\Omega \frac{1}{p(x)}(|\Delta u|^{p(x)} - \lambda|u|^{p(x)}) \, dx - \int_\Omega F(x,u) \, dx$$

$$\geq \frac{1}{p^+} \int_\Omega (|\Delta u|^{p(x)} - \lambda|u|^{p(x)}) \, dx - C_9 \int_\Omega |u|^{\alpha(x)} \, dx - C_{10}\|u\|_{L^1}$$

$$\geq \frac{1}{p^+}\|u\|^{p^-} - C_9 \int_\Omega |u|^{\alpha(x)} \, dx - C_{11}.$$

If $|u|_{\alpha(x)} \leq 1$ then $\int_\Omega |u|^{\alpha(x)} \, dx \leq |u|_{\alpha(x)}^{\alpha^-} \leq 1$. However, if $|u|_{\alpha(x)} > 1$ then $\int_\Omega |u|^{\alpha(x)} \, dx \leq |u|_{\alpha(x)}^{\alpha^+} \leq (\beta_k\|u\|)^{\alpha^+}$. So, we conclude that

$$I(u) \geq \begin{cases} \dfrac{1}{p^+}\|u\|^{p^-} - (C_9 + C_{11}) & \text{if } |u|_{\alpha(x)} \leq 1 \\[2mm] \dfrac{1}{p^+}\|u\|^{p^-} - C_9(\beta_k\|u\|)^{\alpha^+} - C_{11} & \text{if } |u|_{\alpha(x)} > 1 \end{cases}$$

$$\geq \frac{1}{p^+}\|u\|^{p^-} - C_9(\beta_k\|u\|)^{\alpha^+} - C_{12},$$

For $r_k = (C_9\alpha^+\beta_k^{\alpha^+})^{1/(p^- - \alpha^+)}$, it follows that

$$I(u) \geq r_k^{p^-}\left(\frac{1}{p^+} - \frac{1}{\alpha^+}\right) - C_{12}.$$

Since $\beta_k \to 0$ and $p^- \leq p^+ < \alpha^+$, we have $r_k \to +\infty$ as $k \to +\infty$. Consequently,

$$I(u) \to +\infty \quad \text{as } \|u\| \to +\infty, u \in Z_k$$

and the assertion ($\Phi 1$) is true.

(ii) Note that the space Y_k is finite dimension, then all norms are equivalent, it follows that there exists $C_k > 0$, $\|u\| \geq 1$, where $u \in Y_k$, we get

$$\int_\Omega \frac{1}{p(x)}(|\Delta u|^{p(x)} - \lambda|u|^{p(x)})\,dx \leq \frac{1}{p^-}\int_\Omega (|\Delta u|^{p(x)} - \lambda|u|^{p(x)})\,dx$$
$$\leq \frac{1}{p^-}\|u\|^{p^+}$$
$$\leq C_k|u|_{p^+}^{p^+}. \tag{6.25}$$

By (E7), there exist $R_k > 0$, $|t| \geq R_k$, $F(x,t) \geq 2C_k|t|^{p^+}$.

Let $M_k = \max\{0, \inf_{x \in \Omega, |t| \leq R_k} F(x,t)\}$, so, for all $(x,t) \in \Omega \times \mathbb{R}$, we obtain that

$$F(x,t) \geq 2C_k|t|^{p^+} - M_k. \tag{6.26}$$

By (6.25), (6.26), for each $u \in Y_k$, it follows that

$$I(u) = \int_\Omega \frac{1}{p(x)}(|\Delta u|^{p(x)} - \lambda|u|^{p(x)})\,dx - \int_\Omega F(x,u)\,dx \leq -C_k|u|_{p^+}^{p^+} + M_k|\Omega|.$$

For $\rho_k > 0$ large enough, we get $a_k := \max_{u \in Y_k, \|u\| = \rho_k} I(u) \leq 0$.

The assertion ($\Phi 2$) is then satisfied and the proof of Theorem 1.11 is completed. $\square$

Proof of Theorem 6.5. We will apply Theorem 1.6 to the functional I associated with problem (6.9).

Step 1. I has a local linking at 0 with respect to $W^{2,p}(\Omega) \cap W_0^{1,p}(\Omega) = V \oplus W$, according to Remark 6.6.

For $u \in W$, by (E6) and (E11), we have

$$|F(x,t)| \leq \frac{1}{p}\lambda_1|t|^p + C_{13}|t|^\alpha, \tag{6.27}$$

for $(x,t) \in \Omega \times \mathbb{R}$, which implies

$$I(u) \geq \frac{1}{p}\|u\|^p - \frac{\lambda_1}{p}|u|_p^p - C_{13}|u|_\alpha^\alpha.$$

Noting that $\alpha > p$, we have $I(u) \geq 0$ for $u \in W$ with $\|u\| \leq r_1$ where $r_1 > 0$ is small enough.

For $u \in V$ and $\|u\| \leq r_2$, due to (E6) and (6.12) we have

$$
\begin{aligned}
I(u) &= \frac{1}{p} \int_\Omega |\Delta u|^p dx - \int_\Omega F(x, u) dx \\
&= \frac{1}{p} \int_\Omega (|\Delta u|^p - \overline{\lambda}|u|^p) dx - \int_{|u| \leq \delta} \left(F(x, u) - \frac{\overline{\lambda}}{p}|u|^p \right) dx \\
&\quad - \int_{|u| > \delta} \left(F(x, u) - \frac{\overline{\lambda}}{p}|u|^p \right) dx \\
&\geq \frac{1}{p} \left(1 - \frac{\overline{\lambda}}{\check{\lambda}} \right) \|u\|^p - C_{14} \int_\Omega |u|^\alpha dx \\
&\geq \frac{1}{p} \left(1 - \frac{\overline{\lambda}}{\check{\lambda}} \right) \|u\|^p - C_{15}\|u\|^\alpha,
\end{aligned}
$$

where $p < \alpha < p^*$. Since $\alpha > p$, it follows that $I(u) > 0$ for $r_2 > 0$ sufficiently small. Take $r = \min\{r_1, r_2\}$. We know that I has a local linking at 0 with respect to $W^{2,p}(\Omega) \cap W_0^{1,p}(\Omega) = V \oplus W$.

Step 2. I maps bounded sets into bounded sets.

Assume $\|u\| \leq M$, where M is a constant. By (6.27), one has

$$
\begin{aligned}
I(u) &\leq \frac{1}{p}\|u\|^p + \int_\Omega |F(x, u)| dx \\
&\leq \frac{1}{p}\|u\|^p + \frac{1}{p}\lambda_1 |u|_p^p + C_{13}|u|_\alpha^\alpha \\
&\leq \frac{1}{p}M^p + \frac{1}{p}\lambda_1 C_{16} M^p + C_{17} M^\alpha \\
&< \infty,
\end{aligned}
$$

which implies that I maps bounded sets into bounded sets.

Step 3. For every finite dimensional subspace $E \subset W$, $I(u) \to -\infty$, as $\|u\| \to \infty$, $u \in V \oplus E$.

By (E12), for any $R > 0$, there exists a constant $h(R)$ such that $F(x, t) \geq R|t|^p - h(R)$ for all $(x, t) \in \Omega \times \mathbb{R}$. Since $\dim(V \oplus E)$ is of finite, there exists $C_{18} > 0$ such that $\|u\| \leq C_{18}|u|_p$ for all $u \in V \oplus E$, which implies

$$
\begin{aligned}
I(u) &\leq \frac{1}{p}\|u\|^p - R|u|_p^p + h(R)|\Omega| \\
&\leq \left(\frac{1}{p} - RC_{18}^p \right) \|u\|^p + h(R)|\Omega|.
\end{aligned}
$$

Choosing $R > \frac{p}{C_{18}^p}$, we have $I(u) \to -\infty$ as $\|u\| \to \infty$ and $u \in V \oplus E$.

Step 4. I satisfies the $(Ce)^*$ condition.

Consider a sequence $\{u_{\alpha_n}\}$ such that $\{\alpha_n\}$ is admissible and

$$u_{\alpha_n} \in X_{\alpha_n}, \quad \sup I(u_{\alpha_n}) < \infty, \quad (1 + \|u_{\alpha_n}\|)I'_{\alpha_n}(u_{\alpha_n}) \to 0 \quad \text{as } n \to \infty \tag{6.28}$$

Let $u_n := u_{\alpha_n}$ for convenience. Similar to the proof of the Lemma 6.1, we can prove that $\{u_n\}$ has a convergent subsequence. Thus I satisfies the $(Ce)^*$ condition.

Summing up the above, I satisfies all the assumptions of Theorem 1.6. Hence by Theorem 1.6, problem (6.9) has at least one nontrivial solution. The proof of Theorem 6.5 is completed. $\square$

Proof of Theorem 6.6. Choose $h \in C^\infty([0,\infty), \mathbb{R})$ such that $0 \le h(t) \le 1$ for $t \in [0,\infty)$, and for every $\varepsilon > 0$, $h(t) = 1$ for $0 \le t \le \varepsilon/2$, $h(t) = 0$ for $t \ge \varepsilon$. Let $\varphi(u) = h(\|u\|)$. We consider the truncated functional

$$\widetilde{I}(u) = \frac{1}{2}\int_\Omega |\Delta u|^2 \, dx - \varphi(u)\int_\Omega F(x,u) \, dx.$$

where $F(x,t) = \int_0^t f(x,s)\,\mathrm{d}s$. We know that $\widetilde{I} \in C^1(X, \mathbb{R})$. If we can prove that $\widetilde{I}(u)$ admits a sequence of nontrivial weak solutions $\{u_n\}$ with $u_n \to 0$ as $n \to \infty$ in X, Theorem 6.6 holds. Indeed, in this circumstance, it is not hard to see that $I(u) = \widetilde{I}(u)$ for $\|u\| < \varepsilon/2$. So the weak solutions of $\widetilde{I}$ are just weak solutions of problems (6.13). By applying Theorem 1.12 we show $\widetilde{I}(u)$ admits a sequence of nontrivial weak solutions converges to zero in X.

For $\|u\| \ge \varepsilon$, we have $\widetilde{I}(u) = \int_\Omega \frac{1}{2}|\Delta u|^2 \, dx$, which implies $\widetilde{I}(u_n) \to \infty$ as $\|u_n\| \to \infty$. Hence $\widetilde{I}$ is coercive on X. Thus $\widetilde{I}$ is bounded from below and satisfies the (PS) condition. By (E10), it is easy to see that $\widetilde{I}$ is even and $\widetilde{I}(0) = 0$. This shows that (Φ_1) holds.

By (E13), for any $\eta > 0$, there exists $\delta > 0$, such that

$$F(x,t) \ge \frac{1}{2}\eta^{-1}t^2 \text{ for } |t| \le \delta.$$

Given any $k \in \mathbb{N}$, let $E_k := \operatorname{span}\{\varphi_1, \varphi_2, \dots, \varphi_k\}$ be a k-dimensional subspace of X, where φ_i is the eigenfunction and λ_i is eigenvalue of biharmonic operator. Then there exists a constant C_k such that $|u| \le C_k\|u\|$ for $u \in E_k$. Therefore, for any $u \in E_k$ with $\|u\| = \rho \le \min\{\frac{\delta}{C_k}, \frac{\varepsilon}{2}\}$ and ε, η small enough, we have

$$\widetilde{I}(u) = \frac{1}{2}\int_\Omega |\Delta u|^2 \, dx - \varphi(u)\int_\Omega F(x,u) \, dx.$$
$$\le \frac{\rho^2}{2} - \frac{1}{2}\eta^{-1}C_{19}^2\rho^2$$
$$< 0,$$

that is, $\{u \in E_k : \|u\| = \rho\} \subset \{u \in X : \tilde{I}(u) < 0\}$. Since $A = \{u \in E_k : \|u\| = \rho\}$ is a sphere with radius ρ in E_k, a $k - 1$-dimensional subspace of E_k, so $\gamma(A) = k$ by Proposition 1.12. Therefore, $\gamma(\{u \in X : \tilde{I}(u) < 0\}) \geq \gamma(A) = k$. Let $A_k = \{u \in X : \tilde{I}(u) < 0\}$, we have $A_k \in \Gamma_k$, and $\sup_{u \in A_k} \tilde{I}(u) < 0$. This shows (Φ_1) holds. Hence, by Theorem 1.12, we complete the proof of Theorem 6.6. $\qquad\square$

Notes and Comments

In recent years, the study of differential equations and variational problems with $p(x)$-growth conditions had been an interesting topic, which arises from nonlinear electrorheological fluids and elastic mechanics. In that context we refer the reader to Ruzicka [386], Zhikov [385] and the references therein. Recently, Harjulehto *et al.* [420] surveyed the differential equations with non-standard growth and compared results on existence and regularity. Moreover, we point out that elliptic equations involving the non-standard growth are not trivial generalizations of similar problems studied in the constant case since the non-standard growth operator is not homogeneous and, thus, some techniques which can be applied in the case of the constant growth operators will fail in this new situation, such as the Lagrange Multiplier Theorem.

This section is motivated by recent advances in mathematical modeling of non-Newtonian fluids and elastic mechanics, in particular, the electrorheological fluids (smart fluids). This important class of fluids is characterized by the change of viscosity which is not easy and depends on the electric field. These fluids, which are known under the name ER fluids, have many applications in elastic mechanics, fluid dynamics etc.. For more information, the reader can refer to Ruzicka [386]. This section was obtained in [402].

6.5   Two Parameter $p(x)$-Biharmonic Problems

In this section, we consider the fourth-order quasilinear elliptic equation

$$\Delta^2_{p(x)} u + |u|^{p(x)-2} u = \lambda f(x, u) + \mu g(x, u), \quad \text{in } \Omega,$$
$$u = 0, \quad \Delta u = 0, \quad \text{on } \partial\Omega, \tag{6.29}$$

where $\Delta^2_{p(x)} u = \Delta(|\Delta u|^{p(x)-2} \Delta u)$ is the $p(x)$-biharmonic operator of fourth order, $\lambda, \mu \in [0, \infty)$, $\Omega \subset \mathbb{R}^N (N > 1)$ is a nonempty bounded open set with a sufficient smooth boundary $\partial\Omega$. $f, g \colon \Omega \times \mathbb{R} \to \mathbb{R}$ are Carathéodory functions. Next, let $F(x, u) = \int_0^u f(x, s)ds$ and $G(x, u) = \int_0^u g(x, s)ds$. For $p \in C(\overline{\Omega})$, denote $1 < p^- = \min_{x \in \overline{\Omega}} p(x) \leq p^+ = \max_{x \in \overline{\Omega}} p(x) < +\infty$. Moreover,

$$p_2^*(x) = \begin{cases} \dfrac{Np(x)}{N - 2p(x)} & p(x) < \dfrac{N}{2}, \\ \infty & p(x) \geq \dfrac{N}{2}, \end{cases}$$

is the critical exponent just as in many papers. Obviously, $p(x) < p^*(x)$ for all $x \in \overline{\Omega}$. In the sequel, X will denote the Sobolev space $W^{2,p(x)}(\Omega) \cap W_0^{1,p(x)}(\Omega)$.

The energy functional corresponding to problem (6.29) is defined on X as

$$H(u) = \Phi(u) + \lambda\Psi(u) + \mu J(u), \tag{6.30}$$

where

$$\Phi(u) = \int_\Omega \frac{1}{p(x)}(|\Delta u|^{p(x)} + |u|^{p(x)})dx, \tag{6.31}$$

$$\Psi(u) = -\int_\Omega F(x, u)dx, \tag{6.32}$$

$$J(u) = -\int_\Omega G(x, u)dx. \tag{6.33}$$

Let us recall that a weak solution of (6.29) is any $u \in X$ such that

$$\int_\Omega (|\Delta u|^{p(x)-2}\Delta u\Delta v + |u|^{p(x)-2}uv)dx$$

$$= \lambda \int_\Omega f(x, u)vdx + \mu \int_\Omega g(x, u)vdx \quad \text{for all } v \in X.$$

The aim of this section is to prove the following result

Theorem 6.7. *Assume that* $\sup_{(x,s) \in \Omega \times \mathbb{R}} \frac{|f(x,s)|}{1+|s|^{t(x)-1}} < +\infty$, *where* $t \in C(\overline{\Omega})$ *and* $t(x) < p^*(x)$ *for all* $x \in \overline{\Omega}$ *and there exist two positive constants* ϱ, ϑ *and a function* $\gamma(x) \in C(\overline{\Omega})$ *with* $1 < \gamma^- \leq \gamma^+ < p^-$, *such that*

(I1) $F(x, s) > 0$ *for a.e.* $x \in \Omega$ *and all* $s \in]0, \varrho]$;

(I2) *there exist* $p_1(x) \in C(\overline{\Omega})$ *and* $p^+ < p_1^- \leq p_1(x) < p^*(x)$, *such that*

$$\limsup_{s \to 0} \sup_{x \in \Omega} \frac{F(x, s)}{|s|^{p_1(x)}} < +\infty;$$

(I3) $|F(x, s)| \leq \vartheta(1 + |s|^{\gamma(x)})$ *for a.e.* $x \in \Omega$ *and all* $s \in \mathbb{R}$.

Then, there exist an open interval $\Lambda \subseteq (0, +\infty)$ and a positive real number ρ with the following property: for each $\lambda \in \Lambda$ and each function $g(x, s) \colon \Omega \times \mathbb{R} \to \mathbb{R}$ satisfying

$$\sup_{(x,s) \in \Omega \times \mathbb{R}} \frac{|g(x,s)|}{1 + |s|^{p_2(x)-1}} < +\infty,$$

where $p_2 \in C(\overline{\Omega})$ and $p_2(x) < p^*(x)$ for all $x \in \overline{\Omega}$, there exists $\delta > 0$ such that, for each $\mu \in [0, \delta]$, problem (6.29) has at least three weak solutions whose norms in X are less than ρ.

Remark 6.9. The conclusion of Theorem 6.7 gives a precise information about the $p(x)$-biharmonic equation (6.29) with parameter, namely, one can see that (6.29) is stable with respect to small perturbations.

We denote by $W_0^{k,p(x)}(\Omega)$ the closure of $C_0^\infty(\Omega)$ in $W^{k,p(x)}(\Omega)$.

Note that the weak solutions of (6.29) are considered in the generalized Sobolev space

$$X := W^{2,p(x)}(\Omega) \cap W_0^{1,p(x)}(\Omega),$$

equipped with the norm

$$\|u\| = \inf \left\{ \alpha > 0 : \int_\Omega \left(\left| \frac{\Delta u(x)}{\alpha} \right|^{p(x)} + \left| \frac{u(x)}{\alpha} \right|^{p(x)} \right) dx \le 1 \right\}.$$

Remark 6.10. (1) According to [15], the norm $\| \cdot \|_{2,p(x)}$, cited in the preliminaries, is equivalent to the norm $|\Delta \cdot |_{p(x)}$ in the space X. Consequently, the norms $\| \cdot \|_{2,p(x)}, \| \cdot \|$ and $|\Delta \cdot |_{p(x)}$ are equivalent.

(2) By the above remark and Proposition 1.10, there is a continuous and compact embedding of X into $L^{q(x)}(\Omega)$, where $q(x) < p_2^*(x)$ for all $x \in \overline{\Omega}$.

We consider the functional

$$\Phi(u) = \int_\Omega \frac{1}{p(x)} \big(|\Delta u|^{p(x)} + |u|^{p(x)} \big) dx.$$

It is well known that $\Phi(u)$ is well defined and continuous differentiable in X. Now we give the following fundamental proposition.

Proposition 6.2. *For $u \in X$ we have*

(1) $\|u\| < (=; >)1 \Leftrightarrow \Phi(u) < (=; >)1$,
(2) $\|u\| \le 1 \Rightarrow \|u\|^{p^+} \le \Phi(u) \le \|u\|^{p^-}$,
(3) $\|u\| \ge 1 \Rightarrow \|u\|^{p^-} \le \Phi(u) \le \|u\|^{p^+}$, for all $u_n \in X$ we have
(4) $\|u_n\| \to 0 \Leftrightarrow \Phi(u_n) \to 0$,

(5) $\|u_n\| \to \infty \Leftrightarrow \Phi(u_n) \to \infty$.

The proof of this proposition is similar to the proof in [9, Theorem 1.3]. Moreover, the operator $T := \Phi' : X \to X'$ defined as

$$\langle T(u), v \rangle = \int_\Omega (|\Delta u|^{p(x)-2}\Delta u \Delta v + |u|^{p(x)-2}uv)dx \quad \text{for any } u, v \in X,$$

satisfies the assertions of the following theorem.

Theorem 6.8. *The following statements hold:*

(1) T is continuous, bounded and strictly monotone.
(2) T is of (S_+) type.
(3) T is a homeomorphism.

Proof. (1) Since T is the Fréchet derivative of Φ, it follows that T is continuous and bounded. Let us define the sets

$$U_p = \{x \in \Omega : p(x) \geq 2\}, \quad V_p = \{x \in \Omega : 1 < p(x) < 2\}.$$

Using the elementary inequalities [97]

$$|x - y|^\gamma \leq 2^\gamma(|x|^{\gamma-2}x - |y|^{\gamma-2}y)(x - y) \quad \text{if } \gamma \geq 2,$$

$$|x - y|^2 \leq \frac{1}{(\gamma - 1)}(|x| + |y|)^{2-\gamma}(|x|^{\gamma-2}x - |y|^{\gamma-2}y)(x - y) \quad \text{if } 1 < \gamma < 2,$$

for all $(x, y) \in \mathbb{R}^N \times \mathbb{R}^N$, we obtain for all $u, v \in X$ such that $u \neq v$,

$$\langle T(u) - T(v), u - v \rangle > 0,$$

which means that T is strictly monotone.

(2) Let $(u_n)_n$ be a sequence of X such that

$$u_n \rightharpoonup u \text{ weakly in } X \quad \text{and} \quad \limsup_{n \to +\infty}\langle T(u_n), u_n - u \rangle \leq 0.$$

From Proposition 6.2, it suffices to show that

$$\int_\Omega (|\Delta u_n - \Delta u|^{p(x)} + |u_n - u|^{p(x)})dx \to 0. \tag{6.34}$$

In view of the monotonicity of T, we have

$$\langle T(u_n) - T(u), u_n - u \rangle \geq 0,$$

and since $u_n \rightharpoonup u$ weakly in X, it follows that

$$\limsup_{n \to +\infty}\langle T(u_n) - T(u), u_n - u \rangle = 0. \tag{6.35}$$

Put

$$\varphi_n(x) = (|\Delta u_n|^{p(x)-2}\Delta u_n - |\Delta u|^{p(x)-2}\Delta u)(\Delta u_n - \Delta u),$$
$$\psi_n(x) = (|u_n|^{p(x)-2}u_n - |u|^{p(x)-2}u)(u_n - u).$$

By the compact embedding of X into $L^{p(x)}(\Omega)$, it follows that

$$u_n \to u \quad \text{in } L^{p(x)}(\Omega),$$
$$|u_n|^{p(x)-2}u_n \to |u|^{p(x)-2}u \quad \text{in } L^{q(x)}(\Omega),$$

where $1/q(x) + 1/p(x) = 1$ for all $x \in \Omega$. It results that

$$\int_\Omega \psi_n(x)dx \to 0. \tag{6.36}$$

It follows by (6.35) and (6.36) that

$$\limsup_{n \to +\infty} \int_\Omega \varphi_n(x)dx = 0. \tag{6.37}$$

Thanks to the above inequalities,

$$\int_{U_p} |\Delta u_n - \Delta u_k|^{p(x)}dx \le 2^{p^+} \int_{U_p} \varphi_n(x)dx,$$
$$\int_{U_p} |u_n - u_k|^{p(x)}dx \le 2^{p^+} \int_{U_p} \psi_n(x)dx.$$

Then

$$\int_{U_p} (|\Delta u_n - \Delta u|^{p(x)} + |u_n - u|^{p(x)})dx \to 0 \quad \text{as } n \to +\infty. \tag{6.38}$$

On the other hand, in V_p, setting $\delta_n = |\Delta u_n| + |\Delta u|$, we have

$$\int_{V_p} |\Delta u_n - \Delta u|^{p(x)}dx \le \frac{1}{p^- - 1}\int_{V_p} (\varphi_n)^{\frac{p(x)}{2}}(\delta_n)^{\frac{p(x)}{2}(2-p(x))}dx.$$

For $d > 0$, by Young's inequality,

$$d\int_{V_p} |\Delta u_n - \Delta u|^{p(x)}dx \le \int_{V_p} [d(\varphi_n)^{\frac{p(x)}{2}}](\delta_n)^{\frac{p(x)}{2}(2-p(x))}dx,$$
$$\le \int_{V_p} \varphi_n(d)^{\frac{2}{p(x)}}dx + \int_{V_p} (\delta_n)^{p(x)}dx. \tag{6.39}$$

From (6.37) and since $\varphi_n \ge 0$, one can consider that

$$0 \le \int_{V_p} \varphi_n dx < 1.$$

If $\int_{V_p} \varphi_n dx = 0$ then $\int_{V_p} |\Delta u_n - \Delta u|^{p(x)} dx = 0$. If $0 < \int_{V_p} \varphi_n dx < 1$, we choose

$$d = \left(\int_{V_p} \varphi_n(x) dx \right)^{-1/2} > 1,$$

and the fact that $2/p(x) < 2$, inequality (6.39) becomes

$$\int_{V_p} |\Delta u_n - \Delta u|^{p(x)} dx \leq \frac{1}{d} \left(\int_{V_p} \varphi_n d^2 dx + \int_\Omega \delta_n^{p(x)} dx \right),$$

$$\leq \left(\int_{V_p} \varphi_n dx \right)^{1/2} \left(1 + \int_\Omega \delta_n^{p(x)} dx \right).$$

Note that, $\int_\Omega \delta_n^{p(x)} dx$ is bounded, which implies

$$\int_{V_p} |\Delta u_n - \Delta u|^{p(x)} dx \to 0 \quad \text{as } n \to +\infty.$$

A similar method gives

$$\int_{V_p} |u_n - u|^{p(x)} dx \to 0 \quad \text{as } n \to +\infty.$$

Hence, it results that

$$\int_{V_p} (|\Delta u_n - \Delta u|^{p(x)} + |u_n - u|^{p(x)}) dx \to 0 \quad \text{as } n \to +\infty. \tag{6.40}$$

Finally, (6.34) is given by combining (6.38) and (6.40).

(3) Note that the strict monotonicity of T implies its injectivity. Moreover, T is a coercive operator. Indeed, since $p^- - 1 > 0$, for each $u \in X$ such that $\|u\| \geq 1$ we have

$$\frac{\langle T(u), u \rangle}{\|u\|} = \frac{\Phi(u)}{\|u\|} \geq \|u\|^{p^- - 1} \to \infty \quad \text{as } \|u\| \to \infty.$$

Consequently, thanks to Minty-Browder theorem [299], the operator T is an surjection and admits an inverse mapping. It suffices then to show the continuity of T^{-1}. Let $(f_n)_n$ be a sequence of X' such that $f_n \to f$ in X'. Let u_n and u in X such that

$$T^{-1}(f_n) = u_n \quad \text{and} \quad T^{-1}(f) = u.$$

By the coercivity of T, one deducts that the sequence (u_n) is bounded in the reflexive space X. For a subsequence, we have $u_n \rightharpoonup \hat{u}$ in X, which implies

$$\lim_{n \to +\infty} \langle T(u_n) - T(u), u_n - \hat{u} \rangle = \lim_{n \to +\infty} \langle f_n - f, u_n - \hat{u} \rangle = 0.$$

It follows by the second assertion and the continuity of T that

$$u_n \to \hat{u} \quad \text{in } X \quad \text{and} \quad T(u_n) \to T(\hat{u}) = T(u) \quad \text{in } X'.$$

Moreover, since T is an injection, we conclude that $u = \hat{u}$. $\qquad \square$

Now we can give the proof of our main result.

Proof Theorem 6.7. Set $\Phi(u)$, $\Psi(u)$ and $J(u)$ as (6.31), (6.32) and (6.33). So, for each u, $v \in X$, one has

$$\langle \Phi'(u), v \rangle = \int_\Omega (|\Delta u|^{p(x)-2} \Delta u \Delta v + |u|^{p(x)-2} uv) \, dx,$$

$$\langle \Psi'(u), v \rangle = - \int_\Omega f(x, u) v \, dx,$$

$$\langle J'(u), v \rangle = - \int_\Omega g(x, u) v \, dx.$$

From Theorem 6.8, of course, Φ is a continuous Gâteaux differentiable and sequentially weakly lower semicontinuous functional whose Gâteaux derivative admits a continuous inverse on X', moreover, Ψ and J are continuously Gâteaux differentiable functionals whose Gâteaux derivative is compact. Obviously, Φ is bounded on each bounded subset of X under our assumptions.

From Proposition 6.2, we have: if $\|u\| \geq 1$, then

$$\frac{1}{p^+} \|u\|^{p^-} \leq \Phi(u) \leq \frac{1}{p^-} \|u\|^{p^+}. \tag{6.41}$$

Meanwhile, for each $\lambda \in \Lambda$,

$$\lambda \Psi(u) = -\lambda \int_\Omega F(x, u) dx$$

$$\geq -\lambda \int_\Omega \vartheta (1 + |u|^{\gamma(x)}) dx$$

$$\geq -\lambda \vartheta (|\Omega| + |u|_{\gamma(x)}^{\gamma^+})$$

$$\geq -C_2 (1 + |u|_{\gamma(x)}^{\gamma^+})$$

$$\geq -C_3 (1 + \|u\|^{\gamma^+})$$

for any $u \in X$, where C_2 and C_3 are positive constants. Here, we use condition (I3) and (ii) of Proposition 1.9. Combining the two inequalities above, we obtain

$$\Phi(u) + \lambda \Psi(u) \geq \frac{1}{p^+} \|u\|^{p^-} - C_3 (1 + \|u\|^{\gamma^+}),$$

because of $\gamma^+ < p^-$, it follows that

$$\lim_{\|u\| \to +\infty} (\Phi(u) + \lambda \Psi(u)) = +\infty \quad \forall u \in X, \quad \lambda \in [0, +\infty).$$

Then assumption $\lim_{\|u\| \to +\infty}(\Phi(u) + \lambda J(u)) = +\infty$ of Theorem 1.14 is satisfied.

Next, we will prove that the following assumptions is also satisfied. It suffices to verify the conditions of Proposition 1.14. Let $u_0 = 0$, we can easily have

$$\Phi(u_0) = -\Psi(u_0) = 0.$$

Now we claim that $\sup_{u \in \Phi^{-1}(]-\infty, r])} -\Psi(u) < r\frac{-\Psi(u_1)}{\Phi(u_1)}$ is satisfied.

From (12), exist $\eta \in [0,1]$, $C_4 > 0$, such that

$$F(x,s) < C_4|s|^{p_1(x)} < C_4|s|^{p_1^-} \quad \forall s \in [-\eta, \eta], \text{ a.e. } x \in \Omega.$$

Then, from (13), we can find a constant M such that

$$F(x,s) < M|s|^{p_1^-}$$

for all $s \in \mathbb{R}$ and a.e. $x \in \Omega$. Consequently, by the Sobolev embedding theorem ($X \hookrightarrow L^{p_1^-}(\Omega)$ is continuous), we have (for suitable positive constant C_5, C_6)

$$-\Psi(u) = \int_\Omega F(x,u)dx < M \int_\Omega |u|^{p_1^-}\, dx \leq C_5\|u\|^{p_1^-} \leq C_6 r^{p_1^-/p^+},$$

when $\|u\|^{p^+}/p^+ \leq r$. Hence, being $p_1^- > p^+$, it follows that

$$\lim_{r \to 0^+} \frac{\sup_{\|u\|^{p^+}/p^+ \leq r} -\Psi(u)}{r} = 0. \tag{6.42}$$

Let $u_1 \in C^2(\Omega)$ be a function positive in Ω, with $u_1|_{\partial\Omega} = 0$ and $\max_{\overline{\Omega}} u_1 \leq d$. Then, of course, $u_1 \in X$ and $\Phi(u_1) > 0$. In view of (i_1) we also have $-\Psi(u_1) = \int_\Omega F(x, u_1(x))dx > 0$. Therefore, from (6.42), we can find $r \in \left(0, \min\{\Phi(u_1), \frac{1}{p^+}\}\right)$ such that

$$\sup_{\|u\|^{p^+}/p^+ \leq r} (-\Psi(u)) < r\frac{-\Psi(u_1)}{\Phi(u_1)}.$$

Now, let $u \in \Phi^{-1}((-\infty, r])$. Then, $\int_\Omega(|\Delta u|^{p(x)} + |u|^{p(x)})dx \leq rp^+ < 1$ which, by Proposition 6.2, implies $\|u\| < 1$. Consequently,

$$\frac{1}{p^+}\|u\|^{p^+} \leq \int_\Omega \frac{1}{p(x)}(|\Delta u|^{p(x)} + |u|^{p(x)})dx < r.$$

Therefore, we infer that $\Phi^{-1}((-\infty, r]) \subset \{u \in X : \frac{1}{p^+}\|u\|^{p^+} < r\}$, and so

$$\sup_{u \in \Phi^{-1}(]-\infty, r])} -\Psi(u) < r\frac{-\Psi(u_1)}{\Phi(u_1)}.$$

At this point, conclusion follows from Proposition 1.14 and Theorem 1.14.
$\square$

Notes and Comments

Ricceri's three critical points theorem is a powerful tool to study boundary problem of differential equation (see, for example, [391, 421–423]). Particularly, Mihailescu [393] used three critical points theorem of Ricceri [37] to study a particular $p(x)$-Laplacian equation. He proved existence of three solutions for the problem. Liu [394] studied the solutions of the general $p(x)$-Laplacian equations with Neumann or Dirichlet boundary condition on a bounded domain, and obtained three solutions under appropriate hypotheses. Shi [424] generalized the corresponding result of [393]. To our best of knowledge, there is no result of multiple solutions of $p(x)$-biharmonic equation under sublinear condition. This section was obtained in [401].

Bibliography

[1] L. C. Evans, *Partial differential equations*, Graduate Studies in Mathematics, Vol. 19, 2nd edn. American Mathematical Society, Providence, RI (2010), ISBN 978-0-8218-4974-3, doi:10.1090/gsm/019, http://dx.doi.org/10.1090/gsm/019.

[2] R. A. Adams and J. J. F. Fournier, *Sobolev spaces*, Pure and Applied Mathematics (Amsterdam), Vol. 140, 2nd edn. Elsevier/Academic Press, Amsterdam (2003).

[3] D. Gilbarg and N. S. Trudinger, *Elliptic partial differential equations of second order*, Classics in Mathematics. Springer-Verlag, Berlin (2001), ISBN 3-540-41160-7, reprint of the 1998 edition.

[4] H. Brezis, *Functional analysis, Sobolev spaces and partial differential equations*, Universitext. Springer, New York (2011), ISBN 978-0-387-70913-0.

[5] M. Willem, *Minimax theorems*, Progress in Nonlinear Differential Equations and their Applications, Vol. 24. Birkhäuser Boston, Inc., Boston, MA (1996), ISBN 0-8176-3913-6, doi:10.1007/978-1-4612-4146-1, http://dx.doi.org/10.1007/978-1-4612-4146-1.

[6] O. Kováčik and J. Rákosník, On spaces $L^{p(x)}$ and $W^{k,p(x)}$, *Czechoslovak Math. J.* **41(116)**, 4, pp. 592–618 (1991).

[7] S. Samko, Denseness of $C_0^\infty(\mathbf{R}^N)$ in the generalized Sobolev spaces $W^{M,P(X)}(\mathbf{R}^N)$, in *Direct and inverse problems of mathematical physics (Newark, DE, 1997)*, Int. Soc. Anal. Appl. Comput., Vol. 5. Kluwer Acad. Publ., Dordrecht, pp. 333–342 (2000), doi:10.1007/978-1-4757-3214-6_20, http://dx.doi.org/10.1007/978-1-4757-3214-6_20.

[8] X. Fan, J. Shen and D. Zhao, Sobolev embedding theorems for spaces $W^{k,p(x)}(\Omega)$, *J. Math. Anal. Appl.* **262**, 2, pp. 749–760 (2001), doi:10.1006/jmaa.2001.7618, http://dx.doi.org/10.1006/jmaa.2001.7618.

[9] X. Fan and D. Zhao, On the spaces $L^{p(x)}(\Omega)$ and $W^{m,p(x)}(\Omega)$, *J. Math. Anal. Appl.* **263**, 2, pp. 424–446 (2001), doi:10.1006/jmaa.2000.7617, http://dx.doi.org/10.1006/jmaa.2000.7617.

[10] D. Cruz-Uribe, L. Diening and P. Hästö, The maximal operator on weighted variable Lebesgue spaces, *Fract. Calc. Appl. Anal.* **14**, 3, pp. 361–374 (2011), doi:10.2478/s13540-011-0023-7, http://dx.doi.org/10.2478/

s13540-011-0023-7.

[11] X. Fan, Solutions for $p(x)$-Laplacian Dirichlet problems with singular co-
      efficients, *J. Math. Anal. Appl.* **312**, 2, pp. 464–477 (2005), doi:10.1016/j.
      jmaa.2005.03.057, http://dx.doi.org/10.1016/j.jmaa.2005.03.057.

[12] X. Fan and X. Han, Existence and multiplicity of solutions for $p(x)$-
      Laplacian equations in $\mathbf{R}^N$, *Nonlinear Anal.* **59**, 1-2, pp. 173–188
      (2004), doi:10.1016/j.na.2004.07.009, http://dx.doi.org/10.1016/j.na.
      2004.07.009.

[13] D. E. Edmunds, J. Lang and A. Nekvinda, On $L^{p(x)}$ norms, *R. Soc. Lond.*
      Proc. Ser. A Math. Phys. Eng. Sci. **455**, 1981, pp. 219–225 (1999), doi:
      10.1098/rspa.1999.0309, http://dx.doi.org/10.1098/rspa.1999.0309.

[14] D. E. Edmunds and J. Rákosník, Sobolev embeddings with variable expo-
      nent, *Studia Math.* **143**, 3, pp. 267–293 (2000).

[15] A. Zang and Y. Fu, Interpolation inequalities for derivatives in variable
      exponent Lebesgue–Sobolev spaces, *Nonlinear Anal.* **69**, 10, pp. 3629–3636
      (2008), doi:10.1016/j.na.2007.10.001, http://dx.doi.org/10.1016/j.na.
      2007.10.001.

[16] A. El Amrouss, F. Moradi and M. Moussaoui, Existence of solutions for
      fourth-order PDEs with variable exponents, *Electron. J. Differential Equa-*
      tions, pp. No. 153, 13 (2009).

[17] W. Zou and M. Schechter, *Critical point theory and its applications.*
      Springer, New York (2006), ISBN 978-0-387-32965-9; 0-387-32965-X.

[18] I. Ekeland, On the variational principle, *J. Math. Anal. Appl.* **47**, pp. 324–
      353 (1974).

[19] J.-B. Hiriart-Urruty, A short proof of the variational principle for approx-
      imate solutions of a minimization problem, *Amer. Math. Monthly* **90**, 3,
      pp. 206–207 (1983), doi:10.2307/2975554, http://dx.doi.org/10.2307/
      2975554.

[20] G. Cerami, An existence criterion for the critical points on unbounded
      manifolds, *Istit. Lombardo Accad. Sci. Lett. Rend. A* **112**, 2, pp. 332–336
      (1979) (1978).

[21] P. H. Rabinowitz, *Minimax methods in critical point theory with applica-*
      tions to differential equations, CBMS Regional Conference Series in Math-
      ematics, Vol. 65. Published for the Conference Board of the Mathematical
      Sciences, Washington, DC; by the American Mathematical Society, Provi-
      dence, RI (1986), ISBN 0-8218-0715-3.

[22] I. Ekeland, *Convexity methods in Hamiltonian mechanics*, Ergebnisse
      der Mathematik und ihrer Grenzgebiete (3) [Results in Mathemat-
      ics and Related Areas (3)], Vol. 19. Springer-Verlag, Berlin (1990),
      ISBN 3-540-50613-6, doi:10.1007/978-3-642-74331-3, http://dx.doi.org/
      10.1007/978-3-642-74331-3.

[23] S. Luan and A. Mao, Periodic solutions for a class of non-autonomous
      Hamiltonian systems, *Nonlinear Anal.* **61**, 8, pp. 1413–1426 (2005), doi:10.
      1016/j.na.2005.01.108, http://dx.doi.org/10.1016/j.na.2005.01.108.

[24] S. J. Li and M. Willem, Applications of local linking to critical point theory,
      J. Math. Anal. Appl. **189**, 1, pp. 6–32 (1995), doi:10.1006/jmaa.1995.1002,

http://dx.doi.org/10.1006/jmaa.1995.1002.

[25] M. Schechter, *Linking methods in critical point theory*. Birkhäuser Boston, Inc., Boston, MA (1999), ISBN 0-8176-4095-9, doi:10.1007/ 978-1-4612-1596-7, http://dx.doi.org/10.1007/978-1-4612-1596-7.

[26] A. Ambrosetti and A. Malchiodi, *Nonlinear analysis and semilinear elliptic problems*, Cambridge Studies in Advanced Mathematics, Vol. 104. Cambridge University Press, Cambridge (2007).

[27] T. Bartsch, Infinitely many solutions of a symmetric Dirichlet problem, *Nonlinear Anal.* **20**, 10, pp. 1205–1216 (1993).

[28] P. Bartolo, V. Benci and D. Fortunato, Abstract critical point theorems and applications to some nonlinear problems with "strong" resonance at infinity, *Nonlinear Anal.* **7**, 9, pp. 981–1012 (1983).

[29] R. Kajikiya, A critical point theorem related to the symmetric mountain pass lemma and its applications to elliptic equations, *J. Funct. Anal.* **225**, 2, pp. 352–370 (2005), doi:10.1016/j.jfa.2005.04.005, http://dx.doi.org/ 10.1016/j.jfa.2005.04.005.

[30] Z. Liu and Z.-Q. Wang, On Clark's theorem and its applications to partially sublinear problems, *Ann. Inst. H. Poincaré Anal. Non Linéaire* **32**, 5, pp. 1015–1037 (2015), doi:10.1016/j.anihpc.2014.05.002, http://dx.doi.org/ 10.1016/j.anihpc.2014.05.002.

[31] B. Ricceri, A three critical points theorem revisited, *Nonlinear Anal.* **70**, 9, pp. 3084–3089 (2009), doi:10.1016/j.na.2008.04.010, http://dx.doi.org/ 10.1016/j.na.2008.04.010.

[32] G. Bonanno, Some remarks on a three critical points theorem, *Nonlinear Anal.* **54**, 4, pp. 651–665 (2003).

[33] B. Ricceri, Existence of three solutions for a class of elliptic eigenvalue problems, *Math. Comput. Modelling* **32**, 11–13, pp. 1485–1494 (2000), doi:10.1016/S0895-7177(00)00220-X, http://dx.doi.org/ 10.1016/S0895-7177(00)00220-X, nonlinear operator theory.

[34] G. Bonanno, A minimax inequality and its applications to ordinary differential equations, *J. Math. Anal. Appl.* **270**, 1, pp. 210–229 (2002).

[35] B. Ricceri, A further three critical points theorem, *Nonlinear Anal.* **71**, 9, pp. 4151–4157 (2009), doi:10.1016/j.na.2009.02.074, http://dx.doi.org/ 10.1016/j.na.2009.02.074.

[36] B. Ricceri, A general variational principle and some of its applications, *J. Comput. Appl. Math.* **113**, 1-2, pp. 401–410 (2000a), doi:10.1016/ S0377-0427(99)00269-1, http://dx.doi.org/10.1016/S0377-0427(99)00 269-1, fixed point theory with applications in nonlinear analysis.

[37] B. Ricceri, On a three critical points theorem, *Arch. Math. (Basel)* **75**, 3, pp. 220–226 (2000b), doi:10.1007/s000130050496, http://dx.doi.org/10. 1007/s000130050496.

[38] B. Ricceri, A note on the Neumann problem, *Complex Var. Elliptic Equ.* **55**, 5–6, pp. 593–599 (2010), doi:10.1080/17476930903276142, http://dx. doi.org/10.1080/17476930903276142.

[39] G. Bonanno and S. A. Marano, On the structure of the critical set of non-differentiable functions with a weak compactness condition, *Appl. Anal.* **89**,

1, pp. 1–10 (2010), doi:10.1080/00036810903397438, `http://dx.doi.org/`
`10.1080/00036810903397438`.

[40] G. Bonanno and P. Candito, Non-differentiable functionals and applications to elliptic problems with discontinuous nonlinearities, *J. Differential Equations* **244**, 12, pp. 3031–3059 (2008), doi:10.1016/j.jde.2008.02.025, `http://dx.doi.org/10.1016/j.jde.2008.02.025`.

[41] G. Bonanno and G. M. Bisci, Infinitely many solutions for a boundary value problem with discontinuous nonlinearities, *Bound. Value Probl.*, pp. Art. ID 670675, 20 (2009).

[42] G. Bonanno, A critical point theorem via the Ekeland variational principle, *Nonlinear Anal.* **75**, 5, pp. 2992–3007 (2012).

[43] G. Bonanno, Relations between the mountain pass theorem and local minima, *Adv. Nonlinear Anal.* **1**, 3, pp. 205–220 (2012).

[44] A. C. Lazer and P. J. McKenna, Large-amplitude periodic oscillations in suspension bridges: some new connections with nonlinear analysis, *SIAM Rev.* **32**, 4, pp. 537–578 (1990), doi:10.1137/1032120, `http://dx.doi.org/`
`10.1137/1032120`.

[45] Y. An and R. Liu, Existence of nontrivial solutions of an asymptotically linear fourth-order elliptic equation, *Nonlinear Anal.* **68**, 11, pp. 3325–3331 (2008).

[46] A. M. Micheletti and A. Pistoia, Nontrivial solutions for some fourth order semilinear elliptic problems, *Nonlinear Anal.* **34**, 4, pp. 509–523 (1998), doi:10.1016/S0362-546X(97)00596-8, `http://dx.doi.org/10.`
`1016/S0362-546X(97)00596-8`.

[47] A. X. Qian and S. J. Li, Multiple solutions for a fourth-order asymptotically linear elliptic problem, *Acta Math. Sin. (Engl. Ser.)* **22**, 4, pp. 1121–1126 (2006), doi:10.1007/s10114-005-0665-7, `http://dx.doi.org/`
`10.1007/s10114-005-0665-7`.

[48] G. Tarantello, A note on a semilinear elliptic problem, *Differential Integral Equations* **5**, 3, pp. 561–565 (1992).

[49] J. Zhang and S. Li, Multiple nontrivial solutions for some fourth-order semilinear elliptic problems, *Nonlinear Anal.* **60**, 2, pp. 221–230 (2005), doi:10.1016/j.na.2004.07.047, `http://dx.doi.org/10.1016/j.na.2004.07.047`.

[50] X.-L. Liu and W.-T. Li, Existence and multiplicity of solutions for fourth-order boundary value problems with parameters, *J. Math. Anal. Appl.* **327**, 1, pp. 362–375 (2007), doi:10.1016/j.jmaa.2006.04.021, `http://dx.`
`doi.org/10.1016/j.jmaa.2006.04.021`.

[51] G. Bonanno and B. Di Bella, A boundary value problem for fourth-order elastic beam equations, *J. Math. Anal. Appl.* **343**, 2, pp. 1166–1176 (2008), doi:10.1016/j.jmaa.2008.01.049, `http://dx.doi.org/10.1016/j.`
`jmaa.2008.01.049`.

[52] G. A. Afrouzi, S. Heidarkhani and D. O'Regan, Existence of three solutions for a doubly eigenvalue fourth-order boundary value problem, *Taiwanese J. Math.* **15**, 1, pp. 201–210 (2011).

[53] G. Bonanno and B. Di Bella, Infinitely many solutions for a fourth-order elastic beam equation, *NoDEA Nonlinear Differential Equations Appl.* **18**,

3, pp. 357–368 (2011), doi:10.1007/s00030-011-0099-0, `http://dx.doi.org/10.1007/s00030-011-0099-0`.

[54] G. Bonanno, B. Di Bella and D. O'Regan, Non-trivial solutions for nonlinear fourth-order elastic beam equations, *Comput. Math. Appl.* **62**, 4, pp. 1862–1869 (2011), doi:10.1016/j.camwa.2011.06.029, `http://dx.doi.org/10.1016/j.camwa.2011.06.029`.

[55] S. Heidarkhani, Non-trivial solutions for two-point boundary-value problems of fourth-order Sturm–Liouville type equations, *Electron. J. Differential Equations*, pp. No. 27, 9 (2012).

[56] S. Heidarkhani, Existence of solutions for a two-point boundary-value problem of a fourth-order Sturm–Liouville type, *Electron. J. Differential Equations*, pp. No. 84, 15 (2012).

[57] S. M. Khalkhali, S. Heidarkhani and A. Razani, Infinitely many solutions for a fourth-order boundary-value problem, *Electron. J. Differential Equations*, pp. No. 164, 14 (2012).

[58] Y. Song, A nonlinear boundary value problem for fourth-order elastic beam equations, *Bound. Value Probl.*, pp. 2014:191, 11 (2014), doi:10.1186/s13661-014-0191-6, `http://dx.doi.org/10.1186/s13661-014-0191-6`.

[59] G. A. Afrouzi and S. Shokooh, Three solutions for a fourth-order boundary-value problem, *Electron. J. Differential Equations*, pp. No. 45, 11 (2015).

[60] G. Bonanno, A. Chinnì and S. A. Tersian, Existence results for a two point boundary value problem involving a fourth-order equation, *Electron. J. Qual. Theory Differ. Equ.*, pp. No. 33, 9 (2015).

[61] G. Bonanno and B. Di Bella, A fourth-order boundary value problem for a Sturm-Liouville type equation, *Appl. Math. Comput.* **217**, 8, pp. 3635–3640 (2010), doi:10.1016/j.amc.2010.10.019, `http://dx.doi.org/10.1016/j.amc.2010.10.019`.

[62] Y. Wang and Y. Shen, Infinitely many sign-changing solutions for a class of biharmonic equation without symmetry, *Nonlinear Anal.* **71**, 3–4, pp. 967–977 (2009), doi:10.1016/j.na.2008.11.052, `http://dx.doi.org/10.1016/j.na.2008.11.052`.

[63] J. Zhang, Infinitely many nontrivial solutions for a class of biharmonic equations via variant fountain theorems, *Electron. J. Qual. Theory Differ. Equ.*, pp. No. 9, 14 (2011).

[64] W. Wang and P. Zhao, Nonuniformly nonlinear elliptic equations of p-biharmonic type, *J. Math. Anal. Appl.* **348**, 2, pp. 730–738 (2008), doi:10.1016/j.jmaa.2008.07.068, `http://dx.doi.org/10.1016/j.jmaa.2008.07.068`.

[65] P. Candito and R. Livrea, Infinitely many solution for a nonlinear Navier boundary value problem involving the p-biharmonic, *Stud. Univ. Babeş-Bolyai Math.* **55**, 4, pp. 41–51 (2010).

[66] P. Candito and G. Molica Bisci, Multiple solutions for a Navier boundary value problem involving the p-biharmonic operator, *Discrete Contin. Dyn. Syst. Ser. S* **5**, 4, pp. 741–751 (2012), doi:10.3934/dcdss.2012.5.741, `http://dx.doi.org/10.3934/dcdss.2012.5.741`.

[67] C. Li and C.-L. Tang, Three solutions for a Navier boundary value prob-

lem involving the p-biharmonic, *Nonlinear Anal.* **72**, 3–4, pp. 1339–1347 (2010), doi:10.1016/j.na.2009.08.011, `http://dx.doi.org/10.1016/j.na.2009.08.011`.

[68] H. Liu and N. Su, Existence of three solutions for a p-biharmonic problem, *Dyn. Contin. Discrete Impuls. Syst. Ser. A Math. Anal.* **15**, 3, pp. 445–452 (2008).

[69] S. Heidarkhani, Y. Tian and C.-L. Tang, Existence of three solutions for a class of $(p_1, \ldots, p_n)$-biharmonic systems with Navier boundary conditions, *Ann. Polon. Math.* **104**, 3, pp. 261–277 (2012), doi:10.4064/ap104-3-4, `http://dx.doi.org/10.4064/ap104-3-4`.

[70] S. Heidarkhani, Non-trivial solutions for a class of $(p_1, \ldots, p_n)$-biharmonic systems with Navier boundary conditions, *Ann. Polon. Math.* **105**, 1, pp. 65–76 (2012), doi:10.4064/ap105-1-6, `http://dx.doi.org/10.4064/ap105-1-6`.

[71] S. Heidarkhani, Existence of non-trivial solutions for systems of n fourth order partial differential equations, *Math. Slovaca* **64**, 5, pp. 1249–1266 (2014), doi:10.2478/s12175-014-0273-z, `http://dx.doi.org/10.2478/s12175-014-0273-z`.

[72] J. R. Graef, S. Heidarkhani and L. Kong, Multiple solutions for a class of $(p_1, \ldots, p_n)$-biharmonic systems, *Commun. Pure Appl. Anal.* **12**, 3, pp. 1393–1406 (2013), doi:10.3934/cpaa.2013.12.1393, `http://dx.doi.org/10.3934/cpaa.2013.12.1393`.

[73] M. Massar, E. M. Hssini and N. Tsouli, Infinitely many solutions for class of Navier boundary (p, q)-biharmonic systems, *Electron. J. Differential Equations*, pp. No. 163, 9 (2012).

[74] G. Molica Bisci and D. Repovš, Multiple solutions of p-biharmonic equations with Navier boundary conditions, *Complex Var. Elliptic Equ.* **59**, 2, pp. 271–284 (2014), doi:10.1080/17476933.2012.734301, `http://dx.doi.org/10.1080/17476933.2012.734301`.

[75] L. Ding, Multiple solutions for a perturbed Navier boundary value problem involving the p-biharmonic, *Bull. Iranian Math. Soc.* **41**, 1, pp. 269–280 (2015).

[76] G. Barletta and R. Livrea, Infinitely many solutions for a class of differential inclusions involving the p-biharmonic, *Differential Integral Equations* **26**, 9-10, pp. 1157–1167 (2013).

[77] A. Hadjian and S. Shakeri, Multiplicity results for a class of Navier doubly eigenvalue boundary value systems driven by a $(p_1, \ldots, p_n)$-biharmonic operator, *Politehn. Univ. Bucharest Sci. Bull. Ser. A Appl. Math. Phys.* **77**, 1, pp. 63–74 (2015).

[78] J. Liu and X. Shi, Existence of three solutions for a class of quasilinear elliptic systems involving the $(p(x), q(x))$-Laplacian, *Nonlinear Anal.* **71**, 1-2, pp. 550–557 (2009), doi:10.1016/j.na.2008.10.094, `http://dx.doi.org/10.1016/j.na.2008.10.094`.

[79] S. El Manouni and M. Kbiri Alaoui, A result on elliptic systems with Neumann conditions via Ricceri's three critical points theorem, *Nonlinear Anal.* **71**, 5-6, pp. 2343–2348 (2009), doi:10.1016/j.na.2009.01.068, `http://dx.`

doi.org/10.1016/j.na.2009.01.068.

[80] L. Zhang and W. Ge, Solvability of a kind of Sturm–Liouville boundary value problems with impulses via variational methods, *Acta Appl. Math.* **110**, 3, pp. 1237–1248 (2010), doi:10.1007/s10440-009-9504-7, http://dx. doi.org/10.1007/s10440-009-9504-7.

[81] F. Cammaroto, A. Chinnì and B. Di Bella, Multiple solutions for a Neumann problem involving the $p(x)$-Laplacian, *Nonlinear Anal.* **71**, 10, pp. 4486–4492 (2009), doi:10.1016/j.na.2009.03.009, http://dx.doi.org/10. 1016/j.na.2009.03.009.

[82] G. A. Afrouzi and S. Heidarkhani, Multiplicity results for a two-point boundary value double eigenvalue problem, *Ric. Mat.* **59**, 1, pp. 39–47 (2010).

[83] G. A. Afrouzi, S. Heidarkhani and D. O'Regan, Three solutions to a class of Neumann doubly eigenvalue elliptic systems driven by a $(p_1, \ldots, p_n)$-Laplacian, *Bull. Korean Math. Soc.* **47**, 6, pp. 1235–1250 (2010).

[84] S. Heidarkhani and Y. Tian, Multiplicity results for a class of gradient systems depending on two parameters, *Nonlinear Anal.* **73**, 2, pp. 547–554 (2010), doi:10.1016/j.na.2010.03.051, http://dx.doi.org/10.1016/j.na. 2010.03.051.

[85] L. Li and C.-L. Tang, Existence of three solutions for (p, q)-biharmonic systems, *Nonlinear Anal.* **73**, 3, pp. 796–805 (2010), doi:10.1016/j.na.2010. 04.018, http://dx.doi.org/10.1016/j.na.2010.04.018.

[86] J. Sun, H. Chen, J. J. Nieto and M. Otero-Novoa, The multiplicity of solutions for perturbed second-order Hamiltonian systems with impulsive effects, *Nonlinear Anal.* **72**, 12, pp. 4575–4586 (2010), doi:10.1016/j.na. 2010.02.034, http://dx.doi.org/10.1016/j.na.2010.02.034.

[87] J. Chabrowski and J. Marcos do Ó, On some fourth-order semilinear elliptic problems in $\mathbb{R}^N$, *Nonlinear Anal.* **49**, 6, pp. 861–884 (2002), doi:10.1016/S0362-546X(01)00144-4, http://dx.doi.org/10. 1016/S0362-546X(01)00144-4.

[88] A. M. Micheletti and A. Pistoia, Multiplicity results for a fourth-order semilinear elliptic problem, *Nonlinear Anal.* **31**, 7, pp. 895–908 (1998), doi:10.1016/S0362-546X(97)00446-X, http://dx.doi.org/10. 1016/S0362-546X(97)00446-X.

[89] L. Li and S. Heidarkhani, Existence of three solutions to a double eigenvalue problem for the p-biharmonic equation, *Ann. Polon. Math.* **104**, 1, pp. 71–80 (2012), doi:10.4064/ap104-1-5, http://dx.doi.org/10.4064/ ap104-1-5.

[90] G. Talenti, Elliptic equations and rearrangements, *Ann. Scuola Norm. Sup. Pisa Cl. Sci. (4)* **3**, 4, pp. 697–718 (1976).

[91] L. A. Pelete, R. K. A. M. Van der Vorst and V. K. Troĭ, Stationary solutions of a fourth-order nonlinear diffusion equation, *Differentsialnye Uravneniya* **31**, 2, pp. 327–337, 367 (1995).

[92] P. Candito, L. Li and R. Livrea, Infinitely many solutions for a perturbed nonlinear Navier boundary value problem involving the p-biharmonic, *Nonlinear Anal.* **75**, 17, pp. 6360–6369 (2012), doi:10.1016/j.na.2012.07.015,

http://dx.doi.org/10.1016/j.na.2012.07.015.

[93] G. Bonanno and A. Chinnì, Existence and multiplicity of weak solutions for elliptic Dirichlet problems with variable exponent, *J. Math. Anal. Appl.* **418**, 2, pp. 812–827 (2014), doi:10.1016/j.jmaa.2014.04.016, http://dx. doi.org/10.1016/j.jmaa.2014.04.016.

[94] V. D. Rădulescu, Singular phenomena in nonlinear elliptic problems: from blow-up boundary solutions to equations with singular nonlinearities, in *Handbook of differential equations: stationary partial differential equations.* Vol. IV, Handb. Differ. Equ. Elsevier/North-Holland, Amsterdam, pp. 485–593 (2007), doi:10.1016/S1874-5733(07)80010-6, http://dx.doi.org/10. 1016/S1874-5733(07)80010-6.

[95] L. Li, Existence of three solutions for a class of Navier quasilinear elliptic systems involving the $(p_1, \ldots, p_n)$-biharmonic, *Bull. Korean Math. Soc.* **50**, 1, pp. 57–71 (2013), doi:10.4134/BKMS.2013.50.1.057, http:// dx.doi.org/10.4134/BKMS.2013.50.1.057.

[96] È. Mitidieri, A simple approach to Hardy inequalities, *Mat. Zametki* **67**, 4, pp. 563–572 (2000), doi:10.1007/BF02676404, http://dx.doi.org/10. 1007/BF02676404.

[97] J. Simon, Régularité de la solution d'une équation non linéaire dans $\mathbf{R}^N$, in *Journées d'Analyse Non Linéaire (Proc. Conf., Besançon, 1977)*, Lecture Notes in Math., Vol. 665. Springer, Berlin, pp. 205–227 (1978).

[98] X.-L. Fan and Q.-H. Zhang, Existence of solutions for $p(x)$-Laplacian Dirichlet problem, *Nonlinear Anal.* **52**, 8, pp. 1843–1852 (2003), doi:10.1016/S0362-546X(02)00150-5, http://dx.doi.org/ 10.1016/S0362-546X(02)00150-5.

[99] A. Kristály and C. Varga, Multiple solutions for elliptic problems with singular and sublinear potentials, *Proc. Amer. Math. Soc.* **135**, 7, pp. 2121–2126 (2007), doi:10.1090/S0002-9939-07-08715-1, http://dx.doi.org/10. 1090/S0002-9939-07-08715-1.

[100] Y. Wang and Y. Shen, Nonlinear biharmonic equations with Hardy potential and critical parameter, *J. Math. Anal. Appl.* **355**, 2, pp. 649–660 (2009), doi:10.1016/j.jmaa.2009.01.076, http://dx.doi.org/10.1016/ j.jmaa.2009.01.076.

[101] E. Berchio, D. Cassani and F. Gazzola, Hardy–Rellich inequalities with boundary remainder terms and applications, *Manuscripta Math.* **131**, 3–4, pp. 427–458 (2010).

[102] N. T. Chung, Multiple solutions for a fourth order elliptic equation with Hardy type potential, *Acta Univ. Apulensis Math. Inform.*, 28, pp. 115–124 (2011).

[103] H. Xie and J. Wang, Infinitely many solutions for p-harmonic equation with singular term, *J. Inequal. Appl.*, pp. 2013:9, 13 (2013), doi:10.1186/ 1029-242X-2013-9, http://dx.doi.org/10.1186/1029-242X-2013-9.

[104] R. Pei and J. Zhang, Sign-changing solutions for a fourth-order elliptic equation with Hardy singular terms, *J. Appl. Math.*, pp. Art. ID 627570, 6 (2013).

[105] Y. Huang and X. Liu, Sign-changing solutions for p-biharmonic equa-

tions with Hardy potential, *J. Math. Anal. Appl.* **412**, 1, pp. 142–154 (2014), doi:10.1016/j.jmaa.2013.10.044, http://dx.doi.org/10.1016/j. jmaa.2013.10.044.

[106] M. Pérez-Llanos and A. Primo, Semilinear biharmonic problems with a singular term, *J. Differential Equations* **257**, 9, pp. 3200–3225 (2014), doi:10.1016/j.jde.2014.06.011, http://dx.doi.org/10.1016/j.jde.2014. 06.011.

[107] M. Ferrara and G. Molica Bisci, Existence results for elliptic problems with Hardy potential, *Bull. Sci. Math.* **138**, 7, pp. 846–859 (2014), doi:10.1016/j. bulsci.2014.02.002, http://dx.doi.org/10.1016/j.bulsci.2014.02.002.

[108] L. Li, Two weak solutions for some singular fourth order elliptic problems, (2015), in press.

[109] T. Bartsch and Z. Q. Wang, Existence and multiplicity results for some superlinear elliptic problems on $\mathbf{R}^N$, *Comm. Partial Differential Equations* **20**, 9-10, pp. 1725–1741 (1995).

[110] P.-L. Lions, The concentration-compactness principle in the calculus of variations. The limit case. I, *Rev. Mat. Iberoamericana* **1**, 1, pp. 145–201 (1985), doi:10.4171/RMI/6, http://dx.doi.org/10.4171/RMI/6.

[111] Y. Yin and X. Wu, High energy solutions and nontrivial solutions for fourth-order elliptic equations, *J. Math. Anal. Appl.* **375**, 2, pp. 699–705 (2011), doi:10.1016/j.jmaa.2010.10.019, http://dx.doi.org/10.1016/j.jmaa.2010.10.019.

[112] A. Kristály, Bifurcations effects in sublinear elliptic problems on compact Riemannian manifolds, *J. Math. Anal. Appl.* **385**, 1, pp. 179–184 (2012), doi:10.1016/j.jmaa.2011.06.031, http://dx.doi.org/10.1016/j. jmaa.2011.06.031.

[113] B. Ricceri, On an elliptic Kirchhoff-type problem depending on two parameters, *J. Global Optim.* **46**, 4, pp. 543–549 (2010), doi:10.1007/s10898-009-9438-7, http://dx.doi.org/10.1007/s10898-009-9438-7.

[114] D. S. Moschetto, Existence and multiplicity results for a nonlinear stationary Schrödinger equation, *Ann. Polon. Math.* **99**, 1, pp. 39–43 (2010), doi:10.4064/ap99-1-3, http://dx.doi.org/10.4064/ap99-1-3.

[115] B. E. Breckner, D. Repovš and C. Varga, On the existence of three solutions for the Dirichlet problem on the Sierpinski gasket, *Nonlinear Anal.* **73**, 9, pp. 2980–2990 (2010), doi:10.1016/j.na.2010.06.064, http://dx.doi.org/10.1016/j.na.2010.06.064.

[116] F. Cammaroto and L. Vilasi, Multiple solutions for a Kirchhoff-type problem involving the $p(x)$-Laplacian operator, *Nonlinear Anal.* **74**, 5, pp. 1841–1852 (2011), doi:10.1016/j.na.2010.10.057, http://dx.doi.org/10.1016/j.na.2010.10.057.

[117] A. Kristály and D. Repovš, Multiple solutions for a Neumann system involving subquadratic nonlinearities, *Nonlinear Anal.* **74**, 6, pp. 2127–2132 (2011), doi:10.1016/j.na.2010.11.018, http://dx.doi.org/10.1016/j.na.2010.11.018.

[118] C. Ji, Remarks on the existence of three solutions for the $p(x)$-Laplacian equations, *Nonlinear Anal.* **74**, 9, pp. 2908–2915 (2011), doi:10.1016/j.na.

2010.12.013, http://dx.doi.org/10.1016/j.na.2010.12.013.

[119] A. Iannizzotto, Three solutions for a partial differential inclusion via nonsmooth critical point theory, *Set-Valued Var. Anal.* **19**, 2, pp. 311–327 (2011), doi:10.1007/s11228-010-0145-9, http://dx.doi.org/10.1007/s11228-010-0145-9.

[120] L. Yang and H. Chen, Existence and multiplicity of periodic solutions generated by impulses, *Abstr. Appl. Anal.*, pp. Art. ID 310957, 15 (2011), doi:10.1155/2011/310957, http://dx.doi.org/10.1155/2011/310957.

[121] L. Yang, H. Chen and X. Yang, The multiplicity of solutions for fourth-order equations generated from a boundary condition, *Appl. Math. Lett.* **24**, 9, pp. 1599–1603 (2011), doi:10.1016/j.aml.2011.04.008, http://dx.doi.org/10.1016/j.aml.2011.04.008.

[122] A. Kristály and D. Repovš, On the Schrödinger–Maxwell system involving sublinear terms, *Nonlinear Anal. Real World Appl.* **13**, 1, pp. 213–223 (2012), doi:10.1016/j.nonrwa.2011.07.027, http://dx.doi.org/10.1016/j.nonrwa.2011.07.027.

[123] A. Kristály and I.-I. Mezei, Multiple solutions for a perturbed system on strip-like domains, *Discrete Contin. Dyn. Syst. Ser. S* **5**, 4, pp. 789–796 (2012), doi:10.3934/dcdss.2012.5.789, http://dx.doi.org/10.3934/dcdss.2012.5.789.

[124] L. Li and W.-W. Pan, A note on nonlinear fourth-order elliptic equations on $\mathbb{R}^N$, *J. Global Optim.* **57**, 4, pp. 1319–1325 (2013), doi:10.1007/s10898-012-0031-0, http://dx.doi.org/10.1007/s10898-012-0031-0.

[125] G. Kirchhoff, *Mechanik*. Teubner, Leipzig (1883).

[126] C. O. Alves, F. J. S. A. Corrêa and T. F. Ma, Positive solutions for a quasilinear elliptic equation of Kirchhoff type, *Comput. Math. Appl.* **49**, 1, pp. 85–93 (2005).

[127] S. Bernstein, Sur une classe d'équations fonctionnelles aux dérivées partielles, *Bull. Acad. Sci. URSS. Sér. Math. [Izvestia Akad. Nauk SSSR]* **4**, pp. 17–26 (1940).

[128] S. I. Pohožaev, A certain class of quasilinear hyperbolic equations, *Mat. Sb. (N.S.)* **96(138)**, pp. 152–166, 168 (1975).

[129] J.-L. Lions, On some questions in boundary value problems of mathematical physics, in *Contemporary developments in continuum mechanics and partial differential equations (Proc. Internat. Sympos., Inst. Mat., Univ. Fed. Rio de Janeiro, Rio de Janeiro, 1977)*, North-Holland Math. Stud., Vol. 30. North-Holland, Amsterdam-New York, pp. 284–346 (1978).

[130] A. Arosio and S. Panizzi, On the well-posedness of the Kirchhoff string, *Trans. Amer. Math. Soc.* **348**, 1, pp. 305–330 (1996).

[131] M. M. Cavalcanti, V. N. Domingos Cavalcanti and J. A. Soriano, Global existence and uniform decay rates for the Kirchhoff–Carrier equation with nonlinear dissipation, *Adv. Differential Equations* **6**, 6, pp. 701–730 (2001).

[132] P. D'Ancona and S. Spagnolo, Global solvability for the degenerate Kirchhoff equation with real analytic data, *Invent. Math.* **108**, 2, pp. 247–262 (1992), doi:10.1007/BF02100605, http://dx.doi.org/10.1007/BF02100605.

[133] B. Cheng, Nontrivial solutions for Schrödinger–Kirchhoff-type problem in R^N, *Bound. Value Probl.*, pp. 2013:250, 11 (2013).

[134] F. Cammaroto and L. Vilasi, On a Schrödinger–Kirchhoff-type equation involving the $p(x)$-Laplacian, *Nonlinear Anal.* **81**, pp. 42–53 (2013), doi:10. 1016/j.na.2012.12.011, http://dx.doi.org/10.1016/j.na.2012.12.011.

[135] X. Wu, Existence of nontrivial solutions and high energy solutions for Schrödinger–Kirchhoff-type equations in $\mathbf{R}^N$, *Nonlinear Anal. Real World Appl.* **12**, 2, pp. 1278–1287 (2011), doi:10.1016/j.nonrwa.2010.09.023, http://dx.doi.org/10.1016/j.nonrwa.2010.09.023.

[136] S.-J. Chen and L. Li, Multiple solutions for the nonhomogeneous Kirchhoff equation on $\mathbf{R}^N$, *Nonlinear Anal. Real World Appl.* **14**, 3, pp. 1477–1486 (2013), doi:10.1016/j.nonrwa.2012.10.010, http://dx.doi.org/10. 1016/j.nonrwa.2012.10.010.

[137] Y. Huang and Z. Liu, On a class of Kirchhoff type problems, *Arch. Math. (Basel)* **102**, 2, pp. 127–139 (2014), doi:10.1007/s00013-014-0618-4, http://dx.doi.org/10.1007/s00013-014-0618-4.

[138] Q. Li and X. Wu, A new result on high energy solutions for Schrödinger–Kirchhoff type equations in $\mathbb{R}^N$, *Appl. Math. Lett.* **30**, pp. 24–27 (2014), doi:10.1016/j.aml.2013.12.002, http://dx.doi.org/10.1016/j.aml.2013. 12.002.

[139] S. Liang and J. Zhang, Existence of solutions for Kirchhoff type problems with critical nonlinearity in $\mathbb{R}^3$, *Nonlinear Anal. Real World Appl.* **17**, pp. 126–136 (2014), doi:10.1016/j.nonrwa.2013.10.011, http://dx.doi.org/ 10.1016/j.nonrwa.2013.10.011.

[140] W. Liu and X. He, Multiplicity of high energy solutions for superlinear Kirchhoff equations, *J. Appl. Math. Comput.* **39**, 1–2, pp. 473–487 (2012), doi:10.1007/s12190-012-0536-1, http://dx.doi.org/10.1007/ s12190-012-0536-1.

[141] J. Nie and X. Wu, Existence and multiplicity of non-trivial solutions for Schrödinger–Kirchhoff-type equations with radial potential, *Nonlinear Anal.* **75**, 8, pp. 3470–3479 (2012), doi:10.1016/j.na.2012.01.004, http:// dx.doi.org/10.1016/j.na.2012.01.004.

[142] L. Wang, On a quasilinear Schrödinger–Kirchhoff-type equation with radial potentials, *Nonlinear Anal.* **83**, pp. 58–68 (2013), doi:10.1016/j.na.2012.12. 012, http://dx.doi.org/10.1016/j.na.2012.12.012.

[143] X. Wu, High energy solutions of systems of Kirchhoff-type equations in R^N, *J. Math. Phys.* **53**, 6, pp. 063508, 18 (2012), doi:10.1063/1.4729543, http://dx.doi.org/10.1063/1.4729543.

[144] Y. Ye and C.-L. Tang, Multiple solutions for Kirchhoff-type equations in $\mathbb{R}^N$, *J. Math. Phys.* **54**, 8, pp. 081508, 16 (2013), doi:10.1063/1.4819249, http://dx.doi.org/10.1063/1.4819249.

[145] F. Zhou, K. Wu and X. Wu, High energy solutions of systems of Kirchhoff-type equations on $\mathbb{R}^N$, *Comput. Math. Appl.* **66**, 7, pp. 1299–1305 (2013), doi:10.1016/j.camwa.2013.07.028, http://dx.doi.org/10.1016/j.camwa. 2013.07.028.

[146] C. O. Alves and G. M. Figueiredo, Nonlinear perturbations of a periodic

Kirchhoff equation in $\mathbb{R}^N$, *Nonlinear Anal.* **75**, 5, pp. 2750–2759 (2012).

[147] A. Azzollini, The elliptic Kirchhoff equation in $\mathbb{R}^N$ perturbed by a local nonlinearity, *Differential Integral Equations* **25**, 5–6, pp. 543–554 (2012).

[148] A. Bahrouni, Infinitely many solutions for sublinear Kirchhoff equations in $\mathbb{R}^N$ with sign-changing potentials, *Electron. J. Differential Equations*, pp. No. 98, 8 (2013).

[149] C. Chen, H. Song and Z. Xiu, Multiple solutions for p-Kirchhoff equations in $\mathbb{R}^N$, *Nonlinear Anal.* **86**, pp. 146–156 (2013), doi:10.1016/j.na.2013.03.017, http://dx.doi.org/10.1016/j.na.2013.03.017.

[150] C. Chen and Q. Zhu, Existence of positive solutions to p-Kirchhoff-type problem without compactness conditions, *Appl. Math. Lett.* **28**, pp. 82–87 (2014), doi:10.1016/j.aml.2013.10.005, http://dx.doi.org/10.1016/j.aml.2013.10.005.

[151] J. Chen, Multiple positive solutions to a class of Kirchhoff equation on $\mathbb{R}^3$ with indefinite nonlinearity, *Nonlinear Anal.* **96**, pp. 134–145 (2014), doi:10.1016/j.na.2013.11.012, http://dx.doi.org/10.1016/j.na.2013.11.012.

[152] L. Duan and L. Huang, Infinitely many solutions for sublinear Schrödinger-Kirchhoff-type equations with general potentials, *Results Math.* **66**, 1–2, pp. 181–197 (2014), doi:10.1007/s00025-014-0371-9, http://dx.doi.org/10.1007/s00025-014-0371-9.

[153] G. M. Figueiredo, N. Ikoma and J. R. Santos Júnior, Existence and concentration result for the Kirchhoff type equations with general nonlinearities, *Arch. Ration. Mech. Anal.* **213**, 3, pp. 931–979 (2014), doi:10.1007/s00205-014-0747-8, http://dx.doi.org/10.1007/s00205-014-0747-8.

[154] G. M. Figueiredo and J. R. S. Júnior, Multiplicity and concentration behavior of positive solutions for a Schrödinger-Kirchhoff type problem *via* penalization method, *ESAIM Control Optim. Calc. Var.* **20**, 2, pp. 389–415 (2014), doi:10.1051/cocv/2013068, http://dx.doi.org/10.1051/cocv/2013068.

[155] X. He and W. Zou, Ground states for nonlinear Kirchhoff equations with critical growth, *Ann. Mat. Pura Appl. (4)* **193**, 2, pp. 473–500 (2014), doi:10.1007/s10231-012-0286-6, http://dx.doi.org/10.1007/s10231-012-0286-6.

[156] X. He and W. Zou, Existence and concentration behavior of positive solutions for a Kirchhoff equation in $\mathbb{R}^3$, *J. Differential Equations* **252**, 2, pp. 1813–1834 (2012), doi:10.1016/j.jde.2011.08.035, http://dx.doi.org/10.1016/j.jde.2011.08.035.

[157] N. Ikoma, Existence of ground state solutions to the nonlinear Kirchhoff type equations with potentials, *Discrete Contin. Dyn. Syst.* **35**, 3, pp. 943–966 (2015), doi:10.3934/dcds.2015.35.943, http://dx.doi.org/10.3934/dcds.2015.35.943.

[158] G. Li and H. Ye, Existence of positive solutions for nonlinear Kirchhoff type problems in $\mathbb{R}^3$ with critical Sobolev exponent, *Math. Methods Appl. Sci.* **37**, 16, pp. 2570–2584 (2014), doi:10.1002/mma.3000, http://dx.doi.org/10.1002/mma.3000.

[159] Y. Li, F. Li and J. Shi, Existence of positive solutions to Kirchhoff type problems with zero mass, *J. Math. Anal. Appl.* **410**, 1, pp. 361–374 (2014), doi:10.1016/j.jmaa.2013.08.030, http://dx.doi.org/10.1016/j.jmaa.2013.08.030.

[160] S. Liang and S. Shi, Existence of multi-bump solutions for a class of Kirchhoff type problems in $\mathbb{R}^3$, *J. Math. Phys.* **54**, 12, pp. 121510, 20 (2013), doi:10.1063/1.4850835, http://dx.doi.org/10.1063/1.4850835.

[161] J. Wang, L. Tian, J. Xu and F. Zhang, Multiplicity and concentration of positive solutions for a Kirchhoff type problem with critical growth, *J. Differential Equations* **253**, 7, pp. 2314–2351 (2012), doi:10.1016/j.jde.2012.05.023, http://dx.doi.org/10.1016/j.jde.2012.05.023.

[162] A. Azzollini, A note on the elliptic Kirchhoff equation in $\mathbb{R}^N$ perturbed by a local nonlinearity, *Commun. Contemp. Math.* **17**, 4, pp. 1450039, 5 (2015).

[163] Z. Wang and H.-S. Zhou, Positive solution for a nonlinear stationary Schrödinger-Poisson system in $\mathbb{R}^3$, *Discrete Contin. Dyn. Syst.* **18**, 4, pp. 809–816 (2007), doi:10.3934/dcds.2007.18.809, http://dx.doi.org/10.3934/dcds.2007.18.809.

[164] Y. Li, F. Li and J. Shi, Existence of a positive solution to Kirchhoff type problems without compactness conditions, *J. Differential Equations* **253**, 7, pp. 2285–2294 (2012), doi:10.1016/j.jde.2012.05.017, http://dx.doi.org/10.1016/j.jde.2012.05.017.

[165] G. Bianchi, J. Chabrowski and A. Szulkin, On symmetric solutions of an elliptic equation with a nonlinearity involving critical Sobolev exponent, *Nonlinear Anal.* **25**, 1, pp. 41–59 (1995).

[166] M. Reed and B. Simon, *Methods of modern mathematical physics. IV. Analysis of operators.* Academic Press [Harcourt Brace Jovanovich, Publishers], New York–London (1978), ISBN 0-12-585004-2.

[167] J. Jin and X. Wu, Infinitely many radial solutions for Kirchhoff-type problems in $\mathbb{R}^N$, *J. Math. Anal. Appl.* **369**, 2, pp. 564–574 (2010), doi:10.1016/j.jmaa.2010.03.059, http://dx.doi.org/10.1016/j.jmaa.2010.03.059.

[168] A. Azzollini, P. d'Avenia and A. Pomponio, Multiple critical points for a class of nonlinear functionals, *Ann. Mat. Pura Appl. (4)* **190**, 3, pp. 507–523 (2011).

[169] H. Berestycki and P.-L. Lions, Nonlinear scalar field equations. I. Existence of a ground state, *Arch. Rational Mech. Anal.* **82**, 4, pp. 313–345 (1983).

[170] L. Li and J.-J. Sun, Existence and multiplicity of solutions for the Kirchhoff equations with asymptotically linear nonlinearities, *Nonlinear Anal. Real World Appl.* **26**, pp. 391–399 (2015), doi:10.1016/j.nonrwa.2015.07.002, http://dx.doi.org/10.1016/j.nonrwa.2015.07.002.

[171] J. Mawhin and M. Willem, *Critical point theory and Hamiltonian systems*, Applied Mathematical Sciences, Vol. 74. Springer-Verlag, New York (1989), ISBN 0-387-96908-X.

[172] A. Ambrosetti and P. H. Rabinowitz, Dual variational methods in critical point theory and applications, *J. Funct. Anal.* **14**, pp. 349–381 (1973).

[173] B. Cheng and X. Wu, Existence results of positive solutions of Kirchhoff

type problems, *Nonlinear Anal.* **71**, 10, pp. 4883–4892 (2009), doi:10.1016/j.na.2009.03.065, http://dx.doi.org/10.1016/j.na.2009.03.065.

[174] X. He and W. Zou, Infinitely many positive solutions for Kirchhoff-type problems, *Nonlinear Anal.* **70**, 3, pp. 1407–1414 (2009), doi:10.1016/j.na.2008.02.021, http://dx.doi.org/10.1016/j.na.2008.02.021.

[175] T. F. Ma and J. E. Muñoz Rivera, Positive solutions for a nonlinear nonlocal elliptic transmission problem, *Appl. Math. Lett.* **16**, 2, pp. 243–248 (2003), doi:10.1016/S0893-9659(03)80038-1, http://dx.doi.org/10.1016/S0893-9659(03)80038-1.

[176] A. Mao and Z. Zhang, Sign-changing and multiple solutions of Kirchhoff type problems without the P.S. condition, *Nonlinear Anal.* **70**, 3, pp. 1275–1287 (2009), doi:10.1016/j.na.2008.02.011, http://dx.doi.org/10.1016/j.na.2008.02.011.

[177] Z. Zhang and K. Perera, Sign changing solutions of Kirchhoff type problems via invariant sets of descent flow, *J. Math. Anal. Appl.* **317**, 2, pp. 456–463 (2006), doi:10.1016/j.jmaa.2005.06.102, http://dx.doi.org/10.1016/j.jmaa.2005.06.102.

[178] P. H. Rabinowitz, On a class of nonlinear Schrödinger equations, *Z. Angew. Math. Phys.* **43**, 2, pp. 270–291 (1992), doi:10.1007/BF00946631, http://dx.doi.org/10.1007/BF00946631.

[179] Z. Xiu, The existence of a nontrivial solution for a p-Kirchhoff type elliptic equation in $\mathbb{R}^N$, *Abstr. Appl. Anal.* , pp. Art. ID 281949, 6 (2013).

[180] J. Zhang, X. Tang and W. Zhang, Existence of multiple solutions of Kirchhoff type equation with sign-changing potential, *Appl. Math. Comput.* **242**, pp. 491–499 (2014), doi:10.1016/j.amc.2014.05.070, http://dx.doi.org/10.1016/j.amc.2014.05.070.

[181] K. Wu and X. Wu, Infinitely many small energy solutions for a modified Kirchhoff-type equation in $\mathbb{R}^N$, *Comput. Math. Appl.* **70**, 4, pp. 592–602 (2015), doi:10.1016/j.camwa.2015.05.014, http://dx.doi.org/10.1016/j.camwa.2015.05.014.

[182] Q. Li and Z. Yang, Multiple solutions for N-Kirchhoff type problems with critical exponential growth in $\mathbb{R}^N$, *Nonlinear Anal.* **117**, pp. 159–168 (2015), doi:10.1016/j.na.2015.01.005, http://dx.doi.org/10.1016/j.na.2015.01.005.

[183] H. Ye, The existence of least energy nodal solutions for some class of Kirchhoff equations and Choquard equations in $\mathbb{R}^N$, *J. Math. Anal. Appl.* **431**, 2, pp. 935–954 (2015), doi:10.1016/j.jmaa.2015.06.012, http://dx.doi.org/10.1016/j.jmaa.2015.06.012.

[184] J. Nie, Existence and multiplicity of nontrivial solutions for a class of Schrödinger-Kirchhoff-type equations, *J. Math. Anal. Appl.* **417**, 1, pp. 65–79 (2014), doi:10.1016/j.jmaa.2014.03.027, http://dx.doi.org/10.1016/j.jmaa.2014.03.027.

[185] Y. Guo and J. Nie, Existence and multiplicity of nontrivial solutions for p-Laplacian Schrödinger-Kirchhoff-type equations, *J. Math. Anal. Appl.* **428**, 2, pp. 1054–1069 (2015), doi:10.1016/j.jmaa.2015.03.064, http://dx.doi.org/10.1016/j.jmaa.2015.03.064.

[186] S. Liang and J. Zhang, Existence of solutions for Kirchhoff type problems with critical nonlinearity in $\mathbb{R}^N$, *Z. Angew. Math. Phys.* **66**, 3, pp. 547–562 (2015), doi:10.1007/s00033-014-0418-5, http://dx.doi.org/10.1007/s00033-014-0418-5.

[187] Q. Xie and S. Ma, Existence and concentration of positive solutions for Kirchhoff-type problems with a steep well potential, *J. Math. Anal. Appl.* **431**, 2, pp. 1210–1223 (2015), doi:10.1016/j.jmaa.2015.05.027, http://dx.doi.org/10.1016/j.jmaa.2015.05.027.

[188] Q. Zhang and B. Xu, Multiplicity of solutions for a class of semilinear Schrödinger equations with sign-changing potential, *J. Math. Anal. Appl.* **377**, 2, pp. 834–840 (2011), doi:10.1016/j.jmaa.2010.11.059, http://dx.doi.org/10.1016/j.jmaa.2010.11.059.

[189] T. Bartsch, Z.-Q. Wang and M. Willem, The Dirichlet problem for superlinear elliptic equations, in *Stationary partial differential equations*. Vol. II, Handb. Differ. Equ. Elsevier/North-Holland, Amsterdam, pp. 1–55 (2005).

[190] X. H. Tang, Infinitely many solutions for semilinear Schrödinger equations with sign-changing potential and nonlinearity, *J. Math. Anal. Appl.* **401**, 1, pp. 407–415 (2013), doi:10.1016/j.jmaa.2012.12.035, http://dx.doi.org/10.1016/j.jmaa.2012.12.035.

[191] L. Li, V. Rădulescu and D. Repovš, Nonlocal kirchhoff superlinear equations with indefinite nonlinearity and lack of compactness, (2015), in press.

[192] C. Liu, Z. Wang and H.-S. Zhou, Asymptotically linear Schrödinger equation with potential vanishing at infinity, *J. Differential Equations* **245**, 1, pp. 201–222 (2008), doi:10.1016/j.jde.2008.01.006, http://dx.doi.org/10.1016/j.jde.2008.01.006.

[193] J. Sun, H. Chen and J. J. Nieto, On ground state solutions for some nonautonomous Schrödinger-Poisson systems, *J. Differential Equations* **252**, 5, pp. 3365–3380 (2012), doi:10.1016/j.jde.2011.12.007, http://dx.doi.org/10.1016/j.jde.2011.12.007.

[194] K. Perera and Z. Zhang, Nontrivial solutions of Kirchhoff-type problems via the Yang index, *J. Differential Equations* **221**, 1, pp. 246–255 (2006), doi:10.1016/j.jde.2005.03.006, http://dx.doi.org/10.1016/j.jde.2005.03.006.

[195] X.-m. He and W.-m. Zou, Multiplicity of solutions for a class of Kirchhoff type problems, *Acta Math. Appl. Sin. Engl. Ser.* **26**, 3, pp. 387–394 (2010), doi:10.1007/s10255-010-0005-2, http://dx.doi.org/10.1007/s10255-010-0005-2.

[196] Y. Yang and J. Zhang, Positive and negative solutions of a class of nonlocal problems, *Nonlinear Anal.* **73**, 1, pp. 25–30 (2010a), doi:10.1016/j.na.2010.02.008, http://dx.doi.org/10.1016/j.na.2010.02.008.

[197] Y. Yang and J. Zhang, Nontrivial solutions of a class of nonlocal problems via local linking theory, *Appl. Math. Lett.* **23**, 4, pp. 377–380 (2010b), doi:10.1016/j.aml.2009.11.001, http://dx.doi.org/10.1016/j.aml.2009.11.001.

[198] C.-y. Chen, Y.-c. Kuo and T.-f. Wu, The Nehari manifold for a Kirchhoff type problem involving sign-changing weight functions, *J. Differen-*

tial Equations **250**, 4, pp. 1876–1908 (2011), doi:10.1016/j.jde.2010.11.017, http://dx.doi.org/10.1016/j.jde.2010.11.017.

[199] J.-J. Sun and C.-L. Tang, Existence and multiplicity of solutions for Kirchhoff type equations, *Nonlinear Anal.* **74**, 4, pp. 1212–1222 (2011), doi:10.1016/j.na.2010.09.061, http://dx.doi.org/10.1016/j.na.2010.09.061.

[200] J. Sun and S. Liu, Nontrivial solutions of Kirchhoff type problems, *Appl. Math. Lett.* **25**, 3, pp. 500–504 (2012), doi:10.1016/j.aml.2011.09.045, http://dx.doi.org/10.1016/j.aml.2011.09.045.

[201] B. Cheng, New existence and multiplicity of nontrivial solutions for nonlocal elliptic Kirchhoff type problems, *J. Math. Anal. Appl.* **394**, 2, pp. 488–495 (2012), doi:10.1016/j.jmaa.2012.04.025, http://dx.doi.org/10.1016/j.jmaa.2012.04.025.

[202] Z. Wang and H.-S. Zhou, Positive solutions for a nonhomogeneous elliptic equation on $\mathbb{R}^N$ without (AR) condition, *J. Math. Anal. Appl.* **353**, 1, pp. 470–479 (2009), doi:10.1016/j.jmaa.2008.11.080, http://dx.doi.org/10.1016/j.jmaa.2008.11.080.

[203] L. Ding, L. Li and J.-L. Zhang, Positive Solutions for a Nonhomogeneous Kirchhoff Equation with the Asymptotical Nonlinearity in R^3, *Abstr. Appl. Anal.*, pp. Art. ID 710949, 10 (2014), doi:10.1155/2014/710949, http://dx.doi.org/10.1155/2014/710949.

[204] G. Molica Bisci and P. F. Pizzimenti, Sequences of weak solutions for non-local elliptic problems with Dirichlet boundary condition, *Proc. Edinb. Math. Soc. (2)* **57**, 3, pp. 779–809 (2014), doi:10.1017/S0013091513000722, http://dx.doi.org/10.1017/S0013091513000722.

[205] S. Heidarkhani, Infinitely many solutions for systems of n two-point Kirchhoff-type boundary value problems, *Ann. Polon. Math.* **107**, 2, pp. 133–152 (2013), doi:10.4064/ap107-2-3, http://dx.doi.org/10.4064/ap107-2-3.

[206] S. Heidarkhani and J. Henderson, Infinitely many solutions for nonlocal elliptic systems of $(p_1, \ldots, p_n)$-Kirchhoff type, *Electron. J. Differential Equations* , pp. No. 69, 15 (2012).

[207] H. Zhang and F. Zhang, Ground states for the nonlinear Kirchhoff type problems, *J. Math. Anal. Appl.* **423**, 2, pp. 1671–1692 (2015), doi:10.1016/j.jmaa.2014.10.062, http://dx.doi.org/10.1016/j.jmaa.2014.10.062.

[208] L. Duan and L. Huang, Existence of nontrivial solutions for Kirchhoff-type variational inclusion system in $\mathbb{R}^N$, *Appl. Math. Comput.* **235**, pp. 174–186 (2014), doi:10.1016/j.amc.2014.02.070, http://dx.doi.org/10.1016/j.amc.2014.02.070.

[209] D. Lü, Existence and multiplicity results for perturbed Kirchhoff-type Schrödinger systems in $\mathbb{R}^3$, *Comput. Math. Appl.* **68**, 10, pp. 1180–1193 (2014), doi:10.1016/j.camwa.2014.08.020, http://dx.doi.org/10.1016/j.camwa.2014.08.020.

[210] L. Li and X. Zhong, Infinity many small solutions for the kirchhoff equation with local sublinear nonlinearities, *J. Math. Anal. Appl.* (2015), in press.

[211] S. Li, S. Wu and H.-S. Zhou, Solutions to semilinear elliptic problems with combined nonlinearities, *J. Differential Equations* **185**, 1, pp. 200–224

(2002), doi:10.1006/jdeq.2001.4167, http://dx.doi.org/10.1006/jdeq. 2001.4167.

[212] I. Ekeland, Nonconvex minimization problems, *Bull. Amer. Math. Soc. (N.S.)* **1**, 3, pp. 443–474 (1979), doi:10.1090/S0273-0979-1979-14595-6, http://dx.doi.org/10.1090/S0273-0979-1979-14595-6.

[213] Z. Liang, F. Li and J. Shi, Positive solutions to Kirchhoff type equations with nonlinearity having prescribed asymptotic behavior, *Ann. Inst. H. Poincaré Anal. Non Linéaire* **31**, 1, pp. 155–167 (2014), doi:10. 1016/j.anihpc.2013.01.006, http://dx.doi.org/10.1016/j.anihpc.2013. 01.006.

[214] B. Cheng, X. Wu and J. Liu, Multiple solutions for a class of Kirchhoff type problems with concave nonlinearity, *NoDEA Nonlinear Differential Equations Appl.* **19**, 5, pp. 521–537 (2012), doi:10.1007/s00030-011-0141-2, http://dx.doi.org/10.1007/s00030-011-0141-2.

[215] L. Ding, L. Li and J.-L. Zhang, Solutions to Kirchhoff equations with combined nonlinearities, *Electron. J. Differential Equations*, pp. No. 10, 10 (2014).

[216] T. Bartsch and M. Willem, On an elliptic equation with concave and convex nonlinearities, *Proc. Amer. Math. Soc.* **123**, 11, pp. 3555–3561 (1995).

[217] M. Struwe, *Variational methods, Ergebnisse der Mathematik und ihrer Grenzgebiete. 3. Folge. A Series of Modern Surveys in Mathematics [Results in Mathematics and Related Areas. 3rd Series. A Series of Modern Surveys in Mathematics]*, Vol. 34, 4th edn. Springer-Verlag, Berlin (2008), ISBN 978-3-540-74012-4, applications to nonlinear partial differential equations and Hamiltonian systems.

[218] S.-J. Chen and C.-L. Tang, High energy solutions for the superlinear Schrödinger–Maxwell equations, *Nonlinear Anal.* **71**, 10, pp. 4927–4934 (2009), doi:10.1016/j.na.2009.03.050, http://dx.doi.org/10.1016/j.na. 2009.03.050.

[219] Q. Li, H. Su and Z. Wei, Existence of infinitely many large solutions for the nonlinear Schrödinger–Maxwell equations, *Nonlinear Anal.* **72**, 11, pp. 4264–4270 (2010), doi:10.1016/j.na.2010.02.002, http://dx.doi.org/10. 1016/j.na.2010.02.002.

[220] W. Zou, Variant fountain theorems and their applications, *Manuscripta Math.* **104**, 3, pp. 343–358 (2001), doi:10.1007/s002290170032, http://dx. doi.org/10.1007/s002290170032.

[221] C. O. Alves, M. A. S. Souto and S. H. M. Soares, Schrödinger-Poisson equations without Ambrosetti–Rabinowitz condition, *J. Math. Anal. Appl.* **377**, 2, pp. 584–592 (2011).

[222] M.-H. Yang and Z.-Q. Han, Existence and multiplicity results for the nonlinear Schrödinger–Poisson systems, *Nonlinear Anal. Real World Appl.* **13**, 3, pp. 1093–1101 (2012), doi:10.1016/j.nonrwa.2011.07.008, http://dx. doi.org/10.1016/j.nonrwa.2011.07.008.

[223] S. Liu, On superlinear problems without the Ambrosetti and Rabinowitz condition, *Nonlinear Anal.* **73**, 3, pp. 788–795 (2010), doi:10.1016/j.na. 2010.04.016, http://dx.doi.org/10.1016/j.na.2010.04.016.

[224] V. Benci, D. Fortunato, A. Masiello and L. Pisani, Solitons and the electromagnetic field, *Math. Z.* **232**, 1, pp. 73–102 (1999).

[225] A. Azzollini and A. Pomponio, Ground state solutions for the nonlinear Schrödinger-Maxwell equations, *J. Math. Anal. Appl.* **345**, 1, pp. 90–108 (2008).

[226] G. M. Coclite, A multiplicity result for the nonlinear Schrödinger–Maxwell equations, *Commun. Appl. Anal.* **7**, 2-3, pp. 417–423 (2003).

[227] G. M. Coclite, A multiplicity result for the Schrödinger-Maxwell equations with negative potential, *Ann. Polon. Math.* **79**, 1, pp. 21–30 (2002), doi: 10.4064/ap79-1-2, http://dx.doi.org/10.4064/ap79-1-2.

[228] T. D'Aprile and D. Mugnai, Solitary waves for nonlinear Klein-Gordon-Maxwell and Schrödinger-Maxwell equations, *Proc. Roy. Soc. Edinburgh Sect. A* **134**, 5, pp. 893–906 (2004), doi:10.1017/S030821050000353X, http://dx.doi.org/10.1017/S030821050000353X.

[229] H. Kikuchi, On the existence of a solution for elliptic system related to the Maxwell–Schrödinger equations, *Nonlinear Anal.* **67**, 5, pp. 1445–1456 (2007), doi:10.1016/j.na.2006.07.029, http://dx.doi.org/10.1016/j.na.2006.07.029.

[230] D. Ruiz, The Schrödinger-Poisson equation under the effect of a nonlinear local term, *J. Funct. Anal.* **237**, 2, pp. 655–674 (2006), doi:10.1016/j.jfa.2006.04.005, http://dx.doi.org/10.1016/j.jfa.2006.04.005.

[231] A. Salvatore, Multiple solitary waves for a non-homogeneous Schrödinger-Maxwell system in $\mathbb{R}^3$, *Adv. Nonlinear Stud.* **6**, 2, pp. 157–169 (2006).

[232] L. Zhao and F. Zhao, Positive solutions for Schrödinger–Poisson equations with a critical exponent, *Nonlinear Anal.* **70**, 6, pp. 2150–2164 (2009), doi:10.1016/j.na.2008.02.116, http://dx.doi.org/10.1016/j.na.2008.02.116.

[233] L. Zhao and F. Zhao, On the existence of solutions for the Schrödinger-Poisson equations, *J. Math. Anal. Appl.* **346**, 1, pp. 155–169 (2008), doi:10.1016/j.jmaa.2008.04.053, http://dx.doi.org/10.1016/j.jmaa.2008.04.053.

[234] N. Ackermann, A nonlinear superposition principle and multibump solutions of periodic Schrödinger equations, *J. Funct. Anal.* **234**, 2, pp. 277–320 (2006).

[235] T. D'Aprile and D. Mugnai, Non-existence results for the coupled Klein-Gordon-Maxwell equations, *Adv. Nonlinear Stud.* **4**, 3, pp. 307–322 (2004).

[236] L. Li and S.-J. Chen, Infinitely many large energy solutions of superlinear Schrödinger-Maxwell equations, *Electron. J. Differential Equations* , pp. No. 224, 9 (2012).

[237] X. P. Zhu and H. S. Zhou, Existence of multiple positive solutions of inhomogeneous semilinear elliptic problems in unbounded domains, *Proc. Roy. Soc. Edinburgh Sect. A* **115**, 3–4, pp. 301–318 (1990), doi:10.1017/S0308210500020667, http://dx.doi.org/10.1017/S0308210500020667.

[238] A. Ambrosetti and D. Ruiz, Multiple bound states for the Schrödinger–Poisson problem, *Commun. Contemp. Math.* **10**, 3, pp. 391–404 (2008).

[239] G. Cerami and G. Vaira, Positive solutions for some non-autonomous

Schrödinger–Poisson systems, *J. Differential Equations* **248**, 3, pp. 521–543 (2010), doi:10.1016/j.jde.2009.06.017, http://dx.doi.org/10.1016/j.jde.2009.06.017.

[240] H. Zhu, An asymptotically linear Schrödinger–Poisson system on $\mathbb{R}^3$, *Nonlinear Anal.* **75**, 13, pp. 5261–5269 (2012), doi:10.1016/j.na.2012.04.042, http://dx.doi.org/10.1016/j.na.2012.04.042.

[241] H. Zhu, Asymptotically linear Schrödinger–Poisson systems with potentials vanishing at infinity, *J. Math. Anal. Appl.* **380**, 2, pp. 501–510 (2011), doi:10.1016/j.jmaa.2010.09.071, http://dx.doi.org/10.1016/j.jmaa.2010.09.071.

[242] D. G. Costa and H. Tehrani, On a class of asymptotically linear elliptic problems in $\mathbb{R}^N$, *J. Differential Equations* **173**, 2, pp. 470–494 (2001), doi:10.1006/jdeq.2000.3944, http://dx.doi.org/10.1006/jdeq.2000.3944.

[243] L. Jeanjean, On the existence of bounded Palais–Smale sequences and application to a Landesman–Lazer-type problem set on $\mathbf{R}^N$, *Proc. Roy. Soc. Edinburgh Sect. A* **129**, 4, pp. 787–809 (1999), doi:10.1017/S0308210500013147, http://dx.doi.org/10.1017/S0308210500013147.

[244] G. Li and H.-S. Zhou, The existence of a positive solution to asymptotically linear scalar field equations, *Proc. Roy. Soc. Edinburgh Sect. A* **130**, 1, pp. 81–105 (2000), doi:10.1017/S0308210500000068, http://dx.doi.org/10.1017/S0308210500000068.

[245] Z. Liu and Z.-Q. Wang, Existence of a positive solution of an elliptic equation on $\mathbb{R}^N$, *Proc. Roy. Soc. Edinburgh Sect. A* **134**, 1, pp. 191–200 (2004), doi:10.1017/S0308210500003152, http://dx.doi.org/10.1017/S0308210500003152.

[246] C. A. Stuart and H. S. Zhou, Applying the mountain pass theorem to an asymptotically linear elliptic equation on $\mathbf{R}^N$, *Comm. Partial Differential Equations* **24**, 9–10, pp. 1731–1758 (1999), doi:10.1080/03605309908821481, http://dx.doi.org/10.1080/03605309908821481.

[247] X. P. Zhu, A perturbation result on positive entire solutions of a semilinear elliptic equation, *J. Differential Equations* **92**, 2, pp. 163–178 (1991), doi:10.1016/0022-0396(91)90045-B, http://dx.doi.org/10.1016/0022-0396(91)90045-B.

[248] D.-M. Cao and H.-S. Zhou, Multiple positive solutions of nonhomogeneous semilinear elliptic equations in $\mathbf{R}^N$, *Proc. Roy. Soc. Edinburgh Sect. A* **126**, 2, pp. 443–463 (1996), doi:10.1017/S0308210500022836, http://dx.doi.org/10.1017/S0308210500022836.

[249] K.-J. Chen, Exactly two entire positive solutions for a class of nonhomogeneous elliptic equations, *Differential Integral Equations* **17**, 1–2, pp. 1–16 (2004).

[250] Y. Deng, Existence of multiple positive solutions for a semilinear equation with critical exponent, *Proc. Roy. Soc. Edinburgh Sect. A* **122**, 1–2, pp. 161–175 (1992), doi:10.1017/S0308210500021028, http://dx.doi.org/10.1017/S0308210500021028.

[251] A. M. Candela and A. Salvatore, Multiple solitary waves for nonhomogeneous Schrödinger–Maxwell equations, *Mediterr. J. Math.* **3**, 3–4,

pp. 483–493 (2006), doi:10.1007/s00009-006-0092-8, http://dx.doi.org/
10.1007/s00009-006-0092-8.

[252] L. Ding, L. Li and J.-L. Zhang, Multiple solutions for nonhomogeneous
Schrödinger-Poisson systems with the asymptotical nonlinearity in $\mathbb{R}^3$, *Taiwanese J. Math.* **17**, 5, pp. 1627–1650 (2013), doi:10.11650/tjm.17.2013.
2798, http://dx.doi.org/10.11650/tjm.17.2013.2798.

[253] K. Benmlih and O. Kavian, Existence and asymptotic behaviour of standing
waves for quasilinear Schrödinger-Poisson systems in $\mathbb{R}^3$, *Ann. Inst. H.
Poincaré Anal. Non Linéaire* **25**, 3, pp. 449–470 (2008).

[254] K. Yosida, *Functional analysis*, Classics in Mathematics. Springer-Verlag,
Berlin (1995), ISBN 3-540-58654-7, reprint of the sixth (1980) edition.

[255] V. Benci and D. Fortunato, An eigenvalue problem for the Schrödinger-
Maxwell equations, *Topol. Methods Nonlinear Anal.* **11**, 2, pp. 283–293
(1998).

[256] P. A. Markowich, C. A. Ringhofer and C. Schmeiser, *Semiconductor equations*. Springer-Verlag, Vienna (1990), ISBN 3-211-82157-0, doi:10.1007/
978-3-7091-6961-2, http://dx.doi.org/10.1007/978-3-7091-6961-2.

[257] R. Illner, O. Kavian and H. Lange, Stationary solutions of quasi-linear
Schrödinger-Poisson systems, *J. Differential Equations* **145**, 1, pp. 1–16
(1998), doi:10.1006/jdeq.1997.3405, http://dx.doi.org/10.1006/jdeq.
1997.3405.

[258] R. Illner, H. Lange, B. Toomire and P. Zweifel, On quasi-linear
Schrödinger-Poisson systems, *Math. Methods Appl. Sci.* **20**, 14,
pp. 1223–1238 (1997), doi:10.1002/(SICI)1099-1476(19970925)20:14⟨1223::
AID-MMA911⟩3.3.CO;2-F, http://dx.doi.org/10.1002/(SICI)1099-14
76(19970925)20:14<1223::AID-MMA911>3.3.CO;2-F.

[259] T. D'Aprile and J. Wei, Layered solutions for a semilinear elliptic system in
a ball, *J. Differential Equations* **226**, 1, pp. 269–294 (2006), doi:10.1016/j.
jde.2005.12.009, http://dx.doi.org/10.1016/j.jde.2005.12.009.

[260] T. D'Aprile and J. Wei, Solutions en grappe autour des centres harmoniques
d'un système elliptique couplé, *Ann. Inst. H. Poincaré Anal. Non Linéaire*
24, 4, pp. 605–628 (2007), doi:10.1016/j.anihpc.2006.04.003, http://dx.
doi.org/10.1016/j.anihpc.2006.04.003.

[261] D. Ruiz, Semiclassical states for coupled Schrödinger-Maxwell equations:
concentration around a sphere, *Math. Models Methods Appl. Sci.* **15**, 1, pp.
141–164 (2005), doi:10.1142/S0218202505003939, http://dx.doi.org/10.
1142/S0218202505003939.

[262] G. Siciliano, Multiple positive solutions for a Schrödinger-Poisson-Slater
system, *J. Math. Anal. Appl.* **365**, 1, pp. 288–299 (2010), doi:10.1016/j.
jmaa.2009.10.061, http://dx.doi.org/10.1016/j.jmaa.2009.10.061.

[263] M. Ghimenti and A. M. Micheletti, Number and profile of low energy solutions for singularly perturbed Klein-Gordon-Maxwell systems on a Riemannian manifold, *J. Differential Equations* **256**, 7, pp. 2502–2525 (2014),
doi:10.1016/j.jde.2014.01.012, http://dx.doi.org/10.1016/j.jde.2014.
01.012.

[264] L. Ding, L. Li, Y.-J. Meng and C.-L. Zhuang, Existence and asymptotic

behaviour of ground state solution for quasilinear Schrödinger–Poisson systems in $\mathbb{R}^3$, *Topol. Methods Nonlinear Anal.* (2015), in press.

[265] X. He, Multiplicity of solutions for a nonlinear Klein–Gordon–Maxwell system, *Acta Appl. Math.* **130**, pp. 237–250 (2014), doi:10.1007/s10440-013-9845-0, http://dx.doi.org/10.1007/s10440-013-9845-0.

[266] P. Chen and C. Tian, Infinitely many solutions for Schrödinger–Maxwell equations with indefinite sign subquadratic potentials, *Appl. Math. Comput.* **226**, pp. 492–502 (2014), doi:10.1016/j.amc.2013.10.069, http://dx.doi.org/10.1016/j.amc.2013.10.069.

[267] S.-J. Chen and C.-L. Tang, Multiple solutions for nonhomogeneous Schrödinger–Maxwell and Klein–Gordon–Maxwell equations on $\mathbb{R}^3$, *NoDEA Nonlinear Differential Equations Appl.* **17**, 5, pp. 559–574 (2010), doi:10.1007/s00030-010-0068-z, http://dx.doi.org/10.1007/s00030-010-0068-z.

[268] W.-n. Huang and X. H. Tang, The existence of infinitely many solutions for the nonlinear Schrödinger–Maxwell equations, *Results Math.* **65**, 1–2, pp. 223–234 (2014), doi:10.1007/s00025-013-0342-6, http://dx.doi.org/10.1007/s00025-013-0342-6.

[269] Z. Liu, S. Guo and Z. Zhang, Existence and multiplicity of solutions for a class of sublinear Schrödinger–Maxwell equations, *Taiwanese J. Math.* **17**, 3, pp. 857–872 (2013), doi:10.11650/tjm.17.2013.2202, http://dx.doi.org/10.11650/tjm.17.2013.2202.

[270] J. Sun, Infinitely many solutions for a class of sublinear Schrödinger–Maxwell equations, *J. Math. Anal. Appl.* **390**, 2, pp. 514–522 (2012), doi:10.1016/j.jmaa.2012.01.057, http://dx.doi.org/10.1016/j.jmaa.2012.01.057.

[271] Y. Lv, Existence and multiplicity of solutions for a class of sublinear Schrödinger–Maxwell equations, *Bound. Value Probl.*, pp. 2013:177, 22 (2013), doi:10.1186/1687-2770-2013-177, http://dx.doi.org/10.1186/1687-2770-2013-177.

[272] V. Benci and D. Fortunato, Solitary waves of the nonlinear Klein–Gordon equation coupled with the Maxwell equations, *Rev. Math. Phys.* **14**, 4, pp. 409–420 (2002).

[273] V. Benci and D. Fortunato, The nonlinear Klein–Gordon equation coupled with the Maxwell equations, in *Proceedings of the Third World Congress of Nonlinear Analysts, Part 9 (Catania, 2000)*, Vol. 47, pp. 6065–6072 (2001).

[274] A. Azzollini, L. Pisani and A. Pomponio, Improved estimates and a limit case for the electrostatic Klein–Gordon–Maxwell system, *Proc. Roy. Soc. Edinburgh Sect. A* **141**, 3, pp. 449–463 (2011).

[275] D. Mugnai, Solitary waves in abelian gauge theories with strongly nonlinear potentials, *Ann. Inst. H. Poincaré Anal. Non Linéaire* **27**, 4, pp. 1055–1071 (2010), doi:10.1016/j.anihpc.2010.02.001, http://dx.doi.org/10.1016/j.anihpc.2010.02.001.

[276] A. Azzollini and A. Pomponio, Ground state solutions for the nonlinear Klein–Gordon–Maxwell equations, *Topol. Methods Nonlinear Anal.* **35**, 1, pp. 33–42 (2010).

[277] F. Wang, Ground-state solutions for the electrostatic nonlinear Klein–Gordon–Maxwell system, *Nonlinear Anal.* **74**, 14, pp. 4796–4803 (2011), doi:10.1016/j.na.2011.04.050, http://dx.doi.org/10.1016/j.na.2011.04.050.

[278] G. Vaira, Semiclassical states for the nonlinear Klein–Gordon–Maxwell system, *J. Pure Appl. Math. Adv. Appl.* **4**, 1, pp. 59–95 (2010).

[279] P. D. Makita, Nonradial solutions for the Klein-Gordon-Maxwell equations, *Discrete Contin. Dyn. Syst.* **32**, 6, pp. 2271–2283 (2012), doi:10.3934/dcds.2012.32.2271, http://dx.doi.org/10.3934/dcds.2012.32.2271.

[280] D. Cassani, Existence and non-existence of solitary waves for the critical Klein–Gordon equation coupled with Maxwell's equations, *Nonlinear Anal.* **58**, 7-8, pp. 733–747 (2004), doi:10.1016/j.na.2003.05.001, http://dx.doi.org/10.1016/j.na.2003.05.001.

[281] P. C. Carrião, P. L. Cunha and O. H. Miyagaki, Existence results for the Klein–Gordon–Maxwell equations in higher dimensions with critical exponents, *Commun. Pure Appl. Anal.* **10**, 2, pp. 709–718 (2011), doi:10.3934/cpaa.2011.10.709, http://dx.doi.org/10.3934/cpaa.2011.10.709.

[282] P. C. Carrião, P. L. Cunha and O. H. Miyagaki, Positive ground state solutions for the critical Klein–Gordon–Maxwell system with potentials, *Nonlinear Anal.* **75**, 10, pp. 4068–4078 (2012), doi:10.1016/j.na.2012.02.023, http://dx.doi.org/10.1016/j.na.2012.02.023.

[283] F. Wang, Solitary waves for the Klein–Gordon–Maxwell system with critical exponent, *Nonlinear Anal.* **74**, 3, pp. 827–835 (2011), doi:10.1016/j.na.2010.09.033, http://dx.doi.org/10.1016/j.na.2010.09.033.

[284] W. Jeong and J. Seok, On perturbation of a functional with the mountain pass geometry: applications to the nonlinear Schrödinger–Poisson equations and the nonlinear Klein–Gordon–Maxwell equations, *Calc. Var. Partial Differential Equations* **49**, 1-2, pp. 649–668 (2014), doi:10.1007/s00526-013-0595-7, http://dx.doi.org/10.1007/s00526-013-0595-7.

[285] L. Li and C.-L. Tang, Infinitely many solutions for a nonlinear Klein–Gordon–Maxwell system, *Nonlinear Anal.* **110**, pp. 157–169 (2014), doi:10.1016/j.na.2014.07.019, http://dx.doi.org/10.1016/j.na.2014.07.019.

[286] H. Chen and S. Liu, Standing waves with large frequency for 4-superlinear Schrödinger–Poisson systems, *Ann. Mat. Pura Appl. (4)* **194**, 1, pp. 43–53 (2015), doi:10.1007/s10231-013-0363-5, http://dx.doi.org/10.1007/s10231-013-0363-5.

[287] S.-J. Chen and S.-Z. Song, Multiple solutions for nonhomogeneous Klein–Gordon–Maxwell equations on $\mathbf{R}^3$, *Nonlinear Anal. Real World Appl.* **22**, pp. 259–271 (2015), doi:10.1016/j.nonrwa.2014.09.006, http://dx.doi.org/10.1016/j.nonrwa.2014.09.006.

[288] P. L. Cunha, Subcritical and supercritical Klein–Gordon–Maxwell equations without Ambrosetti–Rabinowitz condition, *Differential Integral Equations* **27**, 3-4, pp. 387–399 (2014), http://projecteuclid.org/euclid.die/1391091371.

[289] L. Li, Solitary waves for a superlinear Klein–Gordon–Maxwell system with sign-changing potential, (2015), in press.

[290] L. Ding and L. Li, Infinitely many standing wave solutions for the nonlinear Klein–Gordon–Maxwell system with sign-changing potential, *Comput. Math. Appl.* **68**, 5, pp. 589–595 (2014), doi:10.1016/j.camwa.2014.07.001, http://dx.doi.org/10.1016/j.camwa.2014.07.001.

[291] L. Li, Multiple solutions for superlinear Klein-Gordon-Maxwell system without odd nonlinearity, (2015), in press.

[292] L. Li, Infinitely many solutions for Klein–Gordon–Maxwell systems under the partially sublinear case, (2015), in press.

[293] D. Fortunato, L. Orsina and L. Pisani, Born-Infeld type equations for electrostatic fields, *J. Math. Phys.* **43**, 11, pp. 5698–5706 (2002), doi: 10.1063/1.1508433, http://dx.doi.org/10.1063/1.1508433.

[294] P. d'Avenia and L. Pisani, Nonlinear Klein–Gordon equations coupled with Born–Infeld type equations, *Electron. J. Differential Equations*, pp. No. 26, 13 (2002).

[295] D. Mugnai, Coupled Klein–Gordon and Born–Infeld-type equations: looking for solitary waves, *Proc. R. Soc. Lond. Ser. A Math. Phys. Eng. Sci.* **460**, 2045, pp. 1519–1527 (2004), doi:10.1098/rspa.2003.1267, http://dx.doi.org/10.1098/rspa.2003.1267.

[296] K. Teng and K. Zhang, Existence of solitary wave solutions for the nonlinear Klein–Gordon equation coupled with Born–Infeld theory with critical Sobolev exponent, *Nonlinear Anal.* **74**, 12, pp. 4241–4251 (2011), doi:10.1016/j.na.2011.04.002, http://dx.doi.org/10.1016/j.na.2011.04.002.

[297] W. A. Strauss, Existence of solitary waves in higher dimensions, *Comm. Math. Phys.* **55**, 2, pp. 149–162 (1977).

[298] S.-J. Chen and L. Li, Multiple solutions for the nonhomogeneous Klein-Gordon equation coupled with Born–Infeld theory on $\mathbf{R}^3$, *J. Math. Anal. Appl.* **400**, 2, pp. 517–524 (2013), doi:10.1016/j.jmaa.2012.10.057, http://dx.doi.org/10.1016/j.jmaa.2012.10.057.

[299] E. Zeidler, *Nonlinear functional analysis and its applications. II/B.* Springer-Verlag, New York (1990), ISBN 0-387-97167-X, doi:10.1007/978-1-4612-0985-0, http://dx.doi.org/10.1007/978-1-4612-0985-0, nonlinear monotone operators, Translated from the German by the author and Leo F. Boron.

[300] G. Bonanno and P. F. Pizzimenti, Neumann boundary value problems with not coercive potential, *Mediterr. J. Math.* **9**, 4, pp. 601–609 (2012), doi:10.1007/s00009-011-0136-6, http://dx.doi.org/10.1007/s00009-011-0136-6.

[301] G. Bonanno and A. Sciammetta, Existence and multiplicity results to Neumann problems for elliptic equations involving the p-Laplacian, *J. Math. Anal. Appl.* **390**, 1, pp. 59–67 (2012), doi:10.1016/j.jmaa.2012.01.012, http://dx.doi.org/10.1016/j.jmaa.2012.01.012.

[302] G. Bonanno and G. D'Aguì, Multiplicity results for a perturbed elliptic Neumann problem, *Abstr. Appl. Anal.*, pp. Art. ID 564363, 10 (2010), doi: 10.1155/2010/564363, http://dx.doi.org/10.1155/2010/564363.

[303] G. Bonanno, G. Molica Bisci and V. Rădulescu, Existence of three solutions for a non-homogeneous Neumann problem through Orlicz–Sobolev spaces,

Nonlinear Anal. **74**, 14, pp. 4785–4795 (2011), doi:10.1016/j.na.2011.04.049, http://dx.doi.org/10.1016/j.na.2011.04.049.

[304] P. Candito and G. D'Aguì, Three solutions for a discrete nonlinear Neumann problem involving the p-Laplacian, *Adv. Difference Equ.*, pp. Art. ID 862016, 11 (2010).

[305] A. Chinnì and R. Livrea, Multiple solutions for a Neumann-type differential inclusion problem involving the $p(\cdot)$-Laplacian, *Discrete Contin. Dyn. Syst. Ser. S* **5**, 4, pp. 753–764 (2012).

[306] G. Bonanno and G. D'Aguì, On the Neumann problem for elliptic equations involving the p-Laplacian, *J. Math. Anal. Appl.* **358**, 2, pp. 223–228 (2009), doi:10.1016/j.jmaa.2009.04.055, http://dx.doi.org/10.1016/j.jmaa.2009.04.055.

[307] G. A. Afrouzi, A. Hadjian and S. Heidarkhani, Multiplicity results for a class of two-point boundary value systems investigated via variational methods, *Bull. Math. Soc. Sci. Math. Roumanie (N.S.)* **55(103)**, 4, pp. 343–352 (2012).

[308] A. Kristály, V. D. Rădulescu and C. G. Varga, *Variational principles in mathematical physics, geometry, and economics*, Encyclopedia of Mathematics and its Applications, Vol. 136. Cambridge University Press, Cambridge (2010), ISBN 978-0-521-11782-1, doi:10.1017/CBO9780511760631, http://dx.doi.org/10.1017/CBO9780511760631, qualitative analysis of nonlinear equations and unilateral problems, With a foreword by Jean Mawhin.

[309] W.-W. Pan and L. Li, A Neumann boundary value problem for a class of gradient systems, *Opuscula Math.* **34**, 1, pp. 171–181 (2014), doi:10.7494/OpMath.2014.34.1.171, http://dx.doi.org/10.7494/OpMath.2014.34.1.171.

[310] J. Pipan and M. Schechter, Non-autonomous second order Hamiltonian systems, *J. Differential Equations* **257**, 2, pp. 351–373 (2014), doi:10.1016/j.jde.2014.03.016, http://dx.doi.org/10.1016/j.jde.2014.03.016.

[311] M. Schechter, *Minimax systems and critical point theory*. Birkhäuser Boston, Inc., Boston, MA (2009), ISBN 978-0-8176-4805-3, doi:10.1007/978-0-8176-4902-9, http://dx.doi.org/10.1007/978-0-8176-4902-9.

[312] A. Ambrosetti and V. Coti Zelati, *Periodic solutions of singular Lagrangian systems*, Progress in Nonlinear Differential Equations and their Applications, Vol. 10. Birkhäuser Boston, Inc., Boston, MA (1993).

[313] M. S. Berger and M. Schechter, On the solvability of semilinear gradient operator equations, *Adv. Math.* **25**, 2, pp. 97–132 (1977).

[314] J. Mawhin and M. Willem, Critical points of convex perturbations of some indefinite quadratic forms and semilinear boundary value problems at resonance, *Ann. Inst. H. Poincaré Anal. Non Linéaire* **3**, 6, pp. 431–453 (1986), http://www.numdam.org/item?id=AIHPC_1986__3_6_431_0.

[315] C. Tang, Semi-coercive monotone variational problems on reflexive Banach spaces, *J. Math. Res. Exposition* **17**, 2, pp. 190–192 (1997).

[316] C. Tang, Periodic solutions of non-autonomous second order systems with γ-quasisubadditive potential, *J. Math. Anal. Appl.* **189**, 3, pp. 671–675

(1995), doi:10.1006/jmaa.1995.1044, `http://dx.doi.org/10.1006/jmaa.`
`1995.1044`.

[317] C.-L. Tang, Periodic solutions for nonautonomous second order systems with sublinear nonlinearity, *Proc. Amer. Math. Soc.* **126**, 11, pp. 3263–3270 (1998), doi:10.1090/S0002-9939-98-04706-6, `http://dx.doi.org/10.`
`1090/S0002-9939-98-04706-6`.

[318] C. Tang, Periodic solutions of non-autonomous second order systems, *J. Math. Anal. Appl.* **202**, 2, pp. 465–469 (1996), doi:10.1006/jmaa.1996.0327, `http://dx.doi.org/10.1006/jmaa.1996.0327`.

[319] C.-L. Tang, Existence and multiplicity of periodic solutions for nonautonomous second order systems, *Nonlinear Anal.* **32**, 3, pp. 299–304 (1998), doi:10.1016/S0362-546X(97)00493-8, `http://dx.doi.org/10.`
`1016/S0362-546X(97)00493-8`.

[320] C.-L. Tang and X.-P. Wu, Periodic solutions for a class of nonautonomous subquadratic second order Hamiltonian systems, *J. Math. Anal. Appl.* **275**, 2, pp. 870–882 (2002), doi:10.1016/S0022-247X(02)00442-0, `http://dx.`
`doi.org/10.1016/S0022-247X(02)00442-0`.

[321] C.-L. Tang and X.-P. Wu, Periodic solutions for second order systems with not uniformly coercive potential, *J. Math. Anal. Appl.* **259**, 2, pp. 386–397 (2001), doi:10.1006/jmaa.2000.7401, `http://dx.doi.org/10.1006/jmaa.`
`2000.7401`.

[322] C.-L. Tang and X.-P. Wu, Notes on periodic solutions of subquadratic second order systems, *J. Math. Anal. Appl.* **285**, 1, pp. 8–16 (2003), doi:10.1016/S0022-247X(02)00417-1, `http://dx.doi.org/10.`
`1016/S0022-247X(02)00417-1`.

[323] D. Yanheng and M. Girardi, Periodic and homoclinic solutions to a class of Hamiltonian systems with the potentials changing sign, *Dynam. Systems Appl.* **2**, 1, pp. 131–145 (1993).

[324] F. Antonacci and P. Magrone, Second order nonautonomous systems with symmetric potential changing sign, *Rend. Mat. Appl. (7)* **18**, 2, pp. 367–379 (1998).

[325] F. Antonacci, Periodic and homoclinic solutions to a class of Hamiltonian systems with indefinite potential in sign, *Boll. Un. Mat. Ital. B (7)* **10**, 2, pp. 303–324 (1996).

[326] G. Barletta and R. Livrea, Existence of three periodic solutions for a nonautonomous second order system, *Matematiche (Catania)* **57**, 2, pp. 205–215 (2005) (2002).

[327] Y.-T. Xu and Z.-M. Guo, Existence of periodic solutions to second-order Hamiltonian systems with potential indefinite in sign, *Nonlinear Anal.* **51**, 7, pp. 1273–1283 (2002), doi:10.1016/S0362-546X(01)00895-1, `http://dx.`
`doi.org/10.1016/S0362-546X(01)00895-1`.

[328] W. Zou and S. Li, Infinitely many solutions for Hamiltonian systems, *J. Differential Equations* **186**, 1, pp. 141–164 (2002), doi:10.1016/S0022-0396(02)
00005-0, `http://dx.doi.org/10.1016/S0022-0396(02)00005-0`.

[329] F. Faraci and R. Livrea, Infinitely many periodic solutions for a second-order nonautonomous system, *Nonlinear Anal.* **54**, 3, pp. 417–

429 (2003), doi:10.1016/S0362-546X(03)00099-3, http://dx.doi.org/10. 1016/S0362-546X(03)00099-3.

[330] G. Bonanno and R. Livrea, Periodic solutions for a class of second-order Hamiltonian systems, *Electron. J. Differential Equations*, pp. No. 115, 13 pp. (electronic) (2005).

[331] G. Bonanno and R. Livrea, Multiple periodic solutions for Hamiltonian systems with not coercive potential, *J. Math. Anal. Appl.* **363**, 2, pp. 627–638 (2010), doi:10.1016/j.jmaa.2009.09.025, http://dx.doi.org/10.1016/ j.jmaa.2009.09.025.

[332] M.-Y. Jiang, Periodic solutions of second order superquadratic Hamiltonian systems with potential changing sign. I, *J. Differential Equations* **219**, 2, pp. 323–341 (2005), doi:10.1016/j.jde.2005.06.012, http://dx.doi.org/ 10.1016/j.jde.2005.06.012.

[333] M.-Y. Jiang, Periodic solutions of second order superquadratic Hamiltonian systems with potential changing sign. II, *J. Differential Equations* **219**, 2, pp. 342–362 (2005), doi:10.1016/j.jde.2005.06.011, http://dx.doi.org/ 10.1016/j.jde.2005.06.011.

[334] L. K. Shilgba, Existence result for periodic solutions of a class of Hamiltonian systems with super quadratic potential, *Nonlinear Anal.* **63**, 4, pp. 565–574 (2005), doi:10.1016/j.na.2005.05.018, http://dx.doi.org/10. 1016/j.na.2005.05.018.

[335] F. Faraci and A. Iannizzotto, A multiplicity theorem for a perturbed second-order non-autonomous system, *Proc. Edinb. Math. Soc. (2)* **49**, 2, pp. 267–275 (2006), doi:10.1017/S001309150400149X, http://dx.doi. org/10.1017/S001309150400149X.

[336] L. Xiao and X. H. Tang, Existence of periodic solutions to second-order Hamiltonian systems with potential indefinite in sign, *Nonlinear Anal.* **69**, 11, pp. 3999–4011 (2008), doi:10.1016/j.na.2007.10.032, http://dx.doi. org/10.1016/j.na.2007.10.032.

[337] Y. Ye and C.-L. Tang, Infinitely many periodic solutions of non-autonomous second-order Hamiltonian systems, *Proc. Roy. Soc. Edinburgh Sect. A* **144**, 1, pp. 205–223 (2014), doi:10.1017/S0308210512001461, http://dx.doi. org/10.1017/S0308210512001461.

[338] L. Li and M. Schechter, Existence solutions for second order Hamiltonian systems, *Nonlinear Anal. Real World Appl.* **27**, pp. 283–296 (2016), doi:10.1016/j.nonrwa.2015.08.001, http://dx.doi.org/10.1016/ j.nonrwa.2015.08.001.

[339] C. Li, Z.-Q. Ou and C.-L. Tang, Three periodic solutions for *p*-Hamiltonian systems, *Nonlinear Anal.* **74**, 5, pp. 1596–1606 (2011), doi:10.1016/j.na. 2010.10.030, http://dx.doi.org/10.1016/j.na.2010.10.030.

[340] D. Averna and G. Bonanno, Three solutions for a quasilinear two-point boundary-value problem involving the one-dimensional *p*-Laplacian, *Proc. Edinb. Math. Soc. (2)* **47**, 2, pp. 257–270 (2004).

[341] K. Liao and C.-L. Tang, Existence and multiplicity of periodic solutions for the ordinary *p*-Laplacian systems, *J. Appl. Math. Comput.* **35**, 1–2, pp. 395–406 (2011), doi:10.1007/s12190-009-0364-0, http://dx.doi.org/10.

1007/s12190-009-0364-0.

[342] H. Lü, D. O'Regan and R. P. Agarwal, On the existence of multiple periodic solutions for the vector p-Laplacian via critical point theory, *Appl. Math.* **50**, 6, pp. 555–568 (2005), doi:10.1007/s10492-005-0037-8, http://dx.doi.org/10.1007/s10492-005-0037-8.

[343] X. Lv, S. Lu and P. Yan, Periodic solutions of non-autonomous ordinary p-Laplacian systems, *J. Appl. Math. Comput.* **35**, 1–2, pp. 11–18 (2011), doi:10.1007/s12190-009-0336-4, http://dx.doi.org/10.1007/s12190-009-0336-4.

[344] S. Ma and Y. Zhang, Existence of infinitely many periodic solutions for ordinary p-Laplacian systems, *J. Math. Anal. Appl.* **351**, 1, pp. 469–479 (2009), doi:10.1016/j.jmaa.2008.10.027, http://dx.doi.org/10.1016/j.jmaa.2008.10.027.

[345] Y. Tian and W. Ge, Periodic solutions for second-order Hamiltonian systems with the p-Laplacian, *Electron. J. Differential Equations*, pp. No. 134, 12 (2006).

[346] Y. Tian and W. Ge, Periodic solutions of non-autonomous second-order systems with a p-Laplacian, *Nonlinear Anal.* **66**, 1, pp. 192–203 (2007), doi:10.1016/j.na.2005.11.020, http://dx.doi.org/10.1016/j.na.2005.11.020.

[347] Z. Wang and J. Zhang, Periodic solutions of non-autonomous second order systems with p-Laplacian, *Electron. J. Differential Equations*, pp. No. 17, 12 (2009).

[348] B. Xu and C.-L. Tang, Some existence results on periodic solutions of ordinary p-Laplacian systems, *J. Math. Anal. Appl.* **333**, 2, pp. 1228–1236 (2007), doi:10.1016/j.jmaa.2006.11.051, http://dx.doi.org/10.1016/j.jmaa.2006.11.051.

[349] Y. Zhang and S. Ma, Some existence results on periodic and subharmonic solutions of ordinary P-Laplacian systems, *Discrete Contin. Dyn. Syst. Ser. B* **12**, 1, pp. 251–260 (2009), doi:10.3934/dcdsb.2009.12.251, http://dx.doi.org/10.3934/dcdsb.2009.12.251.

[350] X. Zhang and P. Zhou, An existence result on periodic solutions of an ordinary p-Laplace system, *Bull. Malays. Math. Sci. Soc. (2)* **34**, 1, pp. 127–135 (2011).

[351] G. Bonanno and G. D'Aguì, A Neumann boundary value problem for the Sturm–Liouville equation, *Appl. Math. Comput.* **208**, 2, pp. 318–327 (2009), doi:10.1016/j.amc.2008.12.029, http://dx.doi.org/10.1016/j.amc.2008.12.029.

[352] G. Bonanno and G. Riccobono, Multiplicity results for Sturm–Liouville boundary value problems, *Appl. Math. Comput.* **210**, 2, pp. 294–297 (2009), doi:10.1016/j.amc.2008.12.081, http://dx.doi.org/10.1016/j.amc.2008.12.081.

[353] G. Bonanno and S. M. Buccellato, Two point boundary value problems for the Sturm-Liouville equation with highly discontinuous nonlinearities, *Taiwanese J. Math.* **14**, 5, pp. 2059–2072 (2010).

[354] G. Bonanno and A. Chinnì, Existence of three solutions for a perturbed two-point boundary value problem, *Appl. Math. Lett.* **23**, 7, pp. 807–811 (2010),

doi:10.1016/j.aml.2010.03.015, http://dx.doi.org/10.1016/j.aml.2010.03.015.

[355] H. Chen and J. Sun, An application of variational method to second-order impulsive differential equation on the half-line, *Appl. Math. Comput.* **217**, 5, pp. 1863–1869 (2010), doi:10.1016/j.amc.2010.06.040, http://dx.doi.org/10.1016/j.amc.2010.06.040.

[356] J. Sun, H. Chen and T. Zhou, Multiplicity of solutions for a fourth-order impulsive differential equation via variational methods, *Bull. Aust. Math. Soc.* **82**, 3, pp. 446–458 (2010), doi:10.1017/S0004972710001802, http://dx.doi.org/10.1017/S0004972710001802.

[357] D. Averna and G. Bonanno, A three critical points theorem and its applications to the ordinary Dirichlet problem, *Topol. Methods Nonlinear Anal.* **22**, 1, pp. 93–103 (2003).

[358] G. Bonanno and P. Candito, Infinitely many solutions for a class of discrete non-linear boundary value problems, *Appl. Anal.* **88**, 4, pp. 605–616 (2009), doi:10.1080/00036810902942242, http://dx.doi.org/10.1080/00036810902942242.

[359] G. Bonanno and G. Molica Bisci, Infinitely many solutions for a Dirichlet problem involving the p-Laplacian, *Proc. Roy. Soc. Edinburgh Sect. A* **140**, 4, pp. 737–752 (2010), doi:10.1017/S0308210509000845, http://dx.doi.org/10.1017/S0308210509000845.

[360] G. Bonanno and E. Tornatore, Infinitely many solutions for a mixed boundary value problem, *Ann. Polon. Math.* **99**, 3, pp. 285–293 (2010), doi:10.4064/ap99-3-5, http://dx.doi.org/10.4064/ap99-3-5.

[361] G. Bonanno, G. Molica Bisci and D. O'Regan, Infinitely many weak solutions for a class of quasilinear elliptic systems, *Math. Comput. Modelling* **52**, 1-2, pp. 152–160 (2010), doi:10.1016/j.mcm.2010.02.004, http://dx.doi.org/10.1016/j.mcm.2010.02.004.

[362] L. Ding, L. Li and C. Li, On a p-Hamiltonian system, *Bull. Math. Soc. Sci. Math. Roumanie (N.S.)* **57(105)**, 1, pp. 45–57 (2014).

[363] G. Chen and S. Ma, Infinitely many solutions for resonant cooperative elliptic systems with sublinear or superlinear terms, *Calc. Var. Partial Differential Equations* **49**, 1-2, pp. 271–286 (2014), doi:10.1007/s00526-012-0581-5, http://dx.doi.org/10.1007/s00526-012-0581-5.

[364] S. Ma, Infinitely many solutions for cooperative elliptic systems with odd nonlinearity, *Nonlinear Anal.* **71**, 5–6, pp. 1445–1461 (2009), doi:10.1016/j.na.2008.12.012, http://dx.doi.org/10.1016/j.na.2008.12.012.

[365] W. Zou, Solutions for resonant elliptic systems with nonodd or odd nonlinearities, *J. Math. Anal. Appl.* **223**, 2, pp. 397–417 (1998), doi:10.1006/jmaa.1998.5938, http://dx.doi.org/10.1006/jmaa.1998.5938.

[366] G. Bonanno, G. Molica Bisci and V. Rădulescu, Qualitative analysis of gradient-type systems with oscillatory nonlinearities on the Sierpiński gasket, *Chin. Ann. Math. Ser. B* **34**, 3, pp. 381–398 (2013), doi:10.1007/s11401-013-0772-1, http://dx.doi.org/10.1007/s11401-013-0772-1.

[367] A. Kristály, On a new class of elliptic systems with nonlinearities of arbitrary growth, *J. Differential Equations* **249**, 8, pp. 1917–1928 (2010),

doi:10.1016/j.jde.2010.05.001, `http://dx.doi.org/10.1016/j.jde.2010.` `05.001`.

[368] W. Zou, Multiple solutions for asymptotically linear elliptic systems, *J. Math. Anal. Appl.* **255**, 1, pp. 213–229 (2001), doi:10.1006/jmaa.2000.7236, `http://dx.doi.org/10.1006/jmaa.2000.7236`.

[369] D. G. Costa and C. A. Magalhães, A variational approach to sub-quadratic perturbations of elliptic systems, *J. Differential Equations* **111**, 1, pp. 103–122 (1994), doi:10.1006/jdeq.1994.1077, `http://dx.doi.org/` `10.1006/jdeq.1994.1077`.

[370] V. Benci and P. H. Rabinowitz, Critical point theorems for indefinite functionals, *Invent. Math.* **52**, 3, pp. 241–273 (1979).

[371] S. Ma, Nontrivial solutions for resonant cooperative elliptic systems via computations of the critical groups, *Nonlinear Anal.* **73**, 12, pp. 3856–3872 (2010), doi:10.1016/j.na.2010.08.013, `http://dx.doi.org/10.1016/j.na.` `2010.08.013`.

[372] A. Pomponio, Asymptotically linear cooperative elliptic system: existence and multiplicity, *Nonlinear Anal.* **52**, 3, pp. 989–1003 (2003), doi:10. 1016/S0362-546X(02)00148-7, `http://dx.doi.org/10.1016/S0362-546X` `(02)00148-7`.

[373] W. Zou, S. Li and J. Q. Liu, Nontrivial solutions for resonant cooperative elliptic systems via computations of critical groups, *Nonlinear Anal.* **38**, 2, Ser. A: Theory Methods, pp. 229–247 (1999), doi:10.1016/S0362-546X(98) 00191-6, `http://dx.doi.org/10.1016/S0362-546X(98)00191-6`.

[374] G. H. Fei, Multiple solutions of some nonlinear strongly resonant elliptic equations without the (PS) condition, *J. Math. Anal. Appl.* **193**, 2, pp. 659–670 (1995), doi:10.1006/jmaa.1995.1259, `http://dx.doi.org/10.` `1006/jmaa.1995.1259`.

[375] L. Li and C.-L. Tang, Infinitely many solutions for resonance elliptic systems, *C. R. Math. Acad. Sci. Paris* **353**, 1, pp. 35–40 (2015), doi:10.1016/ j.crma.2014.10.010, `http://dx.doi.org/10.1016/j.crma.2014.10.010`.

[376] L. Boccardo and D. Guedes de Figueiredo, Some remarks on a system of quasilinear elliptic equations, *NoDEA Nonlinear Differential Equations Appl.* **9**, 3, pp. 309–323 (2002).

[377] O. H. Miyagaki and R. S. Rodrigues, On the existence of weak solutions for p, q-Laplacian systems with weights, *Electron. J. Differential Equations*, pp. No. 115, 18 (2008).

[378] A. Kristály, Existence of two non-trivial solutions for a class of quasilinear elliptic variational systems on strip-like domains, *Proc. Edinb. Math. Soc. (2)* **48**, 2, pp. 465–477 (2005), doi:10.1017/S0013091504000112, `http://` `dx.doi.org/10.1017/S0013091504000112`.

[379] F. Cammaroto, A. Chinní and B. Di Bella, Multiple solutions for a quasilinear elliptic variational system on strip-like domains, *Proc. Edinb. Math. Soc. (2)* **50**, 3, pp. 597–603 (2007), doi:10.1017/S0013091505001380, `http://dx.doi.org/10.1017/S0013091505001380`.

[380] C. Li and C.-L. Tang, Three solutions for a class of quasilinear elliptic systems involving the (p, q)-Laplacian, *Nonlinear Anal.* **69**, 10, pp. 3322–

3329 (2008), doi:10.1016/j.na.2007.09.021, http://dx.doi.org/10.1016/ j.na.2007.09.021.

[381] G. A. Afrouzi and S. Heidarkhani, Existence of three solutions for a class of Dirichlet quasilinear elliptic systems involving the $(p_1, \ldots, p_n)$-Laplacian, *Nonlinear Anal.* **70**, 1, pp. 135–143 (2009).

[382] S. Samko, On a progress in the theory of Lebesgue spaces with variable exponent: maximal and singular operators, *Integral Transforms Spec. Funct.* **16**, 5-6, pp. 461–482 (2005), doi:10.1080/10652460412331320322, http:// dx.doi.org/10.1080/10652460412331320322.

[383] L. Diening, P. Harjulehto, P. Hästö and M. Røužička, *Lebesgue and Sobolev spaces with variable exponents*, Lecture Notes in Mathematics, Vol. 2017. Springer, Heidelberg (2011), ISBN 978-3-642-18362-1, doi:10.1007/ 978-3-642-18363-8, http://dx.doi.org/10.1007/978-3-642-18363-8.

[384] X.-L. Fan, $p(x)$-Laplacian equations, in *Topological methods, variational methods and their applications (Taiyuan, 2002)*. World Sci. Publ., River Edge, NJ, pp. 117–123 (2003).

[385] V. V. Zhikov, Averaging of functionals of the calculus of variations and elasticity theory, *Izv. Akad. Nauk SSSR Ser. Mat.* **50**, 4, pp. 675–710, 877 (1986).

[386] M. Ružička, *Electrorheological fluids: modeling and mathematical theory*, Lecture Notes in Mathematics, Vol. 1748. Springer-Verlag, Berlin (2000), ISBN 3-540-41385-5, doi:10.1007/BFb0104029, http://dx.doi.org/10. 1007/BFb0104029.

[387] F. Li, Z. Li and L. Pi, Variable exponent functionals in image restoration, *Appl. Math. Comput.* **216**, 3, pp. 870–882 (2010), doi:10.1016/j.amc.2010. 01.094, http://dx.doi.org/10.1016/j.amc.2010.01.094.

[388] R. A. Mashiyev, Three solutions to a Neumann problem for elliptic equations with variable exponent, *Arab. J. Sci. Eng.* **36**, 8, pp. 1559–1567 (2011), doi:10.1007/s13369-011-0141-x, http://dx.doi.org/ 10.1007/s13369-011-0141-x.

[389] M. Mihăilescu and V. Rădulescu, A multiplicity result for a nonlinear degenerate problem arising in the theory of electrorheological fluids, *Proc. R. Soc. Lond. Ser. A Math. Phys. Eng. Sci.* **462**, 2073, pp. 2625–2641 (2006), doi:10.1098/rspa.2005.1633, http://dx.doi.org/10.1098/ rspa.2005.1633.

[390] G. Anello and G. Cordaro, Existence of solutions of the Neumann problem for a class of equations involving the p-Laplacian via a variational principle of Ricceri, *Arch. Math. (Basel)* **79**, 4, pp. 274–287 (2002).

[391] G. Bonanno and P. Candito, Three solutions to a Neumann problem for elliptic equations involving the p-Laplacian, *Arch. Math. (Basel)* **80**, 4, pp. 424–429 (2003).

[392] G. Dai, Three solutions for a Neumann-type differential inclusion problem involving the $p(x)$-Laplacian, *Nonlinear Anal.* **70**, 10, pp. 3755–3760 (2009), doi:10.1016/j.na.2008.07.031, http://dx.doi.org/10.1016/j.na. 2008.07.031.

[393] M. Mihăilescu, Existence and multiplicity of solutions for a Neumann prob-

lem involving the $p(x)$-Laplace operator, *Nonlinear Anal.* **67**, 5, pp. 1419–1425 (2007), doi:10.1016/j.na.2006.07.027, http://dx.doi.org/10.1016/j.na.2006.07.027.

[394] Q. Liu, Existence of three solutions for $p(x)$-Laplacian equations, *Nonlinear Anal.* **68**, 7, pp. 2119–2127 (2008), doi:10.1016/j.na.2007.01.035, http://dx.doi.org/10.1016/j.na.2007.01.035.

[395] W.-W. Pan, G. A. Afrouzi and L. Li, Three solutions to a $p(x)$-Laplacian problem in weighted-variable-exponent Sobolev space, *An. Ştiinţ. Univ. "Ovidius" Constanţa Ser. Mat.* **21**, 2, pp. 195–205 (2013).

[396] G. Dai, Infinitely many solutions for a $p(x)$-Laplacian equation in $\mathbb{R}^N$, *Nonlinear Anal.* **71**, 3-4, pp. 1133–1139 (2009), doi:10.1016/j.na.2008.11.037, http://dx.doi.org/10.1016/j.na.2008.11.037.

[397] M. M. Rodrigues, Multiplicity of solutions on a nonlinear eigenvalue problem for $p(x)$-Laplacian-like operators, *Mediterr. J. Math.* **9**, 1, pp. 211–223 (2012), doi:10.1007/s00009-011-0115-y, http://dx.doi.org/10.1007/s00009-011-0115-y.

[398] L. Li and S.-J. Chen, Existence and multiplicity of solutions for a class of p-Laplacian equations, *Appl. Math. Lett.* **27**, pp. 59–63 (2014), doi:10.1016/j.aml.2013.07.010, http://dx.doi.org/10.1016/j.aml.2013.07.010.

[399] A. Kristály and C. Varga, On a class of quasilinear eigenvalue problems in $\mathbb{R}^N$, *Math. Nachr.* **278**, 15, pp. 1756–1765 (2005), doi:10.1002/mana.200510339, http://dx.doi.org/10.1002/mana.200510339.

[400] W.-M. Ni and J. Serrin, Nonexistence theorems for quasilinear partial differential equations, in *Proceedings of the conference commemorating the 1st centennial of the Circolo Matematico di Palermo (Italian) (Palermo, 1984)*, 8, pp. 171–185 (1985).

[401] L. Li, L. Ding and W.-W. Pan, Existence of multiple solutions for a $p(x)$-biharmonic equation, *Electron. J. Differential Equations*, pp. No. 139, 10 (2013).

[402] L. Li and C. Tang, Existence and multiplicity of solutions for a class of $p(x)$-biharmonic equations, *Acta Math. Sci. Ser. B Engl. Ed.* **33**, 1, pp. 155–170 (2013), doi:10.1016/S0252-9602(12)60202-1, http://dx.doi.org/10.1016/S0252-9602(12)60202-1.

[403] L. Gasiński and N. S. Papageorgiou, A pair of positive solutions for the Dirichlet $p(z)$-Laplacian with concave and convex nonlinearities, *J. Global Optim.* **56**, 4, pp. 1347–1360 (2013), doi:10.1007/s10898-011-9841-8, http://dx.doi.org/10.1007/s10898-011-9841-8.

[404] L. Gasiński and N. S. Papageorgiou, Anisotropic nonlinear Neumann problems, *Calc. Var. Partial Differential Equations* **42**, 3-4, pp. 323–354 (2011), doi:10.1007/s00526-011-0390-2, http://dx.doi.org/10.1007/s00526-011-0390-2.

[405] M. Růžička, Modeling, mathematical and numerical analysis of electrorheological fluids, *Appl. Math.* **49**, 6, pp. 565–609 (2004), doi:10.1007/s10492-004-6432-8, http://dx.doi.org/10.1007/s10492-004-6432-8.

[406] L. Ying and L. Li, Existence and multiplicity of solutions for $p(x)$-laplacian equations in $\mathbb{R}^n$, *Bull. Malays. Math. Sci. Soc.* (2015), in press.

[407] Y. Liu and Z. Wang, Biharmonic equations with asymptotically linear nonlinearities, *Acta Math. Sci. Ser. B Engl. Ed.* **27**, 3, pp. 549–560 (2007), doi:10.1016/S0252-9602(07)60055-1, http://dx.doi.org/10.1016/S0252-9602(07)60055-1.

[408] Y. Deng and Y. Li, Regularity of the solutions for nonlinear biharmonic equations in $\mathbb{R}^N$, *Acta Math. Sci. Ser. B Engl. Ed.* **29**, 5, pp. 1469–1480 (2009), doi:10.1016/S0252-9602(09)60119-3, http://dx.doi.org/10.1016/S0252-9602(09)60119-3.

[409] Z. Boucheche, R. Yacoub and H. Chtioui, On a bi-harmonic equation involving critical exponent: existence and multiplicity results, *Acta Math. Sci. Ser. B Engl. Ed.* **31**, 4, pp. 1213–1244 (2011), doi:10.1016/S0252-9602(11)60311-1, http://dx.doi.org/10.1016/S0252-9602(11)60311-1.

[410] O. H. Miyagaki and M. A. S. Souto, Superlinear problems without Ambrosetti and Rabinowitz growth condition, *J. Differential Equations* **245**, 12, pp. 3628–3638 (2008), doi:10.1016/j.jde.2008.02.035, http://dx.doi.org/10.1016/j.jde.2008.02.035.

[411] D. G. Costa and C. A. Magalhães, Variational elliptic problems which are nonquadratic at infinity, *Rev. Un. Mat. Argentina* **37**, 1–2, pp. 24–28 (1992) (1991), x Latin American School of Mathematics (Spanish) (Tanti, 1991).

[412] M. Willem and W. Zou, On a Schrödinger equation with periodic potential and spectrum point zero, *Indiana Univ. Math. J.* **52**, 1, pp. 109–132 (2003), doi:10.1512/iumj.2003.52.2273, http://dx.doi.org/10.1512/iumj.2003.52.2273.

[413] M. Schechter and W. Zou, Superlinear problems, *Pacific J. Math.* **214**, 1, pp. 145–160 (2004), doi:10.2140/pjm.2004.214.145, http://dx.doi.org/10.2140/pjm.2004.214.145.

[414] G. Li and C. Yang, The existence of a nontrivial solution to a nonlinear elliptic boundary value problem of p-Laplacian type without the Ambrosetti-Rabinowitz condition, *Nonlinear Anal.* **72**, 12, pp. 4602–4613 (2010), doi:10.1016/j.na.2010.02.037, http://dx.doi.org/10.1016/j.na.2010.02.037.

[415] L. Jeanjean and J. F. Toland, Bounded Palais-Smale mountain-pass sequences, *C. R. Acad. Sci. Paris Sér. I Math.* **327**, 1, pp. 23–28 (1998), doi:10.1016/S0764-4442(98)80097-9, http://dx.doi.org/10.1016/S0764-4442(98)80097-9.

[416] L. Jeanjean, Local conditions insuring bifurcation from the continuous spectrum, *Math. Z.* **232**, 4, pp. 651–664 (1999), doi:10.1007/PL00004774, http://dx.doi.org/10.1007/PL00004774.

[417] Z. Liu and Z.-Q. Wang, On the Ambrosetti-Rabinowitz superlinear condition, *Adv. Nonlinear Stud.* **4**, 4, pp. 563–574 (2004).

[418] P. Drábek and M. Ôtani, Global bifurcation result for the p-biharmonic operator, *Electron. J. Differential Equations*, pp. No. 48, 19 pp. (electronic) (2001).

[419] S. Liu and M. Squassina, On the existence of solutions to a fourth-order quasilinear resonant problem, *Abstr. Appl. Anal.* **7**, 3, pp. 125–133 (2002), doi:10.1155/S1085337502000805, http://dx.doi.org/10.

1155/S1085337502000805.

[420] P. Harjulehto, P. Hästö, Ú. V. Lê and M. Nuortio, Overview of differential equations with non-standard growth, *Nonlinear Anal.* **72**, 12, pp. 4551–4574 (2010), doi:10.1016/j.na.2010.02.033, http://dx.doi.org/10.1016/j.na.2010.02.033.

[421] G. A. Afrouzi and S. Heidarkhani, Three solutions for a Dirichlet boundary value problem involving the *p*-Laplacian, *Nonlinear Anal.* **66**, 10, pp. 2281–2288 (2007).

[422] G. Bonanno and R. Livrea, Multiplicity theorems for the Dirichlet problem involving the *p*-Laplacian, *Nonlinear Anal.* **54**, 1, pp. 1–7 (2003), doi:10.1016/S0362-546X(03)00027-0, http://dx.doi.org/10.1016/S0362-546X(03)00027-0.

[423] G. Bonanno, G. Molica Bisci and V. Rǎdulescu, Multiple solutions of generalized Yamabe equations on Riemannian manifolds and applications to Emden-Fowler problems, *Nonlinear Anal. Real World Appl.* **12**, 5, pp. 2656–2665 (2011), doi:10.1016/j.nonrwa.2011.03.012, http://dx.doi.org/10.1016/j.nonrwa.2011.03.012.

[424] X. Shi and X. Ding, Existence and multiplicity of solutions for a general *p*(*x*)-Laplacian Neumann problem, *Nonlinear Anal.* **70**, 10, pp. 3715–3720 (2009), doi:10.1016/j.na.2008.07.027, http://dx.doi.org/10.1016/j.na.2008.07.027.

Index

Printed in the United States
By Bookmasters